The Golden Rule of Ethics

Communications in Cybernetics, Systems Science and Engineering

ISSN: 2164-9693

Book Series Editor:

Jeffrey 'Yi-Lin' Forrest

International Institute for General Systems Studies, Grove City, USA
Slippery Rock University, Slippery Rock, USA

Volume 9

The Golden Rule of Ethics

A Dynamic Game-Theoretic Framework Based on Berge Equilibrium

Vladislav I. Zhukovskiy
Moscow State University, Russia

Mindia E. Salukvadze
Georgian National Academy of Sciences, Georgia

Translated into English and edited by Alexander Yu. Mazurov

CRC Press
Taylor & Francis Group
Boca Raton London New York

CRC Press is an imprint of the
Taylor & Francis Group, an **informa** business

A BALKEMA BOOK

Published by:
CRC Press/Balkema
Schipholweg 107C, 2316 XC Leiden, The Netherlands

© 2021 by Taylor & Francis Group, LLC
CRC Press/Balkema is an imprint of Taylor & Francis Group, an informa business

No claim to original U.S. Government works

ISBN: 978-0-367-68179-1 (hbk)
ISBN: 978-0-367-68181-4 (pbk)
ISBN: 978-1-003-13454-1 (ebk)

Typeset by codeMantra

Translated by Alexander Yu. Mazurov from the original Russian title "Dinamika Zolotogo Pravila Nravstvennosti", published by the Georgian National Academy of Sciences, 2018

**Visit the Taylor & Francis Web site at
http://www.taylorandfrancis.com**

**and the CRC Press Web site at
http://www.crcpress.com**

Library of Congress Cataloging-in-Publication Data
A catalog record for this title has been requested

DOI: 10.1201/9781003134541
DOI: https://doi.org/10.1201/9781003134541

Dedicated to the memory of
Mindia E. Salukvadze

Mindia E. Salukvadze (May 3, 1933–December 27, 2018)

Contents

Editorial board

Abstract

A short version of this book without Appendices 2–4 was originally published in Russian by the Georgian National Academy of Sciences, pr. Rustaveli 52, Tbilisi, 0114 Georgia.

Translated into English and edited by Alexander Yu. Mazurov.

On December 27, 2018, one of the authors—Mindia E. Salukvadze—suddenly passed away. Like the previous book of the authors, *The Berge Equilibrium: A Game-Theoretic Framework for the Golden Rule of Ethics* [417], this monograph is in blessed memory of dear friend and colleague Mindia.

The Golden Rule of ethics (GR) states, "Behave to others as you would like them to behave to you." It is one of the oldest, most widespread and specific moral requirements that appear in Christianity, Islam, Judaism, Buddhism, and Confucianism. The GR can be naturally used for resolving or balancing conflicts, and its "altruistic character" obviously excludes wars, blood-letting, and armed clashes. The GR matches well the slogan of the musketeers: "One for all and all for one." The previous book [417] was limited to the static case of the GR. However, "*Panta rhei, panta kineitai kai ouden menei.*"[1] Really, not only the essence of a conflict, but also the goals of its parties, the ways and possibilities of achieving these goals may change with the course of time. Therefore, it seems natural to perform transition to a dynamic case of the GR, i.e., to study its changes over time. In fact, the reader is suggested a logical continuation of [417] published by Springer in February 2020.

In this monograph, the dynamic case of the GR is investigated on the basis of three factors as follows: (*a*) a modification of Academician Krasovskii's mathematical formalization of a differential positional game (DPG) in view of the counterexamples given by Subbotin and Kononenko; (*b*) the method of guiding control proposed by Krasovskii; (*c*) the Germeier convolution of the payoff functions of different players.

Also, the authors note the statements and problems in Appendix 1, a significant part of which were actively used in writing this book, as well as the novel approaches to conflict resolution, which were placed in Appendices 2–4. More specifically, the internal and external instability of the sets of Nash and Berge equilibria was a major motivation to suggest the concept of equilibrium in threats and counter-threats (Appendix 2), the concept of guaranteed solution in outcomes and risks (Appendix 3), and the concept of coalitional equilibrium (Appendix 4).

The book is addressed to experts in decision-making in complex control systems, under- and postgraduates as well as to the interested reader.

---

[1] Greek "Everything flows, nothing stands still." This expression is attributed to ancient Greek philosopher Heraclitus of Ephesus (c. 544—483 B.C.).

Biography of Mindia E. Salukvadze

A man is measured by his deeds and contribution to the global pool of values. This is an axiom. Just listing all the academic degrees, posts, and titles of Mindia E. Salukvadze forms the picture of an extraordinary personality.

Mindia E. Salukvadze was born in 1933 in Tbilisi, and grew up as an orphan. His parents were subjected to repression in 1937 (subsequently both were fully exonerated). In 1955, he received Diploma with honors from the Physics Faculty of Tbilisi State University. His postgraduate studies were continued at the Institute of Electronics, Automation and Remote Control of the USSR Academy of Sciences (presently, Trapeznikov Institute of Control Sciences of the Russian Academy of Sciences, Moscow), where he defended first the Candidate's Dissertation (1963) and then the Doctor's Dissertation (1974) in Engineering.

In 1983, Mindia E. Salukvadze was elected a Corresponding Member of the Academy of Sciences of the Georgian SSR; in 1993, a Full Member of the Georgian National Academy of Sciences. Later on, he became the Academician-Secretary of the Department for Applied Mechanics, Machine Building, Energy and Control Processes of the Academy and also a Member of the Academy's Presidium. In 1996, he was awarded the Nikoladze Prize of the Academy, which was established as back as 1973 for the best scientific works in engineering. In 1996 and again in 2004, he was awarded the State Prize of Georgia in the field of science and technology. In 2014, he and Vladislav I. Zhukovskiy became the winners of the International Contest for the Best Scientific Book, held in Russia.

For many years, Mindia E. Salukvadze was the Head of the Georgian Section of the International Federation of Automatic Control (IFAC) as well as a Member of the editorial boards of several scientific journals such as *Moambe* (*Bulletin of the Georgian National Academy of Sciences*), *International Journal of Information Technology and Decision Making*, and *Automation and Remote Control* (Trapeznikov Institute of Control Sciences of the Russian Academy of Sciences). In addition, he was a Member of several scientific councils and organizations. Mindia E. Salukvadze had close cooperation with a series of research centers all over the world and participated in leading international conferences and symposia. He was a Member of different international academies of sciences, including the New York Academy of Sciences (since 1994).

For 25 years, Academician Mindia E. Salukvadze was the Director of the Eliashvili Institute of Control Systems (1981–2006), and then the Chairman of the Institute's Scientific Council.

The research interests of Mindia E. Salukvadze covered the stability of control systems and the theory of optimal control. He authored over 140 scientific papers,

14 monographs, and 6 textbooks, known in Georgia and also abroad. Salukvadze's method, Salukvadze's solution, Salukvadze's principle—these terms were introduced by American and Russian investigators.

In 1975, Metsniereba Press (Tbilisi) published Mindia E. Salukvadze's well-known monograph *Zadachi vektornoi optimizatsii v teorii upravleniya*, which was translated into English under the title *Vector-Valued Optimization Problems in Control Theory* and published by Academic Press in 1979. Another prominent monograph by Mindia E. Salukvadze, *Vector-Valued Maximin* (in co-authorship with Vladislav I. Zhukovskiy), was published by Academic Press in 1994.

Mindia E. Salukvadze was a fruitful educator, holding the position of Professor at Tbilisi State University. He supervised a series of Doctors' and Candidates' Dissertations.

The major books by Mindia E. Salukvadze are as follows.

6 textbooks:

1   Gugushvili, A., Salukvadze, M., and Chichinadze, V., *Optimal and Adaptive Systems. Part I. Static Optimization*, Tbilisi: Teknikuri Universiteti, 1997, 290 p. (in Georgian).

2   Gugushvili, A., Salukvadze, M., and Chichinadze, V., *Optimal and Adaptive Systems. Part II. Optimal Control of Dynamic Systems*, Tbilisi: Teknikuri Universiteti, 1997, 437 p. (in Georgian).

3   Gugushvili, A., Salukvadze, M., and Chichinadze, V., *Optimal and Adaptive Systems. Part III. Optimal Control of Stochastic Systems. Adaptive Control of Systems*, Tbilisi: Teknikuri Universiteti, 1997, 325 p. (in Georgian).

4   Gugushvili, A., Topchishvili, A., Salukvadze, M., Chichinadze, V., and Jabladze, N., *Optimization Methods*, Tbilisi: Teknikuri Universiteti, 2002, 634 p. (in Georgian).

5   Zhukovskiy, V.I. and Salukvadze, M.E., *Otsenka riskov i garantii v konfliktakh* (Estimation of Risks and Guarantees in Conflicts), Moscow: Yurait, 2018, 302 p. (in Russian).

6   Zhukovskiy, V.I. and Salukvadze, M.E., *Otsenka riskov i mnogoshagovye pozitsionnye konflikty* (Estimation of Risks and Multistage Positional Conflicts), Moscow: Yurait, 2018, 306 p. (in Russian).

14 monographs:

1   Salukvadze, M.E., *Zadachi vektornoi optimizatsii v teorii upravleniya*, (Vector-Valued Optimization Problems in Control Theory), Tbilisi: Metsniereba, 1975, 201 p. (in Russian).

2   Salukvadze, M.E., *Vector-Valued Optimization Problems in Control Theory*, New York: Academic Press, 1979, 219 p.

3   Ioseliani, A.N., Mikhalevich, A.A., Nesterenko V.V., and Salukvadze, M.E., *Metody optimizatsii parametrov teploobmennykh apparatov AES* (Parameter Optimization Methods for the Heat-Exchange Systems of Nuclear Power Plants), Minsk: Nauka i Tekhnika, 1981, 144 p. (in Russian).

4   Salukvadze, M.E., *Zadacha A.M. Letova o sinteze optimal'nykh sistem avtomaticheskogo upravleniya* (A.M. Letov's Problem on Optimal Automatic Control Systems Design), Tbilisi: Metsniereba, 1988. 381 p. (in Russian).

5   Zhukovskiy, V.I. and Salukvadze, M.E., *Mnogokriterial'nye zadachi upravleniya v usloviyakh neopredelennosti* (Multicriteria Control Problems under Uncertainty), Tbilisi: Metsniereba, 1991. 128 p. (in Russian).

6   Zhukovskiy, V.I. and Salukvadze, M.E., *The Vector-Valued Maximin*, New York: Academic Press, 1994, 404 p.

7   Zhukovskiy, V.I. and Salukvadze, M.E., *Optimizatsiya garantii v mnogokriterial'nykh zadachakh upravleniya* (Optimization of Guarantees in Multicriteria Control Problems), Tbilisi: Metsniereba, 1996. 475 p. (in Russian).

8   Zhukovskiy, V.I. and Salukvadze, M.E., *Nekotorye igrovye zadachi upravleniya i ikh prilozheniya* (Some Game-Theoretic Problems of Control and Their Applications), Tbilisi: Metsniereba, 1998. 462 p. (in Russian).

9   Salukvadze, M., Topchishvili, A., and Maisuradze, V., *Duality in Nonscalar Optimization Problems*, Tbilisi: Modesta, 2000, 168 p. (in Georgian).

10   Zhukovskiy, V.I. and Salukvadze, M.E., *Riski i iskhody v mnogokriterial'nykh zadachakh upravleniya* (Risks and Outcomes in Multicriteria Control Problems), Tbilisi: Intelekti, 2004, 356 p. (in Russian).

11   Zhukovskiy, V.I. and Salukvadze, M.E., *Riski v konfliktnykh sistemakh upravleniya* (Risks in Conflict Control Systems), Tbilisi: Intelekti, 2008, 456 p. (in Russian).

12   Zhukovskiy, V.I., Salukvadze, M.E., and Beltadze, G.N., *Matematicheskie osnovy Zolotogo pravila nravstvennosti* (Mathematical Foundations of the Golden Rule of Ethics), Tbilisi: the Georgian National Academy of Sciences, 2017, 343 p. (in Russian).

13   Zhukovskiy, V.I. and Salukvadze, M.E., *Dinamika Zolotogo pravila nravstvennosti* (Dynamics of the Golden Rule of Ethics), Tbilisi: the Georgian National Academy of Sciences, 2018, 400 p. (in Russian).

14   Zhukovskiy, V.I. and Salukvadze, M.E., *The Berge Equilibrium: A Game-Theoretic Framework for the Golden Rule of Ethics*, Springer, 2020, 290 p.

Mindia E. Salukvadze organized a series of international scientific conferences and meetings.

Mindia E. Salukvadze was a well-known public figure. At different times, he was elected a Deputy to the Supreme Soviet of the Georgian SSR, Tbilisi Soviet, and district Soviets.

The aforesaid characterizes Mindia E. Salukvadze as a talented researcher, manager, and public figure. His true portrait includes greatheartedness, rarely encountered honesty, and unselfishness. For all of us—his relatives, friends, and colleagues—his decease will always be an irreparable loss.

Introduction

The mathematical theory of games (MTG) resembles an old church organ that periodically produces wonderful spiritual melodies: in recent years, these are a starfall of the Nobel Prizes in Economic Sciences on several issues closely connected with game-theoretic problems. The authors are especially pleased with the research on the mathematical foundations of the Golden Rule of ethics. The first swallow here is our book entitled *The Berge Equilibrium: A Game-Theoretic Framework for the Golden Rule of Ethics* [417], which was published by Springer in February 2020. However, it covered only the static case of the Golden Rule. As a matter of fact, this book further develops the approaches suggested in [417] to the dynamic case of the Golden Rule by taking into account the changing attributes of a conflict over time. In 2019, the Russian original received the first prize at the Faculty of Computational Mathematics and Cybernetics, Moscow State University, in the competition of works contributing to the Development Program of MSU (nomination "Outstanding publications," category "Scientific monographs of outstanding importance for the development of science and education").

In MTG, dynamic conflicts are described by the theory of positional differential games, and Chapters 1 and 2 of this book are devoted to the Golden Rule within this framework. Chapter 1 is of an auxiliary character and considers the general postulates of the Golden Rule and games with nature; in particular, it presents the concept of maximin and, especially, the principle of minimax regret as well as the recently discovered capabilities of this principle in MGT.

In conclusion, note that the studies of the Golden Rule of ethics have not been completed. We are still far from exclaiming "Acta est fabula!"[2] The approach adopted in this book is not the only possible one: even for the single-criterion problem (playing with nature), in addition to the principle of minimax regret put forward below, there exist other principles, classical (such as maximin utility, pessimism–optimism, Bayes–Laplace, and extended minimax) and derivative (such as Laplace–Bayes, Hodges–Lehmann, Germeier, *BL*-criterion, and *P*-criterion [205]). Each of them has certain advantages and shortcomings. Really, nulla lex satis commoda omnibus est.[3] The above-mentioned

---

[2] Latin "The play is over!"

[3] Latin "No law can possibly meet the convenience of every one." A quote from *The History of Rome* XXXIV: 2, by great Roman historian Titus Livius (appr. 59/64 b.c.–17 a.d.). This expression is also attributed to Marcus Porcius Cato (234–149 b.c.), a Roman statesman, orator, and prose writer.

approaches are waiting for thorough study, and the reader would certainly discover many interesting facts getting deeper into them. "On deep paths of mystery unknown creatures leave their spoor."[4]

At the end of the Introduction, we express gratitude to Alexander Yu. Mazurov, Candidate of Sciences (Physics and Mathematics), for his careful translation of the Russian text, editorial changes, permanent feedback and negotiations with the publisher, as well as for his valuable contribution to the English version of the book.

---

[4]A fragment from *Ruslan and Lyudmila*, a poem by Aleksandr S. Pushkin, (1799–1837), a Russian poet, novelist, dramatist, and short-story writer. Considered to be the greatest poet and founder of modern Russian literature.

Compendium

Start from inferiority
in order to reach superiority;
in other words, scratch your heels
rather than the back of your head.

—Kozma Prutkov[1]

Following this aphorism of K. Prutkov, let us start with preliminaries. Along this way, we will however try to forget his 40th aphorism, which says, "Explanatory sentences clarify obscure thoughts." [232, p. 83]. In Section 1.1 of this chapter, the meaning of the Golden Rule of ethics and its connection to philosophy are briefly considered. For a detailed discussion, the interested reader is referred to Chapter 1 of the previous book [417]. In Section 1.2, the key sources and components of the static case of the Golden Rule for "frozen" conflicts [417], including the fundamental concept of Berge equilibrium, are outlined. In Section 1.3, the elementary statement of "struggle with nature" (*single-criterion choice under uncertainty*) is introduced, and our hierarchical approach to constructing a guaranteed (maximin) solution for this statement, in the static and dynamic cases, is presented. As an example, a discrete-time model of optimal advertising is considered. In Section 1.4, the principle of minimax regret and the concept of guaranteed solution in risks are described.

1.1 WHAT IS THE MEANING OF THE GOLDEN RULE OF ETHICS?

Quod tibi fieri non vis, alteri ne feceris.[2]

In this section, The Golden Rule of ethics in different religions as well as its connection with philosophy, ethics, and moral behavior are discussed in detail.

1.1.1 The Golden Rule as a basis of world religions

The Golden Rule of ethics is a general moral rule that can be formulated in the following way: "Behave to others as you would like them to behave to you." The negative statement of this rule is: "Do not do unto others what you don't want others to do unto you."

---

[1] An English translation of a quote from [232, p. 239].

[2] Latin "Do not do unto others what you don't want others to do unto you." A favourite phrase of Roman emperor Marcus Aurelius Severus Alexandrus (209–235 A.D.).

The Golden Rule of ethics is well known in the religious and philosophical teachings of the East and West. Moreover, it underlies many world religions (Abrahamic, Dharmic, Confucianism) and ancient philosophy [70] as well as expresses a fundamental ethical principle. This rule arose in the middle of the first millennium B.C., manifested itself almost simultaneously and independently in various cultures of the East and West, and apparently served as a most vivid embodiment of the humanistic revolution of that period. It received the Golden status in the 18th century in the Western European spiritual tradition.

Types of the Golden Rule

Being an expression of some general philosophical and moral law, the Golden Rule however has different types in different cultures. Scientists and philosophers endeavored to classify the forms of the Golden Rule by ethical or social attributes.

German jurist and philosopher Christian Thomasius identified three forms of the Golden Rule, differentiating the fields of law, policy, and morality. He called them the principles of justum, decorum, and honestum; see below.

- The principle of justum requires that man never does to others what he would not like others to do to him.
- The principle of decorum requires that man does to others what he would like others to do to him.
- The principle of honestum requires that man behaves as he would like others to behave.

Clearly, there are two formal aspects (statements) of the Golden Rule, namely,

- negative, which denies evil ("Do not do to others…");
- positive, which affirms good ("Do to others…").

Russian philosopher V.S. Solovyov called the first (negative) aspect of the Golden Rule the rule of justice, and the second (positive, of Christ) the rule of charity [5, 6].

German philosopher H. Reiner suggested three statements of the Golden Rule, which are similar to the ones of Christian Thomasius and V.S. Solovyov [319]:

- the rule of empathy (Ein-fuhlungsregel), which instructs: "(Do not) do to others what you would (not) like for yourself";
- the rule of autonomy (Autonomieregel), which instructs: "(Do not) do what you would (not) find praiseworthy in others";
- the rule of reciprocity (Gegenseitigkeitsregel), which instructs: "In a way you would (not) like others to behave to you, do (not) behave to them."

The Golden Rule of ethics is not only a topic of academic studies, but also a subject of contemplation for any thinking person. This rule dates back to the middle of the first millennium B.C., the period of humanistic revolution. The Golden status was assigned to it in the 18th century. For details, see Chapter 1 of the book [417].

World religions about the Golden Rule

Those who cannot remember the past
are condemned to repeat it.
—Santayana[3]

Ancient philosophy

In Aristotle's works, the Golden Rule of ethics did not appear in pure formulation. There were, however, many consonant judgments in his ethics. For example, to the question: "How should friends be behaved to?", Aristotle answered: "The way you would like them to behave to you."

In one form or another, the Golden Rule of ethics can be found in Thales of Miletus [176], Hesiod, Socrates [180], Plato [180], Aristotle [7], and Seneca [176, 323, 324].

Abrahamic religions

Judaism

The Five Books of Moses: "*You shall not take vengeance, nor bear any grudge against the children of your people, but you shall love your neighbor as yourself: I am the Lord.* (The Book of Levit, Chapter 19:18). This precept is considered the major commandment of Judaism by Judaic sages. A well-known Judaic parable says: "Once there was a gentile who came before Shammai, and said to him: "Convert me on the condition that you teach me the whole Torah while I stand on one foot." Shammai pushed him aside with the measuring stick he was holding. The same fellow came before Hillel, and Hillel converted him, saying: "That which is despicable to you, do not do to your fellow, this is the whole Torah, and the rest is commentary, go and learn it." [23, 176]. Shammai and Hillel The Elder were most influential rabbis in Jewish history.

Christianity

In the New Testament, this commandment was repeated several times by Jesus Christ.

– The Gospel according to St. Matthew: "*Therefore all things whatsoever ye would that men should do to you, do ye even so to them: for this is the law and the prophets.*" (Chapter 7:12); "*You shall love your neighbor as yourself.*" (Chapter 19:19); "*Jesus said to him, "You shall love the Lord your God with all your heart, with all your soul, and with all your mind." This is the first and great commandment. And the second is like it: "You shall love your neighbor as yourself." On these two commandments hang all the Law and the Prophets.*" (Chapter 22:37–40).

– The Gospel according to St. Mark: "*And to love Him [God] with all the heart, with all the understanding, with all the soul, and with all the strength, and to love one's neighbor as oneself, is more than all the whole burnt offerings and sacrifices.*" (Chapter 12:33).

– The Gospel according to St. Luke: "*And as ye would that men should do to you, do ye also to them likewise.*" (Chapter 6:31).

---

[3]Jorge Agustín Nicolás Ruiz de Santayana y Borrás, well known in the English speaking world as George Santayana (1863–1952), was a Spanish–American philosopher, poet, and humanist who made important contributions to aesthetics, speculative philosophy, and literary criticism.

This rule was repeated by the apostles of Jesus Christ.

- St. Paul's Epistle to the Romans: "*For the commandments, 'You shall not commit adultery,' 'You shall not murder,' 'You shall not steal,' 'You shall not bear false witness,' 'You shall not covet,' and if there is any other commandment, are all summed up in this saying, namely, 'You shall love your neighbor as yourself.'*" (Chapter 13:9).
- St. Paul's Epistle to the Galatians: "*For all the law is fulfilled in one word, even in this: 'You shall love your neighbor as yourself.'*" (Chapter 5:14).
- The Epistle of St. James: "*If you really fulfill the royal law according to the Scripture, 'You shall love your neighbor as yourself,' you do well*;" (Chapter 2:8).
- The Acts of the Apostles: "*For it seemed good to the Holy Spirit, and to us, to lay upon you no greater burden than these necessary things: that you abstain from things offered to idols, from blood, from things strangled, and from sexual immorality. If you keep yourselves from these, you will do well. Farewell.*" (Chapter 15:28,29).

In a Brief addressed to the French bishops on April 6, 1233, Pope Gregory IX recommended: "*Est autem Judaeis a Christianis exhibenda benignitas quam Christianis in Paganismo existentibus cupimus exhiberi.*"[4]

Islam

The Golden Rule of ethics was not mentioned in the Qur'an, but it appeared simultaneously in the positive and negative interpretation in *Sunnah* as one of the sayings of the Prophet Muhammad, who taught the highest principle of faith in this way: "Do to all people what you would like them to do to you, and do not do to others what you would not like for yourself." [176].

Indian religions

Hinduism

Before The Battle of Kurukshetra (around 1500–1000 B.C.), which was described in *Mahabharata*, Vidura admonished his brother, King Dhritarashtra, in the following way: "*One should never do that to another which one regards as injurious to one's own self. This, in brief, is the rule of dharma. Other behavior is due to selfish desires.*" See *Brihaspati, Mahabharata* (Anusasana Parva, Section CXIII, Verse 8).

In *The Upanishads*, a new idea of no difference between the finite and the eternal was introduced. This idea is expressed by the so-called Upadishad unity formula, *Tat Tvam Asi*.[5] This unity formula can be interpreted as the Golden Rule of ethics, as it expresses the desire to see oneself in another person [176].

Chinese philosophy

The Golden Rule appeared in the teachings of Confucius, see *The Analects (Lun Yu)*, Chapter XV, 24: "*Zi Gong [a disciple] asked: "Is there any one word that could guide a person throughout life?" The Master replied: "How about 'reciprocity'! Never impose on others what you would not choose for yourself."*" [176]. Similar principles can be found in Daoism and Mohism.

The Golden Rule and Nature

The reciprocity of the Golden Rule also applies to natural world.

---

[4]Latin "Christians must show towards Jews the same good will which we desire to be shown to Christians in pagan lands."

[5]Sanskrit "Thou art that, or That thou art, or That art thou, or You are that, or That you are, or You're it."

"Knowing the connection of the world, (carelessness is not for his benefit). Look at the exterior (world from analogy with thy own) self; [then] thou wilt neither kill nor destroy (living beings); viz. out of reciprocal regard [well examining] he does no sinful act. What is the characteristic of a sage? Recognising the equality (of all living beings), he appeases himself." (Jainism; see *Holy Akaranga Sutra*, Book 1: Hot and Cold, Third Lesson:128).

Anyone who is going to take a sharp stick and pierce though a chick with it should first try it on himself in order to feel how painful it is. (African traditional religions. A proverb of the Yoruba people, Nigeria).

"If we say that birds, horses, dogs and monkeys are entirely alien to us, we might equally reasonably assert that all savage, black and yellow people are alien to us. And if we consider them aliens, the black and the yellow people may equally reasonably consider us aliens. Who then is our neighbor? To this there is but one answer: do not ask who is your neighbor, but do unto every creature what you desire to have done unto you." (L. Tolstoy; a quote from *The Pathway of Life*).

Criticism of the Golden Rule

German classical philosopher I. Kant formulated a practical imperative close to the famous categorical one in his work *Foundations of the Metaphysics of Morals* (1785), developing it in *Critique of Practical Reason* (1788), as follows:

"Act as if the maxim of your action were to be erected by your will a universal law of nature."

Since man is the subject of possible unconditional goodwill, he is *the superior purpose*. As a result, the supreme principle of ethics can be stated in a different way:

"So act in such a way that you treat humanity, whether in your person in the person of any other, always at the same time as an end and never merely as means." [169].

Discussing the feasibility of this imperative (principle), in a footnote to his second remark, he wrote: "Let it not be thought that the common: quod tibi non vis fieris,[6] could serve here as the rule or principle. For it is only a deduction from the former, though with several limitations; it cannot be a universal law, for it does not contain the principle of duties to oneself, nor of the duties of benevolence to others (for many a one would gladly consent that others should not benefit him, provided only that he might be excused from showing benevolence to them), nor finally that of duties of strict obligation to one another, for on this principle the criminal might argue against the judge who punishes him, and so on." [169].

Only a moral law that is independent of outer reasons makes man truly free. At the same time, for man a moral law is an imperative that categorically commands, since man has needs and is subjected to sensual impulses, i.e., he is capable of maxims that *contradict* the moral law. Imperative means the relation of human will to this law as an obligation, viz., *inner rational coercion* to ethical deeds. This is the concept of duty.

In a secular humanistic form, the Golden Rule of ethics was also stated by J.-P. Sartre in his work *Existentialism Is a Humanism*: "When we say that man chooses himself, we do mean that every one of us must choose himself; but by that we also mean that in choosing for himself he chooses for all men. For in effect, of all the actions a man may take in order to create himself as he wills to be, there is not one which is not creative, at the same time, of an image of man such as he believes he ought to be. To choose between this or that is at the same time to affirm the value of that which is chosen; for we are

---

[6]Latin "Do not do unto others what you don't want others to do unto you." See footnote no. 2 on p. 3 of this book.

unable ever to choose the worse. What we choose is always the better; and nothing can be better for us unless it is better for all." [226].

I.I.2 The Golden Rule and philosophy

> Philosophy is the science
> which considers truth.
> —Aristotle[7]

In this subsection, the connection between the Golden Rule and philosophy is considered.

The philosophical doctrine of the Golden Rule of ethics contains the moral prerequisites for human behavior as follows:

1) Every man strives for good and, above all, for himself.

2) Every man is the cause of his deeds: he makes decisions what to do or not to do, which stems from the conscious and purposeful nature of human activity.

3) The best deeds cannot turn into evil for every man doing them, as the man himself will not have to regret them.

On this basis, the Golden Rule is a mechanism that allows every man to determine the value and/or fallacy of human deeds. At the same time, it cannot make an immoral man moral. The Golden Rule helps every man to adhere to moral categories and to stay in the field of ethics, as well as to maintain self-respect and help others. Therefore, in the course of decision-making the Golden Rule is a requirement applied to oneself.

In the positive statement, the Golden Rule says, "do to others as you would like them to do to you"; in the negative statement, "never do to others what you would not like them to do to you." These postulates contain a holistic and concentrated idea of morality: the attitude to the other as to oneself. The Golden Rule fixes and determines a measure of the "human" in every man, as well as equalizes and similizes people to each other. In Guseinov's opinion, whenever we speak about moral equality we are concerned with only one thing: each individual is worthy of the right for happiness and "the mutual acknowledgement of this right is a prerequisite for moral communication." The Golden Rule demands "from an individual to put himself/herself in place of other individuals and behave unto them as if he/she would be in their place." "The mechanism of the Golden Rule can be defined as assimilation, as a requirement to mentally take the place of another individual." [65, p. 134; 66–71].

The Golden Rule in the positive statement encourages good deeds and sets a high hurdle for moral attitudes towards others, as well as specifies the maximum ethical requirements to human behavior in society. Covering the entire range of human deeds, the Golden Rule serves as the basis for determining and distinguishing between the moral categories of good and evil. J. Korczak wrote, "Many times I thought what "being good" means. To my mind, a good man is a man who has imagination and understands others, who can feel like others do."[8] A quote from his book *Kak lyubit' detei* (How to Love Children, Moscow: Znanie, 1968).

---

[7] Aristotle, Greek Aristoteles, (384–322 B.C.), was an ancient Greek philosopher and scientist. He is considered to be one of the greatest intellectual figures of Western history.

[8] Janusz Korczak, the pen name of Henryk Goldszmit, (1878–1942), was a Polish–Jewish doctor, writer, and child advocate who, in order to maintain his orphanage, refused to escape Nazi-occupied Poland during World War II.

Among the many features of the Golden Rule, note its self-sufficiency, sustainability, and self-fundamentality. In particular, it integrates the categories of chance and necessity, that is, "I want to" and "I have to." Such a union creates essential concepts, and the first among them is freedom. Being combined in the Golden Rule, "I want to" and "I have to" on the one hand admit of one another, and on the other limit one another, establishing a measure to balance each other. Therefore, the Golden Rule is considered the Formula of Freedom. With these categories of chance and necessity, the Golden Rule also solves the dilemma of the ethics of happiness and the ethics of duty, since it requires from every man only what he wants for himself.

At the same time, the Golden Rule is not so elementary and obvious: for implementing this rule, it is necessary to fulfill a number of conditions, at least two:

1) Man should not be abnormal or unhealthy in determining his attitude towards himself and others.

2) Man should be able to mentally put himself in the place of others and thus correct his behavior in a specific situation. Really, often people do harm to others not by malice, but due to his own thoughtlessness.

The Golden Rule is not a high abstract norm, but a very specific principle to help people dealing with temptations and doubts in their real life.

The Golden Rule is not a logical formula, but a pattern of human behavior that does not require any training, education, or special skills. It is present in the everyday experience of every man since birth and is a fundamental value of moral life. Thus, decision-making at all levels should be based on the Golden Rule of ethics.

Note that some other relevant issues, such as the moral decline of modern society, the remedy suggested by the Golden Rule and its role as the key principle of social life, were discussed in detail in Chapter 1 of the book [417].

I.2 STATIC CASE OF THE GOLDEN RULE: SOURCES AND COMPONENTS OF MATHEMATICAL THEORY

> Everything starts with small. In the beginning,
> God created the heaven and the earth.
> —Krotky[9]

As it has been mentioned, the foundations of the mathematical theory of the Golden Rule of ethics suggested by the authors consist of two parts; see [65–71, 173, 417]. The first part (and actually the previous book [417]) was limited to the static case in which conflicts are assumed to be frozen (time-invariant). Dynamic conflicts evolving over time are the subject of this monograph, which is the second part of the mathematical theory.

I.2.1 Abstract of the book *The Berge Equilibrium: A Game-Theoretic Framework for the Golden Rule of Ethics*

Let us discuss in brief the results presented in [417]. The book was written at the junction of two sciences that have little in common with each other (at least, at first glance),

---

[9]Emil Krotky, the pen name of Emanuel Ya. German, (1892–1963), was a Russian and Soviet poet, satirist, and feuilletonist. An English translation of a quote from [21, p. 498].

namely, philosophy (the Golden Rule of ethics) and cybernetics (mathematical theory of noncooperative games). The authors were motivated by The IX Moscow Festival of Science held on October 10, 2014, at Moscow State University. The program of that event in the fundamental library of MSU included lectures by Nobel laureates chemists Kurt Wüthrich (USA) and Jean-Marie Lehn (France) and biochemist Sir Richard Roberts (USA) as well as by RAS Academicians Mikhail Ya. Marov ("The Chelyabinsk meteor") and Lev M. Zelenyi ("Exoplanets: Searching for a second Earth"), Doctors of Sciences Alexander V. Markov ("Why does a human need such a big brain") and Yury I. Aleksandrov ("Neurons, humans, and cultures"). Among the other lecturers, RAS Academician Abdusalam A. Guseinov, Director of the RAS Institute of Philosophy, delivered his talk "The Golden Rule of ethics." Being inspired by the perfectly organized and delivered lecture, one of the authors of this book addressed to the speaker the following somewhat "impudent" question, "Are You interested in a mathematical theory of the Golden Rule?" The answer was affirmative. The fact was that the asker carried in his pocket the Candidate of Sciences Dissertation of Konstantin S. Vaisman, his former postgraduate, who defended it as back as 1995. The dissertation was devoted to our early attempts to study a new solution concept of noncooperative games [354], called the Berge equilibrium. The term "Berge equilibrium" arose as the result of reviewing Claude Berge's book *Théorie générale des jeux á n personnes games* [260], which was originally published in 1957 and then translated into Russian in 1961. To our deep convictions, the Berge equilibrium matches well the main requirements of the Golden Rule. Unfortunately, Vaisman's sudden decease at the age of 35 suspended further development of Berge equilibrium in Russia. At that time, however, the concept of Berge equilibrium was "exported from Russia" by two Algerian postgraduates of V. Zhukovskiy, M. Radjef and M. Larbani. Later on, it was actively used by Western researchers. As was shown by their publications, most of investigations are focused on the properties of Berge equilibrium, the specific features and modifications of this concept, and relations to the Nash equilibrium. It seems that the nascent theory of Berge equilibrium is getting close to becoming a consistent and rigorous mathematical theory. Probably, an intensive accumulation of facts will lead to an evolutionary inner development. At this stage, following tradition, it is necessary to answer two fundamental questions:

1)   Does a Berge equilibrium exist?

2)   How can it be calculated?

In the book [417], the internal instability of the set of Berge equilibria was revealed. To eliminate this negative feature, we suggested a method to construct a Berge equilibrium that is Pareto-maximal with respect to all other Berge equilibria. The method reduces to the calculation of a saddle point for an auxiliary zero-sum two-player game that is effectively designed using the original noncooperative game. Also, we established the existence of such a (Pareto-refined) Berge equilibrium in mixed strategies under standard assumptions of mathematical game theory, i.e., the compact strategy sets and continuous payoff functions of players. This provides the answer to both questions!

Note that much attention in the book [417] was paid to Berge equilibrium in the games under uncertainty as a brand-new line of research.

Finally, in the book [417] some applications to the competitive economic models (the Cournot and Bertrand oligopolies) were considered.

Thus, the book [417] contained five independent parts as follows.

The general philosophical issues related to the Golden Rule of ethics were discussed in Chapter 1, which was written by A. Guseinov [65–71].

A practical design method for the Berge–Pareto equilibrium was introduced and the existence of such an equilibrium in mixed strategies was proved in Chapter 2.

The results of a pioneering research of guaranteed Berge equilibria in conflicts under interval uncertainty were presented in Chapter 3.

The explicit forms of Berge equilibria in the mathematical models of Cournot and Bertrand oligopolies, including their statements under uncertainty, were studied in Chapter 4.

Three new approaches to important problems of mathematical game theory and multicriteria choice were considered in Chapter 5. The first approach ensures payoff increase with simultaneous risk reduction in the Savage–Niehans sense in multicriteria choice problems and noncooperative games. The second approach allows stabilizing coalitional structures in cooperative games without side payments under uncertainty. The third approach serves to combine the "selfish" Nash equilibrium with the "altruistic" Berge equilibrium.

For refining the results mentioned above, we will use the following notations.

$\mathbb{N} = \{1, \ldots, N\}$ as the set of players;

x_i as a pure strategy of player i from a set $X_i \subseteq \mathbf{R}^{n_i}$, where $\mathbf{R}^m$ is the m-dimensional Euclidean space;

$x = (x_1, \ldots, x_N) \in X = \prod_{i \in \mathbb{N}} X_i$ as a strategy profile;

$Y \subseteq \mathbf{R}^m$ as the set of interval uncertainty y;

Y^X as the set of strategic uncertainties $y(x) : X \to Y$;

comp $\mathbf{R}^{n_i}$ as the set of all compact sets in $\mathbf{R}^{n_i}$;

$f_i(x) : X \to \mathbf{R}$ as the payoff function of player i;

$f_i(\cdot) \in C(X)$ as the class of all continuous functions $f_i(x)$ on a set X;

X^B (X^{SB}) as the set of Berge equilibria (Slater-maximal Berge equilibria, respectively);

$(x\|z_i) = (x_1, \ldots, x_{i-1}, z_i, x_{i+1}, \ldots, x_N)$;

$\exists$ as existential quantifier ("exist(s)").

For a noncooperative normal form N-player game

$$\Gamma = \langle \mathbb{N}, \{X_i\}_{i \in \mathbb{N}}, \{f_i(x)\}_{i \in \mathbb{N}} \rangle,$$

a strategy profile $x^B \in X$ is a *Berge equilibrium* if

$$f_i(x\|x_i^B) \leqslant f_i(x^B) \quad \forall x \in X, \ i \in \mathbb{N}. \tag{1.2.1}$$

The set X^B of Berge equilibria is internally unstable, i.e., there may exist $x^{(1)}$, $x^{(2)} \in X^B$ such that $f_i(x^{(1)}) > f_i(x^{(2)})$ ($\forall\, i \in \mathbb{N}$). Due to this fact, we formalize the concept of Slater-maximal Berge equilibrium (weakly efficient Berge equilibrium) by adding Slater maximality (weak efficiency) with respect to the other strategy profiles from X^B. The set of Berge equilibria with the additional requirement of Slater maximality is denoted by X^{SB}. The set X^{SB} has the following properties:

1) $X^{SB} \in$ comp$X$ (possibly, $X^{SB} = \varnothing$) if $X_i \in$ comp $\mathbf{R}^{n_i}$ and $f_i(\cdot) \in C(X)$ $(i \in \mathbb{N})$; see [3, 22, 33, 64, 71, 79, 148, 277, 279, 283, 286, 292, 343, 355, 359, 394].

2)   The existence of x^{SB} is reduced to the existence of a saddle point for a special Germeier convolution of the payoff functions [121].

3)   There exists a SBE in mixed strategies in Γ if $X_i \in$ comp$\mathbf{R}^{n_i}$ and $f_i(\cdot) \in C(X)$ $(i \in \mathbb{N})$ (an analog of Glicksberg's theorem for Nash equilibrium) [120].

4)   Similar results were established in [120] for the game Γ under strategic uncertainty, i.e., for the game

$$\langle \mathbb{N}, \{X_i\}_{i \in \mathbb{N}}, Y^X, \{f_i(x, y)\}_{i \in \mathbb{N}} \rangle.$$

5)   For the Cournot and Bertrand oligopoly models, the cases in which the players gain higher payoffs in a Berge equilibrium than in a Nash equilibrium were identified; see [71, 102, 106–109, 161, 162, 171, 174, 181–185, 191–193].

However, we are still far from exclaiming "*Acta est fabula!*"[10] Our research efforts in [417] were based on the *Non multa sed multum* principle[11] and the *Nune aut nonquam* slogan.[12] The results presented in the book [417] form a part of the mathematical theory of the Golden Rule that describes the static case. Some relevant issues remained untouched, such as risk consideration [12, 16, 47, 56, 58, 83–85, 94–98, 104, 105, 149–154, 157–160, 212, 278, 280, 288–291, 296–301, 320, 372, 378–380, 385, 401], the dynamic case of the Golden Rule (particularly, for the multistage games [103, 215, 216, 275, 276, 308, 309, 341]), and a gamut of other problems arising in the modern theory of differential positional games. Our intention is to cover all these issues below.

Concluding this subsection, let us quote Sir Richard Stone, who believed that with a mathematical description of processes "our decisions may eventually come to rest a little more on knowledge and a little less on guesswork than they do at present."[13]

I.2.2   Berge equilibrium as a fundamental concept of mathematical theory of the Golden Rule

This assertion is illustrated by the following common situation.

Imagine that there are three sellers in a market, namely, a man (husband), his wife, and their son. At their disposal they have resources X_h, X_w, and X_s, respectively, and for gaining some profit they allocate parts of their resources $x_i \in X_i$ $(i = h, w, s)$ to each family member. Which of these values will yield the greatest (possible) profit for each member? Really, the profit (revenues) $P_i(x_h, x_w, x_s)$ $(i = h, w, s)$ directly depends on the chosen values x_h, x_w, and x_s.

The concept of Nash equilibrium has been "reigning" in such decision problems or game situations so far. A Nash equilibrium is a strategy profile $x^e = \left(x_h^e, x_w^e, x_s^e\right)$ that satisfies the three equalities

---

[10]Latin "The play is over!"

[11]Latin "Not many, but much," meaning not quantity but quality.

[12]Latin "Now or never."

[13]Sir John Richard Nicholas Stone, (1913–1991), was a British economist and the father of national income accounting, who in 1984 received the Nobel Prize in Economic Sciences. A quote from *Scientific American*, 1964, vol. 211, no. 3, pp. 168–182.

$$\begin{cases} \max_{x_h} P_h\left(x_h, x_w^e, x_s^e\right) = P_h\left(x_h^e, x_w^e, x_s^e\right), \\ \max_{x_w} P_w\left(x_h^e, x_w, x_s^e\right) = P_w\left(x_h^e, x_w^e, x_s^e\right), \\ \max_{x_s} P_s\left(x_h^e, x_w^e, x_s\right) = P_s\left(x_h^e, x_w^e, x_s^e\right). \end{cases}$$

Here, a selfish nature clearly appears because everyone seeks to increase (maximize) *one's own* profit only, ignoring the interests of the others.

The concept of Berge equilibrium put forward in this book is the exact opposite of Nash equilibrium. A Berge equilibrium is a strategy profile $x^B = \left(x_h^B, x_w^B, x_s^B\right)$ defined by the three equalities

$$\begin{cases} \max_{x_w, x_s} P_h\left(x_h^B, x_w, x_s\right) = P_h\left(x_h^B, x_w^B, x_s^B\right), \\ \max_{x_h, x_s} P_w\left(x_h, x_w^B, x_s\right) = P_w\left(x_h^B, x_w^B, x_s^B\right), \\ \max_{x_h, x_w} P_s\left(x_h, x_w, x_s^B\right) = P_s\left(x_h^B, x_w^B, x_s^B\right). \end{cases}$$

It is precisely these conditions that implement the Golden Rule of ethics, which states, "do to others as you would like them to do to you." According to these conditions, each family member has to maximize the profit of the other members so that they act in the same way, maximizing his/her profit. Thus, choosing $x_h = x_h^B$ the husband does his best to maximize the profit of his wife and son, as dictated by the equalities

$$\max_{x_h, x_s} P_w\left(x_h, x_w^B, x_s\right) = P_w\left(x^B\right), \quad \max_{x_h, x_w} P_s\left(x_h, x_w, x_s^B\right) = P_s\left(x^B\right).$$

The wife and son reciprocate with $x_w = x_w^B$ and $x_s = x_s^B$, respectively, maximizing the husband's profit, that is,

$$\max_{x_w, x_s} P_h\left(x_h^B, x_w, x_s\right) = P_h\left(x^B\right).$$

The wife has the same behavior: choosing $x_w = x_w^B$, she maximizes the profits of her husband and son. Following the ethical lead of the parents who maximize his payoff

$$\max_{x_h, x_w} P_s\left(x_h, x_w, x_s^B\right) = P_s\left(x^B\right),$$

the son also strives to maximize their profits using $x_s = x_s^B$, i.e.,

$$\max_{x_w, x_s} P_h\left(x_h^B, x_w, x_s\right) = P_h\left(x^B\right), \quad \max_{x_h, x_s} P_w\left(x_h, x_w^B, x_s\right) = P_s\left(x^B\right).$$

Thus, in a Berge equilibrium each of the members maximizes the profit of the other two members and receives the same response from them. In other words, the concept of Berge equilibrium matches well the Golden Rule.

As it was evident from Subsections 1.1.1–1.1.5 of the book [417], the Golden Rule features in many fields of human activity. Moreover, in a series of cases the decision-making procedures based on the Golden Rule yield more profitable solutions in competitive economic models than the generally accepted Nash equilibrium; see Chapter 4 of the book [417]. In this respect we fully agree with B. Russel,[14] who admitted that "in all affairs it's a healthy thing now and then to hang a question mark on the things you have long taken for granted."

---

[14]Bertrand Arthur William Russell, (1872–1970), was a British philosopher, logician, social reformer, and Nobel laureate in Literature.

I.3   UNCERTAINTY AND MAXIMIN

Dubia plus torquent mala.[15]

I.3.I   Preliminaries

The genealogical tree of game theory has roots going deep into centuries,[16] powerful trunks and a thick crown in which numerous modern works on game theory are intertwined. A flowering and fruitful trunk—noncooperative games—was cultivated in 1949 by twenty-one-year-old American mathematician John Nash. In his 27 pages-long doctoral dissertation defended at Princeton University, Nash managed to separate out a "new face" of competition and defined a strategy profile, which was later called Nash equilibrium. After 45 years, J. Nash, together with R. Selten and J. Harsanyi, was awarded the Nobel Prize in Economic Sciences "for their pioneering analysis of equilibria in the theory of non-cooperative games." However, the concept of Nash equilibrium has a number of negative properties (internal instability, nonuniqueness, no equivalence, no interchangeability, improvability) [121, Section 2.2.3] and, especially striking, selfishness that permeates it. Really, following the concept of Nash equilibrium, each conflicting party seeks to improve only his result, paying no attention to the interests of others. In this book and [417], we make an attempt to plant a new sprout, dictated by the altruism of the Golden Rule of ethics—the aspiration to help others, sometimes forgetting about oneself. The life-giving rain for this blossom to flourish is triggered by the following factors.

First, an integration of dynamic programming with the Lyapunov function method was proposed by Academician N.N. Krasovskii. As a result, Lyapunov's brilliant idea to perform the stability analysis of the trajectories of a differential equation using only the definiteness of Lyapunov functions was transformed into the ability to find equilibrium strategies (in particular, Berge equilibrium) by the extreme properties of Bellman–Krasovskii functions.

Second, optimal solutions of guaranteeing control problems are unstable with respect to small disturbances and informational errors. In view of this fact, for regularization of optimal solutions, Academician Krasovskii and his followers introduced and developed the ideology of control procedures in which a real object is considered jointly with a similar reference system—guide. The motion of a guide, conceivable or modeled on a computer, acts as an ideal undisturbed process. Actually, this leads to a stabilization problem in a new game-theoretic statement. In the late 1970s, the control concept of differential and evolutionary systems based on a joint consideration of a real controlled object and an auxiliary model system (guide) was further refined. A convenient tool on that way was a uniform description of the dynamics of a model system suggested by Krasovskii. In Chapter 2 of this book, guiding control will be adopted to identify a class of differential positional games for which there exists a Berge equilibrium in a corresponding differential positional game with "separated" dynamics.

Third, due to the conceptual specifics of Berge equilibrium, the Germeier convolution of the players' payoff functions can be successfully applied not only in the static, but also in the dynamic case of the Golden Rule of ethics.

---

[15]Latin "Doubtful ills plague us worst." A quote from *Agamemnon* 480, by Seneca the Younger. In full Lucius Annaeus Seneca, (c. 4 B.C.–65 A.D.), was a Roman philosopher, statesman, orator, and tragedian.
[16]"The need for making decisions under conflict is as old as humanity itself." A quote from [36, p. 10].

And *fourth*, in mathematical models the presence of uncertain factors (uncertainties) without any probabilistic characteristics, just known ranges (e.g., price jumps in a sales market, disruption and (or) variations in the nomenclature of supplies, man-made changes, etc.), and also multistage control (control at discrete time instants) were successfully taken into account. (As a matter of fact, many problems of economic planning, engineering and production control, military science, ecology, medicine are described by difference equations: in practice, information on the state of a process is acquired and the process itself is controlled at discrete time instants.)

Let us briefly discuss the remainder of Section 1.3. It includes four subsections as follows. In Subsection 1.3.2, which has the character of a survey, the uncertainties used in economic problems are classified by types. Next, explicit forms of the maximin in the linear-quadratic positional dynamic problem for the continuous (Subsection 1.3.3) and discrete (Subsection 1.3.4) cases are obtained. In Subsection 1.3.5, a discrete optimization model of advertising strategy is presented.

1.3.2   Uncertainty and its types

L'homme propose et Dieu dispose.[17]

What is uncertainty? How does uncertainty appear in economic systems and decision-making? These questions are discussed below.

Causes of uncertainty

In these matters the only certainty
is that there is nothing certain.
—Pliny the Elder[18]

In the study of any system, including economic ones, the uncertainties affecting it have to be taken into account.

First, this is due to the peculiarities of the evolution of weakly structured systems—the systems described by both qualitative and quantitative characteristics with dominating qualitative, little-known or uncertain parameters.

Second, economic systems are controlled under insufficient knowledge of the state of an external environment, often with large investments of resources. Moreover, a special class of problems is to study economic systems that will operate at their limiting capability, in order to obtain maximum economic or any other benefits.

Third, the need to consider uncertainty becomes vital if separate, often conflicting subsystems are included into a system under study. In this case, an ambiguous solution cannot be found, and some kind of compromise has to be reached accordingly.

Fourth, both in the theory and practice of control, the starting point is some predetermined goals. In other words, for predicting the evolution of complex economic systems, we have to assign plans that are in essence are rather proactive than corrective.

Fifth, deterministic methods are often used in formal modeling of a particular economic system. With such an approach, certainty is introduced into those situations where it does not actually exist. The inaccuracy of setting parameters during calculations

---

[17]French "Man proposes, God disposes." This proverb means people can make plans but whether or not they are successful depends on God.

[18]Gaius Plinius Secundus, (23–79 A.D.), well known as Pliny the Elder, was a Roman writer, natural philosopher, and scientist.

is neglected, or under certain assumptions, inaccurate parameters are replaced by expert appraisals or average values. The resulting violations of equalities, balance relations, etc., make it necessary to vary some parameters for precisely satisfying the given conditions and obtaining an acceptable output. Such situations may occur due to insufficient knowledge of objects and also because of a person or group of persons participating in the control process. The peculiarity of such systems is that a significant part of the information required for their mathematical description exists in the form of beliefs or recommendations of experts [2].

Notion of uncertainty and classification of uncertainty in economic systems

> As far as the laws of mathematics
> refer to reality, they are not certain;
> and as far as they are certain,
> they do not refer to reality.
> —Einstein[19]

The incomplete and/or inaccurate information on the conditions of implementing a chosen strategy is its inherent *uncertainty*. Uncertainty is caused by *embarras du choix*.[20] For an economic system, the concept of uncertainty characterizes a situation in which there is no reliable information about the possible conditions of the internal and external environment, completely or partially. For example, V.V. Cherkasov [244] considered uncertainty to be an incomplete or inaccurate representation of the values of various parameters in the future, caused by various reasons and, above all, incomplete or inaccurate information on the conditions of implementing decision, including costs and results.

Information about the external factors of an economic system is never absolutely sufficient, at least because it comes from the past and the present whereas a desired behavior of the system is oriented towards the future. The smaller the completeness and accuracy of information is and the longer the period for which the behavior of the system is planned, the greater the uncertainty will be.

F. Knight [206] understood a situation of uncertainty as a lack of awareness and the need to act based on opinion rather than knowledge.

Cherkasov interpreted uncertainty as the continuous variability of conditions, a fast and flexible reconfiguration of production, the actions of competitors, market changes, etc. He called uncertainty a most typical cause of risk in management.

There exist various approaches to classify the types of uncertainty. In the roughest classification, two classes are distinguished, namely, "good" uncertainty (some statistical or probabilistic characteristics for unknown factors are available) and "bad" uncertainty (such characteristics cannot be obtained in principle). Note that both types of uncertainty arising in real problems are taken into account using appropriate methods; for example, see [36].

In [201], the following classification of uncertainties was suggested:

– by degree of uncertainty: *probabilistic, linguistic, interval, and complete uncertainty*;

---

[19]Albert Einstein, (1879–1955), was a German-born physicist who developed the special and general theories of relativity and won the Nobel Prize in Physics (1921) for his explanation of the photoelectric effect.
[20]French "difficulty due to not knowing what to prefer, what to choose."

- by the nature of uncertainty: *is parametric, structural, situational, and strategic uncertainty*;
- by the use of information acquired during control: eliminable and ineliminable uncertainty.

V.S. Diev [74] presented more detailed classifications of uncertainties in modern economic systems.

Sources of uncertainty in economic systems

> If the art of war were nothing
> but the art of avoiding risks,
> glory would become
> the prey of mediocre minds...
> I have made all the calculations;
> fate will do the rest.
> —Napoleon[21]

Considering the sources of uncertainty, we will distinguish three interconnected factors that cause uncertainty in economic systems [18].

1) The complexity factor: as a rule, an economic system is a large system that cannot be assigned a complete formal description, as well as a system with a variable structure, a nontrivial hierarchy and internal contradictions that is often controlled using fuzzy criteria.

2) The human factor: human participation is an essential element that determines the behavior of an economic system at different levels and also affects various aspects of its operation. Moreover, the human factor manifests itself in the fact that many concepts, characteristics and parameters of economic behavior are formulated in natural language without an exact formal equivalent, which creates considerable (sometimes insurmountable) difficulties in modeling.

3) The external environment factor: for any economic system, the influence of other (external) systems has to be taken into account, which are often in conflict with the former.

In view of the above factors causing uncertainty in economic systems, we will divide the sources of uncertainty into three groups as follows.

1) Insufficient information about an economic system itself and about the processes running within it. Consequently, full-fledged conclusions or assumptions on the evolution of an economic system and the final results cannot be made. In turn, such a situation may be due to:
   - few data and other reasons that can be partially eliminated by organizing a system of timely and complete information support (for example, in technical systems, state monitoring is performed using information-measuring systems with inevitable errors, and the number of monitored parameters is limited, which do not prevent the appearance of some uncontrolled technical conditions,

---

[21] Napoleon I, French in full Napoléon Bonaparte, (1769–1821), was a French general, first consul (1799–1804), and emperor of the French (1804–1814/15).

possibly causing disasters; in economic systems, the set of possible outcomes is well known, but the probability of a particular outcome can be unknown);

- imperfect tools used to study an economic system, modeling errors, computational complexity, etc.

2)   Accidental or deliberate counteraction of other economic agents. Such counteraction may have the form of violated contractual obligations by suppliers, uncertain demand for products, difficulties in marketing, or the behavior of local and regional authorities, both official and criminal. In addition, there are uncertainties caused by the competitive environment predetermining to a large extent the fate of a particular enterprise (e.g., industrial espionage, the penetration of competitors into trade secrets, and other effects on the internal affairs of a given enterprise).

3)   The effect of random external factors that cannot be predicted due to their unexpectedness. Also, the impossibility of predicting further evolution of processes due to the objectively inaccurate and ambiguous knowledge of the environment at the modern stage of science development. In particular,

- the uncertainties caused by insufficient knowledge of nature (e.g., the exact composition of supply of fish for a given fishing area in a given season is unknown);

- the uncertainties of the natural phenomena themselves (meteorological conditions affecting the average catch of fish, the mobility of supply of fish, etc.).

Thus, uncertainty is associated either with an insufficient amount of necessary information, or with the objective impossibility to acquire it and suggest reliable scenarios for the evolution of economic processes. In any case, the degree of uncertainty is determined by information, its amount, quality, and timeliness.

Approaches to estimate uncertainty in economic systems

> The quest for certainty blocks the search for meaning.
> Uncertainty is the very condition to impel man to unfold his powers.
> —Fromm[22]

Quantitative and qualitative estimation of uncertainty in economic systems includes many factors. The main factors are as follows: fluctuating demand for goods; resource endowment; price variations for raw materials and components; changes in the cost of energy and labor; inflation, etc. Each of the aspects discussed earlier—the randomness of processes under consideration, the distortion and unavailability of information about possible events and processes, and the counteraction of other systems—makes the problem of quantifying uncertainty complicated.

Historically, probabilistic and statistical approaches to the estimation of uncertainty appeared first.

The frequency, or statistical, interpretation of probability is most common. In this case, probability is identified with the relative frequency of occurrence of a mass random event during sufficiently long trials. However, within a limited experiment, only sample estimates of the parameters of the probability density function or its moments can be

---

[22]Erich Fromm, (1900–1980), was a German-born American psychoanalyst and social philosopher who explored the interaction between psychology and society.

obtained. Therefore, in the statistical description of an uncertainty factor y, its actual first and second moments—the mean $M(y)$ and variance $D(y)$—are replaced by some estimates, and their accuracy is contingent on the design of the experiment, the number of trials, the variance of noises, the estimation method selected, etc. The reliability of statistical conclusions based on such estimates considerably depends on the type of postulated distribution laws, being quite sensitive to violation of the initial hypotheses. From this viewpoint, no individual event has a frequency, and therefore it makes no sense to talk about its probability. Hence, *statistical probability can only be used to quantify such alternatives for which statistical information does exist.* The successful operation of a system with the noises and disturbances of statistical nature also depends on the effectiveness of a prediction algorithm, which can be based on the ideas of the least-squares method, Bayesian and non-Bayesian approaches, etc.

If hypotheses proceed from expert appraisals, a subjective interpretation of probability is adopted. Moreover, the probability is interpreted as a "measure of confidence" in one or another value of an uncertainty factor. There exist various approaches to assessing subjective information—the ones suggested by J. Keynes, F. Ramsey, A. Wald, L. Savage, and others [217].

If hypotheses are accepted for purely logical reasons, the matter concerns a logical interpretation of probability. Logical probability describes an experiment-independent degree of confirmation of one judgment by other judgments, i.e., the probability that a judgment is true depends on its validation using other judgments [168].

Also, note that the probabilistic-statistical approach to the estimation of uncertain factors involves the so-called stochastic description [45]. A random factor y is completely stochastically described if its probability density function is defined. Since probability density functions are an exhaustive characteristic of random variables, some researchers consider the situation when the probability density is known to be deterministic. However, when using stochastic models, we inevitably face a number of difficulties connected with the complexity of obtaining probability density functions for model parameters.

In many applications, there is no reason or insufficient information to treat the uncertain factors as random. (For example, repeated experiments with a studied object under the constant action of unaccounted and uncontrolled factors cannot be performed even hypothetically.) This leads to the need to consider the uncertainty of nonstatistical (generally, unknown) character: the only information available about an uncertain factor y is its limitedness. In such conditions, the most common and natural model is the representation of uncertain factors *in interval form, with a specified range of possible values of variables or relations.* Such an interval-probabilistic formalization of uncertainty actually increases the applicability of probability theory in practice by expanding probabilistic observations to interval-probabilistic observations: instead of a single distribution law, which cannot be justified in most real problems due to the absence of sufficient statistics, a whole set (family, class) of similar probability distributions that are consistent with available information is considered.

With the interval-probabilistic approach to the formalization of uncertainty, we may also weaken the assumption on the statistical homogeneity of observed events to construct probabilistic hypotheses, since in most problems of economics and finance the statistical homogeneity of observed events cannot be proved, which makes the use of classical statistical probabilistic hypotheses impossible. In such informational conditions, the point estimation of the parameters of a probability distribution becomes unreasonable. (Of course, an initial data set is assumed to manifest some law expressed in probabilistic or other form). As a result, it turns out that the exact estimates of the parameters of an observed law are unknown; they are blurred, and their admissible values lie in a certain interval.

Due to the high complexity of an object, considerable nonlinearity, difficulties in formalization, the presence of various subjective criteria and constraints, fuzzy models can be used. If information on the model parameters and system requirements is defined by an expert in a natural language (from a mathematical viewpoint, in rather fuzzy terms), then an approach operating the concepts of a linguistic variable and a fuzzy set [164] is adopted, which has been intensively developed in recent years. This approach rests on the premise that the components of human thinking are not numbers, but elements of some fuzzy sets or classes of objects for which the transition from belonging to a class to nonbelonging is not jump-like, but continuous. Traditional analysis methods are of little avail for such systems precisely because of their incapability to capture the fuzziness of human thinking and behavior. The approach based on the theory of fuzzy sets is, in fact, an alternative to the generally accepted quantitative methods of systems analysis. It has three distinguishing features as follows [2].

– Instead of (or, in addition to) numerical variables, fuzzy values and the so-called linguistic variables are used.
– Elementary relations between variables are described by fuzzy statements.
– Complex relations are described by fuzzy algorithms.

In fuzzy modeling, the concept of a fuzzy set is used, i.e., a set Y for which there exists a membership function $\mu(y)$ such that $\mu(y) = 0$ for $y \notin Y$, $\mu(y) = 1$ for $y \in Y$, and $0 < \mu(y) < 1$ for all other y for which the question whether they belong to Y or not has no clear answer. In the latter case, $\mu(y)$ can be interpreted as the degree of membership of y in the set Y. How should an appropriate membership function be constructed and its values be calculated? This issue depends on the specifics of problems, the nature of source data, the way they are processed, etc. In other words, this is a special issue with a particular solution in each case. The methods for constructing membership functions were described well in [2]. If we ignore the form of a membership function, then the situation with an unknown distribution of the parameters of an economic system can be described in terms of fuzzy sets. In this case, the exact value of a parameter is replaced by a certain set of its admissible values, each having some "degree of confidence" of an expert.

Deterministic approaches to consider uncertain factors are important in the practical design of various systems. To guarantee desired properties of a system, developers and engineers often have to focus not on the probabilities of satisfying some requirements or on the fuzzy expectations of given indicators, but on an acceptable behavior of the system in a set of states that are crucial for further operation. For example, in industrial control systems, the deterministic approach is employed to design controllers that ensure a stable system behavior in the presence of uncertainty in the elements of mathematical models due to their imperfections (inaccurate parameters) or external disturbances (uncertain inputs). If there are limits of element uncertainties, then this information is used in feedback controllers. With feedback control, the output parameters obtained under certain assumptions can be adopted for further refining and revising the input indicators, standards, estimates, etc. In economic systems (e.g., enterprises), such controllers often involve technical and economic indicators of system operation. Given the critical values of the factors affecting a system, we may calculate the critical values of indicators and monitor them for identifying a current state of the system in an admissible set of its behavior.

Conclusion

> Where is the beginning of the end
> that comes at the end of the beginning?
> —Kozma Prutkov

Thus, we may summarize the following [75–78].

1) Modern economic systems have a large number of elements and relations between them, a high degree of dynamism, and nonfunctional relations between the elements; such systems are affected by subjective factors due to human participation. More specifically, human beings make decisions during the operation of economic systems. As a result, an economic system evolves under uncertainty, external and internal.

2) The sources of uncertainty in economic systems include insufficient information about economic processes and their conditions; random or deliberate counteraction from other economic agents; random factors that cannot be predicted due to unexpected occurrence; for details, see [207–212, 217–224, 230–256, 226–229].

3) Uncertainty is estimated using the deterministic and probabilistic-statistical approaches as well as the approaches based on the concepts of a linguistic variable and a fuzzy set.

4) Attempts to adopt a particular mathematical framework (interval analysis, statistical methods, deterministic models, etc.) for making decisions in the face of uncertainty can adequately reflect only some types of data in the model, causing the loss of all other types of information. This especially refers to complex hierarchical systems with human beings in their control loops. For example, in deterministic models, the available statistical information on probability distributions of some parameters is not taken into account, and these distributions are replaced by the corresponding average values. Moreover, in this case there is a pressing need for information of a specific type (e.g., on the probability distribution functions). In view of insufficient information for a strict application of probabilistic models and the described difficulties with random variables, the theory of fuzzy sets comes at the forefront here, also because its framework suits well interval values. Consequently, for assessing the economic performance of a system with uncertainty, we should combine various formal approaches (deterministic, statistical, probabilistic, and fuzzy). Studying the quantitative characteristics obtained by heterogeneous formal methods is the most fruitful approach to decision problems under uncertainty.

I.3.3   Maximin in static case

> Maximin is the problem to find the minimum amount of
> fabric required for sewing a maxi skirt.[23]

This subsection is devoted to the single-criterion choice problem under uncertainty, which is described by an ordered triplet $\Gamma_1 = \langle X, Y, f(x, y) \rangle$.

---

[23]An English translation of a joke from a humorous mathematical glossary in [60, p. 204].

Here the choice of a strategy (alternative) x from a set $X \subseteq \mathbf{R}^n$ is in charge of a decision-maker (DM). In economic systems, the role of DMs belongs to the general managers of industrial enterprises and business companies, the heads of states, sellers (suppliers), and buyers (customers); in mechanical control systems, to the captains of ships or aircrafts and the chiefs of control centers. In other words, a DM has right or authority to make decisions, give instructions, and control their implementation. Each DM chooses from a given set of admissible actions, which will be called *strategies*. More specifically, *a strategy is comprehended as a rule that associates with each state of the player's awareness a certain action (behavior) from a set of admissible actions (behaviors) given this awareness.* Consider the case in which the DM's admissible strategies are the elements x of a well-defined set X. For a seller, a strategy is the price of one good; for the general manager of an industrial enterprise, strategies are production output, the amount of raw materials and equipment purchased, investments, innovations and implementation of new technologies, wages reallocation, penalties, bonuses, and other incentive and punishment mechanisms; for the captain of a ship, a strategy is own course (rudder angle, the direction and magnitude of reactive force).

In the single-criterion choice problem under uncertainty Γ_1, the DM's goal is to choose an appropriate strategy $x \in X$ for maximizing the values of a scalar criterion $f(x, y)$ (outcomes). The DM has to consider a possible realization of any uncertainty $y \in Y \subseteq \mathbf{R}^m$ within given limits. The value of $f(x, y)$ may indicate profit or production output. If the criterion $f_1(x, y)$ is associated with total losses or production cost (to be minimized), then the problem Γ_1 should be solved with $f(x, y) = -f_1(x, y)$, since

$$\max_{x \in X} f(x, y) = -\min_{x \in X} f_1(x, y).$$

Now, we proceed to uncertainty. The following situation seems common for almost everybody: it is necessary to reach a place of employment from home. First of all, a person in such conditions (further called passenger) has to decide which means of transportation to use (subway, bus, tramcar, suburban electric train, etc.). Choosing any means of transportation (strategy), passenger inevitably encounters *incomplete and/or inaccurate information*: delays or breakdowns of vehicles, sudden changes of schedule, strikes of drivers, weather fluctuations, crashes on routes, and other uncertainties. As was noted by O. Holmes, "The longing for certainty…is in every human mind. But certainty is generally illusion."[24] At best passenger knows the variation ranges of these factors, without any probabilistic appraisals. Nevertheless, he/she has to make decision anyway! As a matter of fact, the incomplete and/or inaccurate information about the conditions under which his strategy is implemented makes its inherent uncertainty. In the problem Γ_1, denote by y a numerical value of uncertainty and by Y the set of all such values. We assume that the set Y is *a priori* given and non-empty.

Hereinafter, in accordance with the subject matter of this book, the n-dimensional vector x will be called *the DM's strategy* in the problem Γ_1, and $f(x, y)$ will be called his *payoff function*; the value of $f(x, y)$ for a specific pair $(x, y) \in X \times Y$ will be called *an outcome* for the strategy $x \in X$ and uncertainty $y \in Y$.

Interestingly, Γ_1 can be interpreted as a one-player game with nature.

First, we will introduce the concept of a guaranteed solution in outcomes of the problem Γ_1 and also its hierarchical interpretation using a two-level hierarchical game in

---

[24]Oliver Wendell Holmes, Jr., byname The Great Dissenter, (1841–1935), was a justice of the United States Supreme Court, U.S. legal historian and philosopher who advocated judicial restraint.

the case where the interval uncertainty $y \in Y$ in Γ_1 is replaced by the strategic uncertainty $y(x) : X \to Y$, $y(\cdot) \in Y^X$.

Formalization of guaranteed solution in outcomes

The first attempt to solve the problem Γ_1 was undertaken by Wald in 1939; see [351]. It was based on *the maximin principle*, also known as the principle of guaranteed result. Let us formulate this principle in the following way.

Definition 1.3.1 *The guaranteed solution in payoffs (outcomes) of the problem Γ_1 is a pair $(x^g, f^g) \in X \times \mathbf{R}$ determined by the chain of equalities*

$$f^g = \max_{x \in X} \min_{y \in Y} f(x, y) = \min_{y \in Y} f\left(x^g, y\right). \tag{1.3.1}$$

The strategy x^g is called guaranteeing, and the value f^g the guaranteed outcome.

The whole essence of this solution can be explained as follows: choosing and using a strategy x^g, the DM guarantees an outcome f^g under any uncertainty $y \in Y$, since $f^g = \min_{y \in Y} f(x^g, y)$ implies

$$f\left(x^g, y\right) \geqslant f^g \quad \forall y \in Y.$$

The maximin (1.3.1) includes two successive operations, namely,

1) *the inner minimum*, which is intended to find an m-dimensional vector function $y(\cdot) : X \to Y$ such that, for each $x \in X$,

$$f(x, y(x)) = \min_{y \in Y} f(x, y),$$

and hence for each $x \in X$ it follows that

$$f(x, y) \geqslant f(x, y(x)) \quad \forall y \in Y; \tag{1.3.2}$$

2) *the outer maximum*, which is intended to construct a strategy x^g such that

$$\max_{x \in X} f(x, y(x)) = f\left(x^g, y\left(x^g\right)\right) = f^g,$$

and hence

$$f^g = f\left(x^g, y\left(x^g\right)\right) \geqslant f(x, y(x)) \quad \forall x \in X. \tag{1.3.3}$$

In fact, formula (1.3.3) means that among all minima of $f(x, y(x))$ in (1.3.2) for different $x \in X$, we choose the value f^g maximizing $f(x, y(x))$ in x, which is implemented on the strategy x^g.

Remark 1.3.1 Recall that we consider a special class of uncertainties of the form Y^X, which consists of the functions $y(x)$ with the domain X and the codomain Y. (The latter set is yielded by the inner minimum (1.3.2).) The actions of uncertainty are treated as the behavior of another (dummy) player, which has no payoff function and directs every effort to do as much harm to the DM as possible. (This is a strategic uncertainty in the

terminology of Yu.B. Germeier.) The dummy player can use "any conceivable information. In particular, he/she possibly knows the DM's strategy." [179, p. 353]. In this case,
the so-called *informational discrimination of the DM* takes place [179, p. 353].

The inner minimum in (1.3.2) leads to a parametric problem: for each $x \in X$, find
an m-dimensional vector function $y(\cdot) \in Y^X$ such that

$$\min_{y \in Y} f(x, y) = f(x, y(x)). \tag{1.3.4}$$

In this case, the following result should be taken into account.

Proposition 1.3.1 ([10, pp. 17–18; 204, p. 54]) *Let a scalar function $f(x, y)$ be continuous
on $X \times Y$ and also let the sets X and Y be compact. Then*

(a)   *the function*

$$\min_{y \in Y} f(x, y) = f(x, y(x)) \tag{1.3.5}$$

   is continuous on X, and the multivalued mapping

$$Y(x) = \left\{ y^* \in Y \mid f(x, y^*) = \min_{y \in Y} f(x, y) \right\} \quad \forall x \in X,$$

   i.e., $Y(x) : X \to Y$, has a Borel measurable selector $y(x)$.

(b)   *Moreover, if $f(x, y)$ is strictly convex in $y \in Y$ for each $x \in X$ (i.e., for any $y^{(1)}$,
   $y^{(2)} \in Y$, $y^{(1)} \neq y^{(2)}$, and for each $x \in X$, the inequality*

$$f\left(x, \lambda y^{(1)} + (1 - \lambda)y^{(2)}\right) < \lambda f\left(x, y^{(1)}\right) + (1 - \lambda)f\left(x, y^{(2)}\right)$$

   *holds for any constants $\lambda \in (0, 1)$) and the set Y is convex, then the vector function
   $y(x)$ (1.3.5) is continuous on X.*

Corollary 1.3.1 *If a scalar function $f(x, y)$ is continuous on $X \times Y$ and the sets X and Y
are compact, then the function*

$$\max_{x \in X} f(x, y) \tag{1.3.6}$$

is continuous on Y $\left(\text{because } \min_{x \in X}[-f(x, y)] = - \max_{x \in X} f(x, y) \right)$.

The following concepts are well known in game theory and will be used in further
presentation: a) the strategy x^g defined by (1.3.1) is called *the maximin strategy*, and f^g
is called *the maximin*; by analogy, the uncertainty y^0 from

$$f^0 = \min_{y \in Y} \max_{x \in X} f(x, y) = \max_{x \in X} f\left(x, y^0\right)$$

is called *the minimax uncertainty*, and the value f^0 is called *the minimax* in the problem
Γ_1; b) a pair $(x^g, y^0) \in X \times Y$ is *a saddle point in the problem Γ_1* if

$$\max_{x \in X} f\left(x, y^0\right) = f\left(x^g, y^0\right) = \min_{y \in Y} f\left(x^g, y\right), \tag{1.3.7}$$

or equivalently,

$$\min_{y \in Y} \max_{x \in X} f(x, y) = f\left(x^g, y^0\right) = \max_{x \in X} \min_{y \in Y} f(x, y),$$

where x^g is the maximin strategy and y^0 is the minimax uncertainty in the problem Γ_1.

Proposition 1.3.2 *Assume that in the problem* Γ_1 *the sets* X *and* Y *are compact and the function* $f(x, y)$ *is continuous on* X $\times$ Y. *Then this problem has a guaranteed solution in outcomes (payoffs).*

Proof By Proposition 1.3.1 the function $\min\limits_{y\in Y} f(x, y)$ is continuous in $x \in$ X on the compact set X. According to the Weierstrass extreme-value theorem, a continuous function on a compact set X achieves maximum.

Interpretation of maximin within two-level hierarchical game

Consider the following two-player game with a fixed sequence of moves. Assume player 1 (DM) is given priority in actions over player 2. Such a statement with the first move of player 1 describes well, e.g., an interaction of conflicting parties in two-level hierarchical systems. We will also accept the hypothesis that, whenever the outcome depends on the choice of player 2 only, he/she always minimizes the payoff function $f(x, y)$. Player 1 is informed about this behavior.

Then player 1 takes advantage of the first move, reporting his strategy $x \in$ X to player 2. Making the second move in this game, player 2 responds with a counter strategy $y(x) :$ X $\to$ Y that minimizes the function $f(x, y(x))$ for each $x \in$ X. If for each x this minimum is achieved at a unique point $y(x)$, then the best (guaranteed) result of player 1 makes up

$$f^g = \max_{x\in X}\min_{y\in Y} f(x, y) = \max_{x\in X} f(x, y(x)) = f\left(x^g, y\left(x^g\right)\right) = \min_{y\in Y} f\left(x^g, y\right).$$

The sequence of moves of the DM and player 2 is illustrated in Figure 1.3.1.

As a result, the DM prefers the maximin strategy x^g, which yields the guaranteed payoff

$$f^g \leqslant f\left(x^g, y\right) \quad \forall y \in Y.$$

Note that, for all $x \in$ X, this payoff exceeds all other guaranteed payoffs

$$\min_{y\in Y} f(x, y) \leqslant f^g \quad \forall x \in X.$$

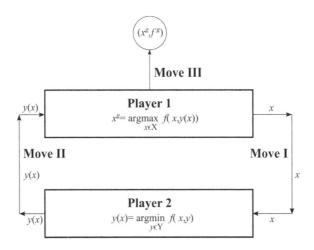

Figure 1.3.1 Sequence of moves.

I.3.4   Maximin in dynamic case: linear-quadratic positional control problem

> Some man married a very skinny woman.
> Being asked why, he said,
> "I have chosen the least evil."
> —Bar Hebraeus[25]

In Subsection 1.3.3, the static statement of the problem Γ_1 has been considered, where all the three elements X, Y, and $f(x, y)$ are assumed to be "frozen" for the entire decision period. However, the mathematical model itself is changing over time: "Everything flows, nothing stands still." In this subsection, the guaranteed solution of the linear-quadratic dynamic statement of the problem Γ_1 is found in explicit form.

First of all, we present *the mathematical model*. Consider a linear-quadratic positional control problem under uncertainty that is described by an ordered quadruple

$$\Gamma_d = \left\langle \Sigma, \mathfrak{A}, \mathcal{Z}_u, \mathcal{J}(U, \mathcal{Z}_u, t_0, x_0) \right\rangle.$$

The controlled system Σ evolves over time $t \in [t_0, \vartheta]$ in accordance with the vector linear differential equation

$$\dot{x} = Ax + u + z, \quad x(t_0) = x_0, \tag{1.3.8}$$

with the following notations: A as a constant matrix of dimensions $n \times n$; $\vartheta > t_0 \geqslant 0$ as a given terminal time instant of the control process; $x(t)$ as the value of the state vector $x \in \mathbf{R}^n$ at a time instant $t \in [t_0, \vartheta]$; $(t, x(t))$ as a pair determining the position at a time instant t; (t_0, x_0) as an initial position; $u \in \mathbf{R}^n$ as a DM's control action; $z \in \mathbf{R}^n$ as an uncertain factor; $\mathfrak{A}$ as the set of DM's positional strategies U. A strategy U will be identified with an n-dimensional vector function $u(t, x) = P(t)x$, where the elements of a matrix $P(t)$ of dimensions $n \times n$ are assumed to be continuous for any $t \in [0, \vartheta]$. (This fact will be indicated by $P(\cdot) \in C^{n \times n}[0, \vartheta]$.) In this case, an appropriate strategy is assigned by choosing a matrix $P(\cdot) \in C^{n \times n}[0, \vartheta]$. In addition, denote by $\mathcal{Z}_u$ the set of uncertainties $\mathcal{Z}_u \div z(t, x, u) = Q(t)x + R(t)u$ for all $Q(\cdot), R(\cdot) \in C^{n \times n}[0, \vartheta]$. Like for the discrete maximin, the special class of uncertainties that depend on the position (t, x) and also on the DM's control action u is dictated by the circumstance that the actions of uncertainties can be treated as the behavior of another (dummy) DM, which has no payoff function and directs every effort to do as much harm to the DM as possible through informational discrimination. Hereinafter, the terminal $\vartheta > 0$ and initial $t_0 \in [0, \vartheta)$ times are assumed to be fixed.

The controlled process has the following dynamics over time. Assume that the DM has chosen and adopted a specific strategy $U \in \mathfrak{A}$, $U \div u(t, x) = P(t)x$, $P(\cdot) \in C^{n \times n}[0, \vartheta]$. Also, let some uncertainty $Z_u \in \mathcal{Z}_u$, $Z_u \div z(t, x, u) = Q(t)x + R(t)u$, have been realized regardless of this choice. Substituting the above strategy $U \div u(t, x) = P(t)x$ and uncertainty $Z_u \div z(t, x, u) = Q(t)x + R(t)u$ into (1.3.8), we obtain a system of linear homogeneous differential equations with continuous in t coefficients in the vector form

$$\dot{x} = Ax + u(t, x) + z(t, x, u(t, x)) = [A + P(t) + Q(t) + R(t)P(t)]\, x, \quad x(t_0) = x_0.$$

---

[25]Bar Hebraeus, Arabic Ibn Al-'Ibri ("Son of the Hebrew"), or Abu al-Faraj, Latin name Gregorius, (1226–1286), was a medieval Syrian scholar noted for his encyclopaedic learning in science and philosophy. An English translation of a quote from [188, p. 23].

This system has a unique and continuous solution $x(t)$ that is extendable to the interval $[t_0, \vartheta]$; see [214, p. 29]. Using this solution, we construct a realization $u[t] = u(t, x(t)) = P(t)x(t)$ of the DM's strategy and a corresponding (independent) realization $z[t] = z(t, x(t), u(t, x(t))) = Q(t)x(t) + R(t)P(t)x(t)$, $t \in [t_0, \vartheta]$, of the uncertainty. Note that each of the three functions $x(t)$, $u[t]$, and $z[t]$ is continuous on $[t_0, \vartheta]$. The resulting triplet

$$(x(t), u[t], z[t] \mid t \in [t_0, \vartheta])$$

will serve as the codomain of the payoff (utility) function of the DM described by the functional

$$\mathcal{J}(U, Z_u, t_0, x_0) = x'(\vartheta)Cx(\vartheta) + \int\limits_{t_0}^{\vartheta} \left(u'[t]Du[t] + z'[t]Lz[t] \right) dt. \tag{1.3.9}$$

A value of (1.3.9) is called *an outcome*. In formula (1.3.9), constant matrices C, D, and L of dimensions $n \times n$ are symmetric. Like before, the prime indicates transposition; $C < 0$ $(>, \leqslant)$ denotes that a quadratic form $x'Cx$ is negative (positive, nonnegative) definite; E_n is an identity matrix and $0_{n \times n}$ is a zero matrix, both of dimensions $n \times n$; 0_n is a zero n-dimensional vector; finally, *Idem* $\{u \to u^g\}$ means the bracketed expression with u replaced by u^g.

At conceptual level, choosing his strategy $U \in \mathfrak{A}$, the DM seeks to maximize the outcome $\mathcal{J}(U, Z_u, t_0, x_0)$ under any *a priori* unpredictable realization of the uncertainty $Z_u \in \mathcal{Z}_u$.

According to Definition 1.3.1, *a pair* $(U^g, \mathcal{J}^g[t_0, x_0])$ *will be called the guaranteed solution in outcomes of the problem* Γ_d *if there exists an uncertainty* $Z_u^g \in \mathcal{Z}_u$ *such that*

$$\mathcal{J}^g[t_0, x_0] = \max_{U \in \mathfrak{A}} \min_{Z_u \in \mathcal{Z}_u} \mathcal{J}(U, Z_u, t_0, x_0) = \min_{Z_u \in \mathcal{Z}_u} \mathcal{J}\left(U^g, Z_u, t_0, x_0 \right)$$
$$= \mathcal{J}\left(U^g, Z_u^g, t_0, x_0 \right). \tag{1.3.10}$$

In this case, U^g *will be called the guaranteeing strategy, and* $\mathcal{J}^g[t_0, x_0]$ *the guaranteed outcome.*

Let us construct the guaranteed solution in outcomes using an appropriate modification of dynamic programming [155, pp. 113–116]. Such an approach gives the following *sufficient conditions for the existence of the guaranteed solution* of the problem Γ_d; see the proposition below. They involve the scalar function

$$W(t, x, u, z, V) = \frac{\partial V}{\partial t} + \left[\frac{\partial V}{\partial x} \right]' (Ax + u + z) + u'Du + z'Lz, \tag{1.3.11}$$

where $\frac{\partial V}{\partial x} = \mathrm{grad}_x V$ is a vector from the space $\mathbf{R}^n$ that consists from the partial derivatives of a scalar function $V(t, x)$ with respect to the coordinates of an n-dimensional vector x.

Proposition 1.3.3 *Let* $V(t, x)$ *be a scalar continuously differentiable function such that*

$$V(\vartheta, x) = x'Cx \quad \forall x \in \mathbf{R}^n, \tag{1.3.12}$$

and let $z(t, x, u, V)$ *and* $u(t, x, V)$ *be n-dimensional vector functions.*

In addition, make the following assumptions.

(a)   *For any $t \in [0, \vartheta]$, $V \in \mathbf{R}$ and $x, u \in \mathbf{R}^n$,*

$$\min_z W(t, x, u, z, V) = W(t, x, u, z(t, x, u, V), V). \tag{1.3.13}$$

(b)   *For all $t \in [0, \vartheta]$, $V \in \mathbf{R}$, and $x \in \mathbf{R}^n$,*

$$\max_u W(t, x, u, z(t, x, u, V), V) = W(t, x, u(t, x, V), z(t, x, u(t, x, V), V), V)$$

$$= W[t, x, V]. \tag{1.3.14}$$

(c)   *For each $x \in \mathbf{R}^n$ and $t \in [0, \vartheta]$,*

$$W[t, x, V(t, x)] \equiv 0. \tag{1.3.15}$$

(d)   *The scalar function $V(t, x)$ and the vector functions $u(t, x, V)$ and $z(t, x, u, V)$ satisfy the relations*

$$u^g(t, x) = u(t, x, V(t, x)) = P^g(t)x \quad \forall x \in \mathbf{R}^n, \ t \in [0, \vartheta],$$
$$z^g(t, x, u) = z(t, x, u, V(t, x)) = Q^g(t)x + R^g(t)u \quad \forall x, u \in \mathbf{R}^n, \ t \in [0, \vartheta],$$

and $P^g(\cdot), Q^g(\cdot), R^g(\cdot) \in C^{n \times n}[0, \vartheta]$.

Then for any initial position $(t_0, x_0) \in [0, \vartheta) \times \mathbf{R}^n$, the guaranteed solution in outcomes of the problem Γ_d has the form

$$U^g \doteq u^g(t, x) = P^g(t)x,$$
$$\mathcal{J}^g[t_0, x_0] = \mathcal{J}(U^g, Z_u^g, t_0, x_0) = V(t_0, x_0), \tag{1.3.16}$$

where $Z_u^g \doteq Q^g(t)x + R^g(t)u$.

Remark 1.3.1 Condition (1.3.13) corresponds to the inner minimum in the chain of equalities (1.3.10), whereas condition (1.3.14) to the outer maximum in the same chain (1.3.10).

Using Proposition 1.3.3, we will find an explicit form of the guaranteed solution in outcomes of the problem Γ_d.

Prior to constructing this solution in the form $(U^g, \mathcal{J}^g[t_0, x_0])$, let us present two auxiliary lemmas as follows.

Lemma 1.3.1 *If symmetric matrices $L > 0$ and $D < 0$ satisfy the relation*

$$L > -D > 0, \tag{1.3.17}$$

then

$$L^{-1} + D^{-1} \leqslant 0. \tag{1.3.18}$$

Proof Since the matrices L and $-D$ are positive definite, from (1.3.17) and [38, p. 89] it follows that

$$L^{-1} \leqslant [-D]^{-1} = -D^{-1},$$

which directly gives (1.3.18).

Note that the inequality $L > -D$ (in equivalent form, $L + D > 0$) means

$$xLx > -xDx \quad \forall x \in \mathbf{R}^n.$$

Lemma 1.3.2 [156, p. 22] *Let $X(t) = e^{A(t-\vartheta)}$ be the fundamental system of solutions for the matrix equation*

$$\dot{X} = AX, X(\vartheta) = E_n, \quad t \in [0, \vartheta],$$

and compile the Cauchy matrix $Y(t) = X(\vartheta)X^{-1}(t) = X^{-1}(t) = e^{A(\vartheta-t)}$. Then the solution of the Riccati matrix equation

$$\dot{\Theta} + \Theta A + A'\Theta - \Theta K \Theta = 0_{n \times n}, \quad \Theta(\vartheta) = C \tag{1.3.19}$$

with symmetric matrices $C < 0$ and $K \leqslant 0$ has the form

$$\Theta(t) = \left[e^{A(\vartheta-t)} \right]' \left\{ C^{-1} + \int_t^\vartheta e^{A(\vartheta-\tau)} K \left[e^{A(\vartheta-\tau)} \right]' d\tau \right\}^{-1} e^{A(\vartheta-t)}, \tag{1.3.20}$$

where $\Theta(t) = \Theta'(t)$, $t \in [0, \vartheta]$.

Proposition 1.3.4 *Consider the problem Γ_d with constant symmetric matrices*

$$L > -D > 0 \quad and \quad C < 0 \tag{1.3.21}$$

of dimensions $n \times n$. For any initial position $(t_0, x_0) \in [0, \vartheta) \times \mathbf{R}^n$, the guaranteed solution in outcomes $(U^g, \mathcal{J}^g[t_0, x_0])$ of this problem is given by

$$U^g \div u^g(t, x) = -D^{-1} \Theta(t) x,$$
$$\mathcal{J}^g[t_0, x_0] = x_0' \Theta(t_0) x_0, \tag{1.3.22}$$

where

$$\Theta(t) = \left[e^{A(\vartheta-t)} \right]' \left\{ C^{-1} + \int_t^\vartheta e^{A(\vartheta-\tau)} \left[D^{-1} + L^{-1} \right] \left[e^{A(\vartheta-\tau)} \right]' d\tau \right\}^{-1} e^{A(\vartheta-t)}$$
$$\forall t \in [t_0, \vartheta]. \tag{1.3.23}$$

Proof We will find the Bellman function in the form $V(t, x) = = x' \Theta(t) x$, $\Theta(t) = \Theta'(t)$. According to (1.3.12),

$$V(\vartheta, x) = x'\Theta(\vartheta)x = x'Cx \quad \forall x \in \mathbf{R}^n \quad \Rightarrow \quad \Theta(\vartheta) = C.$$

Using (1.3.11) we construct

$$W(t, x, u, z, V) = \frac{\partial V}{\partial t} + \left[\frac{\partial V}{\partial x} \right]' (Ax + u + z) + u'Du + z'Lz, \tag{1.3.24}$$

and hence condition (1.3.13) holds if

$$\min_z \widehat{\varphi}_1(t, x, u, z, V) = \widehat{\varphi}_1(t, x, u, z(t, x, u, V), V) \quad \forall t \in [0, \vartheta], \quad x, u \in \mathbf{R}^n, \quad V \in \mathbf{R}, \tag{1.3.25}$$

where $\widehat{\varphi}_1(t, x, u, z, V) = z'\frac{\partial V}{\partial x} + z'Lz$. By (1.3.25) the minimum in (1.3.24) is achieved at $z(t, x, u, V)$ if, for all $t \in [0, \vartheta]$, $x, u \in \mathbf{R}^n$, and $V \in \mathbf{R}$,

$$\left.\frac{\partial \widehat{\varphi}_1(t, x, u, z, V)}{\partial z}\right|_{z(t,x,u,V)} = \frac{\partial V}{\partial x} + 2Lz(t, x, u, V) = 0_n, \qquad (1.3.26)$$

and the Hessian is

$$\frac{\partial^2 \widehat{\varphi}_1(t, x, u, z, V)}{\partial z^2} = 2L > 0.$$

The second requirement is true, because $L > 0$; see (1.3.21). On the other hand, equality (1.3.26) yields

$$z(t, x, u, V) = -\frac{1}{2} L^{-1} \frac{\partial V}{\partial x}; \qquad (1.3.27)$$

in addition,

$$z'(t, x, u, V)Lz(t, x, u, V) + 2z'(t, x, u, V)\frac{\partial V}{\partial x} = -z'(t, x, u, V)Lz(t, x, u, V).$$
$$(1.3.28)$$

In view of (1.3.28) and (1.3.27), the function $W(t, x, u, z(t, x, u, V), V)$ from (1.3.13) will take the form

$$W(t, x, u, z(t, x, u, V), V) = \frac{\partial V}{\partial t} + \left[\frac{\partial V}{\partial x}\right]'(Ax + u) + u'Du - \frac{1}{4}\left[\frac{\partial V}{\partial x}\right]' L^{-1}\frac{\partial V}{\partial x}.$$
$$(1.3.29)$$

Now, we construct $u(t, x, V)$ according to (1.3.14) (of course, taking (1.3.29) into account). The function $u(t, x, V)$ implements the maximum in (1.3.14) if

$$\max_u \widehat{\varphi}_2(t, x, u, V) = \widehat{\varphi}_2(t, x, u(t, x, V), V) \quad \forall t \in [0, \vartheta], \quad x \in \mathbf{R}^n, \quad V \in \mathbf{R},$$
$$(1.3.30)$$

where

$$\widehat{\varphi}_2(t, x, u, V) = u'\frac{\partial V}{\partial x} + u'Du.$$

Then (1.3.30) holds provided that

$$\left.\frac{\partial \widehat{\varphi}_2(t, x, u, V)}{\partial u}\right|_{u(t,x,V)} = \frac{\partial V}{\partial x} + 2Du(t, x, V) = 0_n,$$

$$\left.\frac{\partial^2 \widehat{\varphi}_2(t, x, u, V)}{\partial u^2}\right|_{u(t,x,V)} = 2D < 0.$$

From the first equality above it follows that, first,

$$u(t, x, V) = -\frac{1}{2} D^{-1} \frac{\partial V}{\partial x}; \qquad (1.3.31)$$

and second,

$$u'(t, x, V)\frac{\partial V}{\partial x} + u'(t, x, V)Du(t, x, V) = -u'(t, x, V)Du(t, x, V).$$

Finally, using this equality in combination with (1.3.31), for the function (1.3.29) with $u = u(t, x, V)$ we find

$$W[t, x, V] = \frac{\partial V}{\partial t} + \left[\frac{\partial V}{\partial x}\right]' Ax - \frac{1}{4}\left[\frac{\partial V}{\partial x}\right]'\left[L^{-1} + D^{-1}\right]\frac{\partial V}{\partial x}.$$

According to (1.3.15), we construct $V(t, x) = x'\Theta(t)x$, where $\Theta'(t) = \Theta(t)$. Then (1.3.15) is valid if the symmetric matrix $\Theta(t)$ of dimensions $n \times n$ satisfies the equation

$$\dot{\Theta} + \Theta A + A'\Theta - \Theta[L^{-1} + D^{-1}]\Theta = 0_{n \times n}, \quad \Theta(\vartheta) = C. \tag{1.3.32}$$

Due to the constraints (1.3.21), the condition $C < 0$ and Lemma 1.3.1, we have $L^{-1} + D^{-1} \leqslant 0$. As a result, by Lemma 1.3.2 the matrix $\Theta(t)$—the solution of (1.3.32)—is given by (1.3.20) with $K = L^{-1} + D^{-1}$.

Thus, formula (1.3.23) has been established. On the other hand, (1.3.22) follows from (1.3.31), since

$$V(t, x) = x'\Theta(t)x \quad \Rightarrow \quad \frac{\partial V(t, x)}{\partial x} = 2\Theta(t)x,$$

and by (1.3.16)

$$V(t_0, x_0) = x'_0\Theta(t_0)x_0.$$

1.3.5 Multistage maximin

> Peace is not absence of conflict,
> it is the ability to handle conflict by peaceful means.
> —Reagan[26]

For the difference statement of the linear-quadratic problem from Subsection 1.3.4, the guaranteed solution in outcomes (maximin) is constructed below using an appropriate modification of dynamic programming.

As the mathematical model, let us consider the ordered quadruple

$$\left\langle \sum_0, \mathfrak{A}, \mathcal{Z}_u, \mathcal{J}(U, \mathcal{Z}_u, x_0)\right\rangle, \tag{1.3.33}$$

which will be called *the K-stage positional single-criterion linear-quadratic problem under uncertainty.* We make several assumptions regarding (1.3.33) as follows.

– The controlled system $\sum$ evolves over time in accordance with the vector linear difference equation

$$x(k + 1) = Ax(k) + u + z = f(k, x(k), u, z), \quad x(0) = x_0, \tag{1.3.34}$$

with the following notations: $k = 0, 1, ..., K - 1$ as time instants, i.e., partition points of an entire time interval $[0, K]$ on which *the controlled discrete process $\sum$ is evolving;*

---

[26]Ronald Wilson Reagan, (1911–2004), was the 40th president of the United States (1981–89), "the Great Communicator."

$x(k) \in \mathbf{R}^n$ as the value of *the state vector* x at a time instant $t = k$; $u \in \mathbf{R}^n$ as *a DM's control action*; $z \in \mathbf{R}^n$ as *an uncertain factor*; $(k, x(k))$ as a pair determining the position of (1.3.33) at a time instant k; $(0, x_0)$ as an initial position; A as a constant matrix of dimensions $n \times n$.

- *A DM's positional strategy* $U(k)$ *at a time instant* k is identified with a vector function $u(k, x) = P(k)x$, where $P(k) \in \mathbf{R}^{n \times n}$ is a constant matrix of dimensions $n \times n$. (This fact will be indicated by $U(k) \div u(k, x) = P(k)x$.) Hence, at a time instant k an appropriate strategy is assigned by choosing a specific matrix $P(k) \in \mathbf{R}^{n \times n}$ of dimensions $n \times n$. Thus, an ordered collection

$$U = (U(0), U(1), ..., U(K - 1)) \div (u(0, x), u(1, x), ..., u(K - 1, x))$$
$$= (P(0)x, P(1)x, ..., P(K - 1)x)$$

is a DM's strategy in the problem (1.3.33); the set of all such strategies U will be denoted by $\mathfrak{A}$.

- The set of strategic positional *uncertainties* $Z_u(k)$ *at a time instant* k will be denoted by $\mathcal{Z}_u(k)$. It consists of

$$Z_u(k) \div z(k, x, u) = Q(k)x + R(k)u,$$

where $Q(k), R(k) \in \mathbf{R}^{n \times n}$ are constant matrices of specified dimensions. The special class of uncertainties that depend on the position (k, x) and also on the control action u has been selected due to the reasons discussed in Remark 1.3.1. As a result, the uncertainty in the problem (1.3.33) is described by the ordered collection

$$Z_u = (Z_u(0), Z_u(1), ..., Z_u(K - 1)) \div (z(0, x, u), z(1, x, u), ..., z(K - 1, x, u))$$
$$= (Q(0)x + R(0)u, Q(1)x + R(1)u, ..., Q(K - 1)x + R(K - 1)u);$$

the set of such uncertainties is denoted by $\mathcal{Z}_u$.

The controlled process in the problem (1.3.33) has the following dynamics over time. Assume that the DM has chosen and adopted a specific strategy $U \in \mathfrak{A}$:

$$U \div (u(0, x), u(1, x), ..., u(K - 1, x)) = (P(0)x, P(1)x, ..., P(K - 1)x).$$

Also, let some uncertainty $Z_u \in \mathcal{Z}_u$ have been realized in Σ regardless of this choice:

$$Z_u = (Z_u(0), ..., Z_u(K - 1)) \div (z(0, x, u), ..., z(K - 1, x, u))$$
$$= (Q(0)x + R(0)u, ..., Q(K - 1)x + R(K - 1)u).$$

Substituting the above strategy U and uncertainty Z_u into (1.3.34), we obtain

$$x(1) = Ax_0 + u(0, x_0) + z(0, x_0, u(0, x_0)) = [A + P(0) + Q(0) + R(0)P(0)]x_0$$
$$= f(0, x_0, u(0, x_0), z(0, x_0, u(0, x_0))),$$
$$x(2) = [A + P(1) + Q(1) + R(1)P(1)]x(1)$$
$$= f(1, x(1), u(1, x(1)), z(1, x(1), u(1, x(1)))),$$

$$\vdots$$

$$x(K) = [A + P(K-1) + Q(K-1) + R(K-1)P(K-1)]x(K-1)$$
$$= f(K-1, x(K-1), u(K-1, x(K-1)),$$
$$z(K-1, x(K-1), u(K-1, x(K-1)))).$$

This gives three sequences,

$\{x(k)\}_{k=0}^{K}$,

$\{u[k] = P(k)x(k)\}_{k=0}^{K-1}$, and

$\{z[k] = Q(k)x(k) + R(k)P(k)x(k)\}_{k=0}^{K-1}$,

which form *the criterion* (also called *the payoff* or *utility function of the DM*)

$$\mathcal{J}(U, Z_u, x_0) = x'(K)Cx(K) + \sum_{k=0}^{K-1} \left(u'[k]D(k)u[k] + z'[k]L(k)z[k] \right)$$

$$= \Phi(x(K)) + \sum_{k=0}^{K-1} F(k, x(k, u[k], z[k])). \tag{1.3.35}$$

A value of the function (1.3.35) is called an outcome or DM's payoff. In formula (1.3.35), all matrices C, $D(k)$, $L(k)$, $P(k)$, $Q(k)$, and $R(k)$ of dimensions $n \times n$ are constant, and the matrices C, $D(k)$, and $L(k)$ are symmetric. Recall that the prime indicates transposition; for a symmetric matrix $M \in \mathbf{R}^{n \times n}$, the expression $M > 0$ (< 0) shows that the quadratic form $u'Mu$ is positive (negative, respectively) definite; E_n is an identity matrix of dimensions $n \times n$; 0_n is a zero n-dimensional vector; finally, $Idem\{u \rightarrow u^g\}$ means the bracketed expression with u replaced by u^g. In addition, $detB$ denotes the determinant of a square matrix B.

At conceptual level, choosing his strategy $U \in \mathfrak{A}$, the DM seeks to maximize the outcome $\mathcal{J}(U, Z_u, x_0)$ in the problem (1.3.33) under any realization of the uncertainty $Z_u \in \mathcal{Z}_u$. Definition 1.3.1 naturally leads to the following concept.

A pair $(U^g, \mathcal{J}^g[x_0]) \in \mathfrak{A} \times \mathbf{R}$ will be called the guaranteed solution in outcomes of the problem (1.3.33) if there exists an uncertainty $Z_u^g \in \mathcal{Z}_u$ such that

$$\mathcal{J}^g[x_0] = \max_{U \in \mathfrak{A}} \min_{Z_u \in \mathcal{Z}_u} \mathcal{J}(U, Z_u, x_0) = \min_{Z_u \in \mathcal{Z}_u} \mathcal{J}(U^g, Z_u, x_0) = \mathcal{J}(U^g, Z_u^g, x_0).$$

$$\tag{1.3.36}$$

In this case, U^g will be called the guaranteeing strategy, and $\mathcal{J}^g[x_0]$ the guaranteed outcome.

Remark 1.3.2 The concept of guaranteed solution in outcomes suggests the DM to use the strategy $U^g \in \mathfrak{A}$ in the problem (1.3.33) on two grounds as follows. *First,* the equality

$$\min_{Z_u \in \mathcal{Z}_u} \mathcal{J}(U^g, Z_u, x_0) = \mathcal{J}^g[x_0]$$

implies that $\mathcal{J}(U^g, Z_u, x_0) \geqslant \mathcal{J}^g[x_0]$ under any uncertainty realization of the uncertainty $Z_u \in \mathcal{Z}_u$. In other words, with this strategy the outcome will be not smaller than the guaranteed outcome $\mathcal{J}^g[x_0]$ (the lower bound on $\mathcal{J}(U^g, Z_u, x_0)$ over all $Z_u \in \mathcal{Z}_u$.)

Second, for each strategy $U \in \mathfrak{A}$ the DM will obtain the guaranteed outcome $\min_{Z_u \in \mathcal{Z}_u} \mathcal{J}(U, Z_u, x_0)$, which is not greater than $\mathcal{J}^g[x_0]$.

Now, we introduce *sufficient conditions* for the existence of the multistage maximin (1.3.36) that are based on dynamic programming. At each time instant $k = K$, $K - 1$, $K - 2$, ..., 1, 0, we will use the Bellman function

$$V^{(k)}(x) = x'\Theta(k)x,$$

with a symmetric matrix $\Theta(k) \in \mathbf{R}^{n \times n}$ as well as scalar functions

$$W\left(k, x, u, z, V^{(k+1)}(Ax + u + z)\right)$$
$$= W\left(k, x, u, z, \left(x'A' + u' + z'\right)\Theta(k+1)(Ax+u+z)\right) = W[k, x, u, z, \Theta(k+1)]$$
$$= u'D(k)u + z'L(k)z + \left(x'A' + u' + z'\right)\Theta(k+1)(Ax+u+z)$$
$$(k = K-1, K-2, ..., 1, 0). \tag{1.3.37}$$

Proposition 1.3.5 *Let* $\left\{ V^{(k)}(x) = x'\Theta(k)x \right\}_{k=0}^{K-1}$, $\{u(k, x, \Theta(k+1)) = P(k, \Theta(k+1))x\}_{k=0}^{K-1}$ *and* $\{z(k, x, u, \Theta(k+1)) = Q(k, \Theta(k+1))x + R(k, \Theta(k+1))u\}_{k=0}^{K-1}$ *be three sequences, the first composed of scalar functions and the last two of n-dimensional vector functions, that satisfy the following assumptions:*

$$V^{(K)}(x) = x'Cx \quad \forall x \in \mathbf{R}^n; \tag{1.3.38}$$

for all $x, u \in \mathbf{R}^n$, $\Theta(k+1) \in \mathbf{R}^{n \times n}$, *and* $k = K-1, ..., 1, 0,$

$$\min_z W[k, x, u, z, \Theta(k+1)] = W[k, x, u, z(k, x, u, \Theta(k+1)), \Theta(k+1)]; \tag{1.3.39}$$

for each $x \in \mathbf{R}^n$, $\Theta(k+1) \in \mathbf{R}^{n \times n}$ *and* $k = K-1, ..., 1, 0,$

$$\max_u W[k, x, u, z(k, x, u, \Theta(k+1)), \Theta(k+1)] = W[k, x, u(k, x, \Theta(k+1)),$$
$$z\left(k, x, u(k, x, \Theta(k+1)), \Theta(k+1)\right), \Theta(k+1)]; \tag{1.3.40}$$

for any $x \in \mathbf{R}^n$ *and* $k = K-1, K-2, ..., 1, 0,$

$$V^{(k)}(x) = x'\Theta(k)x = W[k, x, u(k, x, \Theta(k+1)), z(k, x, u(k, x, \Theta(k+1)),$$
$$\Theta(k+1)), \Theta(k+1)]. \tag{1.3.41}$$

Then for any initial state vector $x_0 \in \mathbf{R}^n$, *the guaranteed solution in outcomes* $(U^g, \mathcal{J}^g[x_0])$ *of the problem* (1.3.33) *has the following form:*
  the guaranteed strategy is given by

$$U^g \div \left(u^g[0, x], u^g[1, x], \ldots, u^g[K-1, x]\right),$$

where

$$u^g[k, x] = u(k, x, \Theta(k+1)) \quad (k = 0, 1, \ldots, K-1),$$

and the guaranteed outcome is given by

$$\mathcal{J}^g[x_0] = V^{(0)}(x_0) = x_0'\Theta(0)x_0.$$

(The notations are the same as in (1.3.37). Formula (1.3.41) is used to successively find the matrices $\Theta(k)$ $(k = K-1, \ldots, 1, 0)$.)

Proof This result can be established by a standard procedure; for example, see [19, pp. 366–367].

Remark 1.3.3 For each time instant $(k = 0, 1, \ldots, K-1)$, consider the auxiliary problem

$$\Gamma(k) = \langle \mathfrak{A}, \mathcal{Z}, W[k, x, u, z, \Theta(k+1)] \rangle,$$

where $\mathfrak{A}$ is the set of strategies $U = (U(0), U(1), ..., U(K-1)) \div (u(0, x), u(1, x), ...,$ $u(K-1, x))$ of the form $u(k, x) = P(k)x$; $\mathcal{Z}$ denotes the set of uncertainties $z(k, x, u, \Theta(k+1)) = Q(k)x + R(k)u$; the criterion $W[k, x, u, z, \Theta(k+1)]$ is given by (1.3.37). Then merging the requirements (1.3.39) and (1.3.40) actually means the equalities

$$\max_{u \in \mathfrak{A}} \min_{z \in \mathcal{Z}} W[k, x, u, z, \Theta(k+1)] = \min_{z \in \mathcal{Z}} W[k, x, u(k, x, \Theta(k+1)), z, \Theta(k+1)]$$

$$= W[k, x, u(k, x, \Theta(k+1)),$$

$$z(k, x, u(k, x, \Theta(k+1)), \Theta(k+1)), \Theta(k+1)] = W^g[k, x, \Theta(k+1)]. \tag{1.3.42}$$

In other words, at each time instant $k = K-1, K-2, \ldots, 1, 0$ the DM implements the maximin (1.3.42) in the auxiliary problem $\Gamma(k)$. Consequently, according to Proposition 1.3.5, implementing the local maximin at each time instant $k = K-1, K-2, \ldots, 1, 0$, the DM actually arrives at the global maximin (1.3.36) in the problem (1.3.33).

Taking advantage of Proposition 1.3.5, we will find *an explicit form of the guaranteed solution in outcomes of the problem (1.3.33)*. Before doing it, let us present *three auxiliary results*. Recall that if a quadratic form $z'Gz$ is positive (negative) definite and $G = G' \in \mathbf{R}^{n \times n}$, then all n roots λ_i of the characteristic equation $det[G - \lambda E_n] = 0$ are real and $\lambda_i > 0$ ($\lambda_i < 0$, respectively).

Lemma 1.3.3 *Consider symmetric matrices $L(k-1) > 0$ and $\Theta(k) < 0$ of dimensions $n \times n$. The inequality*

$$L(k-1) + \Theta(k) > 0$$

holds if $\lambda(k-1) > \mu(k)$, where $\lambda(k-1)$ and $-\mu(k)$ are the least roots of the characteristic equations $det[L(k-1) - \lambda E_n] = 0$ and $det[\Theta(k) - \mu E_n] = 0$, respectively.

Proof Let $\lambda(k-1)$ and $-\mu(k)$ be the least roots of the corresponding characteristic equations. In this case,

$$z'L(k-1)z \geqslant \lambda(k-1)z'z, \quad z'\Theta(k)z \geqslant -\mu(k)z'z \quad \forall z \in \mathbf{R}^n,$$

and hence

$$z'[L(k-1) + \Theta(k)]z \geqslant [\lambda(k-1) - \mu(k)]z'z > 0 \quad \forall z \in \mathbf{R}^n \setminus \{0_n\}.$$

Lemma 1.3.4 *Consider symmetric constant matrices $D(k)$, C, $L(k-1)$, and $\Theta(k)$ of dimensions $n \times n$ such that*

$$D(k) < 0, \quad C < 0, \quad L(k-1) > 0, \quad L(k-1) + \Theta(k) > 0. \tag{1.3.43}$$

Then the matrices

$$M(\Theta(k)) = \Theta(k)\left\{\Theta^{-1}(k) - [L(k-1) + \Theta(k)]^{-1}\right\}\Theta(k),$$

$$\Theta(k-1) = A'M(\Theta(k)) \times \left\{M^{-1}(\Theta(k)) - [D(k) + M(\Theta(k))]^{-1}\right\}M(\Theta(k))A \tag{1.3.44}$$

are also symmetric, $M(\Theta(k)) < 0$, and $\Theta(k-1) < 0$ if

$$detA \neq 0. \tag{1.3.45}$$

Proof The symmetry of the matrices $M(\Theta(k))$ and $\Theta(k-1)$ follows from the properties $(AB)' = B'A'$, $\left[A^{-1}\right]' = \left[A'\right]^{-1}$, $A'' = A$ and the two easily checked equalities $M(\Theta(k)) = M'(\Theta(k))$ and $\Theta(k-1) = \Theta'(k-1)$.

The negative definiteness of $M(\Theta(k))$ and $\Theta(k-1)$ is established by the chain of implications

$$\Theta(k) < 0 \Rightarrow \det\Theta(k) \neq 0 \Rightarrow \exists\Theta^{-1}(k) \wedge \Theta^{-1}(k) < 0,$$

$$[L(k-1) + \Theta(k)] > 0 \Rightarrow [L(k-1) + \Theta(k)]^{-1} > 0 \Rightarrow -[L(k-1) + \Theta(k)]^{-1} < 0,$$

$$\Theta^{-1}(k) < 0 \wedge -[L(k-1) + \Theta(k)]^{-1} < 0$$

$$\Rightarrow \Theta^{-1}(k) - [L(k-1) + \Theta(k)]^{-1} < 0 = \{\det\Theta(k) \neq 0\}$$

$$\Rightarrow M(\Theta(k)) = \Theta(k)\left\{\Theta^{-1}(k) - [L(k-1) + \Theta(k)]^{-1}\right\}\Theta(k) < 0;$$

$$D(k) < 0 \wedge M(\Theta(k)) < 0 \Rightarrow [D(k) + M(\Theta(k))] - M(\Theta(k)) = D(k) < 0$$

$$= \{[38, \text{p. } 89]\} \Rightarrow M^{-1}(\Theta(k)) - [D(k) + M(\Theta(k))]^{-1} < 0;$$

$$\det A \neq 0 \wedge M(\Theta(k)) < 0 \Rightarrow \det[AM(\Theta(k))] \neq 0,$$

$$\det[AM(\Theta(k))] \neq 0 \wedge M^{-1}(\Theta(k)) - [D(k) + M(\Theta(k))]^{-1} < 0 \Rightarrow \Theta(k-1)$$

$$= A'M(\Theta(k))\left\{M^{-1}(\Theta(k)) - [D(k) + M(\Theta(k))]^{-1}\right\}M(\Theta(k))A < 0.$$

Corollary 1.3.2 *Under conditions* (1.3.43) *and* (1.3.45) (*see Lemma* 1.3.4), *we have the implication*

$$[\Theta(k) < 0] \Rightarrow [\Theta(k-1) < 0]. \tag{1.3.46}$$

In fact, the validity of (1.3.46) has been demonstrated by Lemma 1.3.4.

Remark 1.3.4 The concept of guaranteed solution in outcomes itself directly leads to the following design method of the guaranteed solution of the problem (1.3.33) using Proposition 1.3.5; see the stages described below. The Bellman functions $V^{(k)}(x)$ have to be constructed as quadratic forms $V^{(k)}(x) = x'\Theta(k)x$, with symmetric matrices $\Theta(k) \in \mathbf{R}^{n\times n}$.

Stage 1 $(k = K)$. From (1.3.38), due to

$$V^{(K)}(x) = x'\Theta(K)x = x'Cx \quad \forall x \in \mathbf{R}^n,$$

find the matrix $\Theta(K) = C$.

Stage 2 $(k = K - 1)$. The function $W[K-1, x, u, z, \Theta(K)]$ (1.3.37) takes the form

$$W[K-1, x, u, z, \Theta(K)] = u'D(K-1)u + z'L(K-1)z$$
$$+ (x'A' + u' + z')\Theta(K)(Ax + u + z). \tag{1.3.47}$$

Checking the condition $L(K-1) + C = L(K-1) + \Theta(K) > 0$, construct $z(K\text{-}1, x, u, \Theta(K))$ according to

$$\min_z W[K-1, x, u, z, \Theta(K)] = W[K-1, x, u, z(K-1, x, u, \Theta(K)), \Theta(K)]$$

$$\forall x, u \in \mathbf{R}^n. \tag{1.3.48}$$

Next, calculate the vector function $u(K - 1, x, \Theta(K))$ according to

$$\max_u W[K - 1, x, u, z(K - 1, x, u, \Theta(K)), \Theta(K)] = W[K - 1, x, u(K - 1, x, \Theta(K)),$$
$$z(K - 1, x, u(K - 1, x, \Theta(K)), \Theta(K)), \Theta(K)] = \overline{W}[K - 1, x] \quad \forall x \in \mathbf{R}^n,$$
(1.3.49)

and find the constant matrix $\Theta(K - 1)$ of dimensions $n \times n$ from the identity

$$x'\Theta(K - 1)x = \overline{W}[K - 1, x] \quad \forall x \in \mathbf{R}^n.$$
(1.3.50)

Thus, Stage 2 yields the n-dimensional vector function

$$u^g[K - 1, x] = u(K - 1, x, \Theta(K) = C) = P(K - 1, \Theta(K))x$$

and also the symmetric matrix $\Theta(K - 1) \in \mathbf{R}^{n \times n}$.

Then, repeating all operations of Stage 2 for $k = K - 2$, obtain the vector function $u^g[K - 2, x] = u(K - 2, x, \Theta(K - 1)) = P(K - 2, \Theta(K - 1))x$ and the matrix $\Theta(K - 2) \in \mathbf{R}^{n \times n}$ of dimensions $n \times n$. And so on, for $k = K - 3, \ldots, 1$.

Finally, repeat the operations of Stage 2 for $k = 0$, replacing $\Theta(K)$ by $\Theta(1)$. For $k = 0$,

$$W[0, x, u, z, \Theta(1)] = u'D(0)u + z'L(0)z + (x'A' + u' + z')\Theta(1)(Ax + u + z).$$

Check the requirement $L(0) + \Theta(1) > 0$ and construct the vector function $z(0, x, u, \Theta(1))$ according to

$$\min_z W[0, x, u, z, \Theta(1)] = W[0, x, u, z(0, x, u, \Theta(1)), \Theta(1)] \quad \forall x, u \in \mathbf{R}^n.$$

Next, find the n-dimensional vector function $u(0, x, \Theta(1))$ according to

$$\max_u W[0, x, u, z(0, x, u, \Theta(1)), \Theta(1)] = W[0, x, u(0, x, \Theta(1)),$$
$$z(0, x, u(0, x, \Theta(1)), \Theta(1)), \Theta(1)] = \overline{W}[0, x] \quad \forall x \in \mathbf{R}^n,$$

and also the matrix $\Theta(0)$ of dimensions $n \times n$ from the identity

$$x'\Theta(0)x = \overline{W}[0, x] \quad \forall x \in \mathbf{R}^n.$$

As a result, the vector function $u^g[0, x] = u(0, x, \Theta(1)) = P(0, \Theta(1))x$ and the constant matrix $\Theta(0)$ of dimensions $n \times n$ are obtained.

Thus, for any initial state vector $x(0) = x_0 \neq 0_n$ in (1.3.34), the guaranteed solution in outcomes $(U^g, \mathcal{J}^g[x_0])$ of the problem (1.3.33) has the explicit form

$$U^g \div \left(u^g[0, x], u^g[1, x], \ldots, u^g[K - 1, x]\right),$$
$$u^g[k, x] = u(k, x, \Theta(k + 1)) = P(k, \Theta(k + 1))x \quad (k = 0, 1, \ldots, K - 1),$$
$$\mathcal{J}^g[x_0] = x_0'\Theta(0)x_0.$$
(1.3.51)

In view of the stages described in Remark 1.3.4, we may formulate the following result.

Proposition 1.3.6 *Consider the problem* (1.3.33) *with*

$$C < 0, \ detA \neq 0, \ D(k) < 0, \ L(k) > 0 \ (k = 0, 1, \ldots, K - 1)$$
(1.3.52)

and let the sequence of matrices $\{\Theta(k)\}_{k=0}^{K}$ constructed by the recursive formulas

$$\Theta(K) = C,$$

$$M(\Theta(K)) = C\left[C^{-1} - (L(K-1) + C)^{-1}\right]C,$$

$$\Theta(K-1) = A'M(\Theta(K))\left\{M^{-1}(\Theta(K)) - [D(K-1) + M(\Theta(K))]^{-1}\right\}$$
$$\times M(\Theta(K))A,$$

$$M(\Theta(K-1)) = \Theta(K-1)\left[\Theta^{-1}(K-1) - (L(K-2) + \Theta(K-1))^{-1}\right]\Theta(K-1),$$

$$\Theta(K-2) = A'M(\Theta(K-1))\left\{M^{-1}(\Theta(K-1)) - [D(K-2)\right.$$
$$\left. + M(\Theta(K-1))]^{-1}\right\}M(\Theta(K-1))A,$$

$$\vdots$$

$$\Theta(k) = A'M(\Theta(k+1))\left\{M^{-1}(\Theta(k+1)) - [D(k) + M(\Theta(k+1))]^{-1}\right\}$$
$$\times M(\Theta(k+1))A,$$

$$M(\Theta(k)) = \Theta(k)\left[\Theta^{-1}(k) - (L(k-1) + \Theta(k))^{-1}\right]\Theta(k),$$

$$\Theta(k-1) = A'M(\Theta(k))\left\{M^{-1}(\Theta(k)) - [D(k-1) + M(\Theta(k))]^{-1}\right\}M(\Theta(k))A,$$

$$\vdots$$

$$\Theta(1) = A'M(\Theta(2))\left\{M^{-1}(\Theta(2)) - [D(1) + M(\Theta(2))]^{-1}\right\}M(\Theta(2))A,$$

$$M(\Theta(1)) = \Theta(1)\left[\Theta^{-1}(1) - (L(0) + \Theta(1))^{-1}\right]\Theta(1),$$

$$\Theta(0) = A'M(\Theta(1))\left\{M^{-1}(\Theta(1)) - [D(0) + M(\Theta(1))]^{-1}\right\}M(\Theta(1))A$$

$$\tag{1.3.53}$$

be such that

$$L(k-1) + \Theta(k) > 0 \quad (k = K, K-1, \ldots, 1). \tag{1.3.54}$$

Then for any initial state vector $x_0 \in \mathbf{R}^n$ in equation (1.3.34), the guaranteed solution in outcomes $(U^g, \mathcal{J}^g[x_0])$ of the problem (1.3.33) has the form

$$U^g \div \left(-[D(0) + M(\Theta(1))]^{-1} M(\Theta(1))Ax, \ldots, \right.$$
$$\left. -[D(K-1) + M(\Theta(K))]^{-1} M(\Theta(K))Ax\right), \tag{1.3.55}$$

$$\mathcal{J}^g[x_0] = x_0'\Theta(0)x_0.$$

Proof According to **Stage 1**, the matrix $\Theta(K)$ is $\Theta(K) = C < 0$ and the Bellman function at the time instant $k = K$ is given by

$$V^{(K)}(x) = x'\Theta(K)x = x'Cx.$$

Following the recommendations of **Stage 2**, we construct the scalar function (1.3.37) for $k = K - 1$

$$W[K-1, x, u, z, \Theta(K)] = u'D(K-1)u + z'L(K-1)z$$
$$+ (x'A' + u' + z')\Theta(K)(Ax + u + z), \qquad (1.3.56)$$

and find $z(K-1, x, u, \Theta(K))$ from (1.3.39), i.e.,

$$\min_{z} W[K-1, x, u, z, \Theta(K)] = Idem[z \to z(K-1, x, u, \Theta(K))].$$

Due to the above explicit form of $W[K-1, x, u, z, \Theta(K)]$, the vector function $z(K-1, x, u, \Theta(K))$ simultaneously minimizes the function

$$\varphi_1(K-1, x, u, z) = z'L(K-1)z + z'\Theta(K)z + 2z'\Theta(K)(Ax + u) \quad \forall x, u \in \mathbf{R}^n.$$

Here, the sufficient conditions can be written as

$$\mathrm{grad}_z \varphi_1(K-1, x, u, z)\big|_{z(K-1, x, u, \Theta(K))} = \frac{\partial \varphi_1(K-1, x, u, z)}{\partial z}\bigg|_{z(K-1, x, u, \Theta(K))}$$
$$= 2[L(K-1) + \Theta(K)]z(K-1, x, u, \Theta(K)) + 2\Theta(K)(Ax + u)$$
$$= 0_n \quad \forall x, u \in \mathbf{R}^n, \qquad (1.3.57)$$

and the Hessian has the form

$$\frac{\partial^2 \varphi_1(K-1, x, u, z)}{\partial z^2} = 2[L(K-1) + \Theta(K)] > 0.$$

The last inequality is immediate from (1.3.54) with $k = K$. On the other hand, condition (1.3.57) implies, first,

$$z(K-1, x, u, \Theta(K)) = -[L(K-1) + \Theta(K)]^{-1}\Theta(K)(Ax + u), \qquad (1.3.58)$$

and second,

$$z'(K-1, x, u, \Theta(K))[L(K-1) + \Theta(K)]z(K-1, x, u, \Theta(K))$$
$$+ 2z'(K-1, x, u, \Theta(K))\Theta(K)(Ax + u)$$
$$= -z'(K-1, x, u, \Theta(K))[L(K-1) + \Theta(K)]z(K-1, x, u, \Theta(K)).$$

Using this relation, equality (1.3.58) and the first row of formula (1.3.44) with $k = K$, we obtain the following chain of equalities from (1.3.56) with $z = = z(K-1, x, u, \Theta(K))$:

$$W[K-1, x, u, z(K-1, x, u, \Theta(K)), \Theta(K)]$$
$$= u'D(K-1)u + (x'A' + u')\Theta(K)(Ax + u)$$
$$- z'(K-1, x, u, \Theta(K))[L(K-1) + \Theta(K)]z(K-1, x, u, \Theta(K))$$
$$= u'D(K-1)u + (x'A' + u')\Theta(K)\Theta^{-1}(K)\Theta(K)(Ax + u)$$
$$- (x'A' + u')\Theta(K)[L(K-1) + \Theta(K)]^{-1}\Theta(K)(Ax + u)$$
$$= u'D(K-1)u + (x'A' + u')\Theta(K)\left\{\Theta^{-1}(K) - [L(K-1) + \Theta(K)]^{-1}\right\}$$
$$\times \Theta(K)(Ax + u)$$
$$= u'D(K-1)u + (x'A' + u')M(\Theta(K))(Ax + u)$$
$$= u'[D(K-1) + M(\Theta(K))]u + 2u'M(\Theta(K))(Ax + u) + x'A'M(\Theta(K))Ax.$$

Now, we get back to (1.3.49), taking into account the formula

$$W[K-1, x, u, z(K-1, x, u, \Theta(K)), \Theta(K)]$$
$$= u'D(K-1)u + (x'A' + u')M(\Theta(K))(Ax + u) = u'[D(K-1) + M(\Theta(K))]u$$
$$+ 2u'M(\Theta(K))(Ax + u) + x'A'M(\Theta(K))Ax.$$

If the maximum in (1.3.49) is achieved at $u = u(K-1, x, \Theta(K))$, then

$$\max_u \varphi_2(K-1, x, u) = \max_u \{u'[D(K-1) + M(\Theta(K))]u + 2u'M(\Theta(K))Ax\}$$
$$\forall x \in \mathbf{R}^n \tag{1.3.59}$$

is also implemented at the same $u = u(K-1, x, \Theta(K))$. The sufficient conditions of this maximum can be written as

$$\left.\frac{\partial \varphi_2(K-1, x, u)}{\partial u}\right|_{u(K-1, x, \Theta(K))} = 2[D(K-1) + M(\Theta(K))]u(K-1, x, \Theta(K))$$
$$+ 2M(\Theta(K))Ax = 0_n \quad \forall x \in \mathbf{R}^n, \tag{1.3.60}$$

$$\frac{\partial^2 \varphi_2(K-1, x, u)}{\partial u^2} = 2[D(K-1) + M(\Theta(K))] < 0.$$

The second requirement is satisfied due to $D(K-1) < 0$ (see (1.3.52) with $k = K - 1$) and $M(\Theta(K)) < 0$. In addition, the matrix $M(\Theta(K))$ has symmetry by Lemma 1.3.4 with $k = K$.

From (1.3.60) it follows that, first,

$$u(K-1, x, \Theta(K)) = -[D(K-1) + M(\Theta(K))]^{-1}M(\Theta(K))Ax; \tag{1.3.61}$$

second,

$$u'(K-1, x, \Theta(K))[D(K-1) + M(\Theta(K))]u(K-1, x, \Theta(K))$$
$$+ 2u'(K-1, x, \Theta(K))M(\Theta(K))Ax = -u'(K-1, x, \Theta(K))[D(K-1)$$
$$+ M(\Theta(K))]u(K-1, x, \Theta(K)) \quad \forall x \in \mathbf{R}^n.$$

In view of this identity, (1.3.61), and the second row of formula (1.3.44), we obtain

$$\overline{W}[K-1, x] = W[K-1, x, u(K-1, x, \Theta(K)),$$
$$z(K-1, x, u(K-1, x, \Theta(K)), \Theta(K)), \Theta(K)]$$
$$= -u'(K-1, x, \Theta(K))[D(K-1) + M(\Theta(K))]u(K-1, x, \Theta(K))$$
$$+ x'A'M(\Theta(K))M^{-1}(\Theta(K))M(\Theta(K))Ax = x'A'M(\Theta(K))\{M^{-1}(\Theta(K))$$
$$- [D(K-1) + M(\Theta(K))]^{-1}\}M(\Theta(K))Ax = x'\Theta(K-1)x = V^{(K-1)}(x).$$
$$\tag{1.3.62}$$

Moreover, by Corollary 1.3.2,

$$[\Theta(K) (= C) < 0] \Rightarrow [\Theta(K-1) < 0],$$

and by Lemma 1.3.4 the matrix $\Theta(K-1)$ is symmetric.

The same considerations can be applied to the case $k = K - 2$ by simply replacing the matrix $\Theta(K)$ with $\Theta(K - 1)$ and the number $K - 1$ with $K - 2$ in all formulas starting from (1.3.56). Following this approach, we establish the analogs of (1.3.61) and (1.3.62),

$$u(K - 2, x, \Theta(K - 1)) = -[D(K - 2) + M(\Theta(K - 1))]^{-1} M(\Theta(K - 1))Ax,$$

and

$$V^{(K-2)}(x) = x'\Theta(K - 2)x,$$

respectively, where the nonnegative definite matrices $M(\Theta(K - 1))$ and $\Theta(K - 2)$ are given by (1.3.45) with $k = K - 1$.

Similar operations should be performed for $k = K - 3, \ldots, 1, 0$. By mathematical induction on k, for each $k = 0, 1, \ldots, K - 1$ we get

$$u(k, x, \Theta(k + 1)) = -[D(k) + M(\Theta(k + 1))]^{-1} M(\Theta(k + 1))Ax, \qquad (1.3.63)$$
$$V^{(k)}(x) = x'\Theta(k)x.$$

Finally, the end of Remark 1.3.4 in combination with the first and second rows of formula (1.3.63) with $k = 0, 1, \ldots, K - 1$ and $k = 0$, respectively, allows us to prove (1.3.55).

Remark 1.3.5 For obtaining the guaranteed solution in outcomes of the linear-quadratic discrete single-criterion problem (1.3.33)–(1.3.36) using Proposition 1.3.6, we have to

1) check the constraints (1.3.52);
2) for $k = K, K - 1, \ldots$, construct the two sequences

   $$\{\Theta(K), \Theta(K - 1), \ldots, \Theta(1), \Theta(0)\},$$

   and

   $$\{M(\Theta(K)), M(\Theta(K - 1)), \ldots, M(\Theta(1)), M(\Theta(0))\}$$

   by the recursive relations (1.3.53);
3) check whether the requirements (1.3.54) are satisfied; if so, analytically design the guaranteed solution in outcomes $(U^g, \mathcal{J}^g[x_0])$ by formulas (1.3.55).

1.3.6 Discrete-time model of optimal advertising

> Advertising may be described as the
> science of arresting the human intelligence
> long enough to get money from it.
> —Leocock[27]

In this subsection, written by Natalia V. Adukova (South Ural State University), dynamic programming is applied to the discrete-time optimal advertising problem.

In many sectors of the economy, firms are competing for a share in the market through advertising. One of the first models that describes an influence of advertising

---

[27]Stephen Butler Leacock, (1869–1944), was an internationally popular Canadian humorist, educator, lecturer, and author of more than 30 books of lighthearted sketches and essays.

costs on the market share of a monopolistic firm was proposed by M.L. Vidale and H.B. Wolfe in 1957; see [347]. In the model, the dynamics of the sales rate $s(t)$ are given by the differential equation

$$\dot{s}(t) = \gamma u(t)\,[m - s(t)] - \delta s(t), \quad s(t_0) = s_0, \tag{1.3.64}$$

where $u(t)$ denotes the advertising effort (a control variable); m is a saturation level of the sales rate; a response constant γ characterizes the effectiveness of advertising; a decay constant δ determines the rate at which consumers are lost due to product obsolescence.

In 1983, another classical model was introduced by S.P. Sethi [327]. Within the framework of the Sethi model, the dynamics of the market share $x(t)$ are described by the nonlinear differential equation

$$\dot{x}(t) = \rho u(t)\sqrt{1 - x(t)} - \delta x(t), \quad x(t_0) = x_0. \tag{1.3.65}$$

In 1989, Sorger constructed an advertising competition duopoly model by combining the Sethi model with the well-known Lancaster model; see [333].

Note that continuous-time models were considered in most works on optimal advertising. Discrete-time statements can be found in a few papers (e.g., [62, 63] and [275]), due to difficulties with explicit solution. At the same time, discrete-time models are preferable in view of the discrete nature of decisions on funding and promotions.

In this subsection, we present a new discrete-time model of optimal advertising. As special cases, it includes the discrete versions of the Vidale–Wolfe model (1.3.64) and the Sethi model (1.3.65). What is important? The new model matches well all desirable properties [305]: the market share has a concave response to advertising and a saturation level.

Mathematical model

Consider the problem of finding the optimal cost of advertising effort for a monopolistic firm that enters the market with a new product (further referred to as the optimal advertising problem). We will use a dynamic model in discrete time.

We introduce the following notations: $t = 1, \ldots, T$ as decision points (time instants of a dynamic process); T as the number of decision points; $x(t)$ as the market share at a time instant t ($x(t) \in [0, 1]$); x_0 as an initial market share; $u(t)$ as the advertising rate (expenditure) at a time instant t; δ as the decay constant of market share in the absence of advertising, $\delta \in [0, 1]$; ρ as the response constant; finally, $\sigma \in (0, 1)$ as the nonlinearity parameter of the model. The parameter ρ determines the effectiveness of advertising, and δ gives the rate at which consumers are lost due to product obsolescence, forgetting, etc.

We will study a generalization of the discrete-time counterpart of the Sethi model [328, 329]. In the model below, the dynamics $x(t)$ are described by the difference equation

$$x(t + 1) = (1 - \delta)x(t) + \rho u(t)[1 - x(t)]^{1-\sigma}, \quad t \in [0; T), \quad x(0) = x_0. \tag{1.3.66}$$

The parameters δ, ρ, and σ and the control variable $u(t)$ are assumed to be such that $x(t) \in [0, 1]$ at each time instant t.

Here the additional term $\rho u(t)[1 - x(t)]^{1-\sigma}$ can be explained as in the Sethi model. On the one hand, advertising effort (i.e., the control variable $u(t)$) has to be applied to the unsold portion of the product, i.e., to the market share $(1 - x(t))$. On the other, the difference between the nonlinear $[1 - x(t)]^{1-\sigma}$ and linear $(1 - x(t))$ terms is represented as

$$[1 - x(t)]^{1-\sigma} - [1 - x(t)] = \sigma x(t)[1 - x(t)] + \frac{\sigma(\sigma + 1)}{2} x^2(t) + \dots,$$

being proportional to $x(t)[1 - x(t)]$ for sufficiently small $x(t)$. Therefore, like in the paper [328],

$$\rho u(t)[1 - x(t)]^{1-\sigma} \approx \rho u(t)(1 - x(t)) + \sigma \rho u(t) x(t)(1 - x(t)).$$

The term $\rho u(t)(1 - x(t))$ is a response to advertising that acts positively on the unsold portion $(1 - x(t))$ of the market. The term $\sigma \rho u(t) x(t)(1 - x(t))$ can be considered an additional process of word-of-mouth communication between consumers from the sold portion $x(t)$ and those from the unsold portion $1 - x(t)$.

The advertising rate $u(t)$ is the control variable of this dynamic system. The control sequence forms a vector

$$u = (u(0), u(1), \dots, u(T - 1)).$$

The control variable $u(t)$ is used to find the optimal advertising expenditure for achieving the maximum profit at the time instant $t = T$.

In the model under study, the discounted profit functional has the form

$$J = \frac{mx(T)}{(1 + r)^T} + \sum_{k=0}^{T-1} \frac{mx(k) - cu^{1/\sigma}(k)}{(1 + r)^k}. \tag{1.3.67}$$

Here m is the revenue potential (a margin per unit product); r denotes the discount rate; c is the parameter characterizing the cost of advertising. Without loss of generality, we will consider the functional J with $c = 1$. This can be done with an appropriate scaling procedure of J.

The term $-\sum_{k=0}^{T-1} \frac{cu^{1/\sigma}(k)}{(1 + r)^k}$ of the profit functional J corresponds to the advertising expenditure. We use the nonlinear advertising expenditure for taking into account a decreasing response of marketing efforts to the firm profit $(0 < \sigma < 1)$. Note that the quadratic advertising expenditure $u^2(k)$ is widespread in the literature; see [287, 331]. We consider a more general case in which the nonlinearity parameter $1/\sigma$ may vary from 1 to ∞. In this case, the problem can be explicitly solved, which is an important argument for such a choice of nonlinearity.

An equivalent approach is to transfer the nonlinearity into the dynamic equation; for example, see [333] and [63]. Then we can adopt the linear function of advertising expenditure in the profit functional. These issues were discussed in detail in [329].

Optimal control design

In what follows, we derive a recursive optimal control formula using an appropriate modification of Bellman's dynamic programming [121]. We construct the sequence of $(T + 1)$ *Bellman functions*

$$V^{(0)}(x(0)), V^{(1)}(x(1)), \dots, V^{(T)}(x(T))$$

and the optimal control sequence

$$u^* = (u^*(0), u^*(1), \dots, u^*(T - 1)).$$

The problem is solved backwards in time, starting from the terminal time instant T. At the time instant T, the Bellman function has the form

$$V^{(T)}(x(T)) = \max_{u(T) \geq} \frac{mx(T) - u^{1/\sigma}(T)}{(1+r)^T} = \frac{mx(T)}{(1+r)^T},$$

because there is no control $u(T)$ at this instant. Recall that $c = 1$. For obtaining a universal recursive formula below, let us write $V^{(T)}(X(T))$ as

$$V^{(T)}(X(T)) = \frac{m}{(1+r)^T}[\alpha_T X(T) + \beta_T],$$

where $\alpha_T = 1$ and $\beta_T = 0$.

The Bellman function at a time instant k is given by

$$V^{(k)}(x(k)) = \max_{u(k) \geq 0} W(k, x(k), u(k)) = \max_{u(k) \geq 0} \left[\frac{mx(k) - u^{1/\sigma}(k)}{(1+r)^k} + V^{(k+1)} \right],$$

where $k = T-1, \ldots, 0$. This definition means that the advertising expenditure at a time instant k is optimized taking into account the sales volume $x(k)$ achieved by this instant. The value $u(k) = u^*(k)$ for which the function $W(k, x(k), u(k))$ achieves maximum is the desired optimal control at the time instant k. Clearly, the value $V^{(0)}(x(0))$ is the maximum value of the profit functional J.

The major result of this subsection is expressed by the two propositions below, which demonstrate that the solution of the optimal advertising problem can be calculated using a recursive procedure.

Proposition 1.3.7 *Let the model parameters $\delta, m, \rho, r,$ and σ be such that $x(t) \in [0, 1]$. Then the solution of the optimal advertising problem exists, and the sequence of Bellman functions $V^{(0)}(x(0)), V^{(1)}(x(1)), \ldots, V^{(T)}(x(T))$ and the optimal control sequence $u^* = (u^*(0), u^*(1), \ldots, u^*(T - 1))$ are calculated by the recursive formulas*

$$V^{(k)}(x(k)) = \frac{m}{(1+r)^T}[\alpha_k x(k) + \beta_k], \tag{1.3.68}$$

$$u^*(k) = \left(\frac{m\sigma\rho\alpha_{k+1}}{(1+r)^{T-k}} \right)^{\frac{\sigma}{1-\sigma}} [1 - x(k)]^\sigma, \tag{1.3.69}$$

where

$$\alpha_k = \alpha_{k+1} \left[(1 - \delta) - (1 - \sigma)\rho \left(\frac{m\sigma\rho\alpha_{k+1}}{(1+r)^{T-k}} \right)^{\frac{\sigma}{1-\sigma}} \right] + (1+r)^{T-k}, \tag{1.3.70}$$

$$\beta_k = \beta_{k+1} + (1 - \sigma)\rho \left(\frac{m\sigma\rho\alpha_{k+1}}{(1+r)^{T-k}} \right)^{\frac{\sigma}{1-\sigma}}, \tag{1.3.71}$$

$\alpha_T = 1$ *and* $\beta_T = 0$. *Here $x(k)$ is given by the linear difference equation*

$$x(k + 1) = \left[1 - \delta - \rho \left(\frac{m\sigma\rho\alpha_{k+1}}{(1+r)^{T-k}} \right)^{\frac{\sigma}{1-\sigma}} \right] x(k) + \rho \left(\frac{m\sigma\rho\alpha_{k+1}}{(1+r)^{T-k}} \right)^{\frac{\sigma}{1-\sigma}}, \tag{1.3.72}$$

which is obtained from (1.3.66) with $u(k) = u^*(k)$.

Proof This proposition will be established by mathematical induction on $k = T - 1, \ldots, 1, 0$.

For $k = T - 1$, the Bellman function can be found from the relation

$$V^{(T-1)}(x(T-1)) = \max_{u(T-1) \geq 0} W^{(T-1)}$$

$$= \max_{u(T-1) \geq 0} \left[\frac{mx(T-1) - u^{1/\sigma}(T-1)}{(1+r)^{T-1}} + V^{(T)} \right].$$

Substituting $x(T)$ from the recursive formula (1.3.66) into $V^{(T)}$ yields

$$W^{(T-1)} = \frac{mx(T-1) - u^{1/\sigma}(T-1)}{(1+r)^{T-1}}$$

$$+ \frac{m}{(1+r)^T} \left((1-\delta)X(T-1) + \rho u(T-1)[1 - x(T-1)]^{1-\sigma} \right).$$

As is easily checked, the function $W^{(T-1)}$ achieves maximum in the variable $u(T-1)$ at

$$u^*(T-1) = \left(\frac{m\rho\sigma}{(1+r)} \right)^{\frac{\sigma}{1-\sigma}} [1 - x(T-1)]^\sigma ,$$

which corresponds to (1.3.69) with $k = T - 1$.

With the optimal control $u^*(T-1)$ at the time instant $(T-1)$, we calculate $W^{(T-1)}$ and obtain the Bellman function

$$V^{(T-1)}(x(T-1)) = \frac{m}{(1+r)^T}$$

$$\times \left[\left((1+r) + (1-\delta) - \rho(1-\sigma) \left(\frac{m\sigma\rho}{1+r} \right)^{\frac{\sigma}{1-\sigma}} \right) x(T-1) + \rho \left(\frac{m\sigma\rho}{1+r} \right)^{\frac{\sigma}{1-\sigma}} \right].$$

Denote

$$\alpha_{T-1} = (1-\delta) - \rho(1-\sigma) \left(\frac{m\sigma\rho}{1+r} \right)^{\frac{\sigma}{1-\sigma}} + (1+r),$$

$$\beta_{T-1} = \rho(1-\sigma) \left(\frac{m\sigma\rho}{1+r} \right)^{\frac{\sigma}{1-\sigma}} .$$

Therefore, the Bellman function at the time instant $T - 1$ can be written as

$$V^{(T-1)}(x(T-1)) = \frac{m}{(1+r)^T} [\alpha_{T-1} x(T-1) + \beta_{T-1}],$$

where α_{T-1} and β_{T-1} are given by (1.3.70) and (1.3.71) with $k = T - 1$. Consequently, the base case has been proved.

Inductive hypothesis. Assume that at the time instant $k + 1$ we have the Bellman function

$$V^{(k+1)}(x(k+1)) = \frac{m}{(1+r)^T} [\alpha_{k+1} x(k+1) + \beta_{k+1}].$$

We should find the Bellman function $V^{(k)}(x(k)) = \max_{u(k) \geq 0} W^{(k)}$ at the time instant k.

Here,

$$W^{(k)} = \frac{mx(k) - u^{1/\sigma}(k)}{(1+r)^k} + V^{(k+1)}(x(k+1)) = \frac{mx(k) - u^{1/\sigma}(k)}{(1+r)^k}$$
$$+ \frac{m}{(1+r)^T}[\alpha_{k+1}x(k+1) + \beta_{k+1}].$$

The stationary point of the function $W^{(k)}$ in the variable $u(k)$ is

$$u^*(k) = \left(\frac{m\sigma\rho\alpha_{k+1}}{(1+r)^{T-k}}\right)^{\frac{\sigma}{1-\sigma}}[1 - x(k)]^\sigma.$$

Because

$$\left(W^{(k)}\right)'' = -\frac{1-\sigma}{(1+r)^k\sigma^2}u^{\frac{1-2\sigma}{\sigma}}(k) < 0,$$

the point $u^*(k)$ corresponds to the maximum of the function $W^{(k)}$. Hence, the optimal positional control $u^*(k)$ is given by (1.3.69).

Now, we can find the Bellman function $V^{(k)}(x(k)) = W^{(k)}\mid_{u(k)=u^*(k)}$ at the time instant k:

$$V^{(k)}(x(k)) = \left\{\left[(1-\delta)\alpha_{k+1} - (1-\sigma)\rho\alpha_{k+1}\left(\frac{m\sigma\rho\alpha_{k+1}}{(1+r)^{T-k}}\right)^{\frac{\sigma}{1-\sigma}} + (1+r)^{T-k}\right]x(k)\right.$$
$$\left. + \beta_{k+1} + \left(\frac{m\sigma\rho\alpha_{k+1}}{(1+r)^{T-k}}\right)^{\frac{\sigma}{1-\sigma}}\right\}\frac{m}{(1+r)^T}.$$

Consequently,

$$V^{(k)}(x(k)) = \frac{m}{(1+r)^T}[\alpha_k x(k) + \beta_k],$$

where α_k and β_k are calculated by formulas (1.3.70) and (1.3.71).

We proceed with some sufficient conditions on the model parameters that guarantee the existence of the above optimal solution. For this, an auxiliary lemma will be needed as follows.

Lemma 1.3.5 *Let the sequence* $\{\alpha_k\}_{k=0}^T$ *be determined by the recursive relation*

$$\alpha_k = \alpha_{k+1}\left[(1-\delta) - (1-\sigma)\rho\left(\frac{m\sigma\rho\alpha_{k+1}}{(1+r)^{T-k}}\right)^{\frac{\sigma}{1-\sigma}}\right] + (1+r)^{T-k}, \quad \alpha_T = 1.$$

If

$$m\sigma\rho^{\frac{1}{\sigma}} < r + \delta, \tag{1.3.73}$$

then the values $b_k = \rho\left(\frac{m\sigma\rho\alpha_k}{(1+r)^{T-k+1}}\right)^{\frac{\sigma}{1-\sigma}}$, $k = 0, 1, \ldots, T$, *satisfy the inequalities*

$$b_k < 1. \tag{1.3.74}$$

Proof By the definition of b_k, conditions (1.3.74) are equivalent to

$$\alpha_k < \frac{(1+r)^{T-k+1}}{m\sigma\rho^{\frac{1}{\sigma}}}.$$

For $k = T$, the inequality obviously holds due to $\delta \in [0, 1]$. Suppose that $\alpha_{k+1} < \frac{(1+r)^{T-k}}{m\sigma\rho^{\frac{1}{\sigma}}}$. Then from the recursive formula for α_k and inequality (1.3.73) it follows that

$$\alpha_k < \alpha_{k+1}(1 - \delta) + (1+r)^{T-k} < \frac{(1+r)^{T-k}}{m\sigma\rho^{\frac{1}{\sigma}}}\left[1 - \delta + m\sigma\rho^{\frac{1}{\sigma}}\right] < \frac{(1+r)^{T-k+1}}{m\sigma\rho^{\frac{1}{\sigma}}}.$$

Proposition 1.3.8 *If inequality* (1.3.73) *holds, then the solution determined by formulas* (1.3.68)–(1.3.72) (*see Proposition* 1.3.7) *exists.*

Proof It suffices to demonstrate that the solution $x(k)$ of the linear difference equation (1.3.72) satisfies $x(k) \in [0, 1]$. We write this equation as

$$x(k + 1) = x(k)(1 - \delta) + [1 - x(k)]b_k, \tag{1.3.75}$$

where $b_k = \rho\left(\dfrac{m\sigma\rho\alpha_k}{(1+r)^{T-k+1}}\right)^{\frac{\sigma}{1-\sigma}}.$

By Lemma 1.3.5, $b_k \in [0, 1]$. Consequently, $(1 - \delta), b_k \in [0, 1]$.

For $k = 0$, the initial value X_0 belongs to the interval $[0, 1]$. Hence, $x(1) = x(0)(1 - \delta) + [1 - x(0)]b_0$ also belongs to the interval $[0, 1]$ as a convex combination of the points $(1 - \delta)$ and b_0.

Assume $x(k) \in [0, 1]$. From equation (1.3.75) it follows that $x(k + 1) \in [0, 1]$ as a convex combination of the points $(1 - \delta)$ and b_0. Proposition 1.3.8 is established by mathematical induction on k.

Thus, we have designed the recursive sequence of optimal controls

$$u^* = (u^*(0), u^*(1), \ldots, u^*(T - 1))$$

and also the sequence of Bellman functions

$$V^{(0)}(x(0)), V^{(1)}(x(1)), \ldots, V^{(T)}(x(T)).$$

The maximum of the profit functional (1.3.67) coincides with the value $V^{(0)}(x(0))$, i.e.,

$$\max_u J(u) = V^{(0)}(x(0)).$$

Algorithm and numerical examples

Using Propositions 1.3.7 and 1.3.8, we introduce the following algorithm of optimal advertising.

Algorithm of optimal advertising

Inputs:
$T \in \mathbb{N}$ as the number of decision points; $x_0 \in [0, 1]$ as an initial market share; $\delta \in [0, 1]$ as the decay constant of market share in the absence of advertising; m as the revenue

potential; ρ as the response constant; r as the discount rate; finally, $\sigma \in (0, 1)$ as the nonlinearity parameter of the model.

Outputs:

$u^* = (u^*(0), u^*(1), \ldots, u^*(T-1))$ as the optimal control sequence (advertising expenditure); $V^{(0)}(x(0)), V^{(1)}(x(1)), \ldots, V^{(T)}(x(T))$ as the sequence of Bellman functions ($V^{(0)}(x(0))$ as the maximum profit achieved by the time instant $k = T$); $x(k)$ as the market share at a time instant k ($k = 1, 2, \ldots, T$).

Step 1. Initialization

Setting of initial values of the input parameters.

Step 2. Verification of x_0, δ, and σ

If $x_0 \notin [0, 1]$, or $\delta \notin [0, 1]$, or $\sigma \notin (0, 1)$, then STOP.

Step 3. Verification of the sufficient condition

If test $:= \frac{1-\delta+m\sigma\rho^{1/\sigma}}{1+r} \geq 1$, then STOP.

Step 3. Calculation of α_k and β_k

The coefficients α_k and β_k are determined by formulas (1.3.70) and (1.3.71) with the initial values $\alpha_T = 1$ and $\beta_T = 0$.

Step 3. Calculation of $x(k)$

The market share $x(k)$ is found from the linear difference equation (1.3.72) with the initial value $x(0) = x_0$.

Step 4. Finding $V(k)$ and u^*

The sequence of Bellman functions $V^{(k)}$ and the optimal control sequence $u^*(k)$ are constructed using formulas (1.3.68) and (1.3.69).

Step 5. End of the algorithm

The algorithm was implemented in Maple as a procedure entitled **OptimalAdvertising**. The input parameters of the procedure are T, x0 $:= x_0$, δ, m, ρ, r, and σ. The procedure returns the vectors U, V, and x and also draws the plots pointU and pointX.

Some examples below demonstrate how to call this procedure.

```
> restart; with(plots):
> T := 24; X0:=0.5; delta:=0.2; m:=0.6; rho:=0.5; r:=0.1;
    sigma:=0.8;
> OptimalAdvertising(T,X0,delta,m,rho,r,sigma);
the total advertising efforts=3.565690
the value of the profit functional=1.459383
the maximum value of the profit functional=1.479754
>display({pointU, polygU}); display({pointX, pointY, polygX,
    polygY});
```

The procedure returns the optimal control sequence U, the sequence of Bellman functions V, and the market share x at each time instant of the dynamic process. Moreover, the procedure calculates the total advertising expenditure sumU = 3.565690. For comparing the market share under the optimal advertising expenditure (the plots pointX, polygX) with its counterpart under the uniform advertising expenditure (the plots pointY, polygY), both graphs are displayed on the same axes.

The graphs obtained for the initial market share $x(0) = 0.3$ and the parameter values $T = 24$, $\delta = 0.2$, $m = 0.6$, $\rho = 0.5$, $r = 0.1$ and different values of σ are shown in

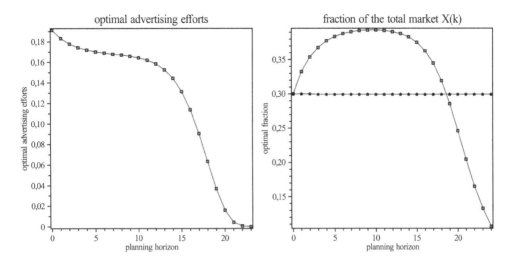

Figure I.3.2 Case $\sigma = 0.9$.

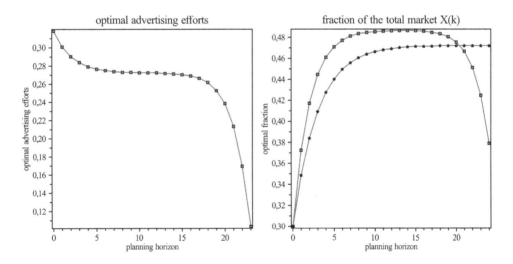

Figure I.3.3 Case $\sigma = 0.5$.

the figures below. More specifically, Figure 1.3.2 corresponds to $\sigma = 0.9$; Figure 1.3.3, to $\sigma = 0.5$; Figure 1.3.4, to $\sigma = 0.1$. Clearly, for greater values of the parameter σ, the values of optimal control (advertising expenditure) are decreasing. As a result, the market share controlled by the firm is being reduced as well.

In addition, note that with an initial market share of 30%, the optimal advertising expenditure achieves maximum at the first time instant, which can be interpreted as a massive advertising campaign in order to increase sales. At subsequent time instants, the advertising expenditure is decreasing to some average steady-state level, finally vanishing in the remaining time instants.

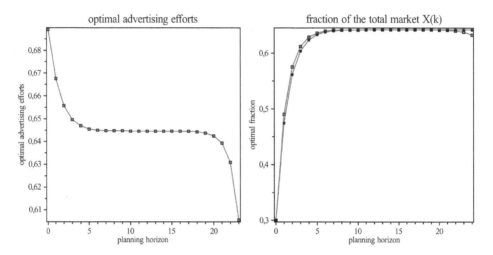

Figure I.3.4   Case $\sigma = 0.1$.

Conclusions

> Sanely applied advertising
> could remake the world.
> —Chase[28]

In Subsection 1.3.6, a discrete-time model of optimal advertising has been proposed. As particular cases, it includes the discrete-time analogs of the classical models of optimal advertising. For example, the case $\sigma = 0.5$ corresponds to the Sethi model [327]; the case $\sigma = 0$, to the Vidale–Wolfe model [347]. Apparently, different values of the parameter σ should be chosen for different markets. The precise tuning of model parameters for a particular market is possible using a sufficient amount of statistical data. The research presented below has not been intended to model any particular market. However, the proposed algorithm can be easily applied to design optimal advertising strategies in real-life markets.

Concerning further extensions of the model, we mention a discrete-time model of optimal advertising in a competitive market, i.e., a game-theoretic statement of the problem. An attempt to construct such a model (as a counterpart of the Sorger game [333]) was undertaken in [184]. Another field of future investigations is to take into account the uncertain factors in the model, e.g., following the approaches [122] and [123].

I.4   MINIMAX REGRET

> Wer wagt, gewinnt.[29]

The epigraph suggests that risk is of definite utility. However, what is risk in a single-criterion choice problem? Why is risk better than maximin? How can risk be described in mathematical terms and what are the properties of risk? All these questions are

---

[28]Stuart Chase, (1888–1985), was an American economist, social theorist, and writer.
[29]German "Those who risk win." This is an analog of the English proverb "Nothing ventured, nothing gained."

considered in Section 1.4. The concept of a guaranteed solution in risks based on the Savage–Niehans minimax regret (risk) functions is formalized and its properties are studied.

1.4.1 Ad narrandum, non ad probandum[30]

Consider the single-criterion choice problem under uncertainty

$$\Gamma_1 = \langle X, Y, f(x, y) \rangle,$$

in which DM's goal is to choose an appropriate strategy $x \in X \subseteq \mathbf{R}^n$ for *maximizing* the values of a scalar criterion $f(x, y)$. Note that the DM has to consider any possible realization of the uncertainty $y \in Y \subseteq \mathbf{R}^m$.

As it has been mentioned in Section 1.3, the guaranteed solution in outcomes of the problem Γ_1 is a pair $(x^g, f^g) \in X \times \mathbf{R}$ given by

$$f^g = \min_{y \in Y} f(x^g, y) = \max_{x \in X} \min_{y \in Y} f(x, y).$$

However, malum consilium est, quod mutari potest.[31] But such attempts are desirable! As a matter of fact, being guided by the maximin principle (see the discussion above), the DM expects a catastrophic situation, i.e., the realization of the worst-case uncertainty: bellum omnium contra omnes.[32] This is a distinguishing feature of the maximin principle! Meanwhile, such a realization is often unlikely. Therefore, in 1951 Savage[33] introduced the principle of minimax regret (risk) as an improvement of the maximin; see [322].

Auxiliary background

Before proceeding to the formalization of the principle of minimax regret (risk), let us dwell on the etymology of the word "risk" and some of its interpretations. The word "risk" in the English language is borrowed from French *risque* and Italian *risco*, both meaning "danger." They go back to Greek *rizikon*, literally meaning "root" but later on used in Latin for "cliff." That is, "to take the risk" is "to go around the cliff." In this sense, note the concept of risk management widely adopted in economics: English *management*

---

[30]Latin "(Is written) to narrate, not to prove." A quote from *Institute of Oratory* X: 1, 13, by Roman rhetorian Marcus Fabius Quintilianus (appr. 35–95 A.D.). He used this phrase to discriminate between the tasks of history and eloquence.

[31]Latin "It is a bad plan that admits of no modification." This phrase is attributed to Publilius Syrus, (flourished 1st century B.C.), Latin mime writer contemporary with Cicero. See [282].

[32]Latin "the war of all against all." In this way, Thomas Hobbes described human existence in the state of nature thought experiment that he conducted in *De Cive* (1642) and *Leviathan* (1651).

[33]Savage, Leonard Jimmy; nee Ogashevich, (1917–1971). He was a American mathematician, economist, and statistician. He began his studies at the University of Michigan, with specialization in chemical engineering. However, he was expelled soon after a fire accident in a chemical laboratory, caused by his poor eyesight. He was allowed to recover only to the specialization in mathematics for physicists. He graduated from the university in 1938. In 1941, he defended his doctoral dissertation. During World War II, he worked as a statistics assistant with J. von Neumann. In the years 1946–1950, he moved to the University of Chicago; then held the position of professor at the Universities of Michigan and Yale. Milton Friedman, Nobel laureate in Economic Sciences, said Savage "one of the few people I have met whom I would unhesitatingly call a genius." In 1977, the Savage Prize was established, which is annually awarded to the authors of two outstanding doctoral dissertations in the field of econometrics and statistics.

originates from Italian *maneggio*, meaning "training of a horse." From this point of view, risk management is the art of "breaking in risk," i.e., understanding its nature, knowing its manners and getting on well with it.

According to the Merriam–Webster Dictionary, risk is: (1) possibility of loss or injury; (2) someone or something that creates or suggests a hazard; (3a) the chance of loss or the perils to the subject matter of an insurance contract; also, the degree of probability of such loss; (3b) a person or thing that is a specified hazard to an insurer; (3c) an insurance hazard from a specified cause or source; (4) the chance that an investment (such as a stock or commodity) will lose value. Thus, on the one hand, risk means a possible hazard; on the other, an action in the hope of a success. The Merriam–Webster Dictionary defines regret as: (1) sorrow aroused by circumstances beyond one's control or power to repair; (2a) an expression of distressing emotion (such as sorrow); (2b) regrets (in plural) a note politely declining an invitation. In the sequel, formalizing the concept of risk for the problems under uncertainty and game-theoretic models, we will simultaneously consider the expected losses from risky decisions and also the possibly favourable actions of the exogenous factors that are not controlled by the decision-maker.

There exist numerous classifications of risks; for example, environmental, political, and technological (anthropogenic) risks are called pure risks. In turn, commercial, investment, and insurance risks are called financial (speculative) risks.

The academic literature describes various mathematical models of risks as well as various methods for measuring risks in quantitative terms. A special place is occupied by stochastic models in which risk is associated with random uncontrolled factors having given distributions. Such models are based on the framework of probability theory and mathematical statistics and have well-known applications, both in engineering and economics.

Consider an illustrative example. Let X be the set of admissible alternatives of a decision-maker, e.g., a collection of investment portfolios. A random event A is that the DM will bear losses from choosing a certain alternative (portfolio) $x \in X$. Then the probability $P(A)$ of this event is a numerical assessment (measure) of the financial risk associated with this alternative (portfolio).

Assessing (measuring) financial risks as the probabilities of adverse events is an inconvenient approach for a financial manager: it forms a probability distribution of losses, but does not evaluate the cost of financial risk.

One of the common methods for determining financial risk is the so-called Value at Risk (VaR). Recently, VaR has become an important tool to control and manage risk in companies of various types. For example, the standard procedure of calculating insurance premiums is based on this risk assessment method.

Consider a mathematical model of risk for an insurance company. Let us analyze a certain period of time, e.g., one year. Insurance premiums are paid at the beginning of the year and, upon deduction of overhead costs, make up the reserve capital K_0 of the insurance company. The insurance payments under all contracts concluded determine the total risk ξ. We define the income of the insurance company as the difference $K_0 - \xi$. Assume that the total risk ξ is a random variable with a given distribution, and the parameters of this distribution are estimated by statistical methods. Note that the hypothesis about *the Gaussian distribution of the random variable* ξ with known parameters, the mean $a = M[\xi]$ and the variance $\sigma = \sqrt{D[\xi]}$, is often accepted. This leads to the well-known *reserve capital formula*

$$K_a = M[\xi] + q_a\sqrt{D[\xi]},$$

where q_a is the quantile of the Gaussian distribution that corresponds to an admissible level of risk $a\%$.

Another method of calculating financial (speculative) risk is Expected Shortfall (ES), which determines the expected amount of losses as a function of the reserve capital K_a.

Since the 1950s, after the appearance of portfolio theory, the risk of financial transactions has been defined using *the standard deviation*. In his paper [306], Nobel laureate Harry Markowitz formalized a bi-criteria mathematical model of an investment portfolio. In this model, the first criterion corresponds to the expected return on the portfolio and the second criterion to its risk, the latter treated as the degree of mutual dependence of assets returns in terms of the covariance matrix composed of the elements $\sigma_{ij} = Cov(\xi_i, \xi_j)$. (Here the random variable ξ_i is the rate of return on asset i.) Within this model of risk, the bi-criteria problem reduces to minimizing a linear-quadratic function.

Nevertheless, the risk models discussed above have encountered many opponents with well-grounded objections. *Their first argument* is connected with the fact that under uncontrolled random factors, the decision-maker will have a subjective attitude towards risk. More specifically, facing the same situation, different people evaluate risk in different ways. Therefore, in the same problem their choices of optimal solutions may not coincide with each other.

On the other hand, measuring risk by the standard deviation may lead to paradoxical results. For example, consider the Sharpe ratio [8] given by $\frac{\sigma[\xi]}{M[\xi]}$, where $M[\xi]$ is the expected value of the random rate of return on a financial transaction and $\sigma[\xi]$ is the standard deviation. Let us study the following optimal choice problem under risk.

A decision-maker can gain 1000 units of income with probability 1 (almost surely). In this case, the value $\sigma[\xi]$ is 0, and hence the Sharpe ratio is $\frac{\sigma[\xi]}{M[\xi]} = 0$. Another alternative is to participate in a lottery (game), where he will gain 1000 units of income or an amount $a > 1000$ equiprobably. Any sane person would prefer to participate in the lottery, despite the fact that in the second case the Sharpe ratio is higher, $\frac{\sigma[\xi]}{M[\xi]} > 0$, due to the appearance of risk.

Note that for stochastic risk modeling, the decision-maker has to know the distribution functions of random variables that are variables in such models. However, the adoption of a statistical hypothesis about the law of distribution may be incorrect. Moreover, the uncontrolled factors can be described by uncertain values, and the decision-maker may know only their ranges.

Finally, the presence of risk in decision-making is not always a negative feature. Numerous examples of salon games (cards, chess, etc.), as well as real problems associated with economic, political, and military conflicts, indicate that the need for making risky choices is often inevitable.

We will consider below some approaches to formalizing optimal decisions in risky conditions under uncertain factors. In the sequel, decision-makers operating in such conditions will be called players. In particular, the decision problem under uncertainty is often called *the game with nature*, despite the fact that this model contains only one rational decision-maker.

Vaque[34]

Let us describe the types of uncertainties in mathematical models of economic problems. Any economic system is subjected to uncontrolled exogenous factors of various kinds: political, social, environmental, technological, and macroeconomic, to name a few.

On the other hand, uncontrolled factors may exist within a given system; some examples include the unpredictable actions of individuals, accidents and diseases, technological failures, etc.

As it has been already mentioned, uncertainties arise during the construction of a mathematical model:

First, the process of problem statement and specification is accompanied by some disagreements between the customer and the developer, due to the insufficient knowledge of the real economy by the latter (on the one hand) and the insufficient mathematical literacy of the former (on the other hand).

Second, the most difficult challenge is to formalize the optimal solution, i.e., to translate into mathematical language the customer's natural ideas about the best solution for the class of problems under study.

Third, there is the adequacy problem of the resulting mathematical model as a description of the real process.

In addition, note the uncertainties that are realized due to the incomplete awareness of decision-makers. In operations research, decisions (strategies) are factors that *are controlled* by players. A real problem also includes *uncontrolled* factors that affect the final outcome. Recall that depending on the awareness of the decision-maker, there are three types of uncontrolled factors: *fixed, random, and uncertain.*

The values of fixed uncontrolled factors are known to the decision-maker and cannot be changed by him. For example, in the optimal control problem of an aircraft flying at a constant altitude above the surface of the Earth, the acceleration of gravity can be considered a fixed uncontrolled factor.

Random uncontrolled factors are random variables that obey a known distribution law with given parameters. The effect of such factors is studied, e.g., in the so-called *statistical games*, in which the decision-maker (statistician) acts as the first player, and the uncertainty is caused not by the actions of a rational opponent, but by objective reality (external environment, commonly called "nature"). Nature is considered the second player, and the corresponding optimal decision problems are called "games with nature." Nature is generally interpreted as a certain disinterested authority, whose behavior is unknown, but not malicious.

Uncertain uncontrolled factors (uncertainties) are variables about which the only available knowledge is the range of admissible values or the class of possible distributions.

I.4.2   Guaranteed solution in risks

> The harm or benefit of an act depend on
> the combination of circumstances.
> —Kozma Prutkov[35]

---

[34] French "uncertainty."
[35] An English translation of a quote from [232, p. 230].

We associate with each uncertainty $y^* \in Y$ in the problem Γ_1 the value $\max\limits_{x \in X} f(x, y^*)$. Therefore, the player determines the maximum outcome under all possible uncertainties $y^* \in Y$. Next, the player constructs the difference between the maximum value of the criterion $f(x, y^*)$ and its value for any strategy $x \in X$, i.e.,

$$\max_{z \in X} f(z, y^*) - f(x, y^*), \tag{1.4.1}$$

where y^* is a fixed uncertainty. Acting in this way, the player numerically assesses his regret from using x instead of $\bar{x} = \arg\max\limits_{x \in X} f(x, y^*)$. Obviously, the regret is 0 in the case of choosing the strategy $\bar{x}$ under the uncertainty y^*. The difference (1.4.1) is called *the Savage–Niehans risk (regret) function* of the player, and its value for a specific pair $(x, y^*) \in X \times Y$ is called **the Savage–Niehans risk** of the player who chooses the strategy $x \in X$ under the uncertainty $y^* \in Y$.

In fact, this risk arises due to the player's incomplete knowledge of the specific realization $y^* \in Y$ of the uncertainty.

Next, the player seeks to choose an appropriate strategy $x \in X$ by *minimizing* the risk (regret). For this purpose, he applies the maximin principle to the problem

$$\langle X, Y, -[\max_{z \in X} f(z, y) - f(x, y)] \rangle.$$

As a result,

$$\min_{x \in X} \max_{y \in Y} \Phi(x, y) = \max_{y \in Y} \Phi(x^r, y),$$

where *the Savage–Niehans risk (regret) function* has the form

$$\Phi(x, y) = \max_{z \in X} f(z, y) - f(x, y). \tag{1.4.2}$$

The corresponding formal definition is as follows.

Definition 1.4.1 *The guaranteed solution in risks of the problem Γ_1 is a pair $(x^r, \Phi^r) \in X \times \mathbf{R}$ that satisfies the chain of equalities*

$$\Phi^r = \min_{x \in X} \max_{y \in Y} \Phi(x, y) = \max_{y \in Y} \Phi(x^r, y), \tag{1.4.3}$$

where the Savage–Niehans risk function is given by

$$\Phi(x, y) = \max_{z \in X} f(z, y) - f(x, y).$$

The strategy x^r is called guaranteeing in risks, whereas the value Φ^r is called the guaranteed risk.

As it has been mentioned earlier, using the function (1.4.1) the DM estimates his risk as the difference between the best-case (maximum) value of the criterion f and the actual one realized. The player's aspiration to minimize this risk is quite natural. According to Definition 1.4.1, choosing the strategy x^r the player obtains his risk $\Phi(x^r, y)$, which cannot *exceed* the guaranteed value Φ^r under any realization of the uncertainty $y \in Y$.

Thus, the solution of the decision problem under uncertainty (the game with nature) can be formalized by a pair that consists of a strategy and a corresponding guarantee.

The strategy $x \in X$ is chosen by the player, and the guarantee is the result obtained by him using this strategy. We have considered two solutions of this kind: the first is based on the maximin principle, introduced by Wald,[36] and the second on the principle of minimax regret (in the Savage–Niehans sense; see Definition 1.4.2).

In final analysis, a player who uses the maximin principle is oriented towards pessimum, the worst-case situation that can only happen. At the same time, the principle of minimax regret expresses the player's orientation towards the best-case situation. In fact, this is the difference between a pessimist and an optimist.[37]

The pessimists and optimists would be united by the concept of a guaranteed solution in which a player increases his payoff simultaneously with reducing his risk. Such a concept is formalized in Section 1.5 of this book.

I.4.3   Properties of risk function

Hier liegt der Hund begraben.
—German proverb[38]

Lemma 1.4.1 *The risk function* $\Phi(x, y)$ *is nonnegative.*
     This fact follows directly from (1.4.2).

Lemma 1.4.2 *Assume that in the problem* Γ_1 *the sets* X *and* Y *are compact and the payoff function* $f(x, y)$ *is continuous on* $X \times Y$. *Then the risk function* $\Phi(x, y)$ *is also continuous on* $X \times Y$.
     Really, the function $\max\limits_{z \in X} f(z, y)$ is continuous on Y, and the difference of continuous functions

$$\Phi(x, y) = \max_{z \in X} f(z, y) - f(x, y)$$

is also continuous.

Proposition 1.4.1 *Assume that in the problem* Γ_1 *the sets* X *and* Y *are compact and the function* $f(x, y)$ *is continuous on* $X \times Y$. *Then there exists the guaranteed solution in risks* (x^r, Φ^r) *of this problem.*

Proof By Lemma 1.4.2 the risk function $\Phi(x, y)$ is continuous on $X \times Y$. Therefore, $\varphi(x, y) = -\Phi(x, y)$ is also continuous on $X \times Y$. In this case, there exists the guaranteed solution in risks (x^r, φ^r) of the problem

$$\langle X, Y, \varphi(x, y) \rangle,$$

---

[36]Abraham Wald, (1902–1950), was a Hungarian mathematician and statistician. He was born in a religious Jewish family. He received home education under the guidance of parents. He continued further studies at the University of Vienna. In 1931, he defended his doctoral dissertation in mathematics. He was forced to emigrate to the United States. During World War II, he used statistical methods for reducing the losses of American military aircrafts. In 1950, while delivering lectures at the invitation of the Indian government, he died as a result of a plane crash in the Nilgiri Mountains.

[37]A pessimist drinks cognac and says: "Yes, it smells like bugs!" An optimist squashes the bug and notices: "Wow! It smells like cognac!"; see [20, p. 276]. A pessimist believes that it cannot be worse; an optimist believes that it can [20, p. 542].

[38]German "That's where the dog lies buried." Close to the English proverb "That's where the shoe pinches!" Used to emphasize the essence of something.

i.e., the pair (x^r, φ^r) satisfies

$$\varphi^r = \max_{x \in X} \min_{y \in Y} \varphi(x, y) = \min_{y \in Y} \varphi(x^r, y). \tag{1.4.4}$$

Due to (1.4.4), passing to the function $\Phi(x, y)$ gives

$$-\Phi^r = \varphi^r = \max_{x \in X} \min_{y \in Y} \varphi(x, y) = \max_{x \in X} \min_{y \in Y} [-\Phi(x, y)] = \min_{y \in Y} \varphi(x^r, y)$$

$$= \min_{y \in Y} [-\Phi(x^r, y)] = -\min_{x \in X} \max_{y \in Y} \Phi(x, y) = -\max_{y \in Y} \Phi(x^r, y),$$

and the conclusion follows.

Remark 1.4.1 According to Lemma 1.4.1, the best-case risk is 0. Under what conditions does this risk exist? Recall the concept of saddle point, as applied to the risk function $\Phi(x, y)$: a pair (x^r, y^o) is *a saddle point* in the problem

$$\langle X, Y, \Phi(x, y) \rangle$$

if

$$\max_{y \in Y} \Phi(x^r, y) = \Phi(x^r, y^o) = \min_{x \in X} \Phi(x, y^o), \tag{1.4.5}$$

or equivalently,

$$\min_{x \in X} \max_{y \in Y} \Phi(x, y) = \Phi(x^r, y^o) = \max_{y \in Y} \min_{x \in X} \Phi(x, y); \tag{1.4.6}$$

in this case, x^r is the minimax strategy, and y^o is the maximin uncertainty. In addition, a well-known result in game theory is *the minimax inequality* [43, p. 31]

$$\max_{y \in Y} \min_{x \in X} \Phi(x, y) \leqslant \min_{x \in X} \max_{y \in Y} \Phi(x, y). \tag{1.4.7}$$

(Though, here the sets X and Y are assumed to be independent.)

Recall that the minimax strategy x^r in the problem

$$G = \langle X, Y, \Phi(x, y) \rangle \tag{1.4.8}$$

is given by

$$\min_{x \in X} \max_{y \in Y} \Phi(x, y) = \max_{y \in Y} \Phi(x^r, y). \tag{1.4.9}$$

Consider the problem (1.4.8) with independent compact sets X and Y and a continuous function $f(x, y)$ on $X \times Y$. (These assumptions guarantee the existence of the maximin and minimax in (1.4.7).) Then the following result is true.

Proposition 1.4.2 *In the problem G, there exists a strategy $x^r \in X$ such that all risks are*

$$\Phi(x^r, y) = 0 \quad \forall y \in Y, \tag{1.4.10}$$

if and only if the risk function $\Phi(x, y)$ (1.4.2) has a saddle point (x^r, y^o). The strategy x^r in (1.4.10) is given by (1.4.9).

Proof *Necessity.* Let the identity (1.4.10) hold for all $y \in Y$. In this case,

$$\max_{y \in Y} \Phi(x^r, y) = 0. \tag{1.4.11}$$

Due to $\Phi(x, y) \geqslant 0$, for any $(x, y) \in X \times Y$ (see Lemma 1.4.1) the relation (1.4.11) directly implies

$$\min_{x \in X} \max_{y \in Y} \Phi(x, y) = \max_{y \in Y} \Phi(x^r, y) = 0.$$

At the same time, by the minimax inequality (1.4.7) and $\Phi(x, y) \geqslant 0$, we have

$$0 \leqslant \max_{y \in Y} \min_{x \in X} \Phi(x, y) = \min_{x \in X} \Phi(x, y^*) \leqslant \min_{x \in X} \max_{y \in Y} \Phi(x, y) = \max_{y \in Y} \Phi(x^r, y) = 0.$$

Here all nonstrict inequalities turn into equalities. Therefore, according to (1.4.6), the pair (x^r, y^*) is a saddle point of the risk function $\Phi(x, y)$.

Sufficiency. Let (x^r, y^o) be a saddle point of $\Phi(x, y)$, i.e., the chain of inequalities (1.4.5) holds. In view of (1.4.2), it takes the form

$$\max_{z \in X} f(z, y) - f(x^r, y) \leqslant \max_{z \in X} f(z, y^o) - f(x^r, y^o) \leqslant \max_{z \in X} f(z, y^o) - f(x, y^o)$$

for any $x \in X$ and $y \in Y$. Applying the strategy $x = \bar{x} = \arg\max_{z \in X} f(z, y^o)$ in the right-hand inequality and taking into account the relations $\max_{z \in X} f(z, y^o) - f(\bar{x}, y^o) = 0$ and $\Phi(x^r, y) = \max_{z \in X} f(z, y) - f(x^r, y) \geqslant 0$ for all $y \in Y$, we arrive in

$$0 \leqslant \Phi(x^r, y) = \max_{z \in X} f(z, y) - f(x^r, y) \leqslant 0 \quad \forall y \in Y,$$

which means that $\Phi(x^r, y) = 0 \quad \forall y \in Y$.

Remark 1.4.2 As it has been mentioned above, for the decision-maker the best-case risk is 0, because the risk function $\Phi(x, y)$ is nonnegative by Lemma 1.4.1. The existence criterion of a strategy x^r implementing zero risk under any uncertainty $y \in Y$ (see Proposition 1.4.2) reduces to the existence of a saddle point of the risk function $\Phi(x, y)$ (minimum over all admissible strategies, maximum over all possible realizations of the uncertainty). In the presence of such a saddle point, the minimax strategy x^r is obtained from the equality

$$\min_{x \in X} \max_{y \in Y} \Phi(x, y) = \max_{y \in Y} \Phi(x^r, y).$$

This strategy x^r always exists if the sets X and Y are compact and the risk function $\Phi(x, y)$ is continuous on $X \times Y$. Then, by Proposition 1.4.2, the risk $\Phi(x^r, y) = 0$ for any uncertainty $y \in Y$.

For example, a saddle point (x^r, y^o) exists if $\Phi(x, y)$ is convex in $x \in X$ for each $y \in Y$, concave in $y \in Y$ for each $x \in X$, and the sets X and Y are convex and compact. The latter requirement is satisfied, e.g., for the sets

$$X = \left\{ x \in \mathbf{R}^n \mid \|x\|^2 \leq \alpha \right\}, \quad Y = \left\{ y \in \mathbf{R}^m \mid \|y\|^2 \leq \beta \right\},$$

where α and β are given positive values. The convexo-concave property of $\Phi(x, y)$ also holds if $\Phi(x, y)$ is linear in x (for each $y \in Y$) and linear in y (for each $x \in X$).

I.4.4   Types of risk functions in noncooperative game under uncertainty (NGU)

> ...nothing whatsoever takes place in the universe in which
> some relation of maximum and minimum does not appear.
>
> —L. Euler[39]

Formalization

As it has been mentioned above, risk is the possibility of a deviation of some charac-teristics from their desired values; risk is measured by the difference between the desired value of a performance criterion of a process and its realized value. According to this approach, for the single-criterion choice problem

$$\Gamma_1 = \langle X, Y, f_1(x, y)\rangle$$

interpreted as a one-player game under an uncertainty $y \in Y$, the measure of risk is *uniquely* defined by the value of the Savage–Niehans risk function

$$\Phi_1(x, y) = \max_{z \in X} f_1(z, y) - f_1(x, y).$$

For the games of two and more players, such a unique definition vanishes, because "our wishes are always limited by someone's capabilities."[40] Consider the design of risk func-tions in the noncooperative N-player game under uncertainty (NGU)

$$\Gamma_N \langle \mathbb{N}, \{X_i\}_{i \in \mathbb{N}}, Y, \{f_i(x, y)\}_{i \in \mathbb{N}}\rangle,$$

where $N \geqslant 2$. In the game Γ_N, the set of players is $\mathbb{N} = \{1, 2, \ldots, N\}$; each player $i \in \mathbb{N}$ chooses his *strategy* $x_i \in X_i \subseteq \mathbf{R}^{n_i}$, thereby yielding *a strategy profile* $x = (x_1, \ldots, x_N) \in X = \prod_{i \in \mathbb{N}} X_i \subseteq \mathbf{R}^n$, where $n = \sum_{i \in \mathbb{N}} n_i$; some uncertainty $y \in Y \subseteq \mathbf{R}^m$ is realized indepen-dently of the actions of all players; *the payoff function* $f_i(x, y)$ of each player i ($i \in \mathbb{N}$) is defined on the direct product $X \times Y$, and its value is called *the payoff* of player i in the game Γ_N.

Now, we construct the risk function of player $i \in \mathbb{N}$ in the game Γ_N. As it has been emphasized earlier, the immediate question is: what is the desired maximum in the risk function of a specific player i? Each player i ($i \in \mathbb{N}$) has a purely subjective idea of "the desired,"[41] which can be reduced to answering the question: which players will assist player i to achieve the maximum $\Phi_1(x, y)$? Therefore, the risk function of player i with the payoff function $f_i(x_1, \ldots, x_i, \ldots, x_N, y)$ may have an alternative and quite natural definition as follows.

Let $\mathbf{K}(i)$ be some coalition (union, alliance) of players from $\mathbb{N}$, *including player i himself.* Then the pure strategy of coalition $\mathbf{K}(i)$ has the form

$$x_{\mathbf{K}(i)} = (x_j, j \in \mathbf{K}(i)) \in X_{\mathbf{K}(i)} = \prod_{j \in \mathbf{K}(i)} X_j.$$

---

[39]Leonhard Euler, (1707–1783), was a Swiss mathematician and physicist. He was recognized as one of the greatest mathematicians of all time. A quote from [240, p. 477].

[40]This phrase is attributed to Klimovich I., (born in 1957), Soviet radio engineer, a columnist of *Literaturnaya gazeta* [20, p. 246].

[41]"For each person, the best thing is what he likes," A English translation of a quote from [113, p. 145].

Denote by $(x||z_{\mathbf{K}(i)}) \in X$ a strategy profile in which the strategies of all players from $\mathbf{K}(i)$ are replaced by the corresponding components of a vector $z_{\mathbf{K}(i)}$.

Define *the risk function of player i* as

$$\Phi_i^{(\mathbf{K}(i))}(x, y) = \max_{z_{\mathbf{K}(i)} \in X_{\mathbf{K}(i)}} f_i(x||z_{\mathbf{K}(i)}, y) - f_i(x, y). \tag{1.4.12}$$

Hereinafter, $\mathbb{N} \backslash \mathbf{K}(i)$ is the countercoalition for $\mathbf{K}(i)$, which includes all players not belonging to $\mathbf{K}(i)$; the strategy of this countercoalition can be written as $x_{\mathbb{N} \backslash \mathbf{K}(i)} = (x_j, j \in \mathbb{N} \backslash \mathbf{K}(i)) \in X_{\mathbb{N} \backslash \mathbf{K}(i)} = \prod_{j \in \mathbb{N} \backslash \mathbf{K}(i)} X_j$; a strategy profile used in the definition of (1.4.12) has the form $x = (x_1, ..., x_N) = (x||x_{\mathbf{K}(i)}) \in X = \prod_{i \in \mathbb{N}} X_i$. Recall that in a strategy profile $(x||z_{\mathbf{K}(i)}) \in X$, the strategies of players j from the coalition $\mathbf{K}(i)$ are $z_j \in X_j$ $(j \in \mathbf{K}(i))$, and the strategies of all other players are $x_k \in X_k$, $k \in \mathbb{N} \backslash \mathbf{K}(i)$.

For the three-player NGU ($\mathbb{N} = \{1, 2, 3\}$), player 1 may have four risk functions as follows:

$$\Phi_1^{(1)}(x, y) = \max_{z_1 \in X_1} f_1(z_1, x_2, x_3, y) - f_1(x, y),$$

$$\Phi_1^{(1,2)}(x, y) = \max_{(z_1, z_2) \in X_1 \times X_2} f_1(z_1, z_2, x_3, y) - f_1(x, y),$$

$$\Phi_1^{(1,3)}(x, y) = \max_{(z_1, z_3) \in X_1 \times X_3} f_1(z_1, x_2, z_3, y) - f_1(x, y),$$

$$\Phi_1^{(1,2,3)}(x, y) = \max_{z \in X} f_1(z, y) - f_1(x, y).$$

Which risk function should player i select? Once again, this issue is subjectively settled by player i, depending on his attitude to the partners. (For the time being, the attitude has not been formalized.) Moreover, chacun a son goût[42] and de gustibus non est disputandum.[43]

For the sake of compactness, further presentation will be restricted mostly to the NGU with two players ($\mathbb{N} = \{1, 2\}$):

$$\langle \{1, 2\}, \{X_i\}_{i=1,2}, Y, \{f_i(x, y)\}_{i=1,2} \rangle. \tag{1.4.13}$$

According to the approach described above, two types of risk functions will be considered:

– for player 1,

$$\Phi_1^{(1)}(x, y) = \max_{z_1 \in X_1} f_1(z_1, x_2, y) - f_1(x_1, x_2, y),$$

$$\Phi_1^{(2)}(x, y) = \max_{z \in X} f_1(z, y) - f_1(x_1, x_2, y); \tag{1.4.14}$$

– for player 2,

$$\Phi_2^{(1)}(x, y) = \max_{z_2 \in X_2} f_2(x_1, z_2, y) - f_2(x_1, x_2, y),$$

$$\Phi_2^{(2)}(x, y) = \max_{z \in X} f_2(z, y) - f_2(x_1, x_2, y). \tag{1.4.15}$$

---

[42] French "to each his own taste."

[43] Latin "In matters of taste, there can be no disputes."

Continuity of risk functions

Lemma 1.4.3 *Assume that in the game* (1.4.13) *the sets* X_i *(i = 1, 2) and Y are compact and the payoff functions* $f_i(x, y)$ *(i = 1, 2) are continuous on* $X_1 \times X_2 \times Y$. *Then all risk functions* $\Phi_i^{(j)}(x, y)$ *(i, j = 1, 2) from* (1.4.14) *and* (1.4.15) *are continuous on* $X_1 \times X_2 \times Y$.

Remark 1.4.3 A result similar to Lemma 1.4.3 applies to the risk function $\Phi_i^{(\mathbf{K}(i))}(x, y)$ $(i \in \mathbb{N})$ defined by (1.4.12):

Assume that in the game Γ_N *the sets* X_i *(i ∈ $\mathbb{N}$) and Y are closed and bounded, and the payoff functions* $f_i(x, y)$ *are continuous on* $\prod_{i \in \mathbb{N}} X_i \times Y$. *Then for any coalitions* $\mathbf{K}(i) \subseteq \mathbb{N}$ *containing player i, the risk function* $\Phi_i^{(\mathbf{K}(i))}(x, y)$ *defined by* (1.4.12) *is continuous on* $\prod_{i \in \mathbb{N}} X_i \times Y$.

In conclusion, note that such an ambiguous formalization of the Savage–Niehans risk function was not thoroughly studied in the literature. In our opinion, this subject needs additional research. Hopefully, the reader will be interested in the corresponding problems.

Chapter 2

Mathematical model of the Golden Rule of ethics in form of differential positional game

> Celui qui croit pouvoir trouver en soi-même
> de quoi se passer de tout le monde se trompe
> fort; mais celui qui croit qu'on ne peut se
> passer de lui se trompe encore davantage.[1]
> —La Rochefoucauld[1]

In Russia, dynamic game-theoretic problems are often studied using the formalization of a differential positional game, proposed by Academician N. Krasovskii [179] is used. However, there are spots on the sun. Krasovskii's student, Academician A. Subbotin, constructed a counterexample in which this formalization leads to the emergence of new motions of a dynamic system with increasing time, and Professor A. Kononenko found a counterexample in which pieces of the initial motions are lost with increasing time. These cases obstruct decision-making (the choice of appropriate strategies by the players), because the outcomes expected by them can be either lost or new ones can appear. The new modification of the original formalization proposed by us (see Part I) can eliminate such drawbacks. With this modification, it is possible to distinguish two fairly general classes of differential positional games in which a Berge equilibrium exists. In Part II, the method of dynamic programming and Poincaré's small parameter method are adopted to establish coefficient criteria for the existence (or nonexistence) of Berge and Nash equilibria in a differential positional game of two players with a small influence of one player on the rate of change of the state vector. Finally, an explicit form of Berge equilibrium in a two-player differential game with interval uncertainty is also found using a suitable version of the dynamic programming method.

PART I. GAMES WITH SEPARATED DYNAMICS

> What is not good to ye, do not make ye to a friend.
> —Clerk (deacon) Joannes[2]

This important proverb was considered the basis for proper behavior, not only by Christians but many other moralists; see Section 1.1. The world would be almost perfect if everyone followed it.

---

[1] French "He who thinks he has the power to content the world greatly deceives himself, but he who thinks that the world cannot be content with him deceives himself yet more." François de La Rochefoucauld, (1613–1680), was a French classical writer; a quote from *Réflexions ou Sentences et Maximes morales* (1665).
[2] A quote from *Izbornik of Sviatoslav* (1073).

In this central part of the book, we endeavor extending the earlier research of the static case of the Golden Rule [417] to the dynamic one. Recall that the fundamental elements of the suggested approach are the Germeier convolution used for the payoff functions of the players, the theory of noncooperative positional differential game of N players in quasimotions, and guiding control.

2.1 MAIN NOTIONS

> Mathematics as an expression of the human mind reflects the active will, the contemplative reason, and the desire for aesthetic perfection. Its basic elements are logic and intuition, analysis and construction, generality and individuality. Though different traditions may emphasize different aspects, it is only the interplay of these antithetic forces and the struggle for their synthesis that constitute the life, usefulness, and supreme value of mathematical science.
> —Courant[3]

2.1.1 Noncooperative positional differential game of N players

The mathematical model of a noncooperative positional differential game of N players (NPDG) is an ordered quadruple

$$\langle \mathbb{N}, \Sigma, \{\mathfrak{A}_i\}_{i\in\mathbb{N}}, \{\mathcal{J}_i\}_{i\in\mathbb{N}} \rangle, \tag{2.1.1}$$

where $\mathbb{N} = \{1, \ldots, N\}$ denotes the set of players; Σ is a controlled dynamic system whose state $x(t)$ evolves over time t in accordance with the vector ordinary differential equation

$$\dot{x} = f(t, x, u_1, \ldots, u_N), \quad x(t_0) = x_0; \tag{2.1.2}$$

$\mathfrak{A}_i$ gives the set of all strategies (actions) U_i of player $i \in \mathbb{N}$; $\mathcal{J}_i$ is the payoff function of player i, and its value (called payoff) is adopted for assessing the performance of player i.

2.1.2 Controlled system

The state $x(t)$ of this game (conflict) at a current time instant t satisfies the vector ordinary differential equation (2.1.2), where $x \in \mathbb{R}^m$ (i.e., the state vector belongs to the m-dimensional real Euclidean space composed of all ordered collections of m real values in the form of columns, with the standard scalar product and Euclidean norm $\|\cdot\|$); the time t varies from $t_0 \geq 0$ to $\vartheta > t_0$, and a current position is $(t, x) \in [t_0, \vartheta] \times \mathbb{R}^m$; the control of player i is $u_i \in Q_i \subset \mathbb{R}^{n_i}$ ($i \in \mathbb{N}$). Now, we introduce the following assumptions.

Conditions 2.1.1 *The components of the m-dimensional vector function $f(t, x, u_1, \ldots, u_N)$ are continuous in all variables. The sets Q_i ($i \in \mathbb{N}$) are closed and bounded, i.e., $Q_i \in$ comp $\mathbb{R}^{n_i}$, and*

$$u = (u_1, \ldots, u_N) \in Q = \prod_{i\in\mathbb{N}} Q_i \subset \mathbb{R}^n \left(n = \sum_{i\in\mathbb{N}} n_i \right).$$

---

[3]Richard Courant, (1888–1972), was a German-born American mathematician, educator, and scientific organizer who made significant advances in the calculus of variations. A quote from *The Australian Mathematics Teacher*, vols. 39–40, Australian Association of Mathematics Teachers, 1983, p. 3.

For each bounded domain G in the space of positions (t, x), *there exists a specific Lipschitz constant* $\lambda(G)$ *such that*

$$\left\| f\left(t^{(1)}, x^{(1)}, u\right) - f\left(t^{(2)}, x^{(2)}, u\right) \right\| \le \lambda(G) \left(\left\| x^{(1)} - x^{(2)} \right\| + \left| t^{(1)} - t^{(2)} \right| \right)$$

for all $(t^{(j)}, x^{(j)}) \in G$ $(j = 1, 2)$, *uniformly in* $u \in Q$.

There exists a $\gamma = const > 0$ *such that, for all* $t \in [t_0, \vartheta]$ *and* $u \in Q$, *the sublinear growth conditions hold*:

$$\|f(t, x, u)\| \le \gamma(1 + \|x\|). \tag{2.1.3}$$

Note that Conditions 2.1.1 guarantee the existence of unique and continuable solutions $x(\cdot) = \{x(t), t_0 \le t \le \vartheta\}$ of the system (2.1.2) on an interval $[t_0, \vartheta]$ for all $x(t_0) = x_0 \in \mathbb{R}^m$ and any Borel measurable n-dimensional vector functions $u(t) \in Q$ $\forall t \in [t_0, \vartheta]$.

2.1.3 Strategies and strategy profiles

The need for a well-timed response to all possible disturbances of a controlled system leads to the principle of feedback control. For the system (2.1.2), this principle can be implemented using the mathematical formalization of players' controls (strategies) in noncooperative differential games suggested by Krasovskii and his followers [178, 179, 233]. Let us formulate several concepts required for further presentation: a strategy of player i, a strategy profile, and the motions of the system (2.1.2) that are induced by them.

Definition 2.1.1 *A strategy* U_i *of player i in the NPDG (2.1.1) is a function* $u_i(t, x)$ *whose values satisfy the inclusion*

$$u_i(t, x) \subseteq Q_i \tag{2.1.4}$$

at each possible position (t, x), *where* Q_i *is a given compact subset of the* n_i-*dimensional Euclidean space* $\mathbb{R}^{n_i}$.

The function $u_i(t, x)$ in (2.1.4) can be multivalued, particularly, $u_i(t, x) = Q_i$. A correspondence between a strategy U_i and a function $u_i(t, x)$ will be denoted by $U_i \div u_i(t, x)$, and the set of all strategies of player i by $\mathfrak{A}_i$. A complete collection of strategies $U = (U_1, \ldots, U_N) \in \mathfrak{A} = \prod_{i \in \mathbb{N}} \mathfrak{A}_i$ will be called a *strategy profile* of the game (2.1.1).

Player i chooses a control u_i. Now, consider the concept of a quasimotion of the system (2.1.2) from an initial position $(t_0, x_0) \in [0, \vartheta] \times \mathbb{R}^m$ that is induced by a fixed strategy U_i of player i $(i \in \mathbb{N})$. The theoretical results presented below were established in [137]. Note that the original concept of a motion of a controlled system (see [178, 179, 233]) has been modified in view of the two counterexamples suggested by Subbotin and Kononenko [137, p. 16]. The first example demonstrated a possible *occurrence of new motions* that were absent in an original pencil of motions and the second example demonstrated a possible *loss of some motions* from that pencil. These negative effects can be avoided using the mathematical formalization of a quasimotion developed by Zhukovskiy in [137]; see that paper for a detailed proof of the facts listed in Subsection 2.1.4 and Section 2.2.

2.1.4   Piecewise continuous step-by-step quasimotions

We will introduce step-by-step quasimotions differing from their counterparts [179, p. 8] by the following properties: *they are piecewise continuous and may have discontinuities of the first kind (finite jumps) at any partition points.*

Step-by-step quasimotions

We define the step-by-step quasimotions (piecewise continuous step-by-step motions) induced by a strategy $U_i \in \mathfrak{A}_i$ and a strategy profile $U = (U_1, \dots, U_N) \in \mathfrak{A}$. Without special mention, let the system (2.1.2) satisfy Conditions 2.1.1.

Consider an initial position $(t_0, x_0) \in [0, \vartheta) \times \mathbb{R}^m$, a value $\alpha \in [0, 1]$ and a chosen strategy $U_i \dot{\div} u_i(t, x)$, $U_i \in \mathfrak{A}_i$. We cover the interval $[t_0, \vartheta]$ by a system Δ of half-intervals $\tau_j \leq t < \tau_{j+1}$ $(j = 0, 1, \dots, r(\Delta) - 1)$, where $\tau_0 \geq t_0$ and $\tau_{r(\Delta)} = \vartheta$. Next, let $u_{\mathbb{N} \setminus \{i\}}(t) = (u_1(t), \dots, u_{i-1}(t), u_{i+1}(t), \dots, u_N(t))$ be Borel measurable vector functions such that, for all $t \in [t_0, \vartheta)$,

$$u_{\mathbb{N} \setminus \{i\}}(t) \in Q_{\mathbb{N} \setminus \{i\}} = \prod_{j \in \mathbb{N}, j \neq i} Q_j.$$

A step-by-step quasimotion of the system (2.1.2) from an initial position $(t_0, x_0) \in [0, \vartheta) \times \mathbb{R}^m$ that is induced by

a) *a strategy $U_i \dot{\div} u_i(t, x) \subseteq Q_i$,*

b) *a partition $\Delta : t_0 \leq \tau_0 < \tau_1 < \dots < \tau_{r(\Delta)} = \vartheta$, and*

c   *a value $\alpha \in [0, 1]$,*

is any function

$$x(\cdot, U_i, \Delta, \alpha) = \{x(t, \tau_0, x_0, \hat{x}_0, u_{\mathbb{N} \setminus \{i\}}(\cdot), U_i, \Delta, \alpha), \ \tau_0 \leq t \leq \vartheta\}$$

that satisfies the quasi step-by-step equation

$$x(t, U_i, \Delta, \alpha) = \hat{x}_i + \int_{\tau_j}^{t} f(\tau, x(\tau, U_i, \Delta, \alpha), u_{\mathbb{N} \setminus \{i\}}(\tau), u_i(\tau_j, \hat{x}_j)) d\tau \qquad (2.1.5)$$

where $\tau_j \leq t < \tau_{j+1}$ $(j = 0, 1, \dots, r(\Delta) - 1)$, subject to the constraint

$$\sum_{j=0}^{r(\Delta)-1} \|\hat{x}_j - x_0(\tau_j, U_i, \Delta, \alpha)\| \leq \alpha, \qquad (2.1.6)$$

where $x_0(\tau_j, U_i, \Delta, \alpha)$ denotes the value at the time instant $t = \tau_j$ of the solution $x(t, U_i, \Delta, \alpha)$ of equation (2.1.5) extended to the right on the interval $\tau_{j-1} \leq t < \tau_j$ $(j = 1, 2, \dots, r(\Delta) - 1)$, i.e.,

$$x_0(\tau_j, U_i, \Delta, \alpha) = \hat{x}_{j-1} + \int_{\tau_{j-1}}^{\tau_j} f(\tau, x(\tau, U_i, \Delta, \alpha), u_{\mathbb{N} \setminus \{i\}}(\tau), u_i(\tau_{j-1}, \hat{x}_{j-1})) d\tau, \quad (2.1.7)$$

$$x(\tau_0, U_i, \Delta, \alpha) = \hat{x}_0, \quad x_0(t_0, U_i, \Delta, \alpha) = x_0.$$

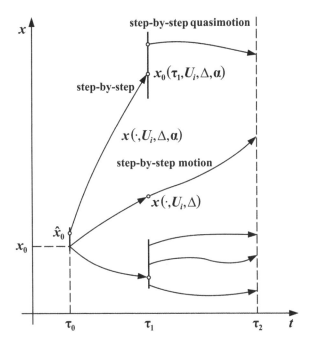

Figure 2.1.1 Step-by-step motion and step-by-step quasimotion.

Unlike the step-by-step motions considered in [179, p. 33], the step-by-step quasimotions may have finite jumps $\|\hat{x}_j - x_0(\tau_j, U_i, \Delta, \alpha)\| \neq 0$ at partition points τ_j. The sum of these jumps must not exceed a given value α; see Figure 2.1.1. This fact will be indicated by a proper notation $x(t, U_i, \Delta, \alpha)$ for quasimotions, which includes α among other variables. For $\alpha = 0$, a step-by-step quasimotion turns into a common step-by-step motion from [179, p. 33].

Denote by $\mathcal{X}(t_0, x_0, U_i)$ the set of all such step-by-step quasimotions of the system (2.1.2) from an initial position (t_0, x_0) that are induced by a fixed strategy U_i, all possible partitions Δ, all values $\alpha \in [0, 1]$ and any Borel measurable functions $u_{\mathbb{N} \setminus \{i\}}(t) \in Q_{\mathbb{N} \setminus \{i\}} = \prod\limits_{s \in \mathbb{N} \setminus \{i\}} Q_s$.

Consider equations (2.1.5) and (2.1.7) with $\tau_j \leq t < \tau_{j+1}$ $(j = 0, 1, \ldots, r(\Delta) - 1)$ and let $u_{\mathbb{N} \setminus \{i\}}(\tau) = u_{\mathbb{N} \setminus \{i\}}(\tau_j, \hat{x}_j)$. Then these equations define a step-by-step quasimotion $x(\cdot, \tau_0, x_0, \hat{x}_0, U, \Delta, \alpha)$ of the system (2.1.2) from an initial position (t_0, x_0) that is induced by a strategy profile $U = (U_1, \ldots, U_N) \in \mathfrak{A}$, where $U = (U_1, \ldots, U_i, \ldots, U_N)$ is a given strategy profile from $\mathfrak{A}$.

Hereinafter, $M_m[t_0, \vartheta]$ will stand for the set of all bounded m-dimensional vector functions $z(t)$, $t \in [t_0, \vartheta]$, with the norm

$$\|z(\cdot)\|_{M_m[t_0, \vartheta]} = \sup_{t_0 \leq t \leq \vartheta} \|z(t)\| .$$

Proposition 2.1.1 *The set $\mathcal{X}(t_0, x_0, U_i)$ of all step-by-step quasimotions $x(\cdot, U_i, \Delta, \alpha)$ of the system (2.1.2) from an initial position (t_0, x_0) that are induced by a fixed strategy U_i, all possible partitions Δ, any values $\alpha \in [0, 1]$, and any Borel measurable functions*

$u_{\mathbb{N}\setminus\{i\}}(t) \in Q_{\mathbb{N}\setminus\{i\}}$ *is bounded by norm in the space* $M_m[t_0, \vartheta]$. *In other words, there exists* $k = const > 0$ *such that, for all* $x(\cdot, U_i, \Delta, \alpha)$,

$$\|x(\cdot, U_i, \Delta, \alpha)\|_{M_m[t_0, \vartheta]} \leq k.$$

In a similar way, the set $\mathcal{X}(t_0, x_0, U)$ of all step-by-step quasimotions $x(\cdot, U, \Delta, \alpha)$ of the system (2.1.2) from an initial position (t_0, x_0) that are induced by a fixed strategy profile $U \in \mathfrak{A}$, all possible partitions Δ, and any values $\alpha \in [0, 1]$ is bounded by norm in the space $M_m[t_0, \vartheta]$.

Quasimotions and their properties

Now, we will define quasimotions as the *continuous* limits (in the space $M_m[t_0, \vartheta]$) of different sequences of step-by-step quasimotions constructed above.

Definition 2.1.2 *A quasimotion* $x[\cdot] = \{x[t, t_0, x_0, U_i], t_0 \leq t \leq \vartheta\}$ *of the system* (2.1.2) *from an initial position* (t_0, x_0) *that is induced by a strategy* $U_i \in \mathfrak{A}_i$ *is any continuous function* $x[\cdot] = \{x[t], t_0 \leq t \leq \vartheta\}$ *on the interval* $[t_0, \vartheta]$ *representing in the metric of the space* $M_m[t_0, \vartheta]$ *the limit of some converging sequence of step-by-step quasimotions (extended to the left to* t_0 *if* $\tau_0^{(r)} > t_0$*) as* $r \to \infty$ *and* $m \to \infty$, *i.e.,*

$$x\left(\cdot, U_i, \Delta^{(r)}, \alpha^{(m)}\right) = \left\{x\left(t, \tau_0^{(r)}, x_0, x_0^{(r)}, u_{\mathbb{N}\setminus\{i\}}^{(r)}(\cdot), \Delta^{(r)}, \alpha^{(m)}\right), t_0 \leq t \leq \vartheta\right\},$$

where

$$diam \, \Delta^{(r)} \to 0,$$

and

$$\left|\tau_0^{(r)} - t_0\right| + \left\|x_0^{(r)} - x_0\right\| \to 0 \text{ as } r \to \infty,$$
$$\alpha^{(m)} \to 0 \text{ as } m \to \infty; \tag{2.1.8}$$

here

$$diam \, \Delta^{(r)} = \max_j \left[\tau_{j+1}^{(r)} - \tau_j^{(r)}\right],$$

$$0 \leq \alpha^{(m)} \leq 1 \quad (r, m = 1, 2, \ldots).$$

Thus, under (2.1.8) this sequence $\{x(\cdot, U_i, \Delta^{(r)}, \alpha^{(m)})\}$ satisfies the condition

$$\sup_{t_0 \leq t \leq \vartheta} \|x[t] - x(t, U_i, \Delta^{(r)}, \alpha^{(m)})\| \to 0.$$

Denote by $\mathcal{X}[t_0, x_0, U_i]$ the pencil of the constructed quasimotions of the system (2.1.2). Different quasimotions of this pencil are implemented with different sequences $\Delta^{(r)}$, $\alpha^{(m)}$ and different $u_{\mathbb{N}\setminus\{i\}}^{(r)}(t) \in Q_{\mathbb{N}\setminus\{i\}}$ that can be used to construct the step-by-step quasimotions converging to the quasimotions $x[\cdot, t_0, x_0, U_i]$.

The quasimotions $x[\cdot, t_0, x_0, U]$ of the system (2.1.2) from an initial position (t_0, x_0) that are induced by a strategy profile $U \in \mathfrak{A}$ are defined by analogy. The pencil of such quasimotions will be denoted by $\mathcal{X}[t_0, x_0, U]$.

Consider the properties of these pencils of quasimotions.

Proposition 2.1.2 *A pencil $\mathcal{X}[t_0, x_0, U_i]$ is a non-empty, bounded by norm and closed subset in the space $C_m[t_0, \vartheta] \subset M_m[t_0, \vartheta]$. The same properties hold for the pencil $\mathcal{X}[t_0, x_0, U]$.*

Proposition 2.1.3 *For any initial position $(t_0, x_0) \in [0, \vartheta) \times \mathbb{R}^m$, any strategy U_i and any strategy profile $U \in \mathfrak{A}$, the cutsets of the corresponding pencils of quasimotions by the hyperplane $t = \vartheta$,*

$$X[\vartheta, t_0, x_0, U_i] = \mathcal{X}[t_0, x_0, U_i] \cap \{t = \vartheta\},$$

and

$$X[\vartheta, t_0, x_0, U] = \mathcal{X}[t_0, x_0, U] \cap \{t = \vartheta\},$$

are non-empty compact sets in $\mathbb{R}^m$.

Proposition 2.1.4 *Whatever the initial position $(t_0, x_0) \in [0, \vartheta) \times \mathbb{R}^m$ and strategy profile $U = (U_1, \ldots, U_i, \ldots, U_N) \in \mathfrak{A}$ are, each quasimotion $x[\cdot, t_0, x_0, U]$ from the pencil $\mathcal{X}[t_0, x_0, U]$ is contained in the pencil $\mathcal{X}[t_0, x_0, U_i]$ $(i \in \mathbb{N})$, i.e.,*

$$\mathcal{X}[t_0, x_0, U] \subset \mathcal{X}[t_0, x_0, U_i] \;\; (i \in \mathbb{N}),$$

and hence

$$X[\vartheta, t_0, x_0, U \div Q] \subset X[\vartheta, t_0, x_0, U_i \div Q_i] \;\; (i \in \mathbb{N}).$$

In what follows, also we will use *the attainability domain* of the system (2.1.2) from an initial position (t_0, x_0), i.e.,

$$X[\vartheta] = X[\vartheta, t_0, x_0, U \div Q] = X[U \div Q] = \bigcup_{U \in \mathfrak{A}} x[\vartheta, t_0, x_0, U].$$

Remark 2.1.1 By Definitions 2.1.1 and 2.1.2 (strategies, strategy profiles, and the quasimotions of the system (2.1.2) from an initial position (t_0, x_0) induced by them),

a)   for the strategy $U_i \div Q_i$ of player i, the pencil of quasimotions is

$$\mathcal{X}[t_0, x_0, U_i \div Q_i] = \{x[t, t_0, x_0, U_i], \; t \in [t_0, \vartheta] \mid \forall \, U_i \in \mathfrak{A}_i\}$$
$$= \bigcup_{U_i \in \mathfrak{A}_i} \{x[t, t_0, x_0, U_i], \; t_0 \leq t \leq \vartheta\};$$

b)   for a strategy profile $U \div Q$, the pencil of quasimotions is

$$\mathcal{X}[t_0, x_0, U \div Q] = \{x[t, t_0, x_0, U], \; t_0 \leq t \leq \vartheta \mid \forall \, U \in \mathfrak{A}\}$$
$$= \bigcup_{U \in \mathfrak{A}} \{x[t, t_0, x_0, U], \; t_0 \leq t \leq \vartheta\}.$$

Completeness of pencil of quasimotions

The following result justifies the transition from the motions of the system (2.1.2) in Krasovskii's sense to the quasimotions.

Theorem 2.1.1 *For any position* $(t^*, x[t^*])$, *where* $t^* \in [t_0, \vartheta]$ *and* $x[\cdot]$ *is some quasimotion from the pencil* $\mathcal{X}[t_0, x_0, U]$,

$$X[\vartheta, t^*, x[t^*], U] \subset X[\vartheta, t_0, x_0, U] \; \forall \, U \in \mathfrak{A}.$$

Thus, in any quasimotion induced by a given strategy profile U from a given initial position, there may not occur or disappear segments of any other quasimotions absent in the original pencil that are induced by the same strategy profile from the same initial position. For the quasimotions formalized in [179], such segments of motions may appear; see below the counterexamples by Kononenko and Subbotin.

The next statement is similar to Theorem 2.1.1.

Proposition 2.1.5 *For any position* $(t^*, x[t^*])$, *where* $t^* \in [t_0, \vartheta]$ *and* $x[\cdot]$ *is an arbitrary quasimotion from the pencil* $\mathcal{X}[t_0, x_0, U_i]$,

$$X[\vartheta, t^*, x[t^*], U_i] \subset X[\vartheta, t_0, x_0, U_i].$$

Corollary 2.1.1 *Let* U_i *be a certain strategy from* $\mathfrak{A}_i$ *and* $x^*[\cdot] = x^*[\cdot, t_0, x_0, U_i]$ *be some quasimotion of the system* (2.1.2) *from an initial position* (t_0, x_0) *that is induced by* U_i. *Consider* $t^* \in [t_0, \vartheta]$ *and let* $x[t] = x[t, t^*, x^*[t^*], U_i]$, $t^* \leq t \leq \vartheta$, *be an arbitrary quasimotion of the system* (2.1.2) *from the initial position* $(t^*, x^*[t^*])$ *that is induced by* U_i. *Then there exists a quasimotion* $\tilde{x}[\cdot] \in \mathcal{X}[t_0, x_0, U_i]$ *such that*

$$\tilde{x}[t] = \begin{cases} x^*[t] & \text{for } t_0 \leq t \leq t^*, \\ x[t] & \text{for } t^* \leq t \leq \vartheta. \end{cases}$$

Consequently, the quasimotion $x[t, t^*, x^*[t^*], U_i]$, $t^* \leq t \leq \vartheta$, is a segment of some quasimotion $\tilde{x}[\cdot]$ from the pencil $\mathcal{X}[t_0, x_0, U_i]$.

Counterexamples by Subbotin and Kononenko

Malum consilium est, quod mutari potest.[4]

As we have mentioned earlier, Krasovskii's concept of motions of a dynamic system is extended to that of quasimotions in view of the two counterexamples suggested by Subbotin and Kononenko.

The first example demonstrates a possible occurrence of new motions [179] that were absent in the original pencil as the game evolved. This fact was first observed by Subbotin as follows.

Counterexample by Subbotin

We know the truth, not only by
the reason, but also by the heart.
—Pascal[5]

Example 2.2.1 Let the controlled system (2.1.2) be given by

$$\dot{x} = u, \;\; 0 \leq t \leq 2, \;\; x[0] = x_0 = 0, \;\; |u| \leq 1. \tag{2.1.9}$$

---

[4]Latin "It is a bad plan that admist of no modification." This phrase is attributed to Publilius Syrus, (flourished 1st century B.C.), Latin mime writer contemporary with Cicero, chiefly remembered for a collection of versified aphorisms.

[5]Blaise Pascal, (1623–1662), was a French mathematician, physician, religious figure, and writer.

Consider the strategy

$$
U \div u(t, x) = \begin{cases}
0 & \text{for } x > 0, \ t \in [0, 1), \\
0 & \text{for } x < 0, \ t \in [0, 1), \\
+1 & \text{for } x = 0, \ t \in [0, 1), \\
0 & \text{for } x = 0, \ t \in [1, 2), \\
+1 & \text{for } x > 0, \ t \in [1, 2), \\
-1 & \text{for } x < 0, \ t \in [1, 2).
\end{cases}
$$

For this system, the pencil of all motions from the initial position $(t_0, x_0) = (0, 0)$ that are induced by the strategy U consists of the two elements, $x^{(1)}[t]$ and $x^{(2)}[t]$, $0 \le t \le 2$; see solid line in Figure 2.1.2. The same strategy U induces three motions from the initial position $(1, 0)$, namely, $x^{(1)}[t]$, $x^{(2)}[t]$, and $x^{(3)}[t] \equiv 0$, $1 \le t \le 2$. Thus, as the current position $(t, x[t])$ is shifting along the motion $x^{(1)}[t]$ or $x^{(2)}[t]$, at the time instant $t = 1$, the original pencil $\mathcal{X}[t_0, x_0, U]$ is augmented by the new motion $x^{(3)}[t]$, $1 \le t \le 2$, which was absent in $\mathcal{X}[t_0, x_0, U]$. Thus, as the game evolves, some segments of the original motions may be lost (see the next example by Kononenko) and new motions may also occur that were absent in the original pencil (the motion $x^{(3)}[t] \equiv 0$ in the current example).

The appearance of new motions (like $x^{(3)}[t] \equiv 0$, $1 \le t \le 2$) is undesired for the players choosing their own strategies. Indeed, a new motion yields new values of the decision criteria adopted by players. As a result, the optimal strategies of players in an initial position (the beginning of the game) may lose optimality as the game evolves if a current position "reaches" the starting point of a new motion. This drawback can be eliminated with a slight modification of the concept of step-by-step motions [179]. More specifically, it is necessary to assume that the step-by-step motions may have finite jumps at partition points.

For instance, in the current example the motion $x[t] \equiv 0$, $0 \le t \le 2$, is the limit of the sequence of step-by-step quasimotions $x^{(k)}[t]$, $0 \le t \le 2$ ($k = 1, 2, \ldots$); see dashed

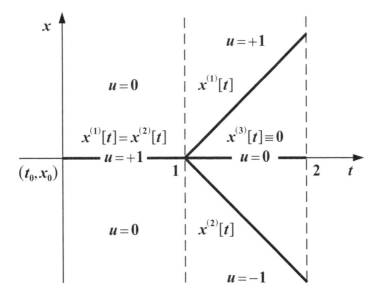

Figure 2.1.2 Pencil of motions from initial position (t_0, x_0).

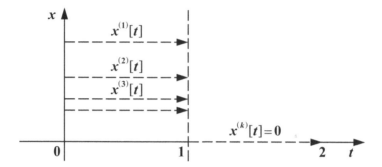

Figure 2.1.3 Step-by-step quasimotions and their limit.

line in Figure 2.1.3. These quasimotions have finite jumps at the point $t = 1$ and are continuous at the other points of the interval $[0, 2]$.

Here the convergence of $x^{(k)}[t]$ to $x[t] \equiv 0$ must be considered in the space $M_m[t_0, \vartheta]$ of all bounded n-dimensional functions with the norm $\|x(\cdot)\|_{M_m[t_0, \vartheta]} = \sup_{t_0 \leq t \leq \vartheta} \|x(t)\|$.
With such a definition, the occurrence of new motions is eliminated and the players can make decisions (choose their strategies) without expecting new motions as the game evolves.

Counterexample by Kononenko

Ex adverso.[6]

The next interesting counterexample was presented by Kononenko [177]. It demonstrates that a segment of a motion (in the sense of [179]) may cease to be a motion as the game evolves.

Example 2.1.2 Let the controlled system (2.1.2) be given by

$$\dot{x} = u, \; 0 \leq t \leq 1, \; x[0] = x_0 = 0, \; |u| \leq 1.$$

Consider the strategy

$$U \div u(t, x) = \begin{cases} +1 & \text{for } x > 0, \; t \in [0, 1), \\ -1 & \text{for } x < 0, \; t \in [0, 1), \\ 0 & \text{for } x = 0, \; t \in [0, \tfrac{1}{2}) \cup (\tfrac{1}{2}, 1), \\ 1 & \text{for } x = 0, \; t = \tfrac{1}{2}. \end{cases} \tag{2.1.10}$$

A motion of this system from the initial position $(t_0, x_0) = (0, 0)$ that is induced by U is $x[t] \equiv 0, 0 \leq t \leq 1$; see bold line in Figure 2.1.4. However, its segment $x^{(3)}[t] \equiv 0, \frac{1}{2} \leq t \leq 1$, is not the motion on the interval $[\frac{1}{2}, 1]$ that starts from the position $(\frac{1}{2}, 0) = (\frac{1}{2}, x[\frac{1}{2}])$. On this interval, the motions are the straight-line segments $x^{(1)}[t] = t - \frac{1}{2}$ and $x^{(2)}[t] = -(t - \frac{1}{2}), t \in [\frac{1}{2}, 1]$, whereas the segment $x^{(3)}[t] \equiv 0, t \in [\frac{1}{2}, 1]$, is not a motion. Therefore, the segment $x[t] \equiv 0 \; (\frac{1}{2} \leq t \leq 1)$ is not the motion induced from the "current" initial position $(\frac{1}{2}, 0)$. This fact can be explained in the following way: the definition introduced

---

[6]Latin "From the opposition."

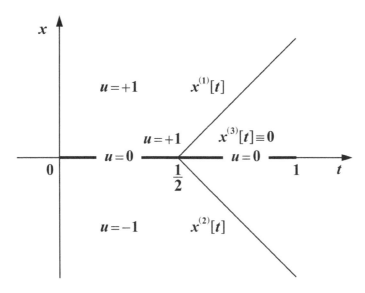

Figure 2.1.4 Lost segment of motion.

in [179, p. 31] suggests no *passage to the limit in the initial time instant*, i.e., for a converging sequence of step-by-step motions, only the initial positions $(t_0, x^{(k)})$ with a fixed time instant t_0 are considered.

In control problems and differential games, this fact may "deteriorate" an optimal strategy. Assume that, for a strategy adopted as the game solution, the performance criterion achieves optimum at the "end" of the motion and "its segment is lost" as the game evolves. Then, starting from some time instant (in Example 2.1.2, $t = \frac{1}{2}$), it is necessary to use another optimal strategy: the previous gives no result because the corresponding motion disappears. For avoiding such "a loss of segments," Kononenko suggested to define the step-by-step motions of system (2.1.2) with the passage to the limit in the initial time instant. That is, each step-by-step motion must be started from an initial position $(t^{(k)}, x^{(k)})$ instead of $(t_0, x^{(k)})$. Then such a step-by-step motion must be extended to the left to t_0 (if $t^{(k)} > t_0$). Due to Conditions 2.1.1, this extension is possible. For designing the motions $x[\cdot, t_0, x_0, U]$, the step-by-step motions $x(t, t^{(k)}, x^{(k)}, U, \Delta^{(k)})$, $t_0 \le t \le \vartheta$, that are extended to the left (if $t^{(k)} > t_0$) or truncated (if $t^{(k)} < t_0$) must be considered in the limit sense under the constraint

$$\lim_{k \to \infty} |t^{(k)} - t_0| = 0. \tag{2.1.11}$$

For instance, the motion $x^{(3)}[t] \equiv 0$, $\frac{1}{2} \le t \le 1$ (see Figure 2.1.4), is the limit of the sequence of the step-by-step motions evolving from $(t^{(k)}, x^{(k)}) = (\frac{1}{2} + \frac{1}{10k}, 0)$, $k = 1, 2, \ldots$.

Consequently, the following result holds for the strategy (2.1.10).

Proposition [177]. *Let $x[t] = x[t, t_0, x_0, U]$, $t_0 \le t \le \vartheta$, be the motion of the system (2.1.2) from an initial position (t_0, x_0) that is induced by the strategy U. Then, for any $t^* \in [t_0, \vartheta)$, the segment $x[t, t_0, x_0, U]$, $t^* \le t \le \vartheta$, is the motion of this system that is induced by U from the initial position $(t^*, x[t^*, t_0, x_0, U])$.*

This explains the use of the limit (2.1.11) in the definition of quasimotions.

2.1.5   Terminal payoff functions

Consider again the NPDG (2.1.1) under Conditions 2.1.1 from Subsection 2.1.2. Let each player i choose his positional strategy $U_i \in \mathfrak{A}_i$ ($i \in \mathbb{N}$) without making coalitions with other players. As a result, a strategy profile $U = (U_1, \dots, U_N) \in \mathfrak{A} = \prod_{i \in \mathbb{N}} \mathfrak{A}_i$ is formed. The strategy profile U induces a pencil $\mathcal{X}[t_0, x_0, U]$ of quasimotions from the initial position (t_0, x_0). The intersection of this pencil with the hyperplane $t = \vartheta$ is a compact subset—the attainability domain $X[\vartheta, t_0, x_0, U \div Q] = X[\vartheta]$ (see Proposition 2.1.3). *The payoff function* of each player $i \in \mathbb{N}$, $\mathcal{J}_i = F_i(x[\vartheta])$, is defined on $X[\vartheta]$. The value of this function is adopted for assessing the performance of player i in the game (2.1.1) and called his *payoff*. Without special mention, accept

Conditions 2.1.2 *In the game* (2.1.1), *the scalar functions* $F_i(x)$ ($i \in \mathbb{N}$) *are continuous in the space* $\mathbb{R}^m$.

In the theory of differential games, the payoff functions of the form $\mathcal{J}_i = F_i(x[\vartheta])$ are said to be *terminal* because they are defined at the "right limits" ($t = \vartheta$) of the quasimotions $x[\cdot] = x[t, t_0, x_0, U]$, $t_0 \leq t \leq \vartheta$.

2.2   ALTERNATIVE AND SADDLE POINT

Ex pose des motifs.[7]

2.2.1   Auxiliary results

Many properties of differential positional games with several players are studied using the theorem of alternative, which was established for motions in [179]. Here are some concepts figuring in this theorem. Under Conditions 2.1.1 and 2.1.2, consider the zero-sum two-player positional differential game

$$\Gamma_a = \langle \{1, 2\}, \Sigma_a, \{\mathfrak{A}, \mathfrak{V}\}, F(x[\vartheta]) \rangle,$$

in which player 1 minimizes, whereas player 2 maximizes a scalar function $F(x[\vartheta])$; a controlled system Σ_a is described by the equation

$$\dot{x} = \bar{f}(t, x, u, v), \quad x[t_0] = x_0.$$

Here $x \in \mathbb{R}^m$ denotes the state vector; $t \in [t_0, \vartheta]$ indicates time; $\vartheta > t_0 \geq 0$ are some constants; the strategy sets of players 1 and 2, $\mathfrak{A}$ and $\mathfrak{V}$, have the following form:

$$\left.\begin{aligned}
\mathfrak{A} &= \{U \div u(t, x) \mid u(t, x) \subseteq L \in comp\ \mathbb{R}^l\}, \\
\mathfrak{V} &= \{V \div v(t, x) \mid v(t, x) \subseteq Q \in comp\ \mathbb{R}^q\}.
\end{aligned}\right\} \tag{2.2.1}$$

We assume that the game Γ_a satisfies the analogs of Conditions 2.1.1 and 2.1.2 as specified below.

Conditions 2.2.1 *The vector function* $\bar{f}(\cdot)$ *is continuous, locally Lipschitzian in* t, x *and* $\left\| \bar{f}(\cdot) \right\| \leq \gamma(1 + \|x\|)$; *moreover, the function* $F(x)$ *is continuous in the space* $\mathbb{R}^m$.

---

[7] French "Exposition of causes."

Consider a set W in the space of all positions (t, x). Such a set W is called *u-stable* if, for any $(t_*, x_*) \in W$, any value $t^* \in (t_*, \vartheta]$ and any vector $v_* \in Q$, the family of all solutions $x(t)$ of the differential contingency equation

$$\dot{x}(t) \in \Phi_u(t, x(t), v_*), \quad x(t_*) = x_*,$$

where

$$\Phi_u(t, x, v) = co \left\{ \bar{f} \in \mathbb{R}^m \mid \bar{f} = \bar{f}(t, x, u, v_*) \; \forall u \in L \right\}$$

and $co\,\{\}$ denotes the convex closed hull of a set $\{\cdot\}$, contains at least one solution $x(t)$, $t_* \leq t \leq \vartheta$, with the inclusion

$$\left(t^*, x\left(t^* \right) \right) \in W.$$

If the equality

$$\min_{u \in L} \max_{v \in Q} z' \bar{f}(t, x, u, v) = \max_{v \in Q} \min_{u \in L} z' \bar{f}(t, x, u, v), \tag{2.2.2}$$

holds for each position $(t, x) \in [0, \vartheta) \times \mathbb{R}^m$ and any vector $x \in \mathbb{R}^m$, then *the saddle point condition in the small game is satisfied* for Γ_a; see [179, p. 56]. Hereinafter, the prime means transposition.

Now, we introduce the concept of an extremal strategy for a set W.

Let (t_*, x_*) be some position. An extremal strategy $U^e \div u^e(t, x)$, $U^e \in \mathfrak{A}$, for W is defined as follows.

If the hyperplane $t = t_*$ does not intersect W in the space of positions, then as $u^e(t_*, x_*)$ we take *any* vector $u \in L$.

If such an intersection occurs, then we take the position $(t_*, w_*) \in W$ that is closest to (t_*, x_*) in the Euclidean metric (any of them if there are several such positions). As $u^e(t_*, x_*)$ take any of the vectors u^* that satisfy the equality,

$$\max_{v \in Q} (x_* - w_*)' \bar{f}\left(t_*, x_*, u^*, v \right) = \min_{u \in L} \max_{v \in Q} (x_* - w_*)' \bar{f}(t_*, x_*, u, v). \tag{2.2.3}$$

2.2.2   Theorem of alternative

Lemma 2.2.1 *Assume that a closed set W is u-stable, a strategy $U^e \div u^e(t, x)$, $U^e \in \mathfrak{A}$, is extremal for W, and an initial position is $(t_*, x_*) \in W$. Then the inclusion*

$$\left(t, x\left[t, t_*, x_*, U^e \right] \right) \in W \; \forall \, t \in [t_*, \vartheta]. \tag{2.2.4}$$

holds for any quasimotion $x[t, t_*, x_*, U^e]$, $t_* \leq t \leq \vartheta$.

Using the proof scheme suggested by Krasovskii [179, pp. 68–70], it is possible to establish the theorem of alternative for the class of quasimotions as follows.

Theorem 2.2.1 *Assume that Conditions 2.2.1 hold for Γ_a and let $M \subset \mathbb{R}^m$ be a given closed set. Whatever the initial position $(t_0, x_0) \in [0, \vartheta) \times \mathbb{R}^m$ is, one of the following situations takes place.*

Either there exists a strategy $U^e \in \mathfrak{A}$ with which the set M is reached at the time instant ϑ for all quasimotions $x[\cdot, t_0, x_0, U^e]$, i.e.,

$$X[\vartheta, t_0, x_0, U^e] \subset M; \tag{2.2.5}$$

or there exist a value $\varepsilon > 0$ and a strategy $V^e \in \mathfrak{V}$ such that, for all quasimotions $x[\cdot, t_0, x_0, V^e]$, at the time instant ϑ the ε-neighborhood of the set M is evaded, i.e.,

$$X[\vartheta, t_0, x_0, V^e] \cap M^\varepsilon = \varnothing, \tag{2.2.6}$$

where

$$M^\varepsilon = \{x \in \mathbb{R}^m \mid \min_{\bar{x} \in M} \|x - \bar{x}\| < \varepsilon\}. \tag{2.2.7}$$

In Theorem 2.2.1, the role of the strategies U^e and V^e is played by the extremal strategies for specially designed stable sets. Moreover, the u-stable set (for the construction of U^e) is terminated to M at the time instant ϑ, whereas the v-stable set (for the construction of V^e) evades the ε-neighborhood of M at the time instant ϑ.

Remark 2.2.1 Interchanging the roles of the strategies U and V gives another alternative as follows. Assume that Conditions 2.2.1 hold and let M be a given closed set. Whatever the initial position $(t_0, x_0) \in [0, \vartheta) \times \mathbb{R}^m$ is,

either there exists a strategy $V^e \in \mathfrak{V}$ such that $x[\vartheta, t_0, x_0, V^e] \in M$ for all quasimotions $x[\cdot, t_0, x_0, V^e] \in \mathcal{X}[t_0, x_0, V^e]$,

or there exists a value $\varepsilon > 0$ and a strategy $U^e \in \mathfrak{A}$ such that

$$X[\vartheta, t_0, x_0, U^e] \cap M^\varepsilon = \varnothing.$$

2.2.3　Minimax, maximin, and saddle point

We get back to the zero-sum two-player positional differential game Γ_a in quasimotions with a scalar terminal payoff function given by

$$\langle \{1, 2\}, \Sigma_a, \{\mathfrak{A}, \mathfrak{V}\}, F(x[\vartheta]) \rangle. \tag{2.2.8}$$

As before, numbers 1 and 2 are associated with players; players 1 minimizes whereas player 2 maximizes $F(x[\vartheta])$ by choosing his strategy $U \in \mathfrak{A}$ and $V \in \mathfrak{V}$, respectively; the strategy sets are the same as in (2.2.1); a controlled system Σ_a satisfies the vector differential equation

$$\dot{x} = \bar{f}(t, x, u, v), \quad x[t_0] = x_0, \tag{2.2.9}$$

and Conditions 2.2.1 are assumed valid.

This game is organized as follows. Player 1 chooses and uses one of his strategies $U \in \mathfrak{A}$ to minimize $F(x[\vartheta])$. Player 2 chooses $V \in \mathfrak{V}$, pursuing the opposite goal. The resulting strategy profile (U, V) induces a pencil $\mathcal{X}[t_0, x_0, U, V]$ of quasimotions $x[t, t_0, x_0, U, V]$, $t_0 \leq t \leq \vartheta$, from the initial position $(t_0, x_0) \in [0, \vartheta) \times \mathbb{R}^m$. Then player 1 loses the value $F(x[\vartheta, t_0, x_0, U, V])$, whereas player 2 gains just as much.

From the view of player 1, the solution of the game (2.2.8) is *the minimax strategy* $U^0 \in \mathfrak{A}$:

$$\min_{U \in \mathfrak{A}} \max_{x[\cdot]} F(x[\vartheta, t_0, x_0, U]) = \max_{x[\cdot]} F(x[\vartheta, t_0, x_0, U^0]) = F_*. \tag{2.2.10}$$

The value F_* is called *the minimax of the game* (2.2.8).

Recall that all the results below were proved in [179].

The minimax strategy possesses two important properties as follows.

Property 2.2.1 *Equality* (2.2.10) *is equivalent to the inequality*

$$\max_{x[\cdot]} F(x[\vartheta, t_0, x_0, U]) \geq \max_{x[\cdot]} F(x[\vartheta, t_0, x_0, U^0]) \ \forall U \in \mathfrak{A}. \tag{2.2.11}$$

Since player 1 seeks to minimize $F(x[\vartheta])$, for a fixed strategy $U \in \mathfrak{A}$ his worst-case loss will be

$$\max_{x[\cdot]} F(x[\vartheta, t_0, x_0, U]) = \max_{x[\cdot] \in \mathcal{X}[t_0, x_0, U]} F(x[\vartheta]) = F(x^*[\vartheta, t_0, x_0, U]).$$

For a fixed strategy U of player 1, this loss will be achieved along a quasimotion $x^*[\cdot, t_0, x_0, U]$ from the pencil $\mathcal{X}[t_0, x_0, U]$ that corresponds to the solution of the optimization problem

$$F(x) \rightarrow \max$$

subject to the constraint

$$x \in X[\vartheta, t_0, x_0, U],$$

where the set $X[\vartheta, t_0, x_0, U]$ is defined by Proposition 2.1.3.

Then, according to inequality (2.2.11), player 1 guarantees the smallest loss among all possible worst-case (maximum) ones using the minimax strategy U^0.

The second equality in (2.2.10) immediately implies

Property 2.2.2

$$F_* \geq F(x[\vartheta, t_0, x_0, U^0])$$

for any quasimotions $x[\cdot]$ *from the pencil* $\mathcal{X}[t_0, x_0, U^0]$.

Therefore, with the strategy $U^0 \in \mathfrak{A}$ player 1 guarantees that his loss will be not greater than F_*.

A direct combination of Properties 2.2.1 and 2.2.2 leads to the following result. Among all possible losses of player 1, the best-case (minimum) one is F_* (2.2.10), which is guaranteed using the minimax strategy U^0. This expresses the principle of maximum guaranteed result [54] for player 1.

In a similar way, for player 2 (maximizing $F(x[\vartheta])$) the solution of the game (2.2.8) is determined by the *maximin strategy* $V^0 \in \mathfrak{V}$:

$$\max_{V \in \mathfrak{V}} \min_{x[\cdot]} F(x[\vartheta, t_0, x_0, V]) = \min_{x[\cdot]} F(x[\vartheta, t_0, x_0, V^0]) = F^*. \tag{2.2.12}$$

The value F^* is called *the maximin of the game* (2.2.8).

Property 2.2.3 *The first equality in* (2.2.12) *is equivalent to the inequality*

$$\min_{x[\cdot]} F(x[\vartheta, t_0, x_0, V^0]) \geq \min_{x[\cdot]} F(x[\vartheta, t_0, x_0, V]) \ \forall V \in \mathfrak{V}.$$

Then the second equality in (2.2.12) gives

Property 2.2.4

$$F^* \leq F(x[\vartheta, t_0, x_0, V^0])$$

for any quasimotions $x[\cdot, t_0, x_0, V^0]$ *from the pencil* $\mathcal{X}[t_0, x_0, V^0]$.

A direct combination of Properties 2.2.3 and 2.2.4 implements the principle of guaranteed result for player 2, maximizing $F(x[\vartheta])$. More specifically, among all worst-case (smallest) payoffs of player 2 the maximum one is guaranteed using the maximin strategy V^0.

A connection between the minimax and maximin of the game (2.2.8) is established by the next result.

Proposition 2.2.1

$$F_* \geq F^*,$$

where the minimax F_* and maximin F^* are given by (2.2.10) and (2.2.12), respectively.

Definition 2.2.1 *A strategy profile $(U^0, V^0) \in \mathfrak{A} \times \mathfrak{V}$ is called [179, p. 48] a saddle point of the differential game (2.2.8) if*

$$
\begin{aligned}
F^* &= \max_{V \in \mathfrak{V}} \min_{x[\cdot]} F\left(x\left[\vartheta, t_0, x_0, V\right]\right) = \min_{x[\cdot]} F\left(x\left[\vartheta, t_0, x_0, V^0\right]\right) \\
&= \max_{x[\cdot]} F\left(x\left[\vartheta, t_0, x_0, U^0\right]\right) = \min_{U \in \mathfrak{A}} \max_{x[\cdot]} F\left(x\left[\vartheta, t_0, x_0, U\right]\right) = F_*.
\end{aligned}
\tag{2.2.13}
$$

The value $F^0 = F_* = F^*$ is called [179, p. 45] *the value of the game* (2.2.8).

Remark 2.2.2 From (2.2.14) and (2.2.13), it follows that

$$
\begin{aligned}
F^0 &= F\left(x\left[\vartheta, t_0, x_0, U^0, V^0\right]\right) = \max_{x[\cdot]} F\left(x\left[\vartheta, t_0, x_0, U^0, V^0\right]\right) \\
&= \min_{x[\cdot]} F\left(x\left[\vartheta, t_0, x_0, U^0, V^0\right]\right) = F_* = F^*
\end{aligned}
$$

for any quasimotions $x[\cdot, t_0, x_0, U^0, V^0]$.

Theorem 2.2.2 *Let Conditions 2.2.1 and equality (2.2.2) hold. Then, for any initial position $(t_0, x_0) \in [0, \vartheta) \times \mathbb{R}^m$, the game (2.2.8) has the maximin V^0 and minimax U^0 strategies forming the saddle point (U^0, V^0); moreover, for each initial position (t_0, x_0) the value F^0 is unique.*

An analogous theorem for motions was proved in [179, pp. 73–77] using the theorem of alternative only. As a matter of fact, the theorem of alternative (see Theorem 2.2.1) is true for the class of quasimotions, and hence the above proof remains in force for quasimotions too.

Remark 2.2.3 Under Conditions 2.2.1, the game (2.2.8) has the maximin V^0 and minimax U^0 strategies for any initial position $(t_0, x_0) \in [0, \vartheta] \times \mathbb{R}^m$.

Interestingly, here the saddle point condition (2.2.2) in the small game is not required.

2.2.4 Properties of saddle point

Property 2.2.5 *A strategy profile $(U^0, V^0) \in \mathfrak{A} \times \mathfrak{V}$ is a saddle point of the game (2.2.8) if and only if*

$$
F\left(x\left[\vartheta, t_0, x_0, U^0\right]\right) \leq F\left(x\left[\vartheta, t_0, x_0, U^0, V^0\right]\right) \leq F\left(x\left[\vartheta, t_0, x_0, V^0\right]\right)
\tag{2.2.14}
$$

for all quasimotions $x\left[\cdot, t_0, x_0, U^0\right]$, $x\left[\cdot, t_0, x_0, U^0, V^0\right]$ and $x\left[\cdot, t_0, x_0, V^0\right]$.

Property 2.2.6 *Let* (U^0, V^0) *be a saddle point of the game* (2.2.8). *Then, for any quasimotions* $x[\cdot, t_0, x_0, U^0, V^0]$ *from the pencil* $\mathcal{X}[t_0, x_0, U^0, V^0]$, *game* (2.2.8) *has the unique value*

$$F^0 = F(x[\vartheta, t_0, x_0, U^0, V^0]) = F_* = F^*.$$

In the game (2.2.8), two saddle points $(U^{(1)}, V^{(1)})$ and $(U^{(2)}, V^{(2)})$ are called *interchangeable* if $(U^{(1)}, V^{(2)})$ and $(U^{(2)}, V^{(1)})$ are also saddle points, and *equivalent* if

$$F\left(x\left[\vartheta, t_0, x_0, U^{(1)}, V^{(1)}\right]\right) = F\left(x\left[\vartheta, t_0, x_0, U^{(2)}, V^{(2)}\right]\right)$$

for any quasimotions $x\left[\cdot, t_0, x_0, U^{(j)}, V^{(j)}\right]$ $(j = 1, 2)$.

Property 2.2.7 *All saddle points of the game* (2.2.8) *are equivalent and interchangeable.*

Remark 2.2.4 The saddle point (U^0, V^0) has been selected as the solution concept for the zero-sum two-player positional differential game (2.2.8) based on the following considerations.

1) The guaranteed values of the payoff function of this game (the minimax F_* (2.2.10) and maximin F^* (2.2.12)) coincide with the value F^0 at a saddle point; see equalities (2.2.13). Hence, the saddle point yields the same guaranteed results for both players.

2) This game has the same value $F^0 = F(x[\vartheta, t_0, x_0, U^0, V^0])$ at any saddle point (the property of equivalence).

3) All saddle points are interchangeable, i.e., it makes no difference to each player which saddle point to use as his strategy. In combination with the opponent's strategy (also from any saddle point), the former strategy forms a saddle point of the game (2.2.8).

Remark 2.2.5 According to Proposition 2.2.1, the inequality $F_* > F^*$ holds if there is no saddle point in the game (2.2.8). An example of such a game was given in [233, pp. 242–244]. This situation may occur if the saddle point condition (2.2.2) in the small game is false. However, even in this case, the existence of a saddle point can be guaranteed by considering the wider class of mixed strategies or using profiles composed of a certain strategy of one player and the counterstrategy of the other. In the class of quasimotions, the existence of a saddle point in the differential game (2.2.8) where the pure-strategy sets $\mathfrak{A}$ and $\mathfrak{V}$ are replaced by the mixed ones or by the strategy–counterstrategy profiles can be established like in [179, pp. 289, 364], using a natural substitution of corresponding motions with quasimotions.

2.2.5   Corollary of theorem of alternative

The theoretical considerations in the next subsection will rely on a corollary of the theorem of alternative. It applies to the game (2.2.8) with a single player only. In this case, the system (2.2.9) takes the form

$$\dot{x} = \bar{f}(t, x, v), \quad x[t_0] = x_0. \tag{2.2.15}$$

Here, as before, $x \in \mathbb{R}^m$ denotes the state vector; $(t_0, x_0) \in [0, \vartheta) \times \mathbb{R}^m$ is a fixed initial position, where $\vartheta = const > t_0 \geq 0$; a strategy V is identified with a function $v(t, x) \subseteq Q$; $\mathfrak{V}$ stands for the set of all such strategies.

Assume that Conditions 2.2.1 hold: the function $\bar{f}(t, x, v)$ is continuous and locally Lipschitzian in x and t; inequality (2.1.3) is valid; the set Q is closed and bounded; finally, the function $F(x)$ is continuous. Consider the attainability domain of the system (2.2.15) from the position (t_0, x_0), i.e., the set

$$X[Q] = X[\vartheta, t_0, x_0, V \div Q] = \bigcup_{V \in \mathfrak{V}} x[\vartheta, t_0, x_0, V].$$

The set $X[Q]$ is closed and bounded (compact) in the space $\mathbb{R}^m$. Moreover, this set consists of the right ends (at $t = \vartheta$) of the quasimotions $x[t, t_0, x_0, V \div Q]$, $t_0 \le t \le \vartheta$, of the system (2.2.15) that are induced by the strategy $V \div Q$ from the position (t_0, x_0).

Now, let M be some closed set in $\mathbb{R}^m$ that has a non-empty intersection with $X[Q]$:

$$M_X = X[Q] \cap M \neq \varnothing. \tag{2.2.16}$$

Then the set M_X is also closed, and the following result holds.

Proposition 2.2.2 *Under Conditions 2.2.1, let the initial position (t_0, x_0) be such that condition (2.2.16) holds for a given closed non-empty set $M \subset \mathbb{R}^m$.*
Then there exists a strategy $V^* \in \mathfrak{V}$ that satisfies the inclusion

$$X[\vartheta, t_0, x_0, V^*] \subseteq M_X. \tag{2.2.17}$$

According to (2.2.17), whatever the closed set M intersecting the attainability domain of the system (2.2.15) is, there surely exists a strategy $V^* \in \mathfrak{V}$ "driving" any quasimotion $x[\cdot, t_0, x_0, V^*]$ of the system (2.2.15) induced by V^* from the position (t_0, x_0) into the set M_X at the time instant ϑ (the end of motion). That is, the inclusion

$$x[\vartheta, t_0, x_0, V^*] \in M_X$$

holds for each quasimotion $x[\cdot, t_0, x_0, V^*] \in \mathcal{X}[t_0, x_0, V^*]$.

Proof Proposition 2.2.2 will be established using Theorem 2.1 (of alternative). To apply the latter, write the system (2.2.15) as

$$\dot{x} = \bar{f}(t, x, v) + 0_n u, \tag{2.2.18}$$

where $u \in [0, 1]$ and 0_n denotes an n-dimensional null vector. Therefore, in the system (2.2.18) the variable u is treated as the control of a dummy player. His strategy U is identified with a scalar function $u(t, x) \in [0, 1]$ for all possible positions $(t, x) \in [0, \vartheta) \times \mathbb{R}^m$. Denote by $\mathfrak{A}$ the set of all such strategies. The boundedness of the set M and Conditions 2.2.1 give all hypotheses for Theorem 2.2.1 (of alternative) applied to the system (2.2.18). For example, we will verify the saddle point condition in the small game. According to (2.2.2), it is necessary to show that, for any $z \in \mathbb{R}^m$ and any positions $(t, x) \in [0, \vartheta) \times \mathbb{R}^m$, there exists a pair $(u^0, v^0) \in [0, 1] \times Q$ such that

$$z'[\bar{f}(t, x, v) + 0_n u^0] \le z'[\bar{f}(t, x, v^0) + 0_n u^0] \le z'[\bar{f}(t, x, v^0) + 0_n u] \tag{2.2.19}$$

for all $u \in [0, 1]$ and $v \in Q$. The relations (2.19) are actually the case: as u^0 we may choose any value from the interval $[0, 1]$, and an appropriate vector v^0 is defined by the equality $\max_{v \in Q} z'\bar{f}(t, x, v) = z'\bar{f}(t, x, v^0)$. The optimum in the left-hand side is achieved due to the continuity of $f(t, x, v)$ in v and the compactness of the set Q. Now, we have every reason to apply Remark 2.2.1 to the system (2.2.18) and the closed set M_X (2.2.16), which describes two possible cases as follows. Given an initial position $(t_0, x_0) \in [0, \vartheta) \times \mathbb{R}^m$, either there exists a strategy $V^* \in \mathfrak{V}$ driving the system (2.2.18) with any quasimotions $x[\cdot, t_0, x_0, V^*]$ into the set M at the time instant ϑ, i.e., $x[\vartheta, t_0, x_0, V^*] \in M$ for any quasimotions $x[\cdot, t_0, x_0, V^*]$ of the system (2.2.18) that are induced by the strategy V^* from the position (t_0, x_0); or the system (2.2.18) evades the ε-neighborhood of the set M_X at the time instant ϑ using some strategy $U^* \in \mathfrak{A}$. But the evasion problem has no solution because the strategy U affects the system (2.2.18) through the null term $0_n u$. Really, for any strategy $U \in \mathfrak{A}$, $\mathcal{X}[t_0, x_0, U] = \mathcal{X}[t_0, x_0, V \div Q]$ by the definition of the pencils of quasimotions for the system (2.2.18). Hence, for a fixed strategy $U \in \mathfrak{A}$, the attainability domain $X[\vartheta, t_0, x_0, U]$ of the system (2.2.18) from the position (t_0, x_0) coincides with the attainability domain $X[\vartheta, t_0, x_0, V \div Q]$ of the system (2.2.18), i.e., $X[\vartheta, t_0, x_0, U] = X[\vartheta, t_0, x_0, V \div Q]$. Therefore, the set M_X cannot be evaded using any strategy $U \in \mathfrak{A}$ and

$$X[\vartheta, t_0, x_0, U] \cap M = X[\vartheta, t_0, x_0, V \div Q] \cap M = M_X \neq \varnothing$$

for all $U \in \mathfrak{A}$. Due to Theorem 2.2.1 (also, see Remark 2.2.1), the system (2.2.18) and hence the system (2.2.15) can be only driven into the set M_X. Consequently, there exists a $V^* \in \mathfrak{V}$ that satisfies (2.2.17).

The next result will be useful for further presentation as well.

Corollary 2.2.1 *Under Conditions 2.2.1, for any point $x^* \in X[\vartheta, t_0, x_0, V \div Q]$ of the attainability domain, there exists a strategy $V^* \in \mathfrak{V}$ such that $x[\vartheta, t_0, x_0, V^*] = x^*$ for all quasimotions $x[\cdot, t_0, x_0, V^*]$ of system (2.2.15) that are induced by V^* from the position (t_0, x_0).*

Since the point x^* represents a closed set in $X[\vartheta, t_0, x_0, V \div Q]$, Corollary 2.2.1 immediately follows from Proposition 2.2.2.

2.3 DIFFERENTIAL GAME WITH SEPARATED DYNAMICS

> Human dignity can be achieved only in the field of ethics,
> and ethical achievement is measured by the degree
> in which our actions are governed by compassion and love,
> not by greed and aggressiveness.
> —Toynbee[8]

In this section, the existence of a Berge equilibrium in a noncooperative positional differential game of N players (NPDG) with separated dynamics will be established. The proof involves, *first*, the Germeier convolution of the payoff functions [213]; *second*, the theory of noncooperative positional differential games, more specifically, the mathematical formalization of quasimotions as described in Sections 2.1 and 2.2; *third*, guiding control suggested in [179].

2.3.1 Preliminaries

Consider the NPDG

$$\langle \mathbb{N}, \{\Sigma_i\}_{i \in \mathbb{N}}, \{\mathfrak{V}_i\}_{i \in \mathbb{N}}, \{F_i(x[\vartheta])\}_{i \in \mathbb{N}} \rangle, \tag{2.3.1}$$

---

[8]Arnold Toynbee, (1889–1975), English historian whose *A Study of History* put forward a philosophy of history, based on an analysis of the cyclical development and decline of civilizations, that provoked much discussion.

where, like in (2.1.1), $\mathbb{N} = \{1, \ldots, N\}$ denotes the set of players and the controlled subsystem Σ_i evolves in accordance with the vector ordinary differential equation

$$\dot{x}_i = f_i(t, x_i, v_i), \ x_i[t_0] = x_i^0 \ (i \in \mathbb{N}). \tag{2.3.2}$$

Here $x_i \in \mathbb{R}^{m_i}$ are components of the state vector $x = (x_1, \ldots, x_N) \in \mathbb{R}^m \ (m = \sum m_i)$; $v_i \in Q_i \in comp \ \mathbb{R}^{n_i}$ gives the control variable of player i; $t \in [t_0, \vartheta]$ indicates time; $\vartheta > t_0 \geq 0$ are some constants. Let the elements of the ordered quadruple (2.3.1) satisfy an analog of Conditions 2.1.1 and 2.1.2 as follows.

Conditions 2.3.1 *The components of the m_i-dimensional vector functions $f_i(t, x_i, v_i)$ are defined and continuous in all variables on the domain $\mathbb{R}^{1+m_i} \times Q_i$, locally Lipschitzian in the first and second variables, i.e.,*

$$\left\| f_i\left(t^{(1)}, x_i^{(1)}, v_i\right) - f_i\left(t^{(2)}, x_i^{(2)}, v_i\right) \right\| \leq \lambda_i(G) \left(\left\| x_i^{(1)} - x_i^{(2)} \right\| + \left| t^{(1)} - t^{(2)} \right| \right)$$

for all $(t^{(j)}, x_i^{(j)}) \in [0, \vartheta) \times G_i \ (j = 1, 2)$, where G_i is any bounded subset in $\mathbb{R}^{m_i}$ and $\lambda_i(G_i) = const > 0$ denotes the Lipschitz constant that depends on the choice of G_i; in addition, each function $f_i(t, x_i, v_i)$ has sublinear growth, i.e., there exists $k_i = const > 0$ such that $\|f_i(t, x_i, v_i)\| \leq k_i(1 + \|x_i\|)$ for any $(t, x_i, v_i) \in [0, \vartheta) \times \mathbb{R}^{m_i} \times Q_i$; finally, the payoff function $F_i(x)$ of player i is continuous in the space $\mathbb{R}^m \ (i \in \mathbb{N})$.

We write the system (2.3.2) as

$$\dot{x} = f(t, x, v), \ x[t_0] = x_0, \tag{2.3.3}$$

where $x = (x_1, \ldots, x_N)$ is the m-dimensional state vector; $v = (v_1, \ldots, v_N) \in Q = \prod_{i \in \mathbb{N}} Q_i \in \mathbb{R}^n \left(n = \sum_{i \in \mathbb{N}} n_i \right)$ denotes the vector of all control variables; finally, $f = (f_1, \ldots, f_N)$ forms the m-dimensional column vector of all functions f_i.

Without special mention, let Conditions 2.3.1 hold for each subsystem $i \ (i \in \mathbb{N})$ in (2.3.2). Hence, these conditions are valid for the entire system (2.3.3), because $\|f\| = \sum_{i \in \mathbb{N}} \|f_i\|$.

Recall that a *strategy* V_i of player i in the game (2.3.1) is identified with an n_i-dimensional vector function $v_i(t, x) \in Q_i \ \forall \ (t, x) \in [0, \vartheta) \times \mathbb{R}^m$, and this fact is written as $V_i \div v_i(t, x)$. Denote by $\mathfrak{V}_i \ (i \in \mathbb{N})$ the set of all such strategies $\{V_i\}$. Also we use the strategy profiles $V = (V_1, \ldots, V_N) \in \mathfrak{V} = \prod_{i \in \mathbb{N}} \mathfrak{V}_i$, and $V \div v(t, x) = (v_1(t, x), \ldots, v_N(t, x)) \in Q$.

The NPDG (2.3.1) describes well, e.g., the competition of two, three, or more producers without direct industrial connections, which explains the separated dynamics in the system (2.3.3). This dynamic game can be also used as a mathematical model of several independent mechanical objects under control, e.g., approaching each other by a given time instant ϑ.

Definition 2.3.1 *A strategy profile $V^B = (V_1^B, \ldots, V_N^B) \in \mathfrak{V}$ is said to be a Berge equilibrium for game (2.3.1) if*

$$\max_{x[\cdot] \in \mathcal{X}[t_0, x_0, V_i^B]} F_i\left(x\left[\vartheta, t_0, x_0, V_i^B\right]\right) \leq \min_{x[\cdot] \in \mathcal{X}[t_0, x_0, V^B]} F_i\left(x\left[\vartheta, t_0, x_0, V^B\right]\right) \ (i \in \mathbb{N}).$$

$$\tag{2.3.4}$$

Remark 2.3.1 Due to Propositions 2.1.4, 2.1.5 and the inclusions

$$\mathcal{X}\left[t_0, x_0, V^B\right] \subset \mathcal{X}\left[t_0, x_0, V_i^B\right] \quad (i \in \mathbb{N}),$$

$$X\left[\vartheta, t_0, x_0, V^B\right] \subset X\left[\vartheta, t_0, x_0, V_i^B\right] \quad (i \in \mathbb{N}),$$

inequalities (2.3.4) are equivalent to

$$F_i\left(x\left[\vartheta, t_0, x_0, V_i^B\right]\right) \leq F_i\left(x\left[\vartheta, t_0, x_0, V^B\right]\right) \quad (i \in \mathbb{N})$$

for any quasimotions

$$x\left[\cdot, t_0, x_0, V_i^B\right]\right) \in \mathcal{X}\left[t_0, x_0, V_i^B\right], \quad x\left[\cdot, t_0, x_0, V^B\right]\right) \in \mathcal{X}\left[t_0, x_0, V^B\right]$$

of the system (2.3.3) that are induced from an initial position (t_0, x_0) by a strategy V_i^B and a strategy profile V^B, respectively. Moreover, the value $F_i(x[\vartheta, t_0, x_0, V^B])$ is *unique* $\forall\, i \in \mathbb{N}$, $\forall x[\cdot, t_0, x_0, V^B] \in \mathcal{X}[t_0, x_0, V^B]$.

Remark 2.3.2 Hereinafter, the notation $V_i \div Q_i$ means that any of the strategies $V_i \in \mathfrak{V}_i$ is used. By the definition of the pencil $\mathcal{X}\left[t_0, x_0, V_i^B\right]$ of quasimotions and Proposition 2.1.4, for all $i \in \mathbb{N}$ the pencil $\mathcal{X}\left[t_0, x_0, V_i^B\right]$ contains all quasimotions $x\left[\cdot, t_0, x_0, V_1 \div Q_1, \ldots, V_{i-1} \div Q_{i-1}, V_i^B \div v_i^B(t, x), V_{i+1} \div Q_{i+1}, \ldots, V_N \div Q_N\right]$ and hence the quasimotion $x\left[\cdot, t_0, x_0, V_1^B, \ldots, V_{i-1}^B, V_i^B, V_{i+1}^B, \ldots, V_N^B\right] = x\left[\cdot, t_0, x_0, V^B\right]$. This guarantees the uniqueness of the value $F_i\left(x\left[\vartheta, t_0, x_0, V^B\right]\right)$ $(i \in \mathbb{N})$ in (2.3.4).

2.3.2 Guiding control

That a Berge equilibrium exists in the game (2.3.1) (in the sense of Definition 2.3.1) is rather a delusive hope. For solving this problem, we will employ the idea of guiding control [179] suggested by Krasovskii for saddle point design in noncooperative positional differential games with stability against noise disturbances. As was noted in [179, p. 248], "…a guide can be considered a certain controller embedded in the control loop for the simulation modeling of a controlled object." Following this approach, for each $i \in \mathbb{N}$ we "augment" the controlled system (2.3.2) with a dynamic system that describes the evolution of the state variable $z_i(t)$ of guide i:

$$\dot{z}_i = f_i(t, z_i, u_i), \ z_i[t_0] = x_i^0 \quad (i \in \mathbb{N}),$$

where $z_i \in \mathbb{R}^{m_i}$ and $u_i \in Q_i$ $(i \in \mathbb{N})$. Like the transition from (2.3.2) to (2.3.3), this yields the vector ordinary differential equation

$$\dot{z} = f(t, z, u), \ z[t_0] = x_0 = (x_1^0, \ldots, x_N^0). \tag{2.3.5}$$

Here the vectors are

$$z = (z_1, \ldots, z_N) \in \mathbb{R}^m, \quad z_i \in \mathbb{R}^{m_i} \ (i \in \mathbb{N}), \quad u = (u_1, \ldots, u_N) \in Q = \prod_{i \in \mathbb{N}} Q_i,$$

and $f = (f_1, \ldots, f_N) \in \mathbb{R}^m$ is a column vector. Based on guiding control, we integrate (2.3.5) with the system (2.3.3). Then the dynamics of the new extended controlled system Σ_e satisfy the following system of ordinary differential equations of order $2m$:

$$\begin{cases} \dot{x}=f(t,x,v),\, x[t_0]=x_0, \\ \dot{z}=f(t,z,u),\, z[t_0]=x_0 \end{cases} \Leftrightarrow \begin{cases} \dot{y}=\bar{f}(t,y,u,v), \\ y[t_0]=y_0=(x_0,x_0) \end{cases}, \qquad (2.3.6)$$

where $y=(x,z)$ and $\bar{f}(t,y,u,v)=(f(t,x,v),f(t,z,u))$ are $2m$-dimensional column vectors. The pencil of all quasimotions $y[t,t_0,y_0,U,V]$, $t_0 \le t \le \vartheta$, of the system (2.3.6) that are induced by the extended strategy profile $(\bar{U},\bar{V}) \in \mathfrak{A} \times \mathfrak{V}$ from an initial position (t_0,y_0) will be denoted by $\mathcal{Y}[t_0,y_0,\bar{U},\bar{V}]$. In the same fashion, $\mathcal{Y}[t_0,y_0,\bar{V}]$ will stand for the pencil of all quasimotions $y[\cdot,t_0,y_0,\bar{V}]$ of this system that is constructed in accordance with the requirements of Subsection 2.1.4 and Definition 2.1.2.

As it has been mentioned earlier, equation (2.3.5) describes the motion (the evolution of the state vector with the course of time t) of a general guide that represents the union of N guides defined by the equations $\dot{z}_i = f_i(t,z_i,u_i)$ ($i \in \mathbb{N}$). During control design, the strategy $\bar{U} \div u(t,x,z) \in Q$ is used to simulate the quasimotion of the controlled system (2.3.5) with the same equation (2.3.5) as (2.3.5), the same initial conditions ($z[t_0] = x_0$) and the same constraints imposed on the control variables ($u_i \in Q_i$).

Thus, adding this guide, we pass from the NPDG (2.3.1) to its extended counterpart—the NPDG

$$\Gamma_e = \langle \mathbb{N}, \Sigma_e \div (2.3.6), \{\mathfrak{A}_i, \mathfrak{V}_i\}_{i\in\mathbb{N}}, \{F_i(x[\vartheta])\}_{i\in\mathbb{N}}\rangle,$$

where, like in (2.3.1), $\mathbb{N} = \{1,\dots,N\}$ denotes the set of players. But, in contrast to (2.3.1), a position of the game Γ_e is represented by a triplet $(t,x,z) \in [0,\vartheta) \times \mathbb{R}^{2m}$, where $(x,z) \in \mathbb{R}^m \times \mathbb{R}^m$ forms the state vector consisting of the m-dimensional vectors $x = (x_1,\dots,x_N)$ and $z = (z_1,\dots,z_N)$, $x_i, z_i \in \mathbb{R}^{m_i}$ ($i \in \mathbb{N}$). Hence, a strategy of player i is $\bar{V}_i \div \bar{v}_i(t,x,z) \in Q_i$ ($i \in \mathbb{N}$); denote by $\mathfrak{V}_i$ the set of all such $\{\bar{V}_i\}$. In a similar fashion, for guide i, $\bar{U}_i \div \bar{u}_i(t,x,z) \in Q_i$ $\forall(t,x,z) \in [0,\vartheta) \times \mathbb{R}^{2m}$, and the set of all $\{\bar{U}_i\}$ will be denoted by $\mathfrak{A}_i$. Note that player i ($i \in \mathbb{N}$) chooses a pair $(\bar{V}_i,\bar{U}_i) \in \mathfrak{V}_i \times \mathfrak{A}_i$. An extended strategy profile in the game Γ_e is a pair $(\bar{V},\bar{U}) \in \mathfrak{V} \times \mathfrak{A}$, where $\bar{V} = (\bar{V}_1,\dots,\bar{V}_N) \in \mathfrak{V} = \prod_{i\in\mathbb{N}} \mathfrak{V}_i$, $\bar{U} = (\bar{U}_1,\dots,\bar{U}_N) \in \mathfrak{A} = \prod_{i\in\mathbb{N}} \mathfrak{A}_i$, $\bar{V}_i \div \bar{v}_i(t,x,z)$, $\bar{V}_i \in \mathfrak{V}_i$ and $\bar{U}_i \div \bar{u}_i(t,x,z)$, $\bar{U}_i \in \mathfrak{A}_i$ ($i \in \mathbb{N}$). In what follows, also we will use the strategy profiles $(\bar{U}_1,\dots,\bar{U}_{i-1},\bar{V}_i,\bar{U}_{i+1},\dots,\bar{U}_N) = (\bar{U} \parallel \bar{V}_i) \in \mathfrak{A}_1 \times \cdots \times \mathfrak{A}_{i-1} \times \mathfrak{V}_i \times \mathfrak{A}_{i+1} \times \cdots \times \mathfrak{A}_N$. In the game Γ_e, each player i chooses his extended strategy $(\bar{V}_i,\bar{U}_i) \in \mathfrak{V}_i \times \mathfrak{A}_i$ ($i \in \mathbb{N}$), which leads to an extended strategy profile $(\bar{V},\bar{U}) \in \mathfrak{V} \times \mathfrak{A}$, where $\bar{V} \div \bar{v}(t,x,z) = (\bar{v}_1(t,x,z),\dots,\bar{v}_N(t,x,z))$ and $\bar{U} \div \bar{u}(t,x,z) = (\bar{u}_1(t,x,z),\dots,\bar{u}_N(t,x,z))$. Then, according to Sections 2.1 and 2.2 of this book, we have to construct the pencils $\mathcal{Y}[t_0,y_0,\bar{V},\bar{U}]$ of all quasimotions $y[t,t_0,y_0,\bar{V},\bar{U}]$, $t_0 \le t \le \vartheta$, of the system (2.3.6) that are induced from an initial position (t_0,x_0,x_0) by an extended strategy profile $(\bar{V},\bar{U})$ (due to Conditions 2.3.1, the system (2.3.6) satisfies all requirements of Conditions 2.1.1). The pencil $\mathcal{Y}[t_0,x_0,x_0,\bar{V},\bar{U}]$ consists of the pairs of quasimotions $x[\cdot,t_0,x_0,\bar{V}]$ and $z[\cdot,t_0,x_0,\bar{U}]$ of the subsystems $\dot{x} = f(t,x,v)$, $x[t_0] = x_0$ and $\dot{z} = f(t,z,u)$, $z[t_0] = x_0$, respectively. Finally, the payoff function of player i is given by a functional $F_i(x[\vartheta])$, where $x[\vartheta] \in X[\vartheta,t_0,x_0,V]$, and the *payoff* of player i is its value $F_i(x[\vartheta])$. Note that the functionals $F_i(x[\vartheta])$ are defined at the right ends of the quasimotions $x[\cdot,t_0,x_0,\bar{V}] \in \mathcal{X}[t_0,x_0,\bar{V}]$ of the system (2.3.3) that are induced from (t_0,x_0) by a collection $\bar{V} \in \mathfrak{V}$, i.e., $x[\vartheta] = x[\vartheta,t_0,x_0,\bar{V}] \in X[\vartheta,t_0,x_0,\bar{V}] = \mathcal{X}[t_0,x_0,\bar{V}] \cap \{t = \vartheta\}$. In addition, the functionals $F_i(z[\vartheta] \parallel x_i[\vartheta])$, where

$$(z[\vartheta] \parallel x_i[\vartheta]) = (z_1[\vartheta,t_0,x_0,\bar{U}_1],\dots,z_{i-1}[\vartheta,t_0,x_0,\bar{U}_{i-1}], x_i[\vartheta,t_0,x_0,\bar{V}_i],$$

and

$$z_{i+1}[\vartheta, t_0, x_0, \bar{U}_{i+1}], \ldots, z_N[\vartheta, t_0, x_0, \bar{U}_N]) \ \forall \bar{U}_j \in \bar{\mathfrak{A}}_j \ (j \in \mathbb{N} \setminus \{i\}),$$

will be considered below. Actually, we will establish the existence of a Berge equilibrium using this extension of the space of positions from (t, x) to (t, x, z), in combination with the Germeier convolution of payoff functions and the theory of noncooperative positional differential games in quasimotions. Once again, in the system (2.3.6) with the extended dynamics of N guides $(i \in \mathbb{N})$, a position is represented by a triplet $(t, x, z) \in [0, \vartheta) \times \mathbb{R}^{2m}$.

For this game, Definition 2.3.1 has the following analog.

Definition 2.3.2 *A collection of strategies* $\bar{V}^B = (\bar{V}^B_1, \ldots, \bar{V}^B_N) \in \bar{\mathfrak{V}} = \prod_{i \in \mathbb{N}} \bar{\mathfrak{V}}_i$ *is said to implement the concept of Berge equilibrium in the game* Γ_e *if*

$$F_i(z[\vartheta, t_0, x_0, \bar{U}] \| x_i[\vartheta, t_0, x_0, \bar{V}^B_i]) \leq F_i(x[\vartheta, t_0, x_0, \bar{V}^B]) \ (i \in \mathbb{N}), \qquad (2.3.7)$$

for any quasimotions $z[\cdot, t_0, x_0, \bar{U}] \in \mathcal{Z}[t_0, x_0, \bar{U}]$ *of system (2.3.5) that are induced from an initial position* (t_0, x_0) *by any admissible collection of strategies* $\bar{U} = (\bar{U}_1, \ldots, \bar{U}_N) \doteq \bar{u}(t, x, z) \in Q \ \forall(t, x, z) \in [0, \vartheta) \times \mathbb{R}^{2m}$ *and for any quasimotions* $x[t, t_0, x_0, \bar{V}^B]$ *of system (2.3.3) that are induced from* (t_0, x_0, x_0) *by a collection of strategies* $\bar{V}^B = (\bar{V}^B_1, \ldots, \bar{V}^B_N) \in \bar{\mathfrak{V}}, \ \bar{V}^B_i \doteq \bar{v}^B_i(t, x, z) \in Q_i \ \forall(t, x, z) \in [0, \vartheta) \times \mathbb{R}^{2m} \ (i \in \mathbb{N})$, *where* $(z \| x_i) = (z_1, \ldots, z_{i-1}, x_i, z_{i+1}, \ldots, z_N)$.
Interestingly, the pencils

$$\mathcal{Y}[t_0, y_0, V^B] \subset \mathcal{Y}_i[t_0, y_0, V^B_i] \ (i \in \mathbb{N}),$$

and hence for "frozen" $(t_0, y_0) = (t_0, x_0, x_0) \in [0, \vartheta) \times \mathbb{R}^{2m}$ and $\bar{V}^B_i \in \bar{\mathfrak{V}}_i$, each player i has a unique payoff $F_i(x[\vartheta, t_0, x_0, \bar{V}^B])$. (This fact directly follows from (2.3.7).)

This section is intended to rigorously demonstrate that such a collection $\bar{V}^B$ exists under Conditions 2.3.1.

2.3.3 Auxiliary noncooperative game

For proving the existence of a collection $\bar{V}^B \in \bar{\mathfrak{V}}$, that implements the concept of Berge equilibrium (or in short, *a Berge equilibrium in the game* Γ_e), consider the auxiliary zero-sum two-player positional differential game

$$\Gamma_a = \langle \{I, II\}, \Sigma \doteq (2.3.6), \bar{\mathfrak{V}}, \bar{\mathfrak{A}}, \varphi(x[\vartheta], z[\vartheta])) = \varphi(y[\vartheta]) \rangle. \qquad (2.3.8)$$

In this game, the strategies $\bar{V}$ of player I (who minimizes a functional $\varphi(x[\vartheta], z[\vartheta])$) are identified with n-dimensional vector functions $\bar{v}(t, x, z) \in Q$ for any $t \in [0, \vartheta)$ and $(x, z) \in \mathbb{R}^{2m}$; denote by $\bar{\mathfrak{V}}$ the set of all $\{\bar{V}\}$. The strategies $\bar{U}$ of player II (who maximizes $\varphi(x[\vartheta], z[\vartheta])$) are introduced by analogy: $\bar{U} \doteq \bar{u}(t, x, z) \in Q$ for all $(t, x, z) \in [0, \vartheta) \times \mathbb{R}^{2m}$ and $\{\bar{U}\} = \bar{\mathfrak{A}}$. These strategies form strategy profiles $(\bar{V}, \bar{U}) \in \bar{\mathfrak{V}} \times \bar{\mathfrak{A}}$. Following Subsection 2.1.4, we will construct the pencils of all quasimotions $\mathcal{X}[t_0, x_0, \bar{V}]$ of the system (2.3.3), $\mathcal{Z}[t_0, x_0, \bar{U}]$ of the system (2.3.5) and $\mathcal{Y}[t_0, x_0, \bar{V}, \bar{U}]$ of the system (2.3.6) that are induced from an initial position (t_0, x_0) by a strategy $\bar{V} \in \bar{\mathfrak{V}}$ of player I, by a strategy $\bar{U} \in \bar{\mathfrak{A}}$ of player II, and by a strategy profile $(\bar{V}, \bar{U})$, respectively. Also the $2m$-dimensional column vector $y = (x, z)$ will be used.

The payoff function of player II (i.e., the loss function of player I) is designed on the basis of the payoff functions $F_i(x[\vartheta])$ of player i in the game Γ_e as follows. Each of the scalar functions $F_i(x)$ ($i \in \mathbb{N}$) is defined on the attainability domain $X[\vartheta, t_0, x_0, \bar{V} \div Q] = X[\bar{V} \div Q]$ of the system (2.3.3), and $X[\bar{V} \div Q]$ coincides with the attainability domain $Z[\vartheta, t_0, x_0, \bar{U} \div Q] = Z[\bar{U} \div Q]$ of the system (2.3.5). Both domains are compact sets in the space $\mathbb{R}^m$. Using $F_i(x)$ ($i \in \mathbb{N}$), we construct in analytic form the scalar function (see (2.3.8))

$$\varphi(y) = \max_{i \in \mathbb{N}}[F_i(z \parallel x_i) - F_i(x)]. \tag{2.3.9}$$

Proposition 2.3.1 *Under Conditions* 2.3.1, *the game* Γ_a *has a saddle point* $(\bar{V}^B, \bar{U}^0) \in \mathfrak{V} \times \mathfrak{A}$ *given by the chain of inequalities*

$$\varphi\left(y\left[\vartheta, t_0, y_0, \bar{V}^B\right]\right) \leq \varphi\left(y\left[\vartheta, t_0, y_0, \bar{V}^B, \bar{U}^0\right]\right) \leq \varphi\left(y\left[\vartheta, t_0, y_0, \bar{U}^0\right]\right) \tag{2.3.10}$$

for any quasimotions $y\left[\cdot, t_0, y_0, \bar{V}^B\right] \in \mathcal{Y}\left[t_0, y_0, \bar{V}^B\right], y\left[\cdot, t_0, y_0, \bar{V}^B, \bar{U}^0\right] \in \mathcal{Y}\left[t_0, y_0, \bar{V}^B, \bar{U}^0\right],$ *and* $y\left[\cdot, t_0, y_0, \bar{U}^0\right] \in \mathcal{Y}\left[t_0, y_0, \bar{U}^0\right]$ *of system* (2.3.6) *that are induced from an initial position* (t_0, x_0) *by a strategy* $\bar{V}^B$ *of player I, by a strategy profile* $(\bar{V}^B, \bar{U}^0)$, *and by a strategy* $\bar{U}^0$ *of player II, respectively.*

Proof By Theorem 2.2.2 there exists a saddle point $(\bar{V}^B, \bar{U}^0) \in \mathfrak{V} \times \mathfrak{A}$ in the game Γ_a for any choice of the initial position $(t_0, y_0) \in [0, \vartheta) \times \mathbb{R}^{2m}$ under Conditions 2.3.1, the continuity of the function $\varphi(x, z)$ on $X[V \div Q] \times Z[U \div Q]$ and the saddle point condition in the small game (an analog of (2.2.2) for the system (2.3.6)). If $F_i(x)$ is continuous, then by [204, p. 54] the function $\varphi(x, z)$ is also continuous in $x \in X[V \div Q]$ and $z \in Z[U \div Q]$. (This follows from the compactness of the sets $X[V \div Q]$ and $Z[U \div Q]$ in the space $\mathbb{R}^m$ and the continuity of $F_i(x)$ ($i \in \mathbb{N}$).) Finally, we check the saddle point condition in the small game: according to (2.2.2), for each position $(t, y) \in [0, \vartheta) \times \mathbb{R}^{2m}$ and any $2m$-dimensional vector $(s_1, s_2) \in \mathbb{R}^m \times \mathbb{R}^m$, this requirement implies the existence of a saddle point $(v^B, u^0) \in Q \times Q$, i.e.,

$$\max_{v \in Q}\left[s_1'f(t, x, v) + s_2'f(t, z, u^0)\right] = s_1'f\left(t, x, v^B\right) + s_2'f\left(t, z, u^0\right)$$

$$= \min_{u \in Q}\left[s_1'f(t, x, v^B) + s_2'f(t, z, u)\right] \quad \forall u, v \in Q. \tag{2.3.11}$$

(As before, the prime indicates transposition and hence s' is an m-dimensional row vector.)

Because the system (2.3.11) has separated dynamics, the relation (2.3.11) is equivalent to the existence of a pair $\left(v^B, u^0\right) \in Q^2$ that satisfies the two equalities

$$\max_{v \in Q} s_1'f(t, x, v) = s_1'f\left(t, x, v^B\right),$$

$$\min_{u \in Q} s_2'f(t, z, u) = s_2'f\left(t, z, u^0\right).$$

Moreover, the existence of (v^B, u^0) for "frozen" $(t, x, z, s_1, s_2) \in [0, \vartheta) \times \mathbb{R}^{4m}$ directly follows from the continuity of the functions $f(t, x, v)$ and $f(t, z, u)$ in the control variables $v \in Q$ and $u \in Q$ and the compactness of Q by the Weierstrass theorem.

Consequently, it has been established that a saddle point $(\bar{V}^B, \bar{U}^0) \in \mathfrak{V} \times \mathfrak{A}$ exists in the game Γ_a under Conditions 2.3.1.

2.3.4 Existence theorem

Theorem 2.3.1 *Let Conditions* 2.3.1 *hold in game* (2.3.1). *For any choice of the initial position* $(t_0, x_0) \in [0, \vartheta) \times \mathbb{R}^m$, *there exists a Berge equilibrium* $\bar{V}^B \in \mathfrak{V}$ *in the noncooperative positional differential game* Γ_e, *i.e., the strategy profile* $\bar{V}^B$ *satisfies* (2.3.7).

Proof By Proposition 2.3.1 the game Γ_a has a saddle point $\left(\bar{V}^B, \bar{U}^0\right) \in \mathfrak{V} \times \bar{\mathfrak{A}}$ for which the chain of inequalities (2.3.10) holds. In the right-hand inequality from (2.3.10), each quasimotion $y\left[\cdot, t_0, y_0, \bar{U}^0\right]$ is formed by two quasimotions $x\left[\cdot, t_0, x_0, \bar{V}\right]$ and $z\left[\cdot, t_0, x_0, \bar{U}^0\right]$ of the systems (2.3.3) and (2.3.5), respectively, for all $\bar{V} \in \mathfrak{V}$. Letting $\bar{V} = \bar{U}^0$ gives $\varphi\left(y\left[\vartheta, t_0, y_0, \bar{U}^0, \bar{U}^0\right]\right) = 0$ for any such quasimotions $x\left[\cdot, t_0, x_0, \bar{U}^0\right]$ and $z\left[\cdot, t_0, x_0, \bar{U}^0\right]$. Really, for each quasimotion $x\left[\cdot, t_0, x_0, \bar{U}^0\right]$, we may find a quasimotion $z\left[\cdot, t_0, x_0, \bar{U}^0\right]$ coinciding with the former at each $t \in [t_0, \vartheta]$. The converse is also true. Note that this result can be obtained using Corollary 2.2.1.

Due to uniqueness, it follows from (2.3.10) and $\varphi\left(y\left[\vartheta, t_0, y_0, \bar{U}^0, \bar{U}^0\right]\right) = 0$ that $\varphi\left(y\left[\vartheta, t_0, y_0, \bar{V}^B, \bar{U}^0\right]\right) \leq 0$. Then, again according to (2.3.10), we have

$$\varphi\left(y\left[\vartheta, t_0, y_0, \bar{V}^B\right]\right) \leqslant 0 \ \forall \ y\left[\cdot, t_0, y_0, \bar{V}^B\right] \in \mathcal{Y}\left[t_0, y_0, \bar{V}^B\right].$$

Each quasimotion $y\left[\cdot, t_0, y_0, V^B\right]$ (see (2.3.6)) is formed by pairs of quasimotions $x\left[\cdot, t_0, x_0, \bar{V}^B\right]$ and $z\left[\cdot, t_0, x_0, \bar{U}\right]$ of the system (2.3.3) that are induced by a collection $\bar{V}^B \in \mathfrak{V}$ and of the system (2.3.5) that are induced by a collection $\bar{U} \in \bar{\mathfrak{A}}$, respectively, from the same initial position (t_0, x_0).

Since $\varphi(x, z) = \max\limits_{i \in \mathbb{N}}[F_i(z \parallel x_i) - F_i(x)] \leq 0$ (see (2.3.10)), for each $i \in \mathbb{N}$ we have

$$F_i\left(z\left[\vartheta, t_0, x_0, \bar{U}\right] \parallel x_i\left[\vartheta, t_0, x_0, \bar{V}_i^B\right]\right) - F_i\left(x\left[\vartheta, t_0, x_0, \bar{V}^B\right]\right) \leqslant 0 \qquad (2.3.12)$$

for any quasimotions $z[\cdot, t_0, x_0, U]$ of the system (2.3.5) for each $\bar{U} \in \bar{\mathfrak{A}}$ and all quasimotions $x\left[\cdot, t_0, x_0, \bar{V}^B\right]$ of the system (2.3.3) that are induced from (t_0, x_0) by a collection $\bar{U}$ and $\bar{V}^B$, respectively. In view of the inclusion $\mathcal{Y}\left[t_0, y_0, \bar{V}^B\right] \subset \mathcal{Y}\left[t_0, y_0, \bar{V}_i^B\right]$ $(i \in \mathbb{N})$ and inequalities (2.3.12) holding for any quasimotions, we can establish,

first, the uniqueness of $F_i\left(x\left[\vartheta, t_0, x_0, \bar{V}^B\right]\right)$ for each $i \in \mathbb{N}$;

second, the validity of (2.3.7) for all $x\left[\cdot, t_0, x_0, \bar{V}_i^B\right]$, $x\left[\cdot, t_0, x_0, \bar{V}^B\right]$, $z\left[\cdot, t_0, x_0, \bar{U}\right]$, $\forall \ \bar{U} \in \bar{\mathfrak{A}}$.

At the end of this proof, we note that the existence theorem is a key element of the theory, its point of support.

PART II. LINEAR-QUADRATIC GAME WITH SMALL INFLUENCE OF ONE PLAYER ON THE RATE OF CHANGE OF STATE VECTOR

> On this shrunken globe, men
> can no longer live as strangers.
> —Stevenson[9]

In this part of Chapter 2, the linear-quadratic differential noncooperative positional game with a small influence of one player on the rate of change of the state vector

---

[9] Adlai E. Stevenson, (1900–1965), was U.S. political leader and diplomat who helped found the United Nations.

is considered. The coefficient criteria for the existence of Berge and Nash equilibria are established and a design procedure for such equilibria is suggested, on the basis of dynamic programming combined with the small parameter method.

Introduction

Those readers who studied Lyapunov's stability theory surely remember algebraic coefficient criteria. The whole idea of such criteria is to establish the stability of an unperturbed motion without solving a system of differential equations using the signs of coefficients and/or relations among them. In this part of Chapter 2, we will suggest a similar approach to equilibrium design in noncooperative linear-quadratic two-player games. More specifically, based on the sign definiteness of the quadratic forms appearing in the payoff functions of players, we will answer two questions as follows.

1)   Do Berge and/or Nash equilibria exist?

2)   How can they be calculated?

In fact, the answers to both questions are concealed in the possibility of judging the existence of a solution for a system of two matrix ordinary differential equations of the Riccati type that is extendable on the time interval of a game. For solving this problem, we will employ dynamic programming, the small parameter method and also Poincaré's theorem on analyticity (conditions under which a solution of a differential equation is analytic with respect to a parameter).

2.4   MATHEMATICAL MODEL

Disjecta membra poetae.[10]

2.4.1   Preliminaries

Consider a noncooperative differential positional linear-quadratic two-player game described by

$$\Gamma_2 = < \{1, 2\}, \Sigma, \{\mathfrak{U}_i\}_{i=1,2}, \{\mathcal{J}_i(U, t_0, x_0)\}_{i=1,2} > .$$

Here $\{1, 2\}$ is the set of players; the n-dimensional state vector $x \in \mathbb{R}^n$ of a controlled dynamic system Σ evolves over time t in accordance with the vector ordinary differential equation

$$\dot{x} = A(t)x + u_1 + \varepsilon u_2, \quad x(t_0) = x_0, \tag{2.4.1}$$

where $t \in [t_0, \vartheta]$ and a terminal time instant $\vartheta > t_0 \geqslant 0$ is fixed; the position of the game Γ_2 at a time instant t is represented by a pair $(t, x) \in [t_0, \vartheta] \times \mathbb{R}^n$, where (t_0, x_0) denotes an initial position; the elements of a system matrix $A(t)$ of dimensions $n \times n$ are assumed to be continuous on $[0, \vartheta]$, and this fact will be indicated by $A(\cdot) \in C^{n \times n}[0, \vartheta]$; $u_i \in \mathbb{R}^n$ gives the control of player i; $\varepsilon > 0$ is a small parameter, and hence Γ_2 belongs to the class of differential positional games with a small influence of player 2 on the rate of change $\dot{x}(t)$ of the state vector $x(t)$.

---

[10]Latin "Scattered fragments."

A strategy U_i of player i is identified with an n-dimensional vector function $u_i(t, x)$ of the form $Q_i(t)x$, where $Q_i(\cdot) \in C^{n \times n}[0, \vartheta]$, and this fact will be indicated by $U_i \div u_i(t, x) = Q_i(t)x$. The set of all such strategies is

$$\mathfrak{U}_i = \left\{ U_i \div Q_i(t)x \;\; \forall Q_i(\cdot) \in C^{n \times n}[0, \vartheta] \right\}.$$

The strategy profile of the game Γ_2 is a pair $U = (U_1, U_2) \in \mathfrak{U} = \mathfrak{U}_1 \times \mathfrak{U}_2$. Therefore, as his strategy player i has to choose a matrix $Q_i(t)$ that is continuous on $[0, \vartheta]$ $(i = 1, 2)$.

A play of the game Γ_2 is organized as follows. Based on his individual considerations (see the payoff function $\mathcal{J}_i(U, t_0, x_0)$ defined below), each player chooses and uses his strategy $U_i^* \div u_i^* = Q_i^*(t)x$ $(i = 1, 2)$. As a result, the system (2.4.1) takes the form

$$\dot{x} = \left[A(t) + Q_1^*(t) + \varepsilon Q_2^*(t) \right] x, \quad x(t_0) = x_0.$$

Such a homogeneous and linear (in the variable x) system with continuous (in the variable t) coefficients has a unique continuous solution $x^*(t)$ that is extendable to $[t_0, \vartheta]$ $\forall t_0 \in [0, \vartheta)$. Using $x^*(t)$ we construct *the realizations* $u_i^*[t] = u_i^*(t, x^*(t)) = Q_i^*(t)x^*(t)$ of the strategies $U_i^* \div Q_i^*(t)x$ $(i = 1, 2)$ chosen by the players. On such a continuous triplet $\{x^*(t), u_1^*[t], u_2^*[t] | t_0 \leqslant t \leqslant \vartheta\}$, *the payoff function of player i* is a priori defined as a quadratic functional

$$\mathcal{J}_i \left(U_1^* U_2^*, t_0, x_0 \right) = \left[x^*(\vartheta) \right]' C_i x^*(\vartheta) + \int_{t_0}^{\vartheta} \left\{ (u_1^*[t])' D_{i1} u_1^*[t] + (u_2^*[t])' D_{i2} u_2^*[t] \right\} dt$$

$$(i = 1, 2); \tag{2.4.2}$$

the value of (2.4.2) is called *the payoff* of player i. In (2.4.2), the prime means transposition, and the matrices C_i and D_{ij} of dimensions $n \times n$ are assumed to be symmetric without loss of generality. Other notations involved include the following: 0_n as a null n-dimensional column vector; $u_i = \left(u_i^{(1)}, \ldots, u_i^{(n)} \right) \in \mathbb{R}^n$ $(i = 1, 2)$; $V = (V_1, V_2)$; E_n and $O_{n \times n}$ as identity and null matrices, respectively, of dimensions $n \times n$; $\det B$ as the determinant of a matrix B of dimensions $n \times n$. In addition, the gradient of a scalar function $W(t, x, u_1, u_2, V)$ with respect to u_i is given by

$$\operatorname{grad}_{u_i} W(t, x, u_1, u_2, V) = \frac{\partial W}{\partial u_i} = \begin{pmatrix} \frac{\partial W}{\partial u_i^{(1)}} \\ \vdots \\ \frac{\partial W}{\partial u_i^{(n)}} \end{pmatrix}.$$

The Hessian of $W(t, x, u_1, u_2, V)$ with respect to the components $u_i \in \mathbb{R}^n$ under fixed values of all other variables is a matrix of dimensions $n \times n$ of the form

$$\frac{\partial^2 W}{\partial u_i^2} = \begin{pmatrix} \frac{\partial^2 W}{\partial u_i^{(1)} \partial u_i^{(1)}} & \cdots & \frac{\partial^2 W}{\partial u_i^{(1)} \partial u_i^{(n)}} \\ \cdots & \cdots & \cdots \\ \frac{\partial^2 W}{\partial u_i^{(n)} \partial u_i^{(1)}} & \cdots & \frac{\partial^2 W}{\partial u_i^{(n)} \partial u_i^{(n)}} \end{pmatrix}.$$

For a constant and symmetric matrix D of dimensions $n \times n$, the inequality $D > 0$ $(< 0, \leqslant 0)$ means that the quadratic form $u_i' D u_i$ is positive definite (negative definite, nonnegative definite, respectively). A direct componentwise verification shows that, for a constant vector $a \in \mathbb{R}^n$,

$$\frac{\partial}{\partial u_i}(u_i' D u_i) = (D + D') u_i,$$

$$\frac{\partial}{\partial u_i}(a' D u_i) = D' a,$$

$$\frac{\partial}{\partial u_i}(a' u_i) = a,$$

$$\frac{\partial^2}{\partial u_i^2}(u_i' D u_i) = D' + D = \{\text{if } D = D'\} = 2D. \tag{2.4.3}$$

For a scalar function $W(t, x, u_i)$,

$$\max_{u_i} W(t, x, u_i) = Idem\{u_i \to u_i(t, x)\}$$

means that

$$\max_{u_i} W(t, x, u_i) = W(t, x, u_i(t, x)) \quad \forall t \in [0, \vartheta], x \in \mathbb{R}^n, \tag{2.4.4}$$

and the identity (2.4.4) holds if

$$\left. \frac{\partial W(t, x, u_i)}{\partial u_i} \right|_{u_i(t,x)} = 0_n, \quad \left. \frac{\partial^2 W(t, x, u_i)}{\partial u_i^2} \right|_{u_i(t,x)} < 0. \tag{2.4.5}$$

2.4.2   Explicit solution of Riccati matrix differential equation

Proposition 2.4.1 *Let a matrix A of dimensions $n \times n$ and also constant and symmetric matrices C and D of dimensions $n \times n$ be such that $A(\cdot) \in C^{n \times n}[0, \vartheta]$ and*

$$C < 0, D < 0.$$

Then the solution $\Theta(t)$ of the Riccati matrix differential equation

$$\dot{\Theta} + \Theta A(t) + A(t)' \Theta - \Theta D^{-1} \Theta = O_{n \times n}, \quad \Theta(\vartheta) = C, \tag{2.4.6}$$

has the form

$$\Theta(t) = [X^{-1}(t)]' \left\{ C^{-1} + \int_t^{\vartheta} X^{-1}(\tau) D^{-1} [X^{-1}(\tau)]' d\tau \right\}^{-1} X^{-1}(t), \tag{2.4.7}$$

where $X(t)$, $0 \le t \le \vartheta$, satisfies the matrix system

$$\dot{X} = A(t)X, \quad X(\vartheta) = E_n. \tag{2.4.8}$$

Proof The matrix linear homogeneous system (2.4.8) with continuous in t coefficients has a solution $X(\cdot) \in C^{n \times n}[0, \vartheta]$ that is extendable to $[0, \vartheta]$; moreover, $\det X(t) \neq 0 \, \forall t \in [0, \vartheta]$, because this matrix of dimensions $n \times n$ represents the fundamental system of solutions for the ordinary differential vector equation $\dot{x} = A(t)x$.

Then two implications are true,

$$[\det X(t) \neq 0 \ \forall t \in [0, \vartheta]] \Rightarrow \left[\exists X^{-1}(t) \ \forall t \in [0, \vartheta]\right]$$

and

$$[X(\vartheta) = E_n] \Rightarrow \left[X^{-1}(\vartheta) = E_n\right].$$

From (2.4.7) it follows that, at $t = \vartheta$,

$$\Theta(\vartheta) = E_n \left\{C^{-1} + O_{n \times n}\right\}^{-1} E_n = C.$$

Once again, due to $\det X(t) \neq 0 \ \forall t \in [0, \vartheta]$ we may write

$$\left[X^{-1}(t)X(t) = E_n\right] \Rightarrow \left[\dot{X}^{-1}(t)X(t) + X^{-1}(t)\dot{X}(t) = O_{n \times n}\right]$$

$$\Rightarrow \left[\dot{X}^{-1}(t) = -X^{-1}(t)A(t)X(t)X^{-1}(t) = -X^{-1}(t)A(t)\right]$$

$$\Rightarrow \left[\frac{d[X^{-1}(t)]'}{dt} = -A'(t)\left[X^{-1}(t)\right]', \ \left[X^{-1}(\vartheta)\right]' = X^{-1}(\vartheta) = E_n\right].$$

As a result,

$$\frac{d[X^{-1}(t)]}{dt} = -X^{-1}(t)A(t), \quad X^{-1}(\vartheta) = E_n,$$

$$\frac{d[X^{-1}(t)]'}{dt} = -A'(t)[X^{-1}(t)]' \quad [X^{-1}(\vartheta)]' = E_n. \tag{2.4.9}$$

Denote by $\{\cdots\}$ the parenthesized expression in (2.4.7). In view of (2.4.9), (2.4.7), and $[X^{-1}(t)]' = [X'(t)]^{-1}$ (see [38, p. 33]), differentiating both sides of (2.4.7) with respect to t gives

$$\frac{d\Theta(t)}{dt} = \left[\frac{d[X^{-1}(t)]'}{dt}\right]\{\cdots\}^{-1}X^{-1}(t)$$

$$+ [X^{-1}(t)]' \left[\frac{d}{dt}\{\cdots\}^{-1}\right]X^{-1}(t) + [X^{-1}(t)]'\{\cdots\}^{-1}\frac{dX^{-1}(t)}{dt}$$

$$= -A'(t)\Theta(t) + [X^{-1}(t)]'\{\cdots\}^{-1}X^{-1}(t)D^{-1}[X^{-1}(t)]'\{\cdots\}^{-1}X^{-1}(t)$$

$$- \Theta(t)A(t) = -A'(t)\Theta(t) + \Theta(t)D^{-1}\Theta(t) - \Theta(t)A(t).$$

The proof of Proposition 2.4.1 is concluded by the two chains of implications

$$[D < 0] \Rightarrow \left[D^{-1} < 0\right] \Rightarrow \left[X^{-1}(\tau)D^{-1}[X^{-1}(\tau)]' < 0 \ \forall \tau \in [0, \vartheta]\right]$$

$$\Rightarrow \left[\int_t^\vartheta X^{-1}(\tau)D^{-1}[X^{-1}(\tau)]'d\tau \leq 0 \ \forall t \in [0, \vartheta]\right],$$

$$\left[C^{-1} < 0 \wedge \int_t^\vartheta X^{-1}(\tau)D^{-1}[X^{-1}(\tau)]'d\tau \leq 0 \ \forall t \in [0, \vartheta]\right]$$

$$\Rightarrow \left[C^{-1} + \int_t^\vartheta X^{-1}(\tau)D^{-1}[X^{-1}(\tau)]'d\tau < 0\right].$$

Remark 2.4.1 Equation (2.4.6) appears if a saddle point $U^0 = (U_1^0, U_2^0) \in \mathfrak{U}$ is designed using dynamic programming:

$$\mathcal{J}\left(U_1, U_2^0, t_0, x_0\right) \le \mathcal{J}\left(U_1^0, U_2^0, t_0, x_0\right) \le \mathcal{J}\left(U_1^0, U_2, t_0, x_0\right)$$

$\forall (t_0, x_0) \in [0, \vartheta) \times \mathbb{R}^n$, $U_i \in \mathfrak{U}_i$ $(i = 1, 2)$, in the zero-sum two-player modification of the game Γ_2 (i.e., the game Γ_2 with $C = C_1 = -C_2$, $D = D_{11} = -D_{22}$, $D_{12} = D_{21} = O_{n \times n}$ and $\mathcal{J} = \mathcal{J}_1 = -\mathcal{J}_2$.) There exist several different types of the solution $\Theta(t)$, $t \in [0, \vartheta]$, of equation (2.4.6), that are reducible to each other. (Recall that the solution $\Theta(t)$ is nonunique.) We have selected (2.4.7) due to its convenience for the small parameter method.

Proposition 2.4.2 *Let $A(\cdot), B(\cdot) \in C^{n \times n}[0, \vartheta]$. Then the solution of the matrix differential equation*

$$\dot{\Theta} + \Theta A(t) + A'(t)\Theta + B(t) = O_{n \times n}, \quad \Theta(\vartheta) = C, \tag{2.4.10}$$

has the form

$$\Theta(t) = [X^{-1}(t)]' \left\{ C + \int_t^\vartheta X'(\tau) B(\tau) X(\tau) d\tau \right\} X^{-1}(t), \tag{2.4.11}$$

where $X(t)$ is the fundamental matrix of solutions for the system

$$\dot{x} = A(t)x, \quad X(\vartheta) = E_n.$$

Proof The matrix system (2.4.10) is linear in x, inhomogeneous and also consists of continuous in $t \in [0, \vartheta]$ coefficients. For any $t_0 \in [0, \vartheta)$, such a system has a unique continuous differentiable solution $\Theta(t)$ that is extendable to the interval $[0, \vartheta]$.

Finally, we will demonstrate that $\Theta(t)$ is given by (2.4.11). Really,

$$[X(\vartheta) = E_n] \Rightarrow [\det X(t) \ne 0 \; \forall t \in [0, \vartheta]] \Rightarrow \left[\exists X^{-1}(t) \; \forall t \in [0, \vartheta]\right].$$

In view of (2.4.9), differentiating both sides of (2.4.11) yields

$$\frac{d\Theta(t)}{dt} = \left[\frac{d[X^{-1}(t)]'}{dt}\right] \{\cdots\} X^{-1}(t) + [X^{-1}(t)]' \left[\frac{d}{dt}\{\cdots\}\right] X^{-1}(t)$$

$$+ [X^{-1}(t)]'\{\cdots\} \frac{dX^{-1}(t)}{dt} = -A'(t)\Theta(t) - B(t) - \Theta(t)A(t).$$

From (2.4.11) it follows that, at $t = \vartheta$, $\Theta(\vartheta) = E_n C E_n = C$.

2.4.3   No maxima in Γ_2

The next result can be used to eliminate the linear-quadratic differential games Γ_2 without any Berge and/or Nash equilibrium, depending on the sign definiteness of the quadratic forms appearing in the integrand of the payoff functions (2.4.2) of the players.

Lemma 2.4.1 *Let the quadratic form $u_1' D_{11} u_1$ in (2.4.2) be positive definite. For any strategy profile $U^* = (U_1^*, U_2^*) \in \mathfrak{U}$, where $U_i^* \div Q_i^*(t)x$ $(i = 1, 2)$ and $Q_i^*(\cdot) \in C^{n \times n}[0, \vartheta]$, any* *initial position* $(t_0, x_0) \in [0, \vartheta) \times \mathbb{R}^n$, $x_0 \neq 0_n$, *and any constant and symmetric matrices* C_1 *and* D_{12}, *there exists a strategy* $\tilde{U}_1 \in \mathfrak{U}_1$, $\tilde{U}_1 \div \tilde{Q}_1(t)x$, *of player 1 such that*

$$\mathcal{J}_1\left(\tilde{U}_1, U_2^*\right) > \mathcal{J}_1\left(U_1^*, U_2^*\right). \tag{2.4.12}$$

Proof Consider some frozen strategy profile from $\mathfrak{U}$,

$$U^* = \left(U_1^*, U_2^*\right) \div \left(Q_1^*(t)x, Q_2^*(t)x\right), \quad Q_i^*(\cdot) \in C^{n \times n}[0, \vartheta] \quad (i = 1, 2),$$

and also some frozen initial position $(t_0, x_0) \in [0, \vartheta) \times [\mathbb{R}^n \setminus \{0_n\}]$.

The proof of Lemma 2.4.1 includes two stages as follows. In the first stage, we will establish the existence of a quadratic form $V(t, x) = x'\Theta(t)x$ for which

$$\mathcal{J}_1\left(U_1^*, U_2^*, t_0, x_0\right) = V(t_0, x_0).$$

In the second stage, we will find a strategy $\tilde{U}_1 \in \mathfrak{U}_1$ of player 1 that satisfies (2.4.12).

First stage We construct the scalar function

$$W(t, x, u_1, u_2, V) = \frac{\partial V}{\partial t} + \left[\frac{\partial V}{\partial x}\right]' (A(t)x + u_1 + \varepsilon u_2) + u_1' D_{11} u_1 + u_2' D_{12} u_2. \tag{2.4.13}$$

For $u_i = Q_i^*(t)x \ (i = 1, 2)$,

$$\begin{aligned} W[t, x, V] &= W(t, x, u_1 = Q_1^*(t)x, u_2 = Q_2^*(t)x, V) \\ &= \frac{\partial V}{\partial t} + \left[\frac{\partial V}{\partial x}\right]' \left(A(t)x + Q_1^*(t)x + \varepsilon Q_2^*(t)x\right) \\ &\quad + \left[Q_1^*(t)x\right]' D_{11} Q_1^*(t)x + \left[Q_2^*(t)x\right]' D_{12} Q_2^*(t)x. \end{aligned}$$

Next, we solve the partial differential equation

$$W[t, x, V] = 0, \quad V(\vartheta, x) = x' C_1 x. \tag{2.4.14}$$

The solution $V = V(t, x)$ is constructed in the class of the quadratic forms $V(t, x) = x\Theta(t)x$ with a matrix $\Theta(\cdot) \in C^{n \times n}[0, \vartheta]$ of dimensions $n \times n$.

Substituting $V(t, x) = x'\Theta(t)x$ into (2.4.14) and collecting like terms at the n-dimensional vector $x \in \mathbb{R}$ give

$$\begin{aligned} W[t, x, V(t, x) = x'\Theta(t)x] = x' &\left\{ \frac{d\Theta(t)}{dt} + [\Theta(t)]' \left[A(t) + Q_1^*(t) + \varepsilon Q_2^*(t)\right] \right. \\ &+ \left[A'(t) + (Q_1^*(t))' + \varepsilon (Q_2^*(t))'\right] \Theta(t) + \left[Q_1^*(t)\right]' D_{11} Q_1^*(t) \\ &+ \left. \left[Q_2^*(t)\right]' D_{12} Q_2^*(t) \right\} x = 0, \end{aligned}$$

$$x'\Theta(\vartheta)x = x' C_1 x.$$

Both of these identities will hold if, for all $(t, x) \in [0, \vartheta] \times \mathbb{R}^n \setminus \{0_n\}$, the matrix $\Theta(t)$ of dimensions $n \times n$ is the solution of the linear inhomogeneous matrix differential equation

$$\dot{\Theta} + \Theta'[A(t) + Q_1^*(t) + \varepsilon Q_2^*(t)] + [A'(t) + (Q_1^*(t))' + \varepsilon(Q_2^*(t))']\Theta + B(t) = O_{n \times n} \tag{2.4.15}$$

with continuous in t elements and the boundary-value condition

$$\Theta(\vartheta) = C_1, \tag{2.4.16}$$

where the matrix

$$B(t) = \left[Q_1^*(t)\right]' D_{11} Q_1^*(t) + \left[Q_2^*(t)\right]' D_{12} Q_2^*(t) \tag{2.4.17}$$

is continuous and symmetric.

By Proposition 2.4.2, the system (2.4.15), (2.4.16) has a unique continuously differentiable solution $\Theta = \Theta^*(t)$ that is extendable to any interval $[t_0, \vartheta] \subset [0, \vartheta]$. Due to the symmetric property of the matrices C and $B(t)$ from (2.4.17) and the explicit form (2.4.11) of $\Theta^*(t)$, the matrix $\Theta^*(t)$ will be symmetric for all $t \in [t, \vartheta]$.

Now, we will construct the realizations of the frozen strategies $U_i^* \div u_i^*(t, x) = Q_i^*(t)x$ along the solution $x^*(t)$ to the vector equation (2.4.1), i.e., we will construct $u_i^*[t] = Q_i^*(t)x^*(t)$, $t \in [t_0, \vartheta]$ $(i = 1, 2)$, where

$$\frac{dx^*(t)}{dt} = A(t)x^*(t) + Q_1^*(t)x^*(t) + \varepsilon Q_2^*(t)x^*(t), \quad x^*(t_0) = x_0.$$

In view of (2.4.14), it follows that

$$W[t, x^*(t), V(t, x^*(t))] = [x^*(t)]'\Theta^*(t)x^*(t)] = W^*[t] = 0 \tag{2.4.18}$$

for all $t \in [t_0, \vartheta]$ along the solution of (2.4.15), (2.4.16), and (2.4.1). Due to (2.4.16), we have $V(\vartheta, x^*(\vartheta)) = [x^*(\vartheta)]'C_1 x^*(\vartheta)$; then integrating both sides of (2.4.18) from t_0 to ϑ gives

$$0 = \int_{t_0}^{\vartheta} W^*[t]dt = \int_{t_0}^{\vartheta} \left\{ \frac{\partial V(t, x)}{\partial t} + \left[\frac{\partial V(t, x)}{\partial x}\right]' \left[A(t)x + Q_1^*(t)x + \varepsilon Q_2^*(t)x\right] \right.$$

$$\left. + \left[Q_1^*(t)\right]' D_{11} Q_1^*(t) + \left[Q_2^*(t)\right]' D_{12} Q_2^*(t) \right\}\Bigg|_{x=x^*(t)} dt$$

$$= \int_{t_0}^{\vartheta} \frac{dV^*(t, x^*(t))}{dt} dt + \int_{t_0}^{\vartheta} \left\{ \left(u_1^*[t]\right)' D_{11} u_1^*[t] + \left(u_2^*[t]\right)' D_{12} u_2^*[t] \right\} dt$$

$$= V\left(\vartheta, x^*(\vartheta)\right) - V(t_0, x_0) + \int_{t_0}^{\vartheta} \left\{ \left(u_1^*[t]\right)' D_{11} u_1^*[t] + \left(u_2^*[t]\right)' D_{12} u_2^*[t] \right\} dt$$

$$= \left[x^*(\vartheta)\right]' C_1 x^*(\vartheta) + \int_{t_0}^{\vartheta} \left\{ \left(u_1^*[t]\right)' D_{11} u_1^*[t] + \left(u_2^*[t]\right)' D_{12} u_2^*[t] \right\} dt - V(t_0, x_0)$$

$$= \mathcal{J}_1(U_1^*, U_2^*, t_0, x_0) - V(t_0, x_0).$$

This directly leads to the equality

$$V(t_0, x_0) = x_0'\Theta^*(t_0)x_0 = \mathcal{J}_1(U^*, t_0, x_0).$$

Second stage Consider the strategy $\tilde{U}_1 \div \tilde{u}_1(t, x) = \beta e'_n x$ of player 1, where e_n is the column vector from $\mathbb{R}^n$ consisting of 1's and a numerical parameter $\beta > 0$ will be determined below. Due to the symmetry of the matrix D_{11} and the condition $D_{11} > 0$,

$$u'_1 D_{11} u_1 \geq \lambda_1 \|u_1\|^2 = \lambda_1 u'_1 u_1 \quad \forall u_1 \in \mathbb{R}^n. \tag{2.4.19}$$

Here $\|\cdot\|$ denotes the Euclidean norm, and $\lambda_1 > 0$ is the smallest root of the characteristic equation $\det[D_{11} - \lambda E_n] = 0$ [38, pp. 88, 109].

We will adopt the matrix $\Theta^*(t)$, $t \in [0, \vartheta]$, of dimensions $n \times n$ obtained in the first stage of solving the problem (2.4.15), (2.4.16). (Note that the elements $\Theta^*(t)$ are continuously differentiable with respect to t.) Taking inequality (2.4.19) into account, also we will use the strategy $U_2^* \div Q_2^*(t)x$ of player 2 chosen in the first stage.

In view of (2.4.19), following (2.4.13) we construct the function

$$\tilde{W}[t, x] = W(t, x, \tilde{u}_1(t, x) = \beta x, u_2^*(t, x) = Q_2^*(t)x, V(t, x) = x'\Theta^*(t)x)$$

$$= \frac{\partial V(t, x)}{\partial t} + \left[\frac{\partial V(t, x)}{\partial x}\right]' \left[A(t)x + \tilde{u}_1(t, x) + \varepsilon u_2^*(t, x)\right]$$

$$+ [\tilde{u}_1(t, x)]' D_{11} \tilde{u}_1(t, x) + [u_2^*(t, x)]' D_{12} u_2^*(t, x) \geq x' \frac{d\Theta^*(t)}{dt} x$$

$$+ 2x'\Theta^*(t)\left[A(t) + \beta E_n + \varepsilon Q_2^*(t)\right]x + x'\lambda_1 \beta^2 e'_n E_n e_n x$$

$$+ x'[Q_2^*(t)]' D_{12} Q_2^*(t)x = x'\left\{\frac{d\Theta^*(t)}{dt} + \Theta^*(t)\left[A(t) + \beta E_n + \varepsilon Q_2^*(t)\right]\right.$$

$$\left. + \left[A'(t) + \beta E_n + \varepsilon[Q_2^*(t)]'\right]\Theta^*(t) + \lambda_1 \beta^2 e'_n e_n E_n + [Q_2^*(t)]' D_{12} Q_2^*(t)\right\}x$$

$$= x' M(t, \beta)x.$$

The parenthesized matrix $M(t, \beta)$ of dimensions $n \times n$ is symmetric and has the form

$$M(t, \beta) = \lambda_1 \beta^2 n E_n + 2\beta \Theta^*(t) + K(t),$$

with the matrix

$$K(t) = \Theta^*(t) + 2\Theta^*(t)\left[A(t) + \varepsilon Q_2^*(t)\right] + [Q_2^*(t)]' D_{12} Q_2^*(t).$$

of dimensions $n \times n$. (Recall that $e'_n e_n = n$.)

The elements of the matrix $M(t, \beta)$ are continuous in $t \in [0, \vartheta]$ and hence uniformly bounded on the compact set $[0, \vartheta]$. The factor β^2 appears in the diagonal elements of the matrix $M(t, \beta)$ only. As before, $\lambda_1 > 0$ is the smallest root of the characteristic equation $\det[D_{11} - \lambda E_n] = 0$ and E_n denotes an identity matrix of dimensions $n \times n$.

Therefore, the constant $\beta = \beta(U_1^*) > 0$ can be chosen sufficiently large so that all leading minors of the matrix $M(t, \beta)$ become positive for all $t \in [0, \vartheta]$ and for all $\beta > \beta(U_1^*)$. By Sylvester's criterion [38, p. 88], the quadratic form $x' M(t, \beta)x$ is positive definite for all $t \in [0, \vartheta]$ and constants $\beta > \beta(U_1^*)$, because the sign of $x' M(t, \beta)x$ is determined by the sign of the quadratic form $\beta^2 \lambda_1 n x' x$.

We fix some constant $\beta^* > \beta(U_1^*)$; then

$$\tilde{W}[t, x] = x'M(t, \beta^*)x > 0 \quad \forall t \in [0, \vartheta], \quad \forall x \in \mathbb{R}^n \setminus \{0_n\}. \tag{2.4.20}$$

Denote by $\tilde{x}(t)$, $t \in [0, \vartheta]$, the solution of the vector equation

$$\dot{x} = A(t)x + \beta^*x + \varepsilon Q_2^*(t)x, \quad x(t_0) = x_0.$$

Since $[x_0 \neq 0_n] \Rightarrow [\tilde{x}(t) \neq 0_n \ \forall t \in [t_0, \vartheta]]$, according to (2.4.20) we have

$$\tilde{W}[t, \tilde{x}(t)] > 0 \quad \forall t \in [t_0, \vartheta].$$

Integrating both sides of this inequality from t_0 to ϑ and using the boundary-value condition $\Theta^*(\vartheta) = C_1$ from (2.4.16) and also $\tilde{u}_1^*[t] = \beta \tilde{x}(t)$, we obtain

$$0 < \int_{t_0}^{\vartheta} \tilde{W}[t, \tilde{x}(t)]dt = \int_{t_0}^{\vartheta} \left\{ \frac{\partial V(t,x)}{\partial t} + [\frac{\partial V(t,x)}{\partial x}]'[A(t)x + \beta^* E_n x + \varepsilon Q_2^*(t)x] \right\}_{x=\tilde{x}(t)} dt$$

$$+ \int_{t_0}^{\vartheta} \left\{ x'\beta^* D_{11}\beta^* x + [Q_2^*(t)]' D_{12} Q_2^*(t)x \right\}_{x=\tilde{x}(t)} dt =$$

$$= \int_{t_0}^{\vartheta} \left\{ \frac{dV(t, \tilde{x}(t))}{dt} \right\} dt + \int_{t_0}^{\vartheta} \left\{ [\tilde{u}_1^*[t]]' D_{11}\tilde{u}_1^*[t] + [u_2^*[t]]' D_{12}u_2^*[t] \right\} dt =$$

$$= \tilde{x}'(\vartheta)C_1\tilde{x}(\vartheta) + \int_{t_0}^{\vartheta} \left\{ [\tilde{u}_1^*[t]]' D_{11}\tilde{u}_1^*[t] + [u_2^*[t]]' D_{12}u_2^*[t] \right\} dt - V(t_0, x_0) =$$

$$= \mathcal{J}_1(\tilde{U}_1, U_2^*, t_0, x_0) - V(t_0, x_0).$$

In combination with $\mathcal{J}_1(U_1^*, U_2^*, t_0, x_0) = V(t_0, x_0)$, this result finally proves Lemma 2.4.1.

Remark 2.4.2 Consider the inner optimization problem in the game Γ_2: find $\max_{U_1 \in \mathfrak{U}_1} \mathcal{J}_1(U_1, U_2^*, t_0, x_0)$ subject to the constraint (2.4.1) with a fixed strategy $U_2^* \in \mathfrak{U}_2$ of player 2 and any $(t_0, x_0) \in [0, \vartheta] \times [\mathbb{R}^n \setminus \{0_n\}]$. In fact, Lemma 2.4.1 states that this maximization problem has no solution for $D_{11} > 0$ and $x_0 \neq 0_n$. Indeed, whatever the strategy $U_1^* \in \mathfrak{U}_1$ of player 1 is, there always exists a strategy $\tilde{U}_1 \in \mathfrak{U}_1$ such that

$$\mathcal{J}_1\left(\tilde{U}_1, U_2^*\right) > \mathcal{J}_1\left(U_1^*, U_2^*\right)$$

for all $(t_0, x_0) \in [0, \vartheta] \times [\mathbb{R}^n \setminus \{0_n\}]$. This result can be used for eliminating the solution concepts of the games Γ_2 that maximize the payoff function of player 1 (e.g., avoiding Nash equilibrium in the game Γ_2 with $D_{11} > 0$). By analogy with Lemma 2.4.1, we may demonstrate that the game Γ_2 with $D_{12} > 0$ has no Berge equilibrium and hence the players should not choose this solution principle for the game Γ_2 with $D_{12} > 0$.

2.5 FORMALIZATION OF EQUILIBRIA AND SUFFICIENT CONDITIONS

Corps de doctrines.[11]

2.5.1 Definitions

Definition 2.5.1 *A strategy profile* $U^e = (U_1^e, U_2^e) \in \mathfrak{U}$ *is a Nash equilibrium in the game* Γ_2 *if, for any initial position* $(t_0, x_0) \in [0, \vartheta) \times [\mathbb{R}^n \setminus \{0_n\}]$,

$$\max_{U_1 \in \mathfrak{U}_1} \mathcal{J}_1 \left(U_1, U_2^e, t_0, x_0 \right) = \mathcal{J}_1 \left(U^e, t_0, x_0 \right),$$

$$\max_{U_2 \in \mathfrak{U}_2} \mathcal{J}_2 \left(U_1^e, U_2, t_0, x_0 \right) = \mathcal{J}_2 \left(U^e, t_0, x_0 \right). \tag{2.5.1}$$

Definition 2.5.2 *A strategy profile* $U^B = (U_1^B, U_2^B) \in \mathfrak{U}$ *is a Berge equilibrium in the game* Γ_2 *if, for any initial posiiton* $(t_0, x_0) \in [0, \vartheta) \times [\mathbb{R}^n \setminus \{0_n\}]$,

$$\max_{U_2 \in \mathfrak{U}_2} \mathcal{J}_1 \left(U_1^B, U_2, t_0, x_0 \right) = \mathcal{J}_1 \left(U^B, t_0, x_0 \right),$$

$$\max_{U_1 \in \mathfrak{U}_1} \mathcal{J}_2 \left(U_1, U_2^B, t_0, x_0 \right) = \mathcal{J}_2 \left(U^B, t_0, x_0 \right). \tag{2.5.2}$$

Remark 2.5.1 Despite the seeming similarity of these two types of equilibria, they have a deep distinction as follows. Unlike Definition 2.5.1 expressing the selfish character of each player (maximization of his own payoff), Definition 2.5.2 postulates altruism, guiding each player towards *the Golden Rule of ethics*—"behave unto the opponent as you would like him to behave unto you."

2.5.2 Sufficient conditions

The sufficient conditions that guarantee the existence of a Nash equilibrium and a Berge equilibrium in the linear-quadratic game under study (see below) are the result of applying dynamic programming to Definitions 2.5.1 and 2.5.2, respectively. They were derived in the book [156, pp. 112, 124].

First, we introduce the two scalar functions

$$W_i(t, x, u_1, u_2, V) = \frac{\partial V_i}{\partial t} + \left[\frac{\partial V_i}{\partial x} \right]' (A(t)x + u_1 + \varepsilon u_2) + u_1' D_{i1} u_1 + u_2' D_{i2} u_2$$

$$(i = 1, 2), \tag{2.5.3}$$

where $V = (V_1, V_2) \in \mathbb{R}^2$.

Nash equilibrium

Proposition 2.5.1 *Let* $V_i^e(t, x)$ $(i = 1, 2)$ *be unique continuously differentiable scalar functions such that*

---

[11] French "Major postulates (of theory)."

1) $V_i^e(\vartheta, x) = x' C_i x \ \forall x \in \mathbb{R}^n.$ $\hspace{5cm}$ (2.5.4)

2)   *Let $u_i^e(t, x, V^e) \ (i = 1, 2)$ be vector functions such that*

$$\max_{u_1} \left\{ W_1 \left(t, x, u_1, u_2^e(t, x, V^e), V^e \right) \right\} = Idem \left\{ u_1 \to u_1^e \left(t, x, V^e \right) \right\},$$

$$\max_{u_2} \left\{ W_2 \left(t, x, u_1^e(t, x, V^e), u_2, V^e \right) \right\} = Idem \left\{ u_2 \to u_2^e \left(t, x, V^e \right) \right\} \hspace{1cm} (2.5.5)$$

for all $(t, x) \in [0, \vartheta] \times [\mathbb{R}^n \setminus \{0_n\}]$ and $V^e = (V_1^e, V_2^e) \in \mathbb{R}^2.$

3)   *Let the functions $V_i^e(t, x) \ (i = 1, 2)$ be the solution for the system of two partial differential equations*

$$W_i(t, x, u_1^e(t, x, V^e), u_2^e(t, x, V^e), V^e) = 0 \ (i = 1, 2) \hspace{2cm} (2.5.6)$$

with the boundary-value conditions (2.5.4) for all $(t, x) \in [0, \vartheta] \times [\mathbb{R}^n \setminus \{0_n\}].$

4)   *Let strategies $U_i^e \div u_i^e(t, x, V^e(t, x)) = u_i^e[t, x]$ be such that $U_i^e \in \mathfrak{U}_i \ (i = 1, 2).$ Then:*

   a)   *the strategy profile $U^e = (U_1^e, U_2^e)$ is a Nash equilibrium in the game Γ_2 (in terms of Definition 2.5.1);*

   b)   *the Nash equilibrium payoffs are*

$$\mathcal{J}_i(U^e, t_0, x_0) = V_i^e(t_0, x_0) \ (i = 1, 2). \hspace{3cm} (2.5.7)$$

Remark 2.5.2 In practice, a Nash equilibrium should be designed by constructing the scalar functions $W_i(t, x, u_1, u_2, V^e)$ (2.5.3) and proceeding with items 1)–4) of Proposition 2.5.1. More specifically, letting $V_i^e(t, x) = x' \Theta_i^e(t) x, \ [\Theta_i^e(t)]' = \Theta_i^e(t)$ $(i = 1, 2)$, we have to perform the following steps.

Step 1 Using (2.5.4), find $\Theta_i^e(\vartheta) = C_i \ (i = 1, 2).$

Step 2 Based on (2.5.5) and (2.4.3)–(2.4.5), construct $u_i^e(t, x, V^e) \ (i = 1, 2).$

Step 3 Find the solution $V_i^e(t, x) \ (i = 1, 2)$ for the system of two partial differential equations (2.5.6) with the boundary-value conditions (2.5.4).

Step 4 Check that $u_i^e[t, x] = u_i(t, x, V^e(t, x)) = Q_i^e(t) x$ and $Q_i^e(\cdot) \in C^{n \times n}[0, \vartheta]$ $(i = 1, 2).$

The resulting pair $U^e = (U_1^e, U_2^e)$ is a Nash equilibrium in the game Γ_2, and the corresponding payoffs of the players are $J_i(U^e, t_0, x_0) = V_i^e(t_0, x_0) \ (i = 1, 2).$

Berge equilibrium

Proposition 2.5.2 *Let $V_i^B(t, x) \ (i = 1, 2)$ be unique continuously differentiable scalar functions such that:*

1) $V_i^B(\vartheta, x) = x' C_i x \ \forall x \in \mathbb{R}^n.$ $\hspace{5cm}$ (2.5.8)

2)   *Let $u_i^B(t, x, V^B)$ $(i = 1, 2)$ be vector functions such that*

$$\max_{u_2} \left\{ W_1 \left(t, x, u_1^B(t, x, V^B), u_2, V^B \right) \right\} = Idem \left\{ u_2 \to u_2^B \left(t, x, V^B \right) \right\},$$

$$\max_{u_1} \left\{ W_2 \left(t, x, u_1, u_2^B(t, x, V^B), V^B \right) \right\} = Idem \left\{ u_1 \to u_1^B \left(t, x, V^B \right) \right\} \qquad (2.5.9)$$

for all $(t, x) \in [0, \vartheta] \times [\mathbb{R}^n \setminus \{0_n\}]$ and $V^B = (V_1^B, V_2^B) \in \mathbb{R}^2$.

3)   *Let the functions $V_i^B(t, x)$ $(i = 1, 2)$ be the solution for the system of two partial differential equations*

$$W_i(t, x, u_1^B(t, x, V^B), u_2^B(t, x, V^B), V^B) = 0 \ (i = 1, 2) \qquad (2.5.10)$$

with the boundary-value conditions (2.5.8) for all $(t, x) \in [0, \vartheta] \times [\mathbb{R}^n \setminus \{0_n\}]$.

4)   *Let the strategies $U_i^B \div u_i^B(t, x, V^B(t, x)) = u_i^B[t, x]$ be such that $U_i^B \in \mathfrak{U}_i$ $(i = 1, 2)$. Then:*

a)   *the strategy profile $U^B = (U_1^B, U_2^B)$ is a Berge equilibrium in the game Γ_2 (in terms of Definition 2.5.2);*

b)   *the Berge equilibrium payoffs are*

$$\mathcal{J}_i(U^B, t_0, x_0) = V_i^B(t_0, x_0) \ (i = 1, 2). \qquad (2.5.11)$$

Remark 2.5.3 Like in the case of Nash equilibrium, a Berge equilibrium should be designed in four steps corresponding to items 1)–4) of Proposition 2.5.2. As the functions $V_i^B(t, x)$ we should choose the quadratic form $V_i^B(t, x) = x'\Theta_i^B(t)x$, where $[\Theta_i^B(t)]' = \Theta_i^B(t)$ for all $t \in [0, \vartheta]$ $(i = 1, 2)$.

2.6   EXPLICIT FORM OF EQUILIBRIA

La charité bien compise commence par soi-même.[12]

Nash equilibrium

Proposition 2.6.1 *Consider the game Γ_2 with the matrices*

$$D_{11} < 0, \ D_{22} < 0, \ C_1 < 0. \qquad (2.6.1)$$

If the system of Riccati matrix equations

$$\begin{cases} \dot{\Theta}_1^e + \Theta_1^e A(t) + A'(t)\Theta_1^e - \Theta_1^e D_{11}\Theta_1^e - \varepsilon^2 \left[\Theta_1^e D_{22}^{-1}\Theta_2^e + \Theta_2^e D_{22}^{-1}\Theta_1^e \right] \\ \quad - \varepsilon^2 \Theta_2^e D_{22}^{-1} D_{12} D_{22}^{-1}\Theta_2^e = O_{n \times n}, \quad \Theta_1^e(\vartheta, \varepsilon) = C_1, \\ \dot{\Theta}_2^e + \Theta_2^e \left[A(t) - D_{11}^{-1}\Theta_1^e \right] + [A'(t) - \Theta_1^e D_{11}^{-1}]\Theta_2^e \\ \quad + \Theta_1^e D_{11}^{-1} D_{12} D_{11}^{-1}\Theta_1^e - \varepsilon^2 \Theta_2^e D_{22}^{-1}\Theta_2^e = O_{n \times n}, \quad \Theta_2^e(\vartheta, \varepsilon) = C_2, \end{cases} \qquad (2.6.2)$$

has a solution $(\Theta_1^e(t), \Theta_2^e(t))$ that is extendable to $[0, \vartheta]$, then in the game Γ_2:

---

[12]French "Charity begins at home."

a)   *the Nash equilibrium is given by*

$$U^e = (U^e_1, U^e_2) \div (-D^{-1}_{11} \Theta^e_1(t, \varepsilon)x, -\varepsilon D^{-1}_{22} \Theta^e_2(t, \varepsilon)x); \tag{2.6.3}$$

b)   *the Nash equilibrium payoffs of the players are*

$$\mathcal{J}_i(U^e, t_0, x_0) = x'_0 \Theta^e_i(t_0, \varepsilon)x_0 \quad (i = 1, 2). \tag{2.6.4}$$

Proof Following Remark 2.5.2, we construct the functions

$$W^e_i(t, x, u_1, u_2, V) = \frac{\partial V_i}{\partial t} + \left[\frac{\partial V_i}{\partial x}\right]' (A(t)x + u_1 + \varepsilon u_2) + u'_1 D_{i1} u_1 + u'_2 D_{i2} u_2$$
$$(i = 1, 2). \tag{2.6.5}$$

Step 1 In view of (2.5.4) and $V^e_i(t, x) = x' \Theta^e_i(t)x$,

$$V^e_i(\vartheta, x) = x' \Theta^e_i(\vartheta, \varepsilon)x = x' C_i x \ \forall x \in \mathbb{R}^n \setminus \{0_n\}, \tag{2.6.6}$$

which gives

$$\Theta^e_i(\vartheta, \varepsilon) = C_i \ (i = 1, 2). \tag{2.6.7}$$

Step 2 Due to (2.5.5),

$$\max_{u_1} \left\{ W_1 \left(t, x, u_1, u^e_2 (t, x, V^e), V^e\right)\right\} = Idem \left\{ u_1 \to u^e_1 \left(t, x, V^e\right)\right\}$$

This equality holds if, according to (2.4.3)–(2.4.5),

$$\frac{\partial W_1(t, x, u_1, u^e_2(t, x, V^e), V^e)}{\partial u_1}\bigg|_{u_1(t,x,V^e)} = \frac{\partial V^e_1}{\partial x} + 2D_{11}u^e_1(t, x, V^e) = 0_n,$$

$$\frac{\partial^2 W_1(t, x, u_1, u^e_2(t, x, V^e), V^e)}{\partial u^2_1}\bigg|_{u_1(t,x,V^e)} = 2D_{11} < 0$$

for any $(t, x) \in [0, \vartheta] \times [\mathbb{R}^n \setminus \{0_n\}]$ and $V^e = (V^e_1, V^e_2) \in \mathbb{R}^2$. By analogy,

$$\frac{\partial W_2(t, x, u^e_1(t, x, V^e), u_2, V^e)}{\partial u_2}\bigg|_{u_2(t,x,V^e)} = \varepsilon \frac{\partial V^e_2}{\partial x} + 2D_{22}u^e_2(t, x, V^e) = 0_n,$$

$$\frac{\partial^2 W_2(t, x, u^e_1(t, x, V^e), u_2, V^e)}{\partial u^2_2}\bigg|_{u_2(t,x,V^e)} = 2D_{22} < 0$$

for all $(t, x) \in [0, \vartheta] \times [\mathbb{R}^n \setminus \{0_n\}]$ and $V^e = (V^e_1, V^e_2) \in \mathbb{R}^2$.

The second and fourth relations are true by (2.6.1). Using the first and third relations, we find

$$u^e_1(t, x, V^e) = -\frac{1}{2}D^{-1}_{11}\frac{\partial V^e_1}{\partial x}, \quad u^e_2(t, x, V^e) = -\frac{\varepsilon}{2}D^{-1}_{22}\frac{\partial V^e_2}{\partial x}. \tag{2.6.8}$$

Step 3 We write the two partial differential equations (2.5.6) with the boundary-value conditions (2.6.7) to find the two scalar functions $V^e_i(t, x)$ $(i = 1, 2)$:

$$0 = W_1^e[t, x, V^e] = W_1\left(t, x, u_1^e\left(t, x, V^e\right), u_2^e\left(t, x, V^e\right), V^e\right)$$

$$= \frac{\partial V_1^e}{\partial t} + \left[\frac{\partial V_1^e}{\partial x}\right]'\left[A(t)x - \frac{1}{2}D_{11}^{-1}\frac{\partial V_1^e}{\partial x}\right] - \frac{\varepsilon^2}{2}\left[\frac{\partial V_1^e}{\partial x}\right]'D_{22}^{-1}\frac{\partial V_2^e}{\partial x}$$

$$+ \frac{1}{4}\left[\frac{\partial V_1^e}{\partial x}\right]'D_{11}^{-1}\frac{\partial V_1^e}{\partial x} + \frac{\varepsilon^2}{4}\left[\frac{\partial V_2^e}{\partial x}\right]'D_{22}^{-1}D_{12}D_{22}^{-1}\frac{\partial V_2^e}{\partial x}$$

$$= \frac{\partial V_1^e}{\partial t} + \left[\frac{\partial V_1^e}{\partial x}\right]'A(t)x - \frac{\varepsilon^2}{2}\left[\frac{\partial V_1^e}{\partial x}\right]'D_{22}^{-1}\frac{\partial V_2^e}{\partial x}$$

$$- \frac{1}{4}\left[\frac{\partial V_1^e}{\partial x}\right]'D_{11}^{-1}\frac{\partial V_1^e}{\partial x} + \frac{\varepsilon^2}{4}\left[\frac{\partial V_2^e}{\partial x}\right]'D_{22}^{-1}D_{12}D_{22}^{-1}\frac{\partial V_2^e}{\partial x},$$

$$W_2^e[t, x, V^e] = W_2\left(t, x, u_1^e\left(t, x, V^e\right), u_2^e\left(t, x, V^e\right), V^e\right)$$

$$= \frac{\partial V_2^e}{\partial t} + \left[\frac{\partial V_2^e}{\partial x}\right]'\left[A(t)x - \frac{1}{2}D_{11}^{-1}\frac{\partial V_1^e}{\partial x} - \frac{\varepsilon^2}{2}D_{22}^{-1}\frac{\partial V_2^e}{\partial x}\right]$$

$$+ \frac{1}{4}\left[\frac{\partial V_1^e}{\partial x}\right]'D_{11}^{-1}D_{21}D_{11}^{-1}\frac{\partial V_1^e}{\partial x} - \frac{\varepsilon^2}{4}\left[\frac{\partial V_2^e}{\partial x}\right]'D_{22}^{-1}\frac{\partial V_2^e}{\partial x} = 0. \quad (2.6.9)$$

In view of (2.4.3) and $V_i^e(t, x) = x'\Theta_i^e x$, we obtain the gradients $\frac{\partial V_i^e}{\partial x} = 2\Theta_i^e x$ and $\frac{\partial V_i^e}{\partial t} = x'\frac{d\Theta_i^e}{dt}x$. Substituting $\frac{\partial V_i^e}{\partial x}$ and $\frac{\partial V_i^e}{\partial t}$ into (2.6.9) and collecting like terms with the pairwise products of the components of the n-dimensional vector x, we arrive at the equations

$$W_1^e[t, x, V^e] = x'\left\{\frac{d\Theta_1^e}{dt} + \Theta_1^e A(t) + A'(t)\Theta_1^e - \Theta_1^e D_{11}^{-1}\Theta_1^e\right.$$

$$\left. + \varepsilon^2[-\Theta_1^e D_{22}^{-1}\Theta_2^e - \Theta_2^e D_{22}^{-1}\Theta_1^e + \Theta_2^e D_{22}^{-1}D_{12}D_{22}^{-1}\Theta_2^e]\right\}x = 0,$$

$$W_2^e[t, x, V^e] = x'\left\{\frac{d\Theta_2^e}{dt} + \Theta_2^e A(t) + A'(t)\Theta_2^e + \Theta_1^e D_{11}^{-1}D_{21}D_{11}^{-1}\Theta_1^e\right.$$

$$\left. - \Theta_2^e D_{11}^{-1}\Theta_1^e - \varepsilon^2\Theta_2^e D_{22}^{-1}\Theta_2^e\right\}x = 0$$

with the boundary-value conditions

$$V_i(\vartheta, x) = x'\Theta_i^e(\vartheta, \varepsilon)x = x'C_i x \quad (i = 1, 2).$$

The identities $W_i^e[t, x, V^e(t, x)] = 0$ for all $(t, x) \in [0, \vartheta] \times [\mathbb{R}^n \setminus \{0_n\}]$ $(i = 1, 2)$ hold if the system of Riccati matrix equations

$$\begin{cases} \dot{\Theta}_1^e + \Theta_1^e A(t) + A'(t)\Theta_1^e - \Theta_1^e D_{11}\Theta_1^e + \varepsilon^2[-\Theta_1^e D_{22}^{-1}\Theta_2^e \\ - \Theta_2^e D_{22}^{-1}\Theta_1^e + \Theta_2^e D_{22}^{-1}D_{12}D_{22}^{-1}\Theta_2^e] = O_{n\times n}, \quad \Theta_1^e(\vartheta, \varepsilon) = C_1, \\ \dot{\Theta}_2^e + \Theta_2^e A(t) + A'(t)\Theta_2^e - \Theta_2^e D_{11}^{-1}\Theta_1^e - \Theta_1^e D_{11}^{-1}\Theta_2^e \\ + \Theta_1^e D_{11}^{-1}D_{21}D_{11}^{-1}\Theta_1^e - \varepsilon^2\Theta_2^e D_{22}^{-1}\Theta_2^e = O_{n\times n}, \quad \Theta_2^e(\vartheta, \varepsilon) = C_2 \end{cases} \quad (2.6.10)$$

has a solution $(\Theta_1^e(t, \varepsilon), \Theta_2^e(t, \varepsilon))$ that is extendable to the interval $[0, \vartheta]$.

Step 4 Using the obtained solutions $(\Theta_1^e(t, \varepsilon), \Theta_2^e(t, \varepsilon))$ and (2.6.8), we find the two n-dimensional vector functions

$$u_1^e[t, x] = u_1(t, x, V^e(t, x)) = -D_{11}^{-1}\Theta_1^e(t, \varepsilon)x,$$

$$u_2^e[t, x] = u_2(t, x, V^e(t, x)) = -\varepsilon D_{22}^{-1}\Theta_2^e(t, \varepsilon)x. \quad (2.6.11)$$

Since $D_{11}^{-1}\Theta_1^e(\cdot, \varepsilon), \varepsilon D_{22}^{-1}\Theta_2^e(\cdot, \varepsilon) \in C^{n \times n}[0, \vartheta]$, the Nash equilibrium in the game Γ_2 will have the form (2.6.3), and the Nash equilibrium payoffs of the players will be given by (2.6.4).

Remark 2.6.1 In the case $D_{11} > 0$ and/or $D_{22} > 0$, by Lemma 2.4.1 at least one of the two maxima from Definition 2.5.1 is not achieved for any $x_0 \neq 0_n$. Really, assume on the contrary that, e.g., in the case $D_{11} > 0$ there exists a strategy $\hat{U}_1 \in \mathfrak{U}_1$ of player 1 such that, for $x_0 \neq 0_n$,

$$\max_{U_1 \in \mathfrak{U}_1} \mathcal{J}_1(U_1, U_2^e, t_0, x_0) = \mathcal{J}_1(\hat{U}_1, U_2^e, t_0, x_0).$$

Then, according to Lemma 2.4.1, for the initial position $(t_0, x_0) \in [0, \vartheta) \times [\mathbb{R}^n \setminus \{0_n\}]$ there also exists a strategy $\tilde{U}_1 \in \mathfrak{U}_1$ for which

$$\mathcal{J}_1(\tilde{U}_1, U_2^*) > \mathcal{J}_1(\hat{U}_1^*, U_2^*),$$

which contradicts the whole essence of the operation $\max_{U_1 \in \mathfrak{U}_1}$. Thus, we have established the following result: if $D_{11} > 0$ and/or $D_{22} > 0$, then there exists no Nash equilibria in the game Γ_2 for any initial position $(t_0, x_0) \in [0, \vartheta) \times [\mathbb{R}^n \setminus \{0_n\}]$.

Berge equilibrium

We will utilize Remark 3.5.3, repeating Steps 1–4 from Remark 3.5.2, with appropriate modifications dictated by Proposition 2.5.2.

Proposition 2.6.2 *Consider the game Γ_2 with the matrices*

$$D_{12} < 0, \ D_{21} < 0, C_2 < 0. \tag{2.6.12}$$

If the system of Riccati matrix equations

$$\begin{cases} \dot{\Theta}_1^B + \Theta_1^B[A(t) - D_{21}^{-1}\Theta_2^B] + [A'(t) - \Theta_2^B D_{21}^{-1}]\Theta_1^B + \\ +\Theta_2^B D_{21}^{-1}D_{11}D_{21}^{-1}\Theta_2^B - \varepsilon^2\Theta_1^B D_{12}^{-1}\Theta_1^e = O_{n \times n}, \ \Theta_1^B(\vartheta, \varepsilon) = C_1, \\ \dot{\Theta}_2^B + \Theta_2^B A(t) + A'(t)\Theta_2^B - \Theta_2^B D_{21}\Theta_2^B + \varepsilon^2[-\Theta_2^B D_{12}^{-1}\Theta_1^B - \\ -\Theta_1^B D_{12}^{-1}\Theta_2^B + \Theta_1^B D_{12}^{-1}D_{22}D_{12}^{-1}\Theta_1^B] = O_{n \times n}, \ \Theta_2^B(\vartheta, \varepsilon) = C_2, \end{cases} \tag{2.6.13}$$

has a solution $(\Theta_1^B(t, \varepsilon), \Theta_2^B(t, \varepsilon))$ that is extendable to $[0, \vartheta]$, then in the game Γ_2:

a) *the Berge equilibrium is given by*

$$U^B = (U_1^B, U_2^B) \div (-D_{21}^{-1}\Theta_2^B(t, \varepsilon)x, -\varepsilon D_{12}^{-1}\Theta_1^B(t, \varepsilon)x); \tag{2.6.14}$$

b) *the Berge equilibrium payoffs of the players are*

$$\mathcal{J}_i(U^B, t_0, x_0) = x_0'\Theta_i^B(t_0, \varepsilon)x_0 \quad (i = 1, 2). \tag{2.6.15}$$

Proof Following Remark 3.5.3, we construct the two scalar functions (2.6.5).
Step 1 In view of (2.5.8) and $V_i^B(t, x) = x'\Theta_i^B(t)x$,

$$\Theta_i^B(\vartheta, \varepsilon) = C_i \quad (i = 1, 2). \tag{2.6.16}$$

Step 2 Due to (2.5.9), using (2.4.3)–(2.4.5) we write

$$\left.\frac{\partial W_1(t, x, u_1^B(t, x, V^B), u_2, V^B)}{\partial u_2}\right|_{u_2(t,x,V^B)} = \varepsilon \frac{\partial V_1}{\partial x} + 2D_{12}u_2^B(t, x, V^B) = 0_n,$$

$$\left.\frac{\partial^2 W_1(t, x, u_1^B(t, x, V^B), u_2, V^B)}{\partial u_2^2}\right|_{u_2(t,x,V^B)} = 2D_{12} < 0 \text{ and}$$

$$\left.\frac{\partial W_2(t, x, u_1, u_2^B(t, x, V^B), V^B)}{\partial u_1}\right|_{u_1(t,x,V^B)} = \varepsilon \frac{\partial V_2}{\partial x} + 2D_{21}u_1^B(t, x, V^B) = 0_n,$$

$$\left.\frac{\partial^2 W_2(t, x, u_1, u_2^B(t, x, V^B), V^B)}{\partial u_1^2}\right|_{u_1(t,x,V^B)} = 2D_{21} < 0$$

for any $(t, x) \in [0, \vartheta] \times [\mathbb{R}^n \setminus \{0_n\}]$ and $V^B = (V_1^B, V_2^B) \in \mathbb{R}^2$.

The second and fourth relations are true by (2.6.12). Using the first and third relations, we find

$$u_1^B(t, x, V^B) = -\frac{1}{2}D_{21}^{-1}\frac{\partial V_2^B}{\partial x}, \quad u_2^B(t, x, V^B) = -\frac{\varepsilon}{2}D_{12}^{-1}\frac{\partial V_1^B}{\partial x}. \tag{2.6.17}$$

Step 3 Substituting (2.6.17) into (2.5.10), we obtain the system of two partial differential equations with boundary conditions (2.5.8):

$$0 = W_1^B[t, x, V^B] = W_1\left(t, x, u_1^B\left(t, x, V^B\right), u_2^B\left(t, x, V^B\right), V^B\right)$$

$$= \frac{\partial V_1^B}{\partial t} + \left[\frac{\partial V_1^B}{\partial x}\right]'\left[A(t)x - \frac{1}{2}D_{21}^{-1}\frac{\partial V_2^B}{\partial x} - \frac{\varepsilon^2}{2}D_{12}^{-1}\frac{\partial V_1^B}{\partial x}\right]$$

$$+ \frac{1}{4}\left[\frac{\partial V_2^B}{\partial x}\right]'D_{21}^{-1}D_{11}D_{21}^{-1}\frac{\partial V_2^B}{\partial x} + \frac{\varepsilon^2}{4}\left[\frac{\partial V_1^B}{\partial x}\right]'D_{12}^{-1}\frac{\partial V_1^B}{\partial x} = \frac{\partial V_1^B}{\partial t}$$

$$+ \left[\frac{\partial V_1^B}{\partial x}\right]'A(t)x + \frac{1}{4}\left[\frac{\partial V_2^B}{\partial x}\right]'D_{21}^{-1}D_{11}D_{21}^{-1}\frac{\partial V_2^B}{\partial x} - \frac{1}{2}\left[\frac{\partial V_1^B}{\partial x}\right]'D_{21}^{-1}\frac{\partial V_2^B}{\partial x}$$

$$- \frac{\varepsilon^2}{4}\left[\frac{\partial V_1^B}{\partial x}\right]'D_{12}^{-1}\frac{\partial V_1^B}{\partial x},$$

$$0 = W_2^B[t, x, V^B] = W_2(t, x, u_1^B(t, x, V^B), u_2^B(t, x, V^B), V^B)$$

$$= \frac{\partial V_2^B}{\partial t} + \left[\frac{\partial V_2^B}{\partial x}\right]'\left[A(t)x - \frac{1}{2}D_{21}^{-1}\frac{\partial V_2^B}{\partial x} - \frac{\varepsilon^2}{2}D_{12}^{-1}\frac{\partial V_1^B}{\partial x}\right] + \frac{1}{4}\left[\frac{\partial V_2^B}{\partial x}\right]'D_{21}^{-1}\frac{\partial V_2^B}{\partial x}$$

$$+ \frac{\varepsilon^2}{4}\left[\frac{\partial V_1^B}{\partial x}\right]'D_{12}^{-1}D_{22}D_{12}^{-1}\frac{\partial V_1^B}{\partial x} = \frac{\partial V_2^B}{\partial t} + \left[\frac{\partial V_2^B}{\partial x}\right]'A(t)x$$

$$- \frac{1}{4}\left[\frac{\partial V_2^B}{\partial x}\right]'D_{21}^{-1}\frac{\partial V_2^B}{\partial x} + \frac{\varepsilon^2}{4}\left[\frac{\partial V_1^B}{\partial x}\right]'D_{12}^{-1}D_{22}D_{12}^{-1}\frac{\partial V_1^B}{\partial x} - \frac{\varepsilon^2}{2}\left[\frac{\partial V_2^B}{\partial x}\right]'D_{12}^{-1}\frac{\partial V_1^B}{\partial x}.$$

Letting $V_i^B(t, x) = x'\Theta_i^B x$ and $\frac{\partial V_i^B}{\partial x} = 2\Theta_i^B x$, we demonstrate that due to (2.6.16) the previous equalities hold if $\Theta_i^B(t, \varepsilon)$ $(i = 1, 2)$ are the solutions of the system (2.6.13). Using the resulting solution $(\Theta_1^B(t, \varepsilon), \Theta_2^B(t, \varepsilon))$ of the system (2.6.13), the explicit form of the functions $V_i^B(t, x) = x'\Theta_i^B x$ and gradients $\frac{\partial V_i^B}{\partial x} = 2\Theta_i^B x$ as well as the inclusions $D_{21}^{-1}\Theta_2^B(\cdot, \varepsilon), \varepsilon D_{12}^{-1}\Theta_1^B(\cdot, \varepsilon) \in C^{n \times n}[0, \vartheta]$, we finally prove (2.6.17). Note that the relations (2.6.15) are true according to (2.5.11).

Remark 2.6.2 By analogy with Remark 2.6.1, we can establish the following result: if $D_{12} > 0$ and/or $D_{21} > 0$, then there are no Berge equilibria in the game Γ_2 for any initial position $(t_0, x_0) \in [0, \vartheta) \times [\mathbb{R}^n \setminus \{0_n\}]$.

2.7 APPLICATION OF SMALL PARAMETER METHOD

> Mais comme tout est compencé dans
> le meilleur des mondes possibles.[13]

2.7.1 Poincaré's theorem

Thus, Propositions 2.6.1 and 2.6.2 have showed that the presence of Berge and/or Nash equilibria is connected with the existence of a solution for the corresponding systems of two matrix ordinary differential equations of the Riccati type that can be extended to the entire interval $[0, \vartheta]$ of the game. As a matter of fact, the existence of solutions in a small left neighborhood $(\vartheta - \delta, \vartheta]$ of the point $t = \vartheta$ is guaranteed by general existence theorems from the theory of ordinary differential equations. The question of the extendability of such solutions to the entire interval $[0, \vartheta]$ of the game remains open. In this section, we will try to answer it using the small parameter method. This method arose in connection with the three-body problem in celestial mechanics; it dates back to J. D'Alembert and was intensively developed starting from the end of the 19th century. Further, from the numerous theoretical results on the small parameter method [200], we will use Poincaré's theorem on the analyticity of solutions with respect to the parameter. It will be formulated for the matrix system of ordinary differential equations

$$\dot{\Theta} = \Xi(t, \Theta, \varepsilon), \quad \Theta(\vartheta, \varepsilon) = C. \tag{2.7.1}$$

The notations are the following: Θ as a matrix of dimensions $n \times n$; $\Xi(t, \Theta, \varepsilon)$ as a matrix of dimensions $n \times n$ whose elements are functions of the variables t, Θ, and ε; ε as a small parameter such that $0 \leqslant \varepsilon \leqslant \varepsilon_0$, where ε_0 is a small number; C as a constant matrix of dimensions $n \times n$; $t \in [0, \vartheta]$ as continuous time. The elements of the matrix $\Xi(t, \Theta, \varepsilon)$ are assumed to be defined and continuous on the domain $G, \varepsilon \in [0, \varepsilon_0]$. Denote by $\Theta = \Theta(t, \varepsilon)$ a solution of (2.7.1) that satisfies the boundary-value condition $\Theta(\vartheta, \varepsilon) = C, (\vartheta, \varepsilon) \in G$. Together with the system (2.7.1), consider the system

$$\dot{\Theta} = \Xi(t, \Theta, 0), \quad \Theta(\vartheta, 0) = C, \tag{2.7.2}$$

which is obtained from (2.7.1) for $\varepsilon = 0$. Let $\Theta = \Theta^{(0)}(t)$ be a solution of (2.7.2) defined on $t \in [0, \vartheta]$ with the same boundary-value condition $\Theta(\vartheta) = C$. For a small value ε, the right-hand sides of these systems are close to each other. Then a natural question is: how do the solutions of the systems (2.7.1) and (2.7.2) differ on the entire interval $[0, \vartheta]$? By the theorem on the continuous dependence of solutions of combined ordinary differential equations on the parameter, generally these solutions are close to each other too.

---

[13]French "There are doubts whether everything is really compensated in the best of all possible worlds"; from A.I. Herzen's letter to N.A. Herzen, June 7, 1851. An ironic combination of two famous quotes from *Des compensations dans les destinées humaines* by French philosopher P.H. Azaïs (1766–1845) and *Candide* by French Enlightenment writer, historian, and philosopher Voltaire (1694–1778).

Moreover, if there exists a unique solution $\Theta^{(0)}(t)$ of the system (2.7.2) and the elements of $\Xi(t, \Theta, \varepsilon)$ are holomorphic (analytic) for $0 \leqslant \varepsilon \leqslant \varepsilon_0$, $\Theta = \Theta^{(0)}(t)$, $t \in [0, \vartheta]$, then for a sufficiently small value ε the solution of (2.7.1) can be written as the series

$$\Theta(t, \varepsilon) = \Theta^{(0)}(t) + \sum_{m=1}^{\infty} \varepsilon^m \Theta^{(m)}(t), \tag{2.7.3}$$

which has uniform convergence on the entire interval $[0, \vartheta]$. This fact is the core of *Poincaré's theorem*.

Among general theorems from the theory of differential equations, also we will employ the theorem on the continuous dependence of solutions on the parameter; see below.

Theorem 2.7.1 *Let the right-hand side of system (2.7.1) be continuously differentiable with respect to the elements of the matrix Θ and also continuous in ε on the domain G. Then for a sufficiently small value $\varepsilon > 0$, the solution $\Theta(t, \varepsilon)$ of system (2.7.1) is well-defined on the same interval $[0, \vartheta]$ as the solution of system (2.7.2).*

2.7.2   Nash equilibrium

In this subsection, we will demonstrate that the existence of a solution $(\Theta_1^e(t, \varepsilon), \Theta_2^e(t, \varepsilon))$ of the system (2.6.2) that is extendable to $[0, \vartheta]$ is a superfluous requirement of Proposition 2.6.1 in the case of a small value $\varepsilon > 0$. In other words, it can be neglected for sufficiently small values $\varepsilon > 0$. More specifically, we will establish the following result.

Proposition 2.7.1 *Consider the game Γ_2 with the matrices*

$$D_{11} < 0, \ D_{22} < 0, C_1 < 0.$$

Then for sufficiently small values $\varepsilon > 0$, the game Γ_2 has the Nash equilibrium (2.6.3) and the corresponding payoffs of the players are given by (2.6.4).

Proof Proposition 2.7.1 can be proved by demonstrating that the system (2.6.2) with sufficiently small values $\varepsilon > 0$ has a solution $(\Theta_1^e(t, \varepsilon), \Theta_2^e(t, \varepsilon))$, $t \in [0, \vartheta]$, that is extendable to $[0, \vartheta]$.

To this end, we will utilize Theorem 2.7.1. For (2.6.2), we construct the null approximation by letting $\varepsilon = 0$. In this case, the system (2.6.2) is decomposed into two subsystems of matrix ordinary differential equations. One of them belongs to the Riccati class, whereas the other is linear in $\Theta_2^{(0)}$:

$$\begin{cases} \dot{\Theta}_1^{(0)} + \Theta_1^{(0)} A(t) + A'(t) \Theta_1^{(0)} - \Theta_1^{(0)} D_{11}^{-1} \Theta_1^{(0)} = O_{n \times n}, \ \Theta_1^{(0)}(\vartheta) = C_1, \\ \dot{\Theta}_2^{(0)} + \Theta_2^{(0)} [A(t) - D_{11}^{-1} \Theta_1^{(0)}] + [A'(t) - \Theta_1^{(0)} D_{11}^{-1}] \Theta_2^{(0)} + \\ + \Theta_1^{(0)} D_{11}^{-1} D_{12} D_{11}^{-1} \Theta_1^{(0)} = O_{n \times n}, \ \Theta_2^{(0)}(\vartheta) = C_2. \end{cases} \tag{2.7.4}$$

For $D_{11} < 0, C_1 < 0$ and $A(\cdot) \in C^{n \times n}[0, \vartheta]$, the solution $\Theta_1^{(0)}(t)$ of the first part of the system (2.7.4) exists, is continuous and extendable to $[0, \vartheta]$, symmetric ($[\Theta_1^{(0)}(t)]' = \Theta_1^{(0)}(t)$) and negative ($\Theta_1^{(0)}(t) < 0$) for all $t \in [0, \vartheta]$ and has the form

$$\Theta_1^{(0)}(t) = \left[X^{-1}(t) \right]' \left\{ C_1^{-1} + \int_t^\vartheta X^{-1}(\tau) D_{11}^{-1} \left[X^{-1}(\tau) \right]' d\tau \right\}^{-1} X^{-1}(t), \tag{2.7.5}$$

where $X(t)$ denotes the fundamental system of solutions for $\dot{x} = A(t)x$, $X[\vartheta] = E_n$; see Proposition 2.4.1. Incorporating this matrix $\Theta_1^{(0)} = \Theta_1^{(0)}(t)$ into the second part of the system, we obtain the following matrix linear inhomogeneous differential equation in $\Theta_2^{(0)}$:

$$\dot{\Theta}_2^{(0)} + \Theta_2^{(0)}\left[A(t) - D_{11}^{-1}\Theta_1^{(0)}(t)\right] + \left[A'(t) - \Theta_1^{(0)}(t)D_{11}^{-1}\right]\Theta_2^{(0)}$$
$$+ \Theta_1^{(0)}(t)D_{11}^{-1}D_{12}D_{11}^{-1}\Theta_1^{(0)}(t) = O_{n\times n} \quad \Theta_1^{(0)}(\vartheta) = C_2. \tag{2.7.6}$$

Since $\Theta_1^{(0)}(\cdot), A(\cdot) \in C^{n\times n}[0, \vartheta]$, for any constant matrix C_2 of dimensions $n \times n$ equation (2.7.6) has a continuous and extendable to $[0, \vartheta]$ solution of the form

$$\Theta_2^{(0)}(t) = \left[X^{-1}(t)\right]'\left\{C_2 + \int_t^\vartheta X'(\tau)B_1(\tau)X(\tau)d\tau\right\}X^{-1}(t), \tag{2.7.7}$$

with the continuous and symmetric matrix

$$B_1(t) = \Theta_1^{(0)}(t)D_{11}^{-1}D_{12}D_{11}^{-1}\Theta_1^{(0)}(t)$$

of dimensions $n\times n$; see Proposition 2.4.2. From (2.7.7) and the symmetry of C_2 and $B(t)$, it follows that (2.7.7) holds for any $t \in [0, \vartheta]$ (like in (2.7.5), $X(t)$ denotes the fundamental matrix). Consequently, the system (2.6.2) with $\varepsilon = 0$ has a continuous and extendable to $[0, \vartheta]$ solution $(\Theta_1^{(0)}(t), \Theta_2^{(0)}(t))$. Therefore, by Theorem 2.7.1 the system (2.6.2) with sufficiently small values $\varepsilon > 0$ also has an extendable to $[0, \vartheta]$ solution $(\Theta_1^e(t, \varepsilon), \Theta_2^e(t, \varepsilon))$. And Proposition 2.7.1 directly follows from Proposition 2.6.1.

2.7.3   Berge equilibrium

Like for Nash equilibrium, we will demonstrate that the existence of a solution $(\Theta_1^B(t, \varepsilon), \Theta_2^B(t, \varepsilon))$ of the system (2.6.13) that is extendable to $[0, \vartheta]$ is a superfluous requirement of Proposition 2.6.2, which can be replaced by the smallness of $\varepsilon > 0$.

Proposition 2.7.2 *Consider the game* Γ_2 *with the matrices*

$$D_{12} < 0, \; D_{21} < 0, C_2 < 0.$$

Then for sufficiently small values $\varepsilon > 0$, *the game* Γ_2 *has the Berge equilibrium (2.6.14) and the corresponding payoffs of the players are given by (2.6.15).*

Proof As before, using Theorem 2.7.1 we will prove that the solution of the system (2.6.13) is extendable to $[0, \vartheta]$. By analogy with Proposition 2.7.1, we construct the null approximation $(\tilde{\Theta}_1^{(0)}(t), \tilde{\Theta}_2^{(0)}(t))$ by letting $\varepsilon = 0$ in (2.6.13). As a result, the system (2.6.13) is decomposed into the two subsystems of matrix nonlinear differential equations

$$\begin{cases} \dot{\tilde{\Theta}}_1^{(0)} + \tilde{\Theta}_1^{(0)}[A(t) - D_{21}^{-1}\tilde{\Theta}_2^{(0)}] + [A'(t) - \tilde{\Theta}_2^{(0)}D_{21}^{-1}]\tilde{\Theta}_1^{(0)} \\ + \tilde{\Theta}_2^{(0)}D_{21}^{-1}D_{11}D_{21}^{-1}\tilde{\Theta}_2^{(0)} = O_{n\times n}, \quad \tilde{\Theta}_1^{(0)}(\vartheta) = C_1, \\ \dot{\tilde{\Theta}}_2^{(0)} + \tilde{\Theta}_2^{(0)}A(t) + A'(t)\tilde{\Theta}_2^{(0)} - \tilde{\Theta}_2^{(0)}D_{21}^{-1}\tilde{\Theta}_2^{(0)} = O_{n\times n}, \quad \tilde{\Theta}_2^{(0)}(\vartheta) = C_2. \end{cases} \tag{2.7.8}$$

For $D_{21} < 0$ and $C_2 < 0$, the solution $\tilde{\Theta}_2^{(0)}(t)$ for the matrix system of Riccati differential equations (the second equation in (2.7.8)) exists, is continuous and extendable to $[0, \vartheta]$, symmetric ($[\tilde{\Theta}_2^{(0)}(t)]' = \tilde{\Theta}_2^{(0)}(t)$) and negative ($\tilde{\Theta}_2^{(0)}(t) < 0$) for all $t \in [0, \vartheta]$ and has the form

$$\tilde{\Theta}_2^{(0)}(t) = \left[X^{-1}(t)\right]' \left\{ C_2^{-1} + \int_t^{\vartheta} X^{-1}(\tau) D_{21}^{-1} \left[X^{-1}(\tau)\right]' d\tau \right\}^{-1} X^{-1}(t). \qquad (2.7.9)$$

Incorporating the solution $\tilde{\Theta}_2^{(0)} = \tilde{\Theta}_2^{(0)}(t)$ into the first part of (2.7.8), we obtain the following matrix linear inhomogeneous ordinary differential equation in $\tilde{\Theta}_1^{(0)}$:

$$\dot{\tilde{\Theta}}_1^{(0)} + \tilde{\Theta}_1^{(0)} \left[A(t) - D_{21}^{-1} \tilde{\Theta}_2^{(0)}(t) \right] + [A'(t) - \tilde{\Theta}_2^{(0)}(t) D_{21}^{-1}] \tilde{\Theta}_1^{(0)}$$
$$+ \tilde{\Theta}_2^{(0)}(t) D_{21}^{-1} D_{11} D_{21}^{-1} \tilde{\Theta}_2^{(0)}(t) = O_{n \times n}, \quad \tilde{\Theta}_1^{(0)}(\vartheta) = C_1.$$

In view of the inclusion $\tilde{\Theta}_2^{(0)}(\cdot), A(\cdot) \in C^{n \times n}[0, \vartheta]$ and Proposition 2.4.2, the explicit solution is given by

$$\tilde{\Theta}_1^{(0)}(t) = \left[X^{-1}(t)\right]' \left\{ C_1 + \int_t^{\vartheta} X'(\tau) B_2(\tau) X(\tau) d\tau \right\} X^{-1}(t), \qquad (2.7.10)$$

with the continuous and symmetric matrix

$$B_2(t) = \tilde{\Theta}_2^{(0)}(t) D_{21}^{-1} D_{11} D_{21}^{-1} \tilde{\Theta}_2^{(0)}(t)$$

of dimensions $n \times n$.

Clearly, the continuous matrix $\tilde{\Theta}_1^{(0)}(t)$ of dimensions $n \times n$ is well-defined for all $t \in [0, \vartheta]$ and symmetric. Hence, for $\varepsilon = 0$ the null approximation ($\tilde{\Theta}_1^{(0)}(t), \tilde{\Theta}_2^{(0)}(t) | t \in [0, \vartheta]$) of the solution ($\Theta_1^B(t, \varepsilon), \Theta_2^B(t, \varepsilon) | t \in [0, \vartheta]$) of the system (2.6.13) is extendable to $[0, \vartheta]$. By Theorem 2.7.1 the system (2.6.13) with sufficiently small values $\varepsilon > 0$ has an extendable to $[0, \vartheta]$ solution ($\Theta_1^B(t, \varepsilon), \Theta_2^B(t, \varepsilon)$). And Proposition 2.7.2 directly follows from Proposition 2.6.2.

2.8 LA FIN COURONNE L'OEUVRE[14]

This section is devoted to the coefficient criteria of the existence (and nonexistence!) of Nash and/or Berge equilibria (in terms of Definitions 2.5.1 and 2.5.2, respectively) in the differential positional linear-quadratic game Γ_2 with a small influence of one player on the rate of change $\dot{x}(t)$ of the state vector $x(t)$. In the game Γ_2, the state vector evolves in accordance with the vector linear differential equation

$$\dot{x} = A(t)x + u_1 + \varepsilon u_2, \quad x(t_0) = x_0,$$

and the payoff function of player i is described by the quadratic functional

$$\mathcal{J}_i(U_1, U_2, t_0, x_0) = x'(\vartheta) C_i x(\vartheta) + \int_{t_0}^{\vartheta} \left\{ u_1'[t] D_{i1} u_1[t] + u_2'[t] D_{i2} u_2[t] \right\} dt \quad (i = 1, 2),$$

---

[14]French "The end crowns the beginning."

where $x, u_i \in \mathbb{R}^n$. As before, the prime indicates transposition. The strategy set of player i has the form

$$\mathfrak{U}_i = \left\{ U_i \div u_i(t, x) | u_i(t, x) = Q_i(t)x \ \forall Q_i(\cdot) \in C^{n \times n}[0, \vartheta] \right\};$$

the game ends at a fixed time instant $\vartheta > t_0 \geq 0$; the symmetric constant matrices C_i and D_{ij} of dimensions $n \times n$ are given; the notation $D > 0 (< 0)$ means that a quadratic form $x'Dx$ is positive definite (negative definite, respectively); $\varepsilon \geq 0$ is a small scalar parameter. The players choose their strategies $U_i \div Q_i(t)x$, find the solution $x(t)$ of the system equation

$$\dot{x} = A(t)x + Q_1(t)x + \varepsilon Q_2(t)x, \quad x(t_0) = x_0,$$

construct the realizations $u_i[t] = Q_i(t)x(t)$ of the chosen strategies U_i and then calculate their payoffs $\mathcal{J}_i(U_1, U_2, t_0, x_0)$ using $x(t)$ and $u_i[t]$. In the noncooperative statement of the game Γ_2, the players have to answer two questions as follows.

1)   Which of the solution concepts (Nash or Berge equilibrium) should they adhere to?

2)   How can these equilibria be constructed?

The answer to the first question is provided by Table 2.8.1. Here NE and BE denote Nash and Berge equilibrium, respectively; $\exists$, $\nexists$, and $\forall$ are the existential, nonexistential, and universal quantifiers, respectively. Propositions 2.7.1 and 2.7.2, as well as Remarks 2.6.1 and 2.6.2, are combined in Table 2.8.1, which presents the coefficient criteria of choosing (or rejecting) Nash and/or Berge equilibrium in the game Γ_2.

For example, if $D_{12} < 0, D_{12} < 0, C_2 < 0$, then there exists a Berge equilibrium; if simultaneously $D_{22} > 0$, then there does not exist a Nash equilibrium (see columns 2 and 7 of Table 2.8.1).

The answer to the second question is based on Poincaré's theorem; see the beginning of Section 2.7. More specifically, we have to consider not only the null term ($\varepsilon = 0$) of the matrix expansion

$$\Theta(t, \varepsilon) = \Theta^{(0)}(t) + \sum_{m=1}^{\infty} \varepsilon^m \Theta^{(m)}(t),$$

but also the subsequent ones $\Theta^{(1)}(t), \Theta^{(2)}(t), \ldots$. This approach will be illustrated by an example of Berge equilibrium design for the game Γ_2: we will find the solution of (2.6.13) and then construct the strategies (2.6.14) and the corresponding payoffs (2.6.15).

Table 2.8.1 Coefficient criteria of equilibrium

	D_{11}	D_{12}	D_{21}	D_{22}	C_1	C_2	NE	BE
1	$D_{11} < 0$	$\forall$	$\forall$	$D_{22} < 0$	$C_1 < 0$	$\forall$	$\exists$	
2	$\forall$	$D_{12} < 0$	$D_{21} < 0$	$\forall$	$\forall$	$C_2 < 0$		$\exists$
3	$D_{11} < 0$	$D_{12} < 0$	$D_{21} < 0$	$D_{22} < 0$	$C_1 < 0$	$C_2 < 0$	$\exists$	$\exists$
4	$D_{11} > 0$	$\forall$	$\forall$	$\forall$	$\forall$	$\forall$	$\nexists$	
5	$\forall$	$D_{12} > 0$	$\forall$	$\forall$	$\forall$	$\forall$		$\nexists$
6	$\forall$	$\forall$	$D_{21} > 0$	$\forall$	$\forall$	$\forall$		$\nexists$
7	$\forall$	$\forall$	$\forall$	$D_{22} > 0$	$\forall$	$\forall$	$\nexists$	

In view of $\Theta_1^B(t, \varepsilon) = \Theta_1^{(0)}(t) + \varepsilon^1\Theta_1^{(1)}(t) + \varepsilon^2\Theta_1^{(2)}(t) + \dots$ and (2.6.13), we have

$$\left(\dot\Theta_1^{(0)} + \varepsilon^1\dot\Theta_1^{(1)} + \varepsilon^2\dot\Theta_1^{(2)} + \dots\right) + \left(\Theta_1^{(0)} + \varepsilon^1\Theta_1^{(1)} + \varepsilon^2\Theta_1^{(2)} + \dots\right)$$
$$\left[A(t) - D_{21}^{-1}\left(\Theta_2^{(0)} + \varepsilon^1\Theta_2^{(1)} + \varepsilon^2\Theta_2^{(2)} + \dots\right)\right]$$
$$+ \left[A'(t) - \left(\Theta_2^{(0)} + \varepsilon^1\Theta_2^{(1)} + \varepsilon^2\Theta_2^{(2)} + \dots\right)D_{21}^{-1}\right]\left(\Theta_1^{(0)} + \varepsilon^1\Theta_1^{(1)} + \varepsilon^2\Theta_1^{(2)} + \dots\right)$$
$$+ \left(\Theta_2^{(0)} + \varepsilon^1\Theta_2^{(1)} + \varepsilon^2\Theta_2^{(2)} + \dots\right)D_{21}^{-1}D_{11}D_{21}^{-1}\left(\Theta_2^{(0)} + \varepsilon^1\Theta_2^{(1)} + \varepsilon^2\Theta_2^{(2)} + \dots\right)$$
$$- \varepsilon^2\left(\Theta_1^{(0)} + \varepsilon^1\Theta_1^{(1)} + \varepsilon^2\Theta_1^{(2)} + \dots\right)D_{12}^{-1}\left(\Theta_1^{(0)} + \varepsilon^1\Theta_1^{(1)} + \varepsilon^2\Theta_1^{(2)} + \dots\right) = O_{n\times n},$$
$$\left(\Theta_1^{(0)}(\vartheta) + \varepsilon^1\Theta_1^{(1)}(\vartheta) + \varepsilon^2\Theta_1^{(2)}(\vartheta) + \dots\right) = C_1,$$
$$\left(\dot\Theta_2^{(0)} + \varepsilon^1\dot\Theta_2^{(1)} + \varepsilon^2\dot\Theta_2^{(2)} + \dots\right) + \left(\Theta_2^{(0)} + \varepsilon^1\Theta_2^{(1)} + \varepsilon^2\Theta_2^{(2)} + \dots\right)A(t)$$
$$+ A'(t)\left(\Theta_2^{(0)} + \varepsilon^1\Theta_2^{(1)} + \varepsilon^2\Theta_2^{(2)} + \dots\right)$$
$$- \left(\Theta_2^{(0)} + \varepsilon^1\Theta_2^{(1)} + \varepsilon^2\Theta_2^{(2)} + \dots\right)D_{21}\left(\Theta_2^{(0)} + \varepsilon^1\Theta_2^{(1)} + \varepsilon^2\Theta_2^{(2)} + \dots\right)$$
$$+ \varepsilon^2\left[-\left(\Theta_2^{(0)} + \varepsilon^1\Theta_2^{(1)} + \varepsilon^2\Theta_2^{(2)} + \dots\right)D_{12}^{-1}\left(\Theta_1^{(0)} + \varepsilon^1\Theta_1^{(1)} + \varepsilon^2\Theta_1^{(2)} + \dots\right)\right.$$
$$- \left(\Theta_1^{(0)} + \varepsilon^1\Theta_1^{(1)} + \varepsilon^2\Theta_1^{(2)} + \dots\right)D_{12}^{-1}\left(\Theta_2^{(0)} + \varepsilon^1\Theta_2^{(1)} + \varepsilon^2\Theta_2^{(2)} + \dots\right)$$
$$+ \left.\left(\Theta_1^{(0)} + \varepsilon^1\Theta_1^{(1)} + \varepsilon^2\Theta_1^{(2)} + \dots\right)D_{12}^{-1}D_{22}D_{12}^{-1}\left(\Theta_1^{(0)} + \varepsilon^1\Theta_1^{(1)} + \varepsilon^2\Theta_1^{(2)} + \dots\right)\right]$$
$$= O_{n\times n},$$
$$\left(\Theta_2^{(0)}(\vartheta) + \varepsilon^1\Theta_2^{(1)}(\vartheta) + \varepsilon^2\Theta_2^{(2)}(\vartheta) + \dots\right) = C_2. \tag{2.8.1}$$

According to the proof of Proposition 2.7.2, the null approximations $\Theta_i^{(0)}(t)$ $(i = 1, 2)$ satisfy the system (2.7.8) and have the explicit forms (2.7.9) and (2.7.10), respectively, where $\tilde\Theta_i^{(0)}(t) = \Theta_i^{(0)}(t) \; \forall[0, \vartheta]$, $(i = 1, 2)$. Equalizing the terms with the factor ε in (2.8.1), we obtain the following system of two matrix linear homogeneous differential equations with time-continuous coefficients to calculate the first approximations:

$$\begin{cases}
\dot\Theta_1^{(1)} + \Theta_1^{(1)}\left[A(t) - D_{21}^{-1}\Theta_2^{(0)}(t)\right] + \left[A'(t) - \Theta_2^{(0)}(t)D_{21}^{-1}\right]\Theta_1^{(1)} \\
\quad + \Theta_2^{(1)}D_{21}^{-1}\left[D_{11}D_{21}^{-1}\Theta_2^{(0)}(t) - \Theta_1^{(0)}(t)\right] + \left[\Theta_2^{(0)}(t)D_{21}^{-1}D_{11} - \Theta_1^{(0)}(t)\right]D_{21}^{-1}\Theta_2^{(1)} \\
\quad = O_{n\times n}, \quad \Theta_1^{(1)}(\vartheta) = O_{n\times n}, \\
\dot\Theta_2^{(1)} + \Theta_2^{(1)}\left[A(t) - D_{21}^{-1}\Theta_2^{(0)}(t)\right] + \left[A'(t) - \Theta_2^{(0)}(t)D_{21}^{-1}\right]\Theta_2^{(1)} = O_{n\times n}, \\
\Theta_2^{(1)}(\vartheta) = O_{n\times n}.
\end{cases}$$

Obviously, it has the trivial solution

$$\Theta_1^{(1)}(t) = \Theta_2^{(1)}(t) = O_{n\times n} \quad \forall t \in [0, \vartheta]. \tag{2.8.2}$$

Now, equalizing the terms with the factor ε^2 in (2.8.1) and using (2.8.2), we derive the following system of two matrix linear inhomogeneous differential equations with time-continuous coefficients to calculate the second approximations $\Theta_1^{(2)}(t)$ and $\Theta_2^{(2)}(t)$:

$$\dot{\Theta}_1^{(2)} + \Theta_1^{(2)} \left[A(t) - D_{21}^{-1} \Theta_2^{(0)}(t) \right] + \left[A'(t) - \Theta_2^{(0)}(t) D_{21}^{-1} \right] \Theta_1^{(2)}$$
$$+ \Theta_2^{(2)} D_{21}^{-1} \left[D_{11} D_{21}^{-1} \Theta_2^{(0)}(t) - \Theta_1^{(0)}(t) \right] + \left[\Theta_2^{(0)}(t) D_{21}^{-1} D_{11} - \Theta_1^{(0)}(t) \right] D_{21}^{-1} \Theta_2^{(2)}$$
$$- \Theta_1^{(0)}(t) D_{12}^{-1} \Theta_1^{(0)}(t) = O_{n\times n},$$
$$\Theta_1^{(2)}(\vartheta) = O_{n\times n}, \tag{2.8.3}$$

$$\dot{\Theta}_2^{(2)} + \Theta_2^{(2)} \left[A(t) - D_{21}^{-1} \Theta_2^{(0)}(t) \right] + \left[A'(t) - \Theta_2^{(0)}(t) D_{21}^{-1} \right] \Theta_2^{(2)}$$
$$+ \Theta_1^{(0)}(t) D_{12}^{-1} D_{22} D_{12}^{-1} \Theta_1^{(0)}(t) - \Theta_2^{(0)}(t) D_{12}^{-1} \Theta_1^{(0)}(t) - \Theta_1^{(0)}(t) D_{12}^{-1} \Theta_2^{(0)}(t) = O_{n\times n},$$
$$\Theta_2^{(2)}(\vartheta) = O_{n\times n}. \tag{2.8.4}$$

We find the explicit-form solution of (2.8.3)–(2.8.4). First, using Proposition 2.4.2 we construct the solution $\Theta_2^{(2)}(t)$ of the second matrix equation from (2.8.3)–(2.8.4). For this purpose, we write the fundamental system of solutions $Y(t)$ for the vector differential equation ($y \in \mathbb{R}^n$)

$$\dot{y} = \left[A(t) - D_{21}^{-1} \Theta_2^{(0)}(t) \right] y, \quad Y(\vartheta) = E_n.$$

According to Proposition 2.4.2, the solution of (2.8.4) takes the form

$$\Theta_2^{(2)}(t) = \left[Y^{-1} \right]' \left\{ \int_t^{\vartheta} Y'(\tau) L(\tau) Y(\tau) d\tau \right\} Y^{-1}(t),$$

where

$$L(t) = \Theta_1^{(0)}(t) D_{12}^{-1} D_{22} D_{12}^{-1} \Theta_1^{(0)}(t) - \Theta_2^{(0)}(t) D_{12}^{-1} \Theta_1^{(0)}(t) - \Theta_1^{(0)}(t) D_{12}^{-1} \Theta_2^{(0)}(t).$$

Substituting $\Theta_2^{(2)} = \Theta_2^{(2)}(t)$ into (2.8.3), we obtain a matrix linear inhomogeneous differential equation with the null boundary-value condition. Its explicit solution $\Theta_1^{(2)}(t)$, like the solution of the second equation from (2.8.3), is found using Proposition 2.4.2. Finally, with the resulting approximations $\Theta_i^{(j)}(t)$ ($j = 0, 1, 2$; $i = 1, 2$), (2.4.14) and (2.4.15), the Berge equilibrium in the game Γ_2 can be written as

$$U^B = \left(U_1^B, U_2^B \right) \div \left(-D_{21}^{-1} \left[\Theta_2^{(0)}(t) + \varepsilon^2 \Theta_2^{(2)}(t) \right] x, -\varepsilon D_{12} \left[\Theta_1^{(0)}(t) + \varepsilon^2 \Theta_1^{(2)}(t) \right] x \right).$$

(The accuracy is up to the second approximation.) The corresponding payoffs of the players are given by

$$\mathcal{J}_1 \left(U^B, t_0, x_0 \right) = x_0' \left[\Theta_1^{(0)}(t_0) + \varepsilon^2 \Theta_1^{(2)}(t_0) \right] x_0,$$
$$\mathcal{J}_2 \left(U^B, t_0, x_0 \right) = x_0' \left[\Theta_2^{(0)}(t_0) + \varepsilon^2 \Theta_2^{(2)}(t_0) \right] x_0.$$

Concluding this section, we suggest that the solution of any game (in particular, Γ_2) should be described by a pair

$$\left(U^S = \left(U_1^S, U_2^S \right), \mathcal{J}^S = \left(\mathcal{J}_1 \left(U^S, t_0, x_0 \right), \mathcal{J}_2 \left(U^S, t_0, x_0 \right) \right) \right).$$

In this case, a strategy profile U^S determines the behavioral rules of the players, and $\mathcal{J}^S$ their payoffs gained.

Chapter 3

Berge equilibria in multistage games

> Celui qui croit pouvoir trouver en soi-même
> de quoi se passer de tout le monde se trompe
> fort; mais celui qui croit qu'on ne peut se
> passer de lui se trompe encore davantage.
> —La Rochefoucauld[1]

The single-criterion multistage problems considered in Chapter 1 are a special case, albeit often encountered in applications. As a rule, decision-making takes place under conflicting interests of several parties. More than 50 years ago, Richard Bellman called such problems N-person games, noting that processes of this kind were of undoubted interest, but unfortunately research in this field was shrouded in obscurity; see [14, p. 252]. Really, in the 1960–1970s, investigations of N-player games dried out, and only in the mid-1980s there was a surge in the practical development of game theory methods, especially in economics and business (micro and macroeconomics, finance, marketing, management, to name a few). Over the past two or three decades, the significance of game theory and also the interest of researchers in it have considerably grown in many branches of economic and social sciences [74]; moreover, one would hardly imagine modern economics without game theory [205, p. 81]. Although the number of monographs, papers, and textbooks on *the theory of N-player games* is now estimated at thousands, this theory has been given a proper assessment just recently. As it has been mentioned, in this field of research the first Nobel Prize in Economic Sciences was awarded in 1994 to J. Nash, J. Harshanyi, and R. Selten "for the pioneering analysis of equilibria in the theory of non-cooperative games." In the succeeding years, a series of Nobel Prizes in Economic Sciences were given to other prominent experts in game theory as follows: in 1996, to J. Mirrlees and W. Vickrey "for their fundamental contributions to the economic theory of incentives under asymmetric information"; in 2001, to G. Akerlof, M. Spence and J. Stiglitz "for their analyses of markets with asymmetric information"; in 2005, to R. Aumann and T. Schelling "for having enhanced our understanding of conflict and cooperation through game-theory analysis"; in 2007, to L. Hurwicz, E. Maskin and R. Myerson "for having laid the foundations of mechanism design theory"; in 2012, to A. Roth and L. Shapley "for the theory of stable allocations and the practice of market design"; in 2014, to J. Tirole "for his analysis of market power and regulation"; in 2016, to O. Hart and B. Holmström "for their contributions to contract theory." Among the prominent results listed, we highlight the research connected with international relations: T. Schelling received the Nobel Prize for a game-theoretic approach to the analysis of

---

[1] French "He who thinks he has the power to content the world greatly deceives himself, but he who thinks that the world cannot be content with him deceives himself yet more." François de La Rochefoucauld, (1613–1680), was a French classical writer; a quote from *Réflexions ou Sentences et Maximes morales* (1665).

various scenarios for the development of relations between the USSR and the USA; R. Aumann,[2] for studying the evolution of conflict over time (repeated games). Many well-known American experts in game theory (J. von Neumann, J. Nash, G. Dantzig, M. Drescher, R. Duncan Luce, H. Raiffa, A. Tucker, L. Shapley, M. Shubik and others) were employees of the RAND Corporation—a think tank created under auspices of the US Air Force in Santa Monica, California, for research on the use of intercontinental ballistic missiles. During World War II, J. von Neumann[3] and M. Flood used game theory methods to develop the optimal atomic bombing strategy for Japan.

In modern economics and business, game theory has a wide variety of applications. At the *macro level*, examples include decision-making on international trade, competition, taxation, or protectionist measures of countries (OPEC-like cartels), with further assessment of expected results or distribution of expected profits. At the *micro level*, as some applications, we mention determining the optimal advertising costs in a competitive market, organizing effective production or choosing the best behavior during tenders (auctions), making alliances of firms to implement joint projects, predicting the behavioral scenarios for competitors, finding mechanisms of inter-regional interactions and income distribution schemes, planning and forecasting the development of firms, determining pricing policies, as well as coordinating the interests and relations of partner companies (owners of assets, employers and workers, etc.). Several applications will be considered at the end of this chapter and also in the Appendix.

Well, Chapter 3 is about conflicts that haunt us throughout life, and also about the theoretical possibility to balance conflicts (at least, at mathematical level). A mathematical model of a multistage conflict is presented, and a new solution based on the concept of Berge equilibrium is suggested. A design procedure of Berge equilibrium using the method of dynamic programming is developed. An explicit form of Berge equilibrium in a series of economic applications is found.

3.1   MATHEMATICAL MODEL OF MULTISTAGE CONFLICT UNDER UNCERTAINTY

> Faber est suae quisque fortunae.[4]

In this section, a multistage noncooperative game representing a mathematical model of the conflict under study is formalized.

---

[2]Robert J. Aumann, (born on June 8, 1930, Frankfurt am Main, Germany), is an outstanding Israeli mathematician. He emigrated to the United States with his family in 1938. He graduated from the City College of New York (B.S., 1950) and the Massachusetts Institute of Technology (S.M., 1952; Ph.D., 1955). He worked as an adviser to the US Agency for Arms Control and Disarmament. In 1956, he moved to Israel, where he was a member of the mathematics faculty at the Hebrew University of Jerusalem until his retirement in 2000.

[3]John von Neumann, (1903–1957), was a Hungarian–German–American mathematician who made an invaluable contribution to functional analysis, set theory, computer science, and economics, and the founder of game theory. He earned a degree in chemical engineering (1925) from the Swiss Federal Institute in Zürich and a doctorate in mathematics (1926) from the University of Budapest. He worked as Privatdozent ("private lecturer") at the Universities of Berlin (1927–1929) and Hamburg (1929–1930). In 1930, he was invited to work at Princeton University, where he held a professorship until his death. He is in game theory community as the author of the fundamental paper *On the theory of strategic games* (1928) and the fundamental monograph *Game Theory and Economic Behavior* (1944) written in collaboration with economist O. Morgenstern.

[4]Latin "Each man is the maker of his own fortune." Attributed to Appius Claudius Caecus, (flourished late 4th century–early 3rd century B.C.), was an outstanding statesman, legal expert, and author of early Rome who was one of the first notable personalities in Roman history.

3.1.1 Preliminaries

Which mythical means were used by Pygmalion to revivify Galatea? We do not know the true answer, but Pygmalion surely was an operations researcher by vocation: since some time his creation had independent life. This idea underlines creative activities in any field, including mathematical modeling. To form an integral entity from a set of odd parts means "to revivify" it in an appropriate sense:

> "She has not yet been born:
> she is music and word,
> and therefore the un-torn,
> fabric of what is stirred."
>   (Mandelshtam)[5]

This section is devoted to the revivification of a conflict.

What is a conflict? Looking aside the common (somewhat criminal) meaning of this word, we will use the following notion from [248, p. 333]: *"conceptually a conflict is any phenomenon that can be considered in terms of its parties, their actions, the outcomes yielded by these actions as well as in terms of its parties interested in these outcomes, including the character of this interest."* As a matter of fact, game theory suggests mathematical models of optimal decision-making under conflict. The logical foundation of game theory is a formalization of three fundamental postulates, namely, the features of a conflict, decision-making rules, and the optimality of solutions. In this book, we are studying only "rigid" conflicts, in which each party is guided by his own motives according to his perception and hence pursues individual goals, *l'esprit les intérêst du clocher*.[6]

A branch of game theory dealing with such rigid conflicts is known as noncooperative games. The noncooperative games described in Chapter 2 possess a series of peculiarities. Let us illustrate them using two simple examples.

Example 3.1.1 Imagine several competing companies (firms) that supply the same product in the market. Product price (hence, the profit of each firm) depends on the total quantity of products launched by them in the market. The goal of each firm is to maximize its profit by choosing an appropriate quantity of supply.

Example 3.1.2 The economic potential of a separate country can be assessed by a special indicator—a function that depends on controllable factors (taxation, financial and economic policy, industrial and agricultural development, foreign supplies, investments, credits, etc.) and also on uncontrollable factors (climate change and environmental disasters, anthropogenic accidents, suddenly sparked wars, etc.). Each country seeks for achieving a maximal economic potential through a reasonable choice of the controllable factors with a proper consideration of the existing economic relations with other countries.

These examples elucidate well the character of noncooperative games.

The *differentia specifica*[7] of such games are the following.

---

[5]Osip E. Mandelshtam, (1891–1938), was a major Russian poet, prose writer, and literary essayist.

[6]French, meaning narrow-mindedness and a lack of understanding or even interest in the world beyond one's own town's boundaries.

[7]Latin, meaning a feature by which two subclasses of the same class of named objects can be distinguished.

First, the decision-making process involves several parties (decision-makers, e.g., sellers or governments), which are often called **players** in game theory. Note that *a priori* they are competitors: *quilibet (quisque) fortunae suae faber*.[8]

Second, each player has an individual goal (profit or economic potential maximization) and the goals are bound to each other: *tout s'enchaine, tout se lie dans ce monde*.[9] A dazzling success of one party may turn out a disaster for another party.

Third, each player uses his own tools for achieving his goal (for sellers, the quantity of products supplied; for a country, the controllable factors in Example 3.1.2); in game theory, the controllable factors of each player are called his *strategies*, and a specific strategy chosen by a player is his decision or action in a noncooperative game.

Let us pay attention to three important circumstances.

First, quantitative analysis in any field requires an appropriate mathematical model; this fully applies to noncooperative games. In the course of mathematical modeling, a researcher inevitably faces the risks of going deep into details ("not seeing the wood for the trees") and presenting the phenomenon under study in a rough outline ("throwing away the baby with the bathwater"). The mathematical model of a noncooperative game often includes the following elements:

- the set of players;
- for each player, the set of his strategies;
- for each player, a scalar functional defined on the set of players' strategies. The value of this functional is the degree to which a given player achieves his goal under given strategies. In game theory, the functional is called the ***payoff function*** (or utility function) of a given player.

Second, as it has been underlined in Chapter 1, "Everything flows, nothing stands still." Therefore, in some cases, the mathematical models of noncooperative games have to take into account the dynamics of conflict over time. The conflicts studied in this chapter are assumed to be multistage in the following sense: either they structurally consist of several steps (also called stages in game theory), or the control process itself is partitioned into a number of successive steps at which the corresponding discrete-time process implements successive transitions of the controlled system from one state to another.

Third, "many intricate phenomena become clear naturally if treated in terms of game theory." [41, p. 97]. Following these *ex cathedra*[10] pronouncements by Russian game theory master N. Vorobiev, we are employing the framework of noncooperative games in this book.

A series of conventional requirements have been established for a game-theoretic model (of course, including a sufficient adequacy to the conflict under consideration) as follows.

First, the model must incorporate all interested parties of the conflict (*players* in the terminology of game theory).

Second, the model must specify possible actions of all parties (the *strategies* of players).

---

[8]Latin "Every man is the artisan of his own fortune." An outstanding Russian satirist, M.E. Saltykov-Shchedrin, repeatedly used this phrase in *Letters to Auntie* II, *Motley Letters* V, and *The Well-Meant Speeches* VIII.

[9]French, meaning that all things in the world are interconnected.

[10]Latin "From the chair," used with regard to statements made by people in positions of authority.

Third, the model must describe the interests of all parties (for each player and each admissible collection of actions chosen by all players, the model must assign a value called the *payoff* of this player).

The main challenges of game theory [43] are:

1) the design of optimality principles;

2) the proof of existence of *optimal actions* for players;

3) the calculation of optimal actions.

Different game-theoretic concepts of optimality often reflect intuitive ideas of *profitability, stability, and equitability*, rarely with an appropriate axiomatic characterization. Therefore, in most cases the notion of optimality in game theory (an optimal solution of a game) is not unique, prior, or absolute.

We will make the *normative*[11] approach to noncooperative games the cornerstone of this book: it will be established which behavior of the players should be considered optimal (rational or reasonable).

Depending on the feasibility of joint actions among the players (coordination of their individual actions), the games are classified as noncooperative, cooperative, and coalitional.

In the *noncooperative statement* of a game (simply called a noncooperative game, see above), each player chooses his action (strategy) in order to achieve the best individual result for himself/herself/itself without any coordination with other players: *chacun pour soi, chacun chez soi*.[12]

The *cooperative statement* of a game (cooperative game) is opposite to the noncooperative one. Here all players jointly choose their strategies in a coordinate way and, in some cases, even share the results (their payoffs). *Alle für einen, einer für alle*.[13]

According to the *coalitional statement* of a game (coalitional game), all players are partitioned into pairwise disjoint groups (coalitions) so that the members of each coalition act cooperatively, whereas all coalitions play a noncooperative game with each other.

3.1.2 Elements of the mathematical model

Ad disputandum[14]

Consider several subsystems that are interconnected with each other. In economics, these can be industrial enterprises or sectors, countries, sellers in a market, productions of every sort and kind with the same type of products, and other economic systems (called firms in [124, p. 28]). In ecology, industrial enterprises with the same purification and treatment facilities, populations of different species (e.g., predators and preys) under competitive exploitation, epidemics propagation, and control. In the mechanics of controlled systems, a group of controlled objects (aircrafts, missiles) that seek for approaching each other or capturing an evader.

---

[11]There also exist other approaches to conflict analysis: *descriptive*, which is to find the resulting collections of players' actions (the so-called strategy profiles) in a given conflict; *constructive*, which is to implement the desired (e.g., optimal) strategies in a given conflict; *predictive*, which is to forecast the actual result (outcome) of a given conflict.

[12]French "Every man for himself, every man to himself."

[13]German "One for all and all for one."

[14]Latin "For discussion."

Each subsystem is controlled by a supervisor (further called player), who undertakes certain actions for achieving his goal based on available information. In social and economic systems, the role of players belongs to the general managers of industrial enterprises and business companies, the heads of states, sellers (suppliers), and buyers (customers). In mechanical control systems, the captains of ships or aircrafts and the chiefs of control centers.

Assume that, due to *a priori* conditions, the players have to follow the "Help yourself" slogan. This leads to the noncooperative statement of their interaction.

As an example, consider a simplified mathematical model of competition among N firms in a market.

Example 3.1.3 There are $N \geqslant 2$ competing firms (players) that supply an infinitely divisible good of the same type (flour, sugar, etc.) in a market. The cost of one good for firm i is $c_i > 0$, $i \in \mathbf{N} = \{1, ..., N\}$. Let the number of market participants be sufficiently small so that the prices for goods directly depend on the quantity supplied by each firm. More specifically, denote by K the total supply of goods in the market; then the price p of one good can be calculated as $p = \max\{a - Kb, 0\}$, where $a > 0$ gives the constant price of one good without any supply in the market and $b > 0$ is the coefficient of elasticity that characterizes the price drop in response to one good supply. Here a natural assumption is $c_i < a$, $i \in \mathbf{N}$, since otherwise the activity of firms makes no economic sense. In addition, the production capacities of the players are unlimited and they sell the goods at the price p.

Suppose the firms operate in stable (not extreme) conditions and hence their behavior is to increase profits. Denote by x_i the quantity of goods supplied by firm i ($i \in \mathbf{N}$). Then the total supply of goods in the market makes up

$$K = \sum_{i=1}^{N} x_i,$$

and, the profit of firm i is described by the function

$$f_i(x) = p x_i - c_i x_i \quad (i \in \mathbf{N}),$$

where (as before) p gives the price of one good.

Another reasonable hypothesis is $a - Kb > 0$, as otherwise $p = 0$ and production yields no benefit for all firms (the profits become negative, $f_i(x) = -c_i x_i < 0$, $i \in \mathbf{N}$). In this case, the function

$$f_i(x) = \left[a - b \sum_{k \in \mathbf{N}} x_k \right] x_i - c_i x_i \quad (i \in \mathbf{N})$$

is the profit of firm i.

Therefore, in Example 3.1.3 the players are the competing firms and the action (strategy) of each player $i \in \mathbf{N}$ consists in choosing the quantity $x_i \in X_i = [0, +\infty)$ of its goods supplied in the market. Making its choice, each player i seeks to maximize its profit $f_i(x)$ (payoff) given the supplied quantities $x = (x_1, ..., x_N)$ of all players.

Idée générale

Now, we clearly outline the framework of the noncooperative game studied in this chapter.

To do it, let us answer Quintilian's questions *"Quis? Quid? Ubi? Quibus auxiliis? Cur? Quomodo? Quando?"*[15]

Prior to giving detailed answers, we discuss a major aspect—time.

Time

In Chapter 3 of this book, the operation of a controlled discrete system (or several interconnected systems of this type) at discrete time instants $0, 1, 2, \ldots, K$, where K is a finite number, is considered. Denote by $\mathbf{K} = \{0, 1, 2, \ldots, K\}$ the set of all time instants.

Throughout this book, we use the term "multistage" subject to different elements of the theory, meaning that either a conflict has a multistage structure, or the control process itself is partitioned into a number of successive stages that generally correspond to different time instants. Thus, a discrete control process implementing successive transitions of a controlled discrete system or a set of such systems from one state to another is called a *multistage* process. In many applications, the multistage character of models naturally follows from the essence of processes under study (for example, the problem to determine the optimal dimensions of different stages of a multistage rocket, or the problem to find a most cost-effective flight mode of an aircraft). At the same time, the multistage character can be artificially introduced into a model in order to apply the method of dynamic programming. As another example, consider the problem on the maximum product of nonnegative numbers whose sum does not exceed a given value $a > 0$; in this problem, steps are introduced [19, pp. 61–63] in an artificial way, describing transition from the product of k numbers to that of $k + 1$ ones. The problems with K steps are also called K-step problems; e.g., two-step in the case of $K = 2$. Time sampling is also motivated by the use of difference approximation schemes for solving continuous problems on computers.

Here are the answers to Quintilian's seven questions.

Quis? (Who?)

In fact, a leading part in noncooperative games is assigned to the **players**. As mentioned earlier, players can be the general managers of industrial enterprises and business companies, the heads of states, sellers (suppliers) and buyers (customers), the captains of ships or aircrafts, and so on, i.e., those who have right or authority to make decisions, give instructions, and control their implementation (interestingly, some people considering themselves to be (fairly!) serious strongly object to such a game-theoretic interpretation of their activity). Each player has a corresponding number: $1, 2, \ldots, i, \ldots, N$. Denote by $\mathbf{N} = \{1, 2, \ldots, N\}$ the set of all players and *let the set* $\mathbf{N}$ *be finite*. Note that games with an infinite number of players (called nonatomic games) are also studied in game theory [223]. Players may form groups, i.e., coalitions $\mathbf{K} \subseteq \mathbf{N}$. *A coalition is any subset* $\mathbf{K} = \{i_1, \ldots, i_k\}$ *of the player set* $\mathbf{N}$. In particular, possible coalitions are singletons (the noncoalitional statement of the game) and the entire set $\mathbf{N}$ (the cooperative statement of the game). A partition of the set $\mathbf{N}$ into pairwise disjoint subsets forming $\mathbf{N}$ in union is a *coalitional structure* of the game:

$$\mathcal{P} = \left\{ \mathbf{K}_1, \mathbf{K}_2, \ldots, \mathbf{K}_l \mid \mathbf{K}_i \bigcap \mathbf{K}_j = \varnothing \ (i, j = 1, \ldots, l; \ i \neq j), \ \bigcup_{i=1}^{l} \mathbf{K}_i = \mathbf{N} \right\}.$$

---

[15] Latin "Who? What? Where? Who helped? Why? How? When?"; a well-known system of seven questions for crime investigation suggested by Roman rhetorician Quintilian, Latin in full Marcus Fabius Quintilianus, (appr. 35–100 A.D.).

For example, in the noncooperative three-player game ($\mathbf{N} = \{1, 2, 3\}$), there exist five possible coalitional structures given by $\mathcal{P}_1 = \{\{1\}, \{2\}, \{3\}\}$, $\mathcal{P}_2 = \{\{1, 2\}, \{3\}\}$, $\mathcal{P}_3 = \{\{1\}, \{2, 3\}\}$, $\mathcal{P}_4 = \{\{1, 3\}, \{2\}\}$, $\mathcal{P}_5 = \{\{1, 2, 3\}\}$.

For compact notation, we will sometimes consider the two-player games only, letting $\mathbf{N} = \{1, 2\}$.

In Example 3.1.3, the players are the general managers of competing firms.

Controlled system

An important class of multistage systems consists of discrete controlled systems, hereinafter denoted by $\overline{\Sigma}$. The state of $\overline{\Sigma}$ depends on control subjects, which have been called players in the previous section. As $\overline{\Sigma}$ the controlled system itself is often comprehended. In what follows, we consider the systems $\overline{\Sigma}$ possibly containing other controlled systems as their subsystems, which generally compete and affect each other. This can be a market of sellers or buyers, a set of productions, ministries or countries, a group of satellites, airplanes, or ships.

A system $\overline{\Sigma}$ evolves in the course of time, which is its distinctive feature: *Tempora mutantur et nos mutamur in illis*.[16] Further analysis will be confined to the case in which the dynamics of $\overline{\Sigma}$ over time are described by a system of difference equations, i.e., equations incorporating finite differences of desired functions. These equations have a deep analogy [52] with ordinary differential equations, and the role of the differentiation operator is played by the difference operator $\Delta x(k) = x(k + 1) - x(k)$.

At each time instant $k \in \mathbf{K}$, the state of a controlled system $\overline{\Sigma}$ is modeled by the n-dimensional column vector $x(k) = (x_1(k), \ldots, x_n(k))$. Generally, this vector is composed of parameter values that characterize the current *state* of $\overline{\Sigma}$ at a time instant k. For a group of enterprises $i \in \mathbf{N}$, possible examples of parameters include the production output of each enterprise i, quantitative assessments of its labor and natural resources, assets, inventory, etc. For a group of moving objects $i \in \mathbf{N}$, the coordinates and velocities of each object i.

Formally speaking, the state of a controlled system at a time instant k is not identical with the value of the vector $x(k)$ that characterizes this state. (Like a citizen is not identical with his passport data.) In view of this fact, by "the state $x(k)$" we will mean the state of a controlled system corresponding to the value $x(k)$ of the state variable x.

Let the evolution of a controlled system $\overline{\Sigma}$ be described by a vector difference equation

$$x(k + 1) = f(k, x(k), u_1, \ldots, u_N), \qquad x(t_0) = x_0, \tag{3.1.1}$$

where the components of the vector function $f(k, x(k), u_1, \ldots, u_N)$ are assumed to be continuous. Using equation (3.1.1) and the values of the vectors $x(k)$ and $u_i = u_i[k] \in \mathbf{R}^{m_i}$ at time instant k, we determine the state of $\overline{\Sigma}$ at time instant $k + 1$. The initial position (x_0, t_0) is such that $t_0 \in \{0, \ldots, K - 1\}$ and $x_0 \in \mathbf{R}^n$.

In the theory of differential games, the vector $x(k)$ is known as the phase or state vector of a system $\overline{\Sigma}$ at a time instant k. Following the terminology of differential games, a pair (k, x), $k \in \mathbf{K}$, will be called a *position* of a corresponding game, and a pair $(k, x(k))$ a *current position* at a time instant k. Then $(t_0, x(t_0))$ gives an *initial position*, i.e., the position at the beginning of the game—the time instant t_0.

---

[16]Latin "Times are changed, we also are changed with them." This phrase is often mis-attributed to Ovid, Latin in full Publius Ovidius Naso, (43 B.C.–17 A.D.), Roman poet. Actually, it is a variant of his phrases that appeared in the 16th century Germany.

Quid? (What?)

Each player chooses and then uses his *strategy. A strategy is comprehended as a rule that associates each state of the player's awareness with a certain action (behavior) from a set of admissible actions (behaviors) given this awareness.* For the head of a state, this is a direction of strategic development. In a sector composed of several industrial enterprises, a strategy of each general manager is the output of his enterprise, the price of products, the amount of raw materials and equipment purchased, supply contracts, investments, innovations and implementation of new technologies, wages reallocation, penalties, bonuses and other incentive and punishment mechanisms. For a seller, a strategy is the price of one good; for the captain of a ship, own course (rudder angle, the direction and magnitude of reactive force).

Thus, the action of each player consists in choosing and using his individual strategy, which gives an answer to the question *Quid?* Speaking formally, a strategy of player i in the game Γ_3 is x_i, and the strategy set of this player (the set of all strategies) is denoted by X_i.

In the mathematical model studied in this chapter, the actions of player i ($i \in \mathbf{N}$) will consist in an appropriate choice of the parameter $u_i \in \mathbf{R}^{m_i}$. Thus, **the control action** $u_i \in \mathbf{R}^{m_i}$ of player i on the controlled system $\overline{\Sigma}$ is formed by player i.

A rigorous formalization of strategies should also contain the description of information available to each player at each time instant $k = 0, 1, \ldots, K - 1$. In the theory of differential games, such information often includes the state vector $x(k)$ realized at a current time instant k. Such an approach identifies a strategy with a function that depends the position (k, x) of a game for any $k = 0, 1, \ldots, K - 1$ and $x \in \mathbf{R}^n$. (In the terminology of control theory, this type of strategy is implemented using *the feedback principle* and called *positional*).

Now, let with each player $i \in \mathbf{N}$ and each position $(k, x) \in \{0, 1, \ldots, K - 1\} \times \mathbf{R}^n$ some subset $\mathcal{U}_i(k, x) \subseteq \mathbf{R}^{m_i}$ be associated. Following [179], the strategy of player i will be conceptually formalized in the following way.

Definition 3.1.1 *a) A positional strategy $U_i(k)$ of player i at a time instant $k \in \{0, 1, \ldots, K - 1\}$ is identified with an m_i-dimensional vector function $u_i(k, x)$ such that*

$$u_i(k, x) \in \mathcal{U}_i(k, x) \quad \forall x \in \mathbf{R}^n;$$

the fact of identification will be denoted by $U_i(k) \div u_i(k, x)$, and the sets of strategies of player i at a time instant k by $\mathfrak{A}_i(k)$.

b) A positional strategy of player i is an ordered collection

$$U_i = (U_i(0), U_i(1), \ldots, U_i(K - 1)) \in \prod_{k=0}^{K-1} \mathfrak{A}_i(k) = \mathfrak{A}_i;$$

c) a strategy profile U is an ordered collection of strategies of all players, i.e.,

$$U = (U_1, \ldots, U_N) \in \prod_{i \in \mathbf{N}} \mathfrak{A}_i = \mathfrak{A}.$$

Remark 3.1.1 If a game has a given initial position (t_0, x_0) with an initial time instant $t_0 \in \{0, 1, \ldots, K - 1\}$, then in Definition 3.1.1, item b), 0 should be replaced by t_0, which gives

$$U_i = (U_i(t_0), U_i(t_0 + 1), \ldots, U_i(K - 1)) \in \mathfrak{A}_i = \prod_{k=t_0}^{K-1} \mathfrak{A}_i(k).$$

Ubi? (Where?)

Here the answer is short: in the conflict, more precisely put, in its mathematical model described by the noncooperative game. In Example 3.1.3, this is the market of goods.

Quibus auxiliis? Quomodo? (Who helped? How?)

Actually, the players affect the conflict using their positional strategies, which is the answer to both questions.

Now, let an initial position (t_0, x_0), where $t_0 \in \{0, 1, \ldots, K-1\}$ and $x_0 \in \mathbf{R}^n$, be given or chosen somehow. *A play of the multistage game* (its evolution over time) is organized as follows. At the initial time instant t_0, player i knows the value sets $\mathcal{U}_i(t_0, x) \subseteq \mathbf{R}^{m_i}$ of the control action u_i of player i.

Next, each player i ($i \in \mathbf{N}$) chooses and adopts a specific strategy $U_i(t_0) \div u_i(t_0, x) \in \mathcal{U}_i(t_0, x)$, $U_i(t_0) \in \mathfrak{A}_i(t_0)$, seeking to maximize his payoff function; see formula (3.1.4) below. This forms a strategy profile of the game:

$$U(t_0) = (U_1(t_0), U_2(t_0), \ldots, U_N(t_0)) \div u(t_0, x)$$
$$= (u_1(t_0, x), u_2(t_0, x), \ldots, u_N(t_0, x)) \quad \forall x \in \mathbf{R}^n.$$

The state vector

$$x(t_0 + 1) = f(t_0, x_0, u(t_0, x_0))$$

at the time instant $t_0 + 1$ is calculated using the strategy profile $U(t_0) \div u(t_0, x)$ and the system (3.1.1) with $k = t_0$. As a result, the position $(t_0 + 1, x(t_0 + 1))$ is realized. At the time instant $k = t_0 + 1$, all players again choose and adopt their strategies $U_i(t_0 + 1) \div u_i(t_0 + 1, x)$, $U_i(t_0 + 1) \in \mathfrak{A}_i(t_0 + 1)$, $i \in \mathbf{N}$, thereby forming a strategy profile

$$U(t_0 + 1) = (U_1(t_0 + 1), U_2(t_0 + 1), \ldots, U_N(t_0 + 1)) \div u(t_0 + 1, x)$$
$$= (u_1(t_0 + 1, x), u_2(t_0 + 1, x), \ldots, u_N(t_0 + 1, x)).$$

And the state vector

$$x(t_0 + 2) = f(t_0 + 1, x(t_0 + 1), u(t_0 + 1, x(t_0 + 1)))$$

at the time instant t_0+2 is calculated using the strategy profile $U(t_0+1)$, the system (3.1.1) with $k = t_0 + 1$, and the previous state vector $x(t_0 + 1)$.

This procedure is repeated for all other time instants by analogy.

For the time instant $k = K - 1$, the strategy profile

$$U(K - 1) = (U_1(K - 1), \ldots, U_N(K - 1)) \div u(K - 1, x)$$
$$= (u_1(K - 1, x), \ldots, u_N(K - 1, x)),$$
$$U_i(K - 1) \in \mathfrak{A}_i(K - 1) \quad (i \in \mathbf{N}),$$

chosen by the players at the time instant $k = K - 1$, the system (3.1.1) with $k = K - 1$, and the previous state vector $x(K - 1)$ finally lead to the state vector

$$x(K) = f(K - 1, x(K - 1), u(K - 1, x(K - 1))).$$

Consequently, we obtain the discrete trajectory

$$x_0, \ x(t_0 + 1), \ \ldots, x(K) \tag{3.1.2}$$

of the difference system (3.1.1) that is induced step-by-step by the strategy profile

$$U \in \mathfrak{A}, \quad U = (U_1, \ldots, U_N), \quad U_i = (U_i(t_0), \ldots, U_i(K-1)),$$
$$U_i(k) \div u_i(k, x), \quad U_i(k) \in \mathfrak{A}_i(k).$$

The realizations of the strategies adopted by the players on the trajectory (3.1.2) are

$$u_i[t_0] = u_i(t_0, x_0), u_i[t_0 + 1] = u_i(t_0 + 1, x(t_0 + 1)), \ldots, u_i[K-1]$$
$$= u_i(K-1, x(K-1)) \quad (i \in \mathbf{N}), \tag{3.1.3}$$
$$u[k] = (u_1[k], \ldots, u_N[k]) \quad (k = t_0, t_0 + 1, \ldots, K-1).$$

In Example 3.1.3, the firms choose the quantities of their goods supplied in the market as their strategies. The resulting situation in the market is the strategy profile in the corresponding noncooperative game.

Cur? (Why?)

The answer is: in order to assess the performance of each player. The noncooperative game (the mathematical model of a conflict accepted in our book) incorporates the *payoff function* of player i ($i \in \mathbf{N}$). The value of this function (called *payoff* or *outcome* in game theory) is a numerical assessment of the desired performance. In Example 3.1.3, the payoff function of player i has the form

$$f_i(x, y) = \left[a - b \sum_{k \in \mathbf{N}} x_k \right] x_i - c_i x_i.$$

It characterizes the profit of firm i in the single-step game. The following circumstances should be taken into account while assessing the performance of each player in a noncooperative game.

First, the design of payoff functions (performance assessment criteria) is rather difficult and often suffers from subjectivism: "*Nous ne désirerions guère de choses avec ardeur, si nous connaissions parfaitement ce que nous désirons.*"[17] [187, p. 55].

Sometimes, the goal consists in higher profits or lower cost; in other cases, in smaller environmental impact. There may exist other goals too. As a rule, in a noncooperative game these criteria represent scalar functions defined on the set of all admissible strategy profiles. For the sake of definiteness, assume that each player seeks to *increase* his payoff function as much as possible.

Second, according to the noncoalitional statement of the game, the players act in an isolated way and do not form coalitions. Being guided by the *Suum cuique* slogan,[18] each player chooses his strategy by maximizing his own payoffs.

As a result, each player endeavors to implement his cherished dream: "*Chacun produit selon ses facultés et recoit selon ses besoins.*"[19]

Third, the decision-making process in the noncooperative game is organized as follows. Each player chooses and then uses his strategy, which yields a strategy profile of the game. The payoff function of each player is defined on the set of all admissible strategy profiles. The value of this function (**payoff**) is a numerical assessment for the player's performance.

Now, we present the dynamic (multistage) conflict considered in this subsection.

---

[17] French "We should earnestly desire but few things if we clearly knew what we desired." A quote from *Réflexions ou Sentences et Maximes morales* by F. de La Rochefoucauld.

[18] Latin "To each his own," or "May all get their due"; also, see the epigraph to Subsection 2.2.1.

[19] French "From each according to his ability, to each according to his needs."

Depending on the goals of players, there are two classes of dynamic games [1, p. 23]— *games of kind* and *games of degree*. The former games have a finite number of outcomes (in a special case, just two outcomes). For example, in pursuit–evasion games with many pursuers, only two outcomes are possible: either the evader has successfully escaped, or has been captured by at least one pursuer. The latter games (of degree) have a continuum of possible outcomes. In such games, the performance of player i is formalized using the sequences (3.1.2) and (3.1.3) and *the payoff function of player i* is generally described by

$$\mathcal{J}_i(U_1, \ldots, U_N, t_0, x_0) = \mathcal{J}_i(U, t_0, x_0) = \Phi_i(x(K)) + \sum_{k=t_0}^{K-1} F_i(k, x(k), u[k]) \quad (i \in \mathbf{N}),$$

(3.1.4)

where the scalar function $\Phi_i(x)$ and $F_i(k, x, u)$ are assumed continuous. In the theory of differential games, *the term* $\Phi_i(x(K))$ is called *terminal*, and *the term* $\sum_{k=t_0}^{K} F_i(k, x(k), u_1[k], \ldots, u_N[k])$ *integral*.

Interestingly, for $N = 1$ ($\mathbf{N} = \{1\}$), the expressions (3.1.4) and (3.1.1) with $U_1 \in \mathfrak{A}_1$ lead to the well-known Boltz optimization problem of variational calculus in the multistage statement with positional control. If the game involves N persons, we use the term "N-player games."

At conceptual level, during the decision-making process in the noncoalitional game player i chooses his strategy $U_i \in U_i$ so that:

1) this choice occurs simultaneously for all N players;

2) no agreements or coalitions among the players are allowed and they do not exchange information *during the game*.

Quando? (When?)
The answer to the last question of Quantilian's system is shortest: at the time of decision-making in the conflict (within its mathematical model—the noncooperative game) through an appropriate choice of strategies by the players.

In principle, a conflict can be treated as a certain controlled system, the "black box" in which the players input their strategies and receive their payoffs at the output. This is a standard approach to "instantaneous, single-step, static" noncooperative games in general game theory [42]. However, in most applications (particularly, in economics and the mechanics of controlled systems), a controlled system itself undergoes some changes with the course of time, and the players are able to vary their strategies during the whole conflict. The games whose state evolves in time are called *dynamic*.

3.1.3 Multistage noncooperative game

An ordered quadruple

$$\langle \mathbf{N}, \Sigma \div (3.1.1), \{\mathfrak{A}_i\}_{i \in \mathbf{N}}, \{\mathcal{J}_i(U_1, \ldots, U_N, t_0, x_0) \div (3.1.4)\}_{i \in \mathbf{N}} \rangle \qquad (3.1.5)$$

forms *a multistage positional noncooperative N-player game* with an initial position (t_0, x_0). In the game (3.1.5), the notations are the following:

$\mathbf{N} = \{1, 2, \ldots, N\}$ as the set of players;

Σ as a controlled discrete system that evolves at discrete time instants $k = 0, 1, \ldots, K - 1$ in accordance with the vector difference equation (3.1.1) with the initial condition $x(t_0) = x_0$; the entire operation of the system $\sum$ from t_0 to K is partitioned into intervals $[k, k + 1)$, $t_0 \in \{0, 1, \ldots, K - 1\}$, and hence the game is called multistage (or K-stage);

$\mathfrak{A}_i$ as the set of all positional strategies of player i given by

$$U_i = (U_i(t_0), U_i(t_0 + 1), \ldots, U_i(K - 1)), \quad U_i(k) \in \mathfrak{A}_i(k);$$

in Definition 2.1.1, a strategy of player i at a time instant k is identified with an m_i-dimensional vector function $U_i(k) \div u_i(k, x) \in \mathcal{U}_i(k, x) \ \forall x \in \mathbf{R}^n$ that depends on the position (k, x); this fact explains the terms "positional game" and "positional strategy" used below;

$\mathcal{J}_i(U_1, U_2, \ldots, U_N, t_0, x_0)$ as the payoff function of player i (see (3.1.4)), whose value (*payoff*) actually assesses the performance of player i in the game (3.1.5) (the greater the payoff is, so much the better for player i); the payoff function (3.1.5) is formed on the $N + 1$ sequences

$$\{x(k) \mid k = t_0, t_0 + 1, \ldots, K\}, \quad \{u_i[k] \mid k = t_0, \ldots, K - 1\} \quad (i \in \mathbf{N})$$

and is determined by the strategy profile $U = (U_1, U_2, \ldots, U_N)$ chosen by the players in the noncooperative game under consideration; see Section 3.2.

A generally accepted solution concept of noncooperative games is the so-called Nash equilibrium.[20] For the game (3.1.5), this concept is reduced to the following.

Definition 3.1.2 *A strategy profile* $U^e \in \mathfrak{A}$,

$$U^e = (U_1^e, \ldots, U_N^e), \quad U_i^e = (U_i^e(t_0), \ldots, U_i^e(K - 1)),$$

$$U_i^e(k) \div u_i^e(k, x) \in \mathcal{U}_i(k, x) \quad \forall x \in \mathbf{R}^n \quad (k = 0, 1, \ldots, K - 1) \quad (i \in \mathbf{N}),$$

is said to be a Nash equilibrium in game (3.1.5) *if for any initial position* (t_0, x_0), *where* $t_0 \in \{0, \ldots, K - 1\}$ *and* $x_0 \in \mathbf{R}^n$, *and each* $i \in \mathbf{N}$,

$$\max_{U_i \in \mathfrak{A}_i} \mathcal{J}_i(U^e \parallel U_i, t_0, x_0) = \mathcal{J}_i(U^e, t_0, x_0),$$

where $(U^e \parallel U_i) = (U_1^e, \ldots, U_{i-1}^e, U_i, U_{i+1}^e, \ldots, U_N^e)$. *The N-dimensional vector*

$$\mathcal{J}(U^e, t_0, x_0) = (\mathcal{J}_1(U^e, t_0, x_0), \ldots, \mathcal{J}_N(U^e, t_0, x_0))$$

is called the Nash equilibrium payoff in game (3.1.5) *with a fixed initial position* (t_0, x_0), *and the pair* $(U^e, \mathcal{J}(U^e, t_0, x_0))$ *is called the Nash equilibrium solution of the game* (3.1.5).

---

[20]John Forbes Nash, Jr., Born June 13, 1928, in Bluefield, West Virginia. He successfully graduated from the Carnegie Institute of Technology (now, Carnegie Mellon University) with bachelor's and master's degrees in mathematics. Richard Duffin, Nash's undergraduate advisor at the Carnegie Institute of Technology, gave him a brief characteristic, "He is a mathematical genius." In 1948, he started postgraduate study at Princeton University, where he was particularly influenced by International Economy, the faculty course of J. von Neumann, and by the famous book *Theory of Games and Economic Behavior* (1944), written by von Neumann together with O. Morgenstern. In 1949, he presented thesis on equilibrium solutions of noncooperative games; after 45 years—in 1994—he was awarded the Noble Prize in Economic Sciences for that research. From 1951 to 1959, he worked in Cambridge at Massachusetts Institute of Technology (MIT). In 1958, *Fortune* called Nash "America's brilliant young star of the 'new mathematics.'" In 1959, moved to California for RAND Corporation and became a leading expert of the Cold War. Since 1959, he suffered from a mental disorder (completely overcame the disease by 1980, to great astonishment of doctors). Since 1980 again worked at Princeton University as a consulting professor. He died in a car crash on May 24, 2015, at the age of 86. Throughout the world, Nash is well known by R. Howard's movie *A Beautiful Mind* (2001, featuring R. Crowe) based on S. Nasar's book *Beautiful Mind: The Life of Mathematical Genius and Nobel Laureate John Nash*.

Remark 3.1.2 Players are recommended to use a Nash equilibrium $(U^e, \mathcal{J}(U^e, t_0, x_0))$, because with their strategies from the profile $U^e = \left(U_1^e, \ldots, U_N^e \right)$ the players gain their payoffs $\mathcal{J}_i(U^e, t_0, x_0)$ $(i \in \mathbf{N})$. According to Definition 3.1.2, a Nash equilibrium U^e is invariant with respect to (does not depend on!) the choice of the initial position (t_0, x_0), whereas the Nash equilibrium payoffs $\mathcal{J}_i(U^e, t_0, x_0)$ $(i \in \mathbf{N})$ of the players generally differ for different initial positions (t_0, x_0).

Remark 3.1.3 The positive and negative properties of Nash equilibrium were discussed in detail in the book [121, pp. 169–183].

As for the negative properties, we note an important counterexample [156, p. 115] with coefficient constraints under which the linear-quadratic differential two-player game has no Nash equilibrium. In 1994, this fact actually motivated K.S. Vaisman and V.I. Zhukovskiy to formalize the concept of Berge equilibrium, replacing U^e and U_i^e in Definition 3.1.2 by U^B and U_i^B, respectively [24–32, 57, 375, 376].

Definition 3.1.3 *A strategy profile* $U^B \in \mathfrak{A}$,

$$U^B = \left(U_1^B, \ldots, U_N^B \right), \quad U_i^B = \left(U_i^B(t_0), \ldots, U_i^B(K-1) \right),$$

$$U_i^B(k) \div u_i^B(k, x) \in \mathcal{U}_i(k, x) \quad \forall x \in \mathbf{R}^n \quad (k = 0, 1, \ldots, K-1) \quad (i \in \mathbf{N}),$$

is said to be a Berge equilibrium in game (3.1.5) *if for any initial position* (t_0, x_0), *where* $t_0 \in \{0, \ldots, K-1\}$ *and* $x_0 \in \mathbf{R}^n$, $x_0 \neq 0_n$, *and each* $i \in \mathbf{N}$,

$$\max_{U \in \mathfrak{A}} \mathcal{J}_i(U \parallel U_i^B, t_0, x_0) = \mathcal{J}_i(U^B, t_0, x_0),$$

where $(U \parallel U_i^B) = \left(U_1, \ldots, U_{i-1}, U_i^B, U_{i+1}, \ldots, U_N \right)$. *The N-dimensional vector*

$$\mathcal{J}(U^B, t_0, x_0) = \left(\mathcal{J}_1(U^B, t_0, x_0), \ldots, \mathcal{J}_N(U^B, t_0, x_0) \right)$$

is called the Berge equilibrium payoff in game (3.1.5) *with a fixed initial position* (t_0, x_0), *and the pair* $(U^B, \mathcal{J}(U^B, t_0, x_0))$ *is called the Berge equilibrium solution of the game* (3.1.5).

Remark 3.1.4 Players are recommended to use a Berge equilibrium $(U^B, \mathcal{J}(U^B, t_0, x_0))$, because with their strategies from the profile $U^B = \left(U_1^B, \ldots, U_N^B \right)$ the players gain their payoffs $\mathcal{J}_i(U^B, t_0, x_0)$ $(i \in \mathbf{N})$. According to Definition 3.1.3, a Berge equilibrium U^B is invariant with respect to the choice of the initial position (t_0, x_0), whereas the Berge equilibrium payoffs $\mathcal{J}_i(U^B, t_0, x_0)$ $(i \in \mathbf{N})$ $(i \in \mathbf{N})$ of the players generally differ for different initial positions (t_0, x_0).

3.2   PRO ET CONTRA[21] OF BERGE EQUILIBRIUM

> Many intricate phenomena become clear naturally
> if treated in terms of game theory.
> —Vorobiev[22]

The properties making the concept of Berge equilibrium more and more widespread as well as the properties drawing just criticism against it are considered.

---

[21] Latin "For and against."

[22] Nikolay N. Vorobiev, (1925–1995), was a Russian mathematician, expert in the field of algebra, mathematical logic, and probability theory, as well as the founder of the largest national school of game theory. A quote from [40, p. 97]

3.2.1 Classical noncooperative game as single-stage statement of game (3.1.5)

By the number of steps, all games of the form (3.1.5) are divided into multistage and single-stage ones. (The terms "multistep" and "single-step" games are also conventional in the literature [73, p. 29].) The classical mathematical theory of noncooperative games [41] deals mainly with single-stage games. Note that any sequence of moves can be represented as an instantaneous choice of strategy. For example, a chess play of 40 moves can be described by a list of moves for each player (his/her strategy). However, if the game (3.1.5) is formally limited to only one step, then the resulting single-step noncooperative N-player game N will be completely determined by three components as follows:

– the set of players;
– the sets of players' strategies;
– the payoff functions of players. (The initial position is assumed to be "frozen.")

Thus, in the general theory of static games, a noncooperative N-player game is defined by an ordered triplet

$$\langle \mathbf{N}, \{\mathcal{U}_i\}_{i\in\mathbf{N}}, \{\mathcal{J}_i(u_1, \dots, u_N)\}_{i\in\mathbf{N}} \rangle, \tag{3.2.1}$$

with the following notations:

$\mathbf{N} = \{1, 2, \dots, N\}$ as the set of players;

$\mathcal{U}_i = \{u_i \mid u_i \in \mathcal{U}_i \subset \mathbf{R}^{m_i}\}$ as the strategy set of player i, where $\mathcal{U}_i$ is a given subset of $\mathbf{R}^{m_i}$ (a common approach is to consider compact sets $\mathcal{U}_i$ in the space $\mathbf{R}^{m_i}$); a strategy u_i of player i is identified with a point u_i from the set $\mathcal{U}_i$, and hence $u_i \in \mathcal{U}_i$, where u_i is an element of the strategy set $\mathcal{U}_i$ of player i;

$\mathcal{J}_i(u_1, \dots, u_N)$ as the payoff function of player i, which is defined on the Cartesian product $\mathcal{U} = \prod_{i\in\mathbf{N}} \mathcal{U}_i$ and is assumed to be continuous; then $\mathcal{U}$ is the set of all strategy profiles $u = (u_1, \dots, u_N)$ and each player seeks to maximize his payoff (the value of his payoff function) in current conditions.

Once again, we emphasize that in the game (3.2.1), the players choose their strategies simultaneously and no coalitions are allowed. (This explains well the term "noncooperative game.")

The concept of Berge equilibrium in the game (3.2.1) is implemented in the following way.

Definition 3.2.1 *A strategy profile* $u^B = \left(u_1^B, \dots, u_N^B\right) \in \mathcal{U}$ *is called a Berge equilibrium in game* (3.2.1) *if*

$$\mathcal{J}_i(u^B) = \max_{u\in\mathcal{U}} \mathcal{J}_i(u \| u_i^B) \quad \forall i \in \mathbf{N}, \tag{3.2.2}$$

where $(u \| u_i^B) = (u_1, \dots, u_{i-1}, u_i^B, u_{i+1}, \dots, u_N)$ *and* $\mathcal{U} = (\mathcal{U}_1 \times \dots \times \mathcal{U}_N)$.

The concept of Berge equilibrium situation is gradually becoming central to the mathematical theory of games. It was proposed at V.I. Zhukovskiy's seminar devoted to the critical analysis of C. Berge's book [260], or rather of its Russian translation appeared in 1961. Since then, this concept has become so widespread that many journals on game theory or operations research regularly publish papers on Berge equilibrium. A significant number of such research works are devoted to the existence of Berge

equilibria, to the methods for calculating the payoffs of different players in Berge equilibria (or their estimates), as well as to the methods for the exact and approximate design
of Berge equilibria. To find a Berge equilibrium in a game and the value of the payoff
functions in the Berge equilibrium means *to completely solve* the game.

For the first time, apparently, the formal concept of Berge equilibrium was suggested
by Zhukovskiy, Salukvadze and Vaisman in the paper [354] and the preprint [368]
published in 1994. The later intensive investigations in this field were mainly focused
on the properties of Berge equilibria, the structure of their sets, possible modifications
and generalizations of this concept and the explicit calculation of Berge equilibria for
different classes of games. Particular attention was paid to existence theorems; see brief
surveys in the book [91, pp. 207–209] and the paper [377]. In the end of the 20th century,
the study of noncooperative games with uncertain factors was initiated [91, 112, 123,
133, 134]. Interestingly, the payoffs and risks of players in quantitative terms were first
considered in the book [92].

From the entire gamut of all these questions, we single out:

- the positive and negative properties of Berge equilibria (Subsections 3.2.2 and 3.2.3,
  respectively);
- the sufficient conditions of existence (Theorems 3.4.1 and 3.4.2).

3.2.2   Positive properties of Berge equilibrium

In this subsection, the properties facilitating a widespread use of Berge equilibrium are
discussed.

Stability of Berge equilibrium

> [Paradoxes of the infinite arise] only when we attempt,
> with our finite minds, to discuss the infinite,
> assigning to it those properties
> which we give to the finite and limited.
> —Galileo[23]

Property 3.2.1 *A Berge equilibrium* $u^B = (u_1^B, ..., u_N^B)$ *of the noncooperative game* (3.2.1) *is
stable against the deviations of all players except for player i* ($\forall i \in \mathbf{N}$), *because for all* $u_j \in \mathcal{U}_j$
($j \in \mathbf{N} \backslash \{i\}$, $i \in \mathbf{N}$) *the payoff of each player i under any deviations of such a coalition of
the other* ($N - 1$) *players from the Berge equilibrium does not exceed his payoff in the
Berge equilibrium.* As a matter of fact, the concept of Berge equilibrium matches well
the slogan of the musketeers: "One for all and all for one."

Compactness of the set of Berge equilibria

> The notion of infinity is our greatest friend;
> it is also the greatest enemy of our peace in mind.
> —Pierpont[24]

---

[23]Galileo, in full Galileo Galilei, (1564–1642), was an Italian natural philosopher, astronomer, and mathematician who made fundamental contributions to the sciences of motion, astronomy, and strength of materials and
to the development of the scientific method.

[24]James P. Pierpont, (1866–1938), was an American mathematician. He is known for research in the field of
real and complex variable functions.

It is shown below that the set of Berge equilibria is closed and bounded.

Thus, we consider the single-step mathematical model of a conflict in the form of a noncooperative N-player game, $N \geqslant 2$, described by an ordered triplet

$$\Gamma = \langle \{\mathbb{N}\}, \{\mathcal{U}_i\}_{i \in \mathbb{N}}, \{f_i(u)\}_{i \in \mathbb{N}} \rangle. \tag{3.2.3}$$

Here $\mathbb{N} = \{1, 2, \ldots, N\}$ denotes the set of players; each of N players chooses his strategy (action) $u_i \in \mathcal{U}_i \subseteq \mathbb{R}^{n_i}$ (throughout the book, the symbol $\mathbb{R}^k$, $k \geqslant 1$, indicates the k-dimensional Euclidean space whose elements are ordered sets of k real values in the form of columns, with the standard scalar product and the Euclidean norm); such a choice yields a *strategy profile* $u = (u_1, \ldots, u_N) \in \mathcal{U} = \prod_{i \in \mathbb{N}} \mathcal{U}_i \subseteq \mathbb{R}^n$ $\left(n = \sum_{i \in \mathbb{N}} n_i\right)$; a payoff function $f_i(u)$ defined on the set $\mathcal{U}$ numerically assesses the performance of player i ($i \in \mathbb{N}$); let $(u \| z_i) = (u_1, \ldots, u_{i-1}, z_i, u_{i+1}, \ldots, u_N)$ and $f = (f_1, \ldots, f_N)$.

Recall that a pair $\left(u^B, f^B\right) = \left(\left(u_1^B, \ldots, u_N^B\right), \left(f_1\left(u^B\right), \ldots, f_N\left(u^B\right)\right)\right) \in \mathcal{U} \times \mathbb{R}^N$ is called *a Berge equilibrium* in game (3.2.3) if

$$\max_{u \in \mathcal{U}} f_i\left(u \| u_i^B\right) = f_i\left(u^B\right) \quad (i \in \mathbb{N}).$$

Similarly, a strategy profile u^B composed of such pairs is also called *a Berge equilibrium* in the game (3.2.3).

Property 3.2.2 *If in the game Γ the sets are closed and bounded, $\mathcal{U}_i \in \text{comp } \mathbb{R}^{n_i}$, and the payoff functions are continuous, $f_i(\cdot) \in C(\mathcal{U})$ ($i \in \mathbb{N}$), then the set $\mathcal{U}^B$ of all Berge equilibria u^B in this game represents a compact set in $\mathcal{U} = \mathcal{U}_1 \times \ldots \times \mathcal{U}_N$, perhaps empty.*

Proof Consider the noncooperative N-player game (3.2.3), in which a Berge equilibrium $u^B = (u_1^B, \ldots, u_N^B) \in \mathcal{U}$ is given by the inequalities

$$\mathcal{J}_i\left(u \| u_i^B\right) \leqslant \mathcal{J}_i\left(u^B\right) \quad \forall u \in \mathcal{U} \ (i \in \mathbf{N}),$$

and the set of all Berge equilibria has the form

$$\mathcal{U}^B = \left\{ u^B \in \mathcal{U} \mid \max_{u \in \mathcal{U}} \mathcal{J}_i\left(u \| u_i^B\right) = \mathcal{J}_i\left(u^B\right), \ i \in \mathbf{N} \right\}.$$

The set $\mathcal{U}^B$ is bounded as a subset of the compact set $\mathcal{U} = \mathcal{U}_1 \times \ldots \times \mathcal{U}_N$ representing the product of compact sets.

Let us prove that $\mathcal{U}^B$ is closed. Consider an arbitrary infinite sequence $\{u^{(k)}\}_0^\infty$ of points from $\mathcal{U}^B$. Since $\{u^{(k)}\}_0^\infty \subset \mathcal{U}^B \subseteq \mathcal{U}$ and $\mathcal{U}$ is compact, there exists a subsequence $\{u^{(k_l)}\}_0^\infty \subset \{u^{(k)}\}_0^\infty$ and a strategy profile $u^* \in \mathcal{U}$ such that

$$\lim_{l \to \infty} u^{(k_l)} = u^*. \tag{3.2.4}$$

(This relation holds componentwise.) The set $\mathcal{U}^B$ is closed if $u^* \in \mathcal{U}^B$. We will establish this fact by contradiction. Assume on the contrary that a strategy profile u^* is not a Berge equilibrium in the game (3.2.3), i.e., $u^* \notin \mathcal{U}^B$. Due to (3.2.2), there exist a number $j \in \mathbf{N}$ and a strategy profile $\bar{u} \in \mathcal{U}$ such that

$$\mathcal{J}_j(\bar{u} \| u_j^*) > \mathcal{J}_j(u^*). \tag{3.2.5}$$

By the continuity of $\mathcal{J}_j(u)$ on $\mathcal{U}$ (hence, the continuity of $\mathcal{J}_j(u\|\bar{u}_j^*)$ on $\mathcal{U}_{\mathbf{N}\setminus\{i\}}$), there exists a sufficiently small value $\delta > 0$ such that, for all strategy profiles $u \in \mathcal{U}$ satisfying

$$\left\| u^* - u \right\| < \delta$$

and consequently

$$\left\| u^*_{\mathbf{N}\setminus\{i\}} - u_{\mathbf{N}\setminus\{i\}} \right\| < \delta,$$

it follows that

$$\mathcal{J}_j\left(u\|\bar{u}_j\right) > \mathcal{J}_j(u).$$

Now, we choose a sufficiently large integer $l^* > 0$ so that

$$\left\| u^* - u^{(k_l)} \right\| < \delta$$

for $l \geqslant l^*$. For such values $u^{(k_l)}$,

$$\mathcal{J}_j\left(\bar{u}\|u_j^{k_j}\right) > \mathcal{J}_j\left(u^{(k_l)}\right) \quad \forall l \geqslant l^*.$$

This inequality contradicts the inclusion $u^{(k_l)} \in \mathcal{U}^B$ ($l \geqslant l^*$), and hence the strategy profile under consideration is a Berge equilibrium in the game (3.2.3), i.e.,

$$\mathcal{J}_j\left(u^{(k_l)}\|u_j\right) \leqslant \mathcal{J}_j\left(u^{(k_l)}\right) \quad \forall u_j \in \mathcal{U}_j \ (j \in \mathbf{N}).$$

The closedness and boundedness of the set $\mathcal{U}^B$ of Berge equilibria imply its compactness.

Corollary 3.2.1 *Under the hypotheses of Property 3.2.2, the set*

$$\mathcal{J}\left(\mathcal{U}^B\right) = \left\{ \mathcal{J}\left(u^B\right) \in \mathbf{R}^N \mid u^B \in \mathcal{U}^B \right\} \tag{3.2.6}$$

is a compact (perhaps, empty) set, where $\mathcal{U}^B$ denotes the set of Berge equilibria u^B in the game (3.2.3).

This result is immediate from Property 3.2.2 and the continuity of the vector function $\mathcal{J}(u)$ in $u \in \mathcal{U}$. (For a continuous mapping, the image of a compact set is another compact set.)

Remark 3.2.1 The compactness of the set $\mathcal{U}^B$ of Berge equilibria u^B can be used (at least, theoretically) for choosing an appropriate Berge equilibrium with additional good properties. For this purpose, consider the auxiliary N-criteria choice problem

$$\left\langle \mathcal{U}^B, \{\mathcal{J}_i(u)\}_{i \in \mathbf{N}} \right\rangle, \tag{3.2.7}$$

where $\mathcal{U}^B$ is the set of alternatives that form Berge equilibria u^B; $\mathcal{J}_i(u)$ is the ith criterion ($i \in \mathbf{N}$) that coincides with the payoff function $\mathcal{J}_i(u)$ of player i in the game (3.2.3).

Now, in the game (3.2.3), we try to identify a Berge equilibrium u^* that is *unimprovable* with respect to the set $\mathcal{U}^B$, i.e., there does not exist another strategy profile $u^B \in \mathcal{U}^B$ in which the payoffs of all players simultaneously satisfy the inequality

$$\mathcal{J}_i(u^B) > \mathcal{J}_i(u^*)$$

for all $i \in \mathbf{N}$. To state it differently, there does not exist a Berge equilibrium in the game (3.2.3) that strictly improves all outcomes $\mathcal{J}_i(u^*)$ $(i \in \mathbf{N})$. In this case, the problem is to construct the vector maximum in the problem (3.2.7); then the compactness of the set $\mathcal{U}^B$ together with the continuity of $\mathcal{J}_i(u)$ $i \in \mathbf{N}$ guarantees the existence of u^*. Note that u^* can be found, e.g., from the condition

$$\max_{u \in \mathcal{U}^e} \sum_{i=1}^{N} \alpha_i \mathcal{J}_i(u) = \sum_{i=1}^{N} \alpha_i \mathcal{J}_i(u^*),$$

where $\alpha_i > 0$, $i \in \mathbf{N}$, are some constants.

Actually, the unimprovable Berge equilibrium can be designed in the following way. *First*, find the ideal point

$$\mathcal{J}^* = (\mathcal{J}_1^*, ..., \mathcal{J}_N^*),$$

where $\mathcal{J}_i^* = \max\limits_{u \in \mathcal{U}^B} \mathcal{J}_i(u)$ $(i \in \mathbf{N})$; generally speaking, the vector $\mathcal{J}^*$ is not achieved at an equilibrium u^B, but all players would like to be as close to it as possible.

Second, design the equilibrium $u^* \in \mathcal{U}^B$ using the equality

$$\min_{u \in \mathcal{U}^B} \sum_{i \in \mathbf{N}} \left[\mathcal{J}_i^* - \mathcal{J}_i(u) \right]^2 = \sum_{i \in \mathbf{N}} \left[\mathcal{J}_i^* - \mathcal{J}_i(u^*) \right]^2;$$

it has been established that u^* implements the vector maximum (in the Pareto sense) in the N-criteria choice problem (3.2.7).

The existence of such a $u^* \in \mathcal{U}^B$ follows from the continuity of $\mathcal{J}_i(u)$ $(i \in \mathbf{N})$ and the compactness of $\mathcal{U}^B$ by the Weierstrass extreme-value theorem.

Corollary 3.2.2 *Under the hypotheses of Property 3.2.2, in the game Γ with $\mathcal{U}^B \neq \varnothing$ there exists a Berge equilibrium that is Pareto-maximal with respect to all $u^B \in \mathcal{U}^B$.*

Really, due to the compact character of $\mathcal{U}^B$, the belonging $f_i(\cdot) \in C(\mathbf{X})$ $(i \in \mathbb{N})$ and Property 3.2.2, the N-criteria choice problem

$$\left\langle \mathcal{U}^B, \{f_i(u)\}_{i \in \mathbb{N}} \right\rangle$$

has a Pareto-maximal alternative $u^B \in \mathcal{U}^B$, i.e., for all $u \in \mathcal{U}^B$ the system of N inequalities

$$f_i(u) \geqslant f_i\left(u^B\right) \quad (i \in \mathbb{N}),$$

with at least one strict inequality, is inconsistent.

Connection between BE and saddle point

Property 3.2.3 *In the antagonistic case of game (3.2.3), i.e., the corresponding zero-sum two-player game with $\mathbf{N} = \{1, 2\}$ and $\mathcal{J}_1(u) = -\mathcal{J}_2(u) = \mathcal{J}(u)$, the Berge equilibrium $u^B = (u_1^B, u_2^B)$ coincides with the saddle point $\mathcal{J}(u_1, u_2)$, that is,*

$$\mathcal{J}\left(u_1^B, u_2\right) \leqslant \mathcal{J}\left(u_1^B, u_2^B\right) \leqslant \mathcal{J}\left(u_1, u_2^B\right) \quad \forall u_i \in \mathcal{U}_i \quad (i = 1, 2).$$

Proof Really, for the antagonistic case, inequalities (3.2.2) are reduced to

$$\mathcal{J}_1\left(u_1^B, u_2\right) = \mathcal{J}\left(u_1^B, u_2\right) \leqslant \mathcal{J}\left(u_1^B, u_2^B\right) = \mathcal{J}_1\left(u_1^B, u_2^B\right) \quad \forall u_2 \in \mathcal{U}_2,$$

$$\mathcal{J}_2\left(u_1, u_2^B\right) = -\mathcal{J}\left(u_1, u_2^B\right) \leqslant -\mathcal{J}\left(u_1^B, u_2^B\right) = \mathcal{J}_2\left(u_1^B, u_2^B\right) \quad \forall u_1 \in \mathcal{U}_1.$$

Remark 3.2.3 Property 3.2.3 underlines that the concept of Berge equilibrium is rather broad: it contains saddle point as a particular case. Saddle point as a solution of zero-sum two-player games would hardly raise any questions!

3.2.3   Negative properties of Berge equilibrium

Internal instability of Berge equilibrium

There are spots on the sun, however. We will identify such spots for the concept of Berge equilibrium in the game (3.2.3) using an illustrative example as follows.

Example 3.2.1 Consider the game (3.2.3) with two players ($\mathbf{N} = \{1, 2\}$), the sets

$$\mathcal{U}_1 = \mathcal{U}_2 = [-1, 1]$$

and the payoff functions

$$\mathcal{J}_1(u) = -u_2^2 + 2u_1 u_2, \quad \mathcal{J}_2(u) = -u_1^2 + 2u_1 u_2.$$

Denote this game by $\Gamma_{(3.2.1)}$. Obviously, the set of all strategy profiles in the game $\Gamma_{(3.2.1)}$ is $u = (u_1, u_2) \in \mathcal{U} = [-1, 1]^2$.

Conditions (3.2.2) determining a Berge equilibrium u^B in this game can be written as

$$-u_2^2 + 2u_1^B u_2 \leqslant -\left[u_1^B\right]^2 + 2u_1^B u_2^B \quad \forall u_2 \in \mathcal{U}_2,$$

$$-u_1^2 + 2u_1 u_2^B \leqslant -\left[u_1^B\right]^2 + 2u_1^B u_2^B \quad \forall u_1 \in \mathcal{U}_1,$$

which is equivalent to

$$-\left[u_2 - u_1^B\right]^2 \leqslant -\left[u_1^B - u_2^B\right]^2 \quad \forall u_2 \in [-1, 1],$$

$$-\left[u_1 - u_2^B\right]^2 \leqslant -\left[u_2^B - u_1^B\right]^2 \quad \forall u_1 \in [-1, 1].$$

Hence, the Berge equilibria $u^B = (u_1^B, u_2^B)$ in the game $\Gamma_{(3.2.1)}$ are given by the equalities

$$u_1^B = u_2^B = \alpha = const \in [-1, 1],$$

and the payoffs of both players at these strategy profiles $u^B = (u_1^B, u_2^B) = (\alpha, \alpha)$ are

$$\mathcal{J}_i(u^B) = \alpha^2 \quad (i = 1, 2)$$

for any constants $\alpha \in [-1, 1]$.

Thus, any point of the segment AB (Figure 3.2.1) is a Berge equilibrium, and the corresponding payoffs form the segment CD of the bisecting line (Figure 3.2.2).

This example reflects well the following properties of Berge equilibria.

Feature 3.2.1 *A Berge equilibrium u^B may be nonunique.*

In Figure 3.2.1 there exists a continuum of such strategy profiles—any point $u^B = (\alpha, \alpha)$ of the segment. (For each constant $\alpha \in [-1, 1]$, the strategy profile (α, α) is a Berge equilibrium.) Recall that according to Property 3.2.2, the set of Berge equilibria is compact. The table of payoffs is presented below.

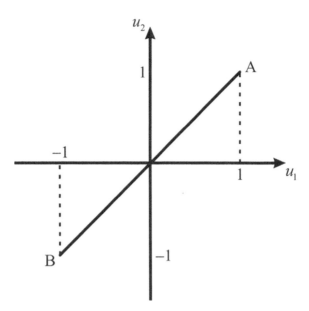

Figure 3.2.1 Berge equilibria.

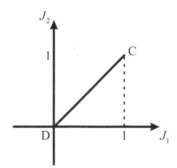

Figure 3.2.2 Payoffs in Berge equilibria.

Table 3.2.1 Payoffs in different strategy profiles.

Profile / Payoff	$(0, 0)$	$\left(\frac{1}{4}, \frac{1}{4}\right)$	$\left(\frac{3}{4}, \frac{3}{4}\right)$	$\left(\frac{1}{4}, \frac{3}{4}\right)$	$(1, 1)$	$\left(\frac{1}{4}, \frac{1}{3}\right)$
$\mathcal{J}_1(u)$	0	$\frac{1}{16}$	$\frac{9}{16}$	$\frac{5}{16}$	1	$\frac{5}{48}$
$\mathcal{J}_2(u)$	0	$\frac{1}{16}$	$\frac{9}{16}$	$-\frac{3}{16}$	1	$\frac{1}{48}$

Using Table 3.2.1, we will reveal a series of other features of Berge equilibria.

One of them concerns *the internal instability of the set of Berge equilibria*. First, let us introduce some concepts from the theory of multicriteria choice problems [213, p. 20].

A set $\overline{U}$ of alternatives u is *internally stable* if there do not exist two elements $u^{(1)}$ and $u^{(2)}$ from the set $\overline{U}$ such that $\mathcal{J}_i(u^{(1)}) > \mathcal{J}_i(u^{(2)})$ $(i \in \mathbf{N})$.

A set $\overline{\mathcal{U}}$ is *internally unstable* if in $\overline{\mathcal{U}}$ there are at least two alternatives $u^{(1)}$ and $u^{(2)}$ that satisfy the strict inequalities $\mathcal{J}_i(u^{(1)}) > \mathcal{J}_i(u^{(2)})$, $i \in \mathbf{N}$.

Feature 3.2.2 *The set of Berge equilibria is internally unstable.*

According to Table 3.2.1, the two Berge equilibria $u^{(1)} = (0,0)$ and $u^{(2)} = (1,1)$ satisfy the strict inequalities

$$0 = \mathcal{J}_i\left(u^{(1)}\right) = \mathcal{J}_i(0,0) < \mathcal{J}_i(1,1) = \mathcal{J}_i\left(u^{(2)}\right) = 1 \quad (i = 1, 2).$$

The absence of internal instability is closely connected with another feature as follows.

Feature 3.2.3. *An equilibrium is generally improvable*, i.e., for an equilibrium there may exist another strategy profile (not necessarily an equilibrium) in which the payoff of each player is greater than in the equilibrium.

For example, in the game $\Gamma_{(3.2.1)}$ consider a Nash equilibrium $u^{(1)} = (0,0)$ and another strategy profile of the form $u^* = \left(\frac{1}{4}, \frac{1}{3}\right)$, which is not a Nash equilibrium. Obviously,

$$\mathcal{J}_1\left(u^{(1)}\right) = 0 < \frac{1}{18} = \mathcal{J}_1\left(u^*\right), \quad \mathcal{J}_2\left(u^{(1)}\right) = 0 < \frac{5}{48} = \mathcal{J}_2\left(x^*\right).$$

Another feature concerns *the absence of interchangeability and equivalence*. What is remarkable, both properties hold for the saddle points of zero-sum two-player games. Let $u^{(1)} = \left(u_1^{(1)}, u_2^{(1)}\right) \in \mathcal{U} = \mathcal{U}_1 \times \mathcal{U}_2$ and $u^{(2)} = \left(u_1^{(2)}, u_2^{(2)}\right) \in \mathcal{U}$ be any saddle points of the zero-sum two-player game

$$\Gamma_2 = \langle \mathcal{U}_1, \mathcal{U}_2, \mathcal{J}(u_1, u_2) \rangle,$$

i.e.,

$$\begin{aligned}
\mathcal{J}\left(u_1, u_2^{(1)}\right) &\leqslant \mathcal{J}\left(u^{(1)}\right) \leqslant \mathcal{J}\left(u_1^{(1)}, u_2\right), \\
\mathcal{J}\left(u_1, u_2^{(2)}\right) &\leqslant \mathcal{J}\left(u^{(2)}\right) \leqslant \mathcal{J}\left(u_1^{(2)}, u_2\right)
\end{aligned} \tag{3.2.8}$$

for all $u_i \in \mathcal{U}_i$ ($i = 1, 2$). Then inequalities (3.2.8) imply the chain of inequalities

$$\mathcal{J}\left(u_1^{(1)}, u_2^{(1)}\right) \leqslant \mathcal{J}\left(u_1^{(1)}, u_2^{(2)}\right) \leqslant \mathcal{J}\left(u_1^{(2)}, u_2^{(2)}\right) \leqslant \mathcal{J}\left(u_1^{(2)}, u_2^{(1)}\right) \leqslant \mathcal{J}\left(u_1^{(1)}, u_2^{(1)}\right),$$

and hence all inequalities here turn into equalities, in particular,

$$\mathcal{J}\left(u^{(1)}\right) = \mathcal{J}\left(u_1^{(1)}, u_2^{(2)}\right) = \mathcal{J}\left(u_1^{(2)}, u_2^{(1)}\right) = \mathcal{J}\left(u^{(2)}\right). \tag{3.2.9}$$

In view of (3.2.8) and (3.2.9), we have

$$\begin{aligned}
\mathcal{J}\left(u_1, u_2^{(1)}\right) &\leqslant \mathcal{J}\left(u_1^{(2)}, u_2^{(1)}\right) \leqslant \mathcal{J}\left(u_1^{(2)}, u_2\right), \\
\mathcal{J}\left(u_1, u_2^{(2)}\right) &\leqslant \mathcal{J}\left(u_1^{(1)}, u_2^{(2)}\right) \leqslant \mathcal{J}\left(u_1^{(1)}, u_2\right)
\end{aligned} \tag{3.2.10}$$

for any $u_i \in \mathcal{U}_i$ ($i = 1, 2$). The relations (3.2.10) mean that the strategy profiles $(u_1^{(2)}, u_2^{(1)})$ and $(u_1^{(1)}, u_2^{(2)})$ are saddle points in the zero-sum two-player game Γ_2. Therefore, we have established that in the game Γ_2,

a)   all saddle points are *equivalent*, i.e., the payoffs $\mathcal{J}(u_1, u_2)$ at these points coincide with each other; see the chain of equalities (3.2.9);

b)   all saddle points are *interchangeable*, i.e., if $(u_1^{(1)}, u_2^{(1)})$ and $(u_1^{(2)}, u_2^{(2)})$ are two saddle points, then $(u_1^{(1)}, u_2^{(2)})$ and $(u_1^{(2)}, u_2^{(1)})$ are saddle points as well; this fact follows from (3.2.8)–(3.2.10).

These properties are fruitful for both players in the zero-sum two-player game: with the property of interchangeability, each of the players may use *his strategy from any saddle point* without agreeing with the other; with the property of equivalence, the corresponding payoffs of the players *coincide with each other*.

Unfortunately, this does not apply to general noncooperative games!

Feature 3.2.4 *A general noncooperative game does not satisfy the properties of equivalence and interchangeability.*

According to Table 3.2.1, we have:

a) $\mathcal{J}_1(0, 0) = 0 \neq \mathcal{J}_1(1, 1) = 1$, and hence the payoffs of player 1 in the equilibria $(0, 0)$ and $(1, 1)$ are different (no equivalence);

b) $\mathcal{J}_1\left(\frac{1}{4}, \frac{3}{4}\right) = -\frac{3}{16} < \mathcal{J}_1\left(\frac{1}{4}, \frac{1}{4}\right) = \frac{1}{16}$, and hence $\left(\frac{1}{4}, \frac{3}{4}\right)$ is not a Berge equilibrium, although $\left(\frac{1}{4}, \frac{1}{4}\right)$ and $\left(\frac{3}{4}, \frac{3}{4}\right)$ are actually Berge equilibria (no interchangeability).

The absence of interchangeability reduces the noncooperative character of a game. Namely, if an equilibrium is not unique, then *before the game starts the players have to agree* which specific equilibrium (out of many) will they prefer.

Remark 3.2.4 From the viewpoint of such an agreement, the concept of Berge equilibrium as an optimality principle for games can be given a quite natural interpretation as follows. The participants in a conflict (players) get together before the start of the game, somehow choosing a Berge equilibrium acceptable for everyone. Having concluded an optional agreement to prefer this equilibrium, the players separate and everyone makes the final choice of his strategy, assuming that the other parties to the conflict will comply with the terms of the agreement. The strategy profile chosen at the beginning is a Berge equilibrium if none of the players benefits by a unilateral deviation from it. Thus, Berge equilibria are precisely those strategy profiles that have stability against the deviations of individual players (Property 3.2.1). A player deviating from the equilibrium receives the status of The Widow of a noncommissioned Officer, "who flagged herself."[25] To tell the truth, other players can suffer in this case as well.

On the one hand, a Berge equilibrium is stable only against the deviations of individual players. Two or more players, united into a coalition, can increase their payoffs (Feature 3.2.3), in some cases also reducing the payoffs of the other parties to the conflict. That is, the Berge equilibrium as an optimality principle neglects a possible formation of coalitions.

On the other hand, a Berge equilibrium cannot be the result of isolated actions of different players. In the interpretation given above, the players are assumed to be able to meet before the start of the game for choosing some acceptable strategy profile. Such an

---

[25]See *The Inspector-General*, a comedy in five acts, by Nikolay V. Gogol, (1809–1852), Ukrainian–Russian humorist, dramatist, and novelist who much contributed to the great 19th-century tradition of Russian realism.

opportunity is a fundamentally important aspect for noncooperative games with the concept of Berge equilibrium. Thus, the Berge equilibrium is generally not an individual way to choose solution. For implementing it, the players have to agree about their adherence to an equilibrium in general, and in the case of nonuniqueness, about the choice of a specific equilibrium from the resulting set of such strategy profiles.

The choice of Berge equilibrium as the solution of a particular game is more or less satisfactory if the Berge equilibrium $u^B = \left(u_1^B, ..., u_N^B \right)$ is unique and strict, i.e., for any $i \in \mathbf{N}$ and $u_i \neq u_i^B$, $u_i \in \mathcal{U}_i$, we have

$$\mathcal{J}_i \left(u^B \right) > \mathcal{J}_i \left(u_1, ..., u_{i-1}, u_i^B, u_{i+1}, ..., u_N \right) \qquad (i \in \mathbf{N}).$$

However, even in this case there may exist more beneficial choices than the Berge equilibrium, for all players simultaneously.

Remark 3.2.5 Internal instability and improvability are negative properties. How can they be eliminated?

First, the players should use only those Berge equilibria that simultaneously implement the vector maximum in the N-criteria choice problem

$$\left\langle \mathcal{U}^B, \{ \mathcal{J}_i(u) \}_{i \in \mathbf{N}} \right\rangle.$$

(In the game $\Gamma_{(3.2.1)}$, such Berge equilibria are the strategy profiles $u^B = (1, 1)$ and $u^B = (-1, -1)$.) But such a coincidence is a rare phenomenon in noncooperative games. As is indicated by the available literature on noncooperative games, there exist just three classes of games [195, 213] in which a Berge equilibrium simultaneously implements the vector maximum.

Second, alternative solution concepts of noncooperative games differing from the Berge equilibrium can be suggested, for example, the equilibrium in threats and counterthreats or the active equilibrium [156]. They implement the vector maximum and have the advantages of equilibrium (see Properties 3.2.1–3.2.2), yet not suffering from the drawbacks of improvability and internal instability.

Feature 3.2.5 *A noncooperative game may have no Nash equilibrium*; see Example 3.2.2 below. In this case, the Nash equilibrium as a solution concept makes no sense, and other types of equilibrium should be chosen (in particular, the Berge equilibrium).

Example 3.2.2 Consider a noncooperative two-player game in which the strategy set of player i is $\mathcal{U}_i = \mathbf{R}^{m_i}$ $(i = 1, 2)$, and the payoff function of player i has the linear-quadratic form:

$$\mathcal{J}_i (u_1, u_2) = u_1' A_{i1} u_1 + u_2' A_{i2} u_2 + 2a_{i1}' u_1 + 2a_{i2}' u_2 \quad (i = 1, 2); \tag{3.2.11}$$

here A_{ij} are constant symmetric matrices of compatible dimensions and the m_i-dimensional vectors a_{ji} have constant components $(i, j = 1, 2)$. Denote this game by $\Gamma_{(3.2.2)}$. Recall that for a matrix M, the notation $M > 0$ means that the quadratic form $z' M z$ is positive definite; as before, the prime indicates transposition.

Proposition 3.2.1 *If*

$$A_{11} > 0 \quad or \ (and) \ A_{22} > 0,$$

then the game $\Gamma_{(3.2.2)}$ *has no Nash equilibrium.*

Proof A Nash equilibrium $u^e = (u_1^e, u_2^e) \in \mathbf{R}^{m_1 + m_2}$ in the game $\Gamma_{(3.2.2)}$ is given by the two inequalities

$$\mathcal{J}_1\left(u_1, u_2^e\right) \leqslant \mathcal{J}_1\left(u^e\right) \qquad \forall u_1 \in \mathcal{U}_1,$$
$$\mathcal{J}_2\left(u_1^e, u_2\right) \leqslant \mathcal{J}_2\left(u^e\right) \qquad \forall u_2 \in \mathcal{U}_2.$$

Hence, there exists no Nash equilibrium in the game $\Gamma_{(3.2.2)}$ if for *each* strategy profile $u = (u_1, u_2) \in \mathbf{R}^{m_1+m_2}$ one (or both) of the following conditions hold: either there is a strategy $u_1^* \in \mathbf{R}^{m_1}$ such that

$$\mathcal{J}_1\left(u_1^*, u_2\right) > \mathcal{J}_1\left(u_1, u_2\right), \qquad (3.2.12)$$

or (and) there is a strategy $u_2^* \in \mathbf{R}^{m_2}$ such that

$$\mathcal{J}_2\left(u_1, u_2^*\right) > \mathcal{J}_2(u).$$

Now, let (u_1, u_2) be an arbitrary strategy profile from $\mathbf{R}^{m_1} \times \mathbf{R}^{m_2}$ and $A_{11} > 0$. We construct a strategy $u_1^* \in \mathbf{R}^{m_1}$ satisfying (3.2.12), i.e., due to the explicit form (3.2.11) of $\mathcal{J}_1(u)$,

$$\left(u_1^*\right)' A_{11} u_1^* + u_2' A_{12} u_2 + 2a_{11}' u_1^* + 2a_{12}' u_2 > u_1' A_{11} u_1 + u_2' A_{12} u_2 + 2a_{11}' u_1 + 2a_{12}' u_2.$$

Then condition (3.2.12) holds if there is a strategy u_1^* such that

$$\left(u_1^*\right)' A_{11} u_1^* + 2a_{11}' u_1^* > u_1' A_{11} u_1 + 2a_{11}' u_1.$$

Since $A_{11} > 0$,

$$u_1' A_{11} u_1 \geqslant \lambda u_1' u_1 \qquad \forall u_1 \in \mathbf{R}^{m_1}, \qquad (3.2.13)$$

where λ is the least root of the equation $det\left[A_{11} - \mu E_{m_1}\right] = 0$; in addition, $\lambda > 0$. In view of (3.2.13), inequality (3.2.12) is valid if

$$\lambda(u_1^*)' u_1^* + 2a_{11}' u_1^* > u_1' A_{11} u_1 + 2a_{11}' u_1 = \alpha. \qquad (3.2.14)$$

Let $u_1^* = \beta e_{m_1}$, where a positive constant β will be specified below and e_{m_1} denotes an m_1-dimensional vector consisting of ones. Substituting $u_1^* = \beta e_{m_1}$ into (3.2.14), we arrive at condition (3.2.12) if β satisfies the inequality

$$\lambda m_1 \beta^2 + 2\beta a_1' e_{m_1} > \alpha,$$

where $a_1 = a_{11}$. This is reduced to

$$\lambda m_1 \left(\beta + \frac{a_1' e_{m_1}}{\lambda m_1}\right)^2 > \alpha + \frac{(a_1' e_{m_1})^2}{\lambda m_1},$$

which holds for all

$$\beta > \left|\sqrt{\frac{|\alpha|}{\lambda m_1} + \frac{(a_1' e_{n_1})^2}{\lambda^2 m_1^2}}\right| + \left|\frac{a_1' e_{m_1}}{\lambda m_1}\right|.$$

Thus, for any constants β specified above, the strategy $u_1^* = \beta e_{m_1}$ guarantees (3.2.14) and consequently (3.2.12).

Using the same considerations together with the condition $A_{22} > 0$, we easily establish the existence of a strategy $u_2^* = \beta e_{m_2}$ for which

$$\mathcal{J}_2\left(u_1, u_2^*\right) > \mathcal{J}_2(u_1, u_2).$$

Remark 3.2.6 By analogy, we may demonstrate that for $A_{12} > 0$ or (and) $A_{21} > 0$, the game $\Gamma_{(3.2.2)}$ has no Berge equilibrium regardless of the parameters A_{ii}, a_{i1} and a_{i2}.

Concluding this analysis of negative properties of Berge equilibria, let us quote F. de Larochefoucault: "Other shortcomings, if used skillfully, sparkle brighter than any virtues." [186, p. 52].

No guaranteed individual rationality

> Among the splendid generalizations effected by modern mathematics,
> there is none more brilliant or more inspiring or more fruitful,
> and none more commensurate with the limitless immensity of being itself,
> than that which produced the great concept designated ...
> hyperspace or multidimensional space.
> —Keyser[26]

A Nash equilibrium has the property of individual rationality whereas a Berge equilibrium generally not; see an illustrative example below, which was suggested by Vaisman in 1994. It is also established that there may exist a Berge equilibrium in which at least one player gains a smaller payoff than the maximin.

Besides the absence of internal stability, another negative property of a Berge equilibrium is the following.

Property 3.2.4 *A Berge equilibrium may not satisfy the individual rationality conditions*, as opposed to the Nash equilibrium x^e in the game Γ_2. (Under the assumptions $X_i \in$ comp $\mathbb{R}^{n_i}$ and $f_i(\cdot) \in C(X)$ ($i \in \mathbb{N}$), the game Γ_2 (the game Γ with $\mathbb{N} = \{1, 2\}$) satisfies the inequalities

$$f_1\left(x^e\right) \geqslant \max_{x_1 \in X_1} \min_{x_2 \in X_2} f_1(x_1, x_2), \quad f_2\left(x^e\right) \geqslant \max_{x_2 \in X_2} \min_{x_1 \in X_1} f_2(x_1, x_2),$$

known as the individual rationality conditions.)

Example 3.2.3 Consider a noncooperative two-player game of the form

$$\Gamma_2' = \Big\langle \{1, 2\}, \{X_1 = (-\infty, +\infty), X_2 = [-1, +1]\},$$

$$\left\{f_1(x) = -4x_1^2 + 2x_1 x_2 + x_2^2, f_2(x) = -(x_1 - 1)^2 + 5\right\}\Big\rangle,$$

where $x = (x_1, x_2)$. A Berge equilibrium $x^B = \left(x_1^B, x_2^B\right)$ in the game Γ_2' is defined by the two equalities

$$\max_{x_2 \in X_2} f_1\left(x_1^B, x_2\right) = f_1\left(x^B\right), \quad \max_{x_1 \in X_1} f_2\left(x_1, x_2^B\right) = f_2\left(x^B\right).$$

The second inequality holds only for the strategy $x_1^B = 1$. Due to the strong convexity of $f_1(x)$ in x_2 (which follows from $\frac{\partial^2 f_1(x_1^B, x_2)}{\partial x_2^2}|_{x_2} = 2 > 0$), the maximum of the function

$$f_1\left(x_1^B, x_2\right) = -4 + 2x_2 + x_2^2$$

---

[26]Cassius Jackson Keyser, (1862–1947), was an American mathematician of pronounced philosophical inclinations. A quote from *On Mathematics and Mathematicians*, R.E. Moritz, Ed., New York: Dover, 1958, pp. 360–361.

is achieved on the boundary of X_2, more specifically, at the point $x_2^B = 1$. Thus, the game Γ_2' has the Berge equilibrium $x^B = (1, 1)$, and the corresponding payoff is $f_1(x^B) = f_1(1, 1) = -1$.

Now, we find $\max\limits_{x_1 \in X_1} \min\limits_{x_2 \in X_2} f_1(x_1, x_2)$ in two stages as follows. In the *first* stage, we construct a scalar function $x_2(x_1)$ that implements the *inner minimum*

$$\min_{x_2 \in X_2} f_1(x_1, x_2) = f_1(x_1, x_2(x_1)) \quad \forall x_1 \in X_1.$$

By the strong convexity of $f_1(x_1, x_2)$ in x_2,

$$\left.\frac{\partial f_1(x_1, x_2)}{\partial x_2}\right|_{x_2(x_1)} = 2x_1 + 2x_2(x_1) = 0,$$

which gives the unique solution $x_2(x_1) = -x_1$ and $f_1[x_1] = f_1(x_1, x_2(x_1)) = -5x_1^2$.

In the *second* stage, we construct *the outer maximum*, i.e., find

$$\max_{x_1 \in X_1} f_1[x_1] = \max_{x_1 \in \mathbb{R}} f_1(x_1, x_2(x_1)) = \max_{x_1 \in \mathbb{R}}\left[-5x_1^2\right] = 0.$$

As a result,

$$f_1(x^B) = -1 < 0 = \max_{x_1 \in \mathbb{R}} f_1(x_1, x_2(x_1)) = \max_{x_1 \in \mathbb{R}} \min_{x_2 \in [-1, +1]} f_1(x_1, x_2),$$

which establishes that the individual rationality property may fail for a Berge equilibrium.

Remark 3.2.7 The individual rationality condition is a requirement for a "good" solution of both noncooperative and cooperative games: each player can guarantee the maximin individually, i.e., by his own maximin strategy, regardless of the behavior of the other players. However, in a series of applications (especially for the linear-quadratic statements of the game), maximin often does not exist. Such games were studied in the books [86, pp. 95–97, 110–116, 120; 156, pp. 124–131].

If the game (3.2.3) has maximins, Vaisman suggested to incorporate the individual rationality property into the definition of a Berge equilibrium. Such equilibria are called *Berge–Vaisman equilibria*.

3.3 SUFFICIENT CONDITIONS

> Trace everything back to the beginning
> and you will understand a lot.
> —Kozma Prutkov[27]

This epigraph expresses the core of Bellman's principle of optimality, which is used in Section 3.4 to establish the existence of a Berge equilibrium in the multistage conflict (3.1.5).

3.3.1 Multistage two-player game

Due to the cumbersome form of mathematical relations of dynamic programming, we will consider the K-stage noncooperative *two-player game* only, which is described by an ordered quadruple

---

[27]An English translation of a quote from [232, p. 13].

$$\Gamma(t_0, x_0) = \left\langle \{1, 2\}, \sum o, \{\mathfrak{A}_i\}_{i=1,2}, \{\mathcal{J}_i(U, t_0, x_0)\}_{i=1,2} \right\rangle. \tag{3.3.1}$$

The game (3.3.1) has the following notations:

1 and 2 as the numbers of players;

$\sum$ as a controlled system evolving over time in accordance with the vector difference equation

$$x(k+1) = f(k, x(k), u_1, u_2), \quad x(t_0) = x_0,$$
$$(k = t_0, t_0 + 1, \ldots, K-1), \tag{3.3.2}$$

where the vector $x(k) \in \mathbf{R}^n$ characterizes the state of the controlled system $\sum$ at given time instants $k = 0, 1, \ldots, K$; $t_0 \in \{0, 1, \ldots, K-1\}$ as the initial time instant of the game; $(k, x(k))$ as the position of the game (3.3.1) at a time instant $k = t_0, t_0 + 1, \ldots, K-1$; finally, (t_0, x_0) as an initial position.

We formalize the strategy U_i of player i ($i = 1, 2$) as follows. In (3.3.2), the m_i-dimensional vector u_i is a control action of player i. For each position $(k, x) \in \{0, 1, \ldots, K-1\} \times \mathbf{R}^n$ and each player ($i = 1, 2$), let $\mathcal{U}_i(k, x) \subseteq \mathbf{R}^{m_i}$ be *a priori* defined subset in the state space $\mathbf{R}^{m_i}$. A strategy $U_i(k)$ of player i *at a time instant* k is identified with an m_i-dimensional vector function $u_i(k, x) \in \mathcal{U}_i(k, x)$ $\forall x \in \mathbf{R}^n$; hereinafter, the fact of such identification will be denoted by $U_i(k) \div u_i(k, x)$. The set of all strategies of player i at a time instant k is denoted by $\mathfrak{A}_i(k)$. Then *a strategy profile* $U(k)$ of the game (3.3.1) *at a time instant* k is a collection

$$U(k) = (U_1(k), U_2(k)) \in \mathfrak{A}_1(k) \times \mathfrak{A}_2(k) = \mathfrak{A}(k).$$

We associate with the game (3.3.1) with the initial position (t_0, x_0) the family of games

$$\Gamma(k, x(k)) = \langle \{1, 2\}, \sum o, \{\mathfrak{A}_i(k)\}_{i=1,2}, \{\mathcal{J}_i(U, k, x(k))\}_{i=1,2} \rangle,$$

where, in contrast to $\Gamma(t_0, x_0)$, the initial position (t_0, x_0) is replaced by $(k, x(k)) \in \{t_0, t_0 + 1, \ldots, K-1\} \times \mathbf{R}^n$. Then a sequence

$$U_i \div (u_i(k, x), u_i(k+1, x(k+1)), \ldots, u_i(K-1, x(K-1))) \quad (i = 1, 2)$$

is called *a strategy of player* i *in the game* $\Gamma(k, x(k))$; here

$$x(k+1) = f(k, x, u_1(k, x), u_2(k, x)),$$
$$x(k+2) = f(k+1, x(k+1), u_1(k+1, x(k+1)), u_2(k+1, x(k+1))),$$

$$\vdots$$

$$x(K) = f(K-1, x(K-1), u_1(K-1, x(K-1)), u_2(K-1, x(K-1))), \tag{3.3.3}$$

$$U_i(j) \div u_i(j, x) \in \mathcal{U}_i(j, x) \quad \forall x \in \mathbf{R}^n,$$

and $U_i(j)$ is a strategy chosen by player i from the set $\mathfrak{A}_i(j)$ at a time instant $j \in \{k, k+1, \ldots, K-1\}$. The set of all strategies U_i in the game $\Gamma(k, x(k))$ is denoted by $\mathfrak{A}_i(k)$, and the set of all strategy profiles $U = (U_1, U_2)$ by $\mathfrak{A}(k)$.

A play of the game (3.3.1) is organized as follows. Assume an initial position $(t_0, x_0) \in \{0, \ldots, K-1\} \times \mathbf{R}^n$ is fixed. being guided by their individual interests (see below), at each stage $j = t_0, t_0 + 1, \ldots, K-1$ both players choose their strategies $U_i(j) \div u_i(j, x) \in \mathcal{U}_i(j, x)$ $\forall x \in \mathbf{R}^n$. Then the strategy of player i takes the form

$$U_i \div (u_i(t_0, x), u_i(t_0+1, x(t_0+1)), \ldots, u_i(K-1, x(K-1))) \quad (i = 1, 2),$$

where $x = x_0 = x(t_0), x(t_0 + 1), \ldots, x(K - 1), x(K)$ are determined by (3.3.3). As a result, we obtain

first, the discrete sequence of values of the state vector,

$$x(t_0) = x_0, x(t_0 + 1), \ldots, x(K),$$

and *second*, the corresponding sequence of *realizations* of the strategies chosen by players 1 and 2,

$$u_i[t_0] = u_i(t_0, x(t_0)), u_i[t_0 + 1] = u_i(t_0 + 1, x(t_0 + 1)), \ldots, u_i[K - 1]$$
$$= u_i(K - 1, x(K - 1)) \quad (i = 1, 2).$$

On the three sequences

$$\{x(j) \,|\, j = t_0, \ldots, K\}, \quad \{u_i[j] \,|\, j = t_0, \ldots, K - 1\} \quad (i = 1, 2),$$

we define *the payoff function of player i* as the functional

$$\mathcal{J}_i(U, t_0, x_0) = \Phi_i(x(K)) + \sum_{j=t_0}^{K-1} F_i(j, x(j), u_1[j], u_2[j]) \quad (i = 1, 2), \tag{3.3.4}$$

whose values are called *the payoff of player i*.

At conceptual level, the goal of player i is *to maximize* his payoff with an appropriate choice of the strategy $U_i \in \mathfrak{A}_i$.

Consider this problem within the framework of *noncooperative games*, in which the players are not allowed to form coalitions for joint strategies. As it has been mentioned, a solution concept of noncooperative games is the Berge equilibrium. Recall that *a strategy profile $U^B = \left(U_1^B, U_2^B\right) \in \mathfrak{A}$, where*

$$U_i^B \doteq \left(u_i^B(t_0, x), u_i^B\left(t_0 + 1, x^B(t_0 + 1)\right), \ldots, u_i^B\left(K - 1, x^B(K - 1)\right)\right) \quad (i = 1, 2),$$

and $x^B(t_0 + 1), \ldots, x^B(K - 1)$ are given by (3.3.3) for $x = x_0$, $x(r) = x^B(r)$, $u_i(j, x) = u_i^B(j, x^B(j))$ $(j = t_0, t_0 + 1, \ldots, K - 1;\ r = t_0 + 1, \ldots, K;\ i = 1, 2)$, is called a Berge equilibrium in game (3.3.1) with an initial position (t_0, x_0), where $t_0 \in \{0, 1, \ldots, K - 1\}$ and $x_0 \in \mathbf{R}^n$, if

$$\mathcal{J}_1\left(U_1^B, U_2, t_0, x_0\right) \leqslant \mathcal{J}_1\left(U^B, t_0, x_0\right) \quad \forall U_2 \in \mathfrak{A}_2,$$
$$\mathcal{J}_2\left(U_1, U_2^B, t_0, x_0\right) \leqslant \mathcal{J}_2\left(U^B, t_0, x_0\right) \quad \forall U_1 \in \mathfrak{A}_1;$$

in this case, the values $\mathcal{J}_i\left(U^B, t_0, x_0\right)$ $(i = 1, 2)$ are called the Berge equilibrium payoffs in game (3.3.1).

Let U^B be a Berge equilibrium in the game (3.3.1) for any initial position $(t_0, x_0) \in \{0, \ldots, K - 1\} \times \mathbf{R}^n$ and also let $\mathcal{J}\left(U^B, t_0, x_0\right) = \left(\mathcal{J}_1\left(U^B, t_0, x_0\right), \mathcal{J}_2\left(U^B, t_0, x_0\right)\right)$ be the corresponding payoffs of the players. Then the pair $\left(U^B, \mathcal{J}\left(U^B, t_0, x_0\right)\right)$ forms *a Berge equilibrium solution* of the game (3.3.1). Such an equilibrium solution will be designed using the method of dynamic programming; see the remainder of Chapter 3. For avoiding cumbersome expressions, Subsections 3.3.2 and 3.3.3 will be confined to the case of the corresponding single-player game, i.e., the multistage control problem with $x \in \mathbf{R}^n$ and $\mathcal{U}_1 = \mathcal{U} = \mathbf{R}^m$.

3.3.2   Brief introduction to dynamic programming

> After all, mathematics started as an experimental field.
> If we do not wish to suffer the usual atrophy of armchair philosophers,
> we must occasionally roll up our sleeves and do some spadework.
> With the aid of dynamic programming and digital computers
> we can methodically engage in mathematical experimentation.
> —Bellman and Dreyfus[28]

Preliminaries

It is difficult to disagree with Bellman's assessment of dynamic programming given in 1962; see the epigraph above. Almost 50 years after its appearance, dynamic programming can be considered one of the most widespread methods for solving optimization problems, along with Pontryagin's maximum principle [215]. The history of dynamic programming is closely connected with the name of Bellman,[29] who "made the biggest contribution to the development of this method and published his results during the 1950s in more than 600 articles and books" [238, p. 24]. The pioneering works were [258, 259]. The emergence of dynamic programming was motivated by a number of interesting and important types of human activity that appeared after World War II, which are conveniently interpreted as multistage decision processes. It was found that the mathematical issues arising in the course of their study go far beyond the scope of classical analysis and therefore require new methods. Following the recognition of these facts, a series of new mathematical theories and methods were created in the middle of the 20th century. Among them, one of the first places was occupied by dynamic programming, which represented a new approach "based on using functional equations and the optimality principle taking into account the capabilities provided by the development of computer technology" [15, p. 17].

Dynamic programming, sometimes called dynamic planning, is a special mathematical method for an optimal (best) planning of controlled processes. By "controlled" we mean processes whose evolution can be affected to some degree.

The first publications on dynamic programming [258, 259] were devoted to multistage controlled processes. These are processes with a sequential transition of an

---

[28] A quote from the book [15, p. 19].

[29] Richard Ernest Bellman was born on August 26, 1920, in New York. He was an American mathematician, one of the leading experts in the field of computer technology. While at school, he achieved the highest level in mathematics among the students of New York. In 1941, he graduated from Brooklyn College, where he received a bachelor's degree in mathematics, and immediately continued his postgraduate studies at Johns Hopkins University in Baltimore. At the beginning of the war with Germany, he took a number of military posts, continuing to actively engage in mathematics. In December 1944, he was drafted into the US army and assigned to the famous Manhattan project in Los Alamos, where he was involved in a number of problems of theoretical physics, until his dismissal in 1946. Then, joining Princeton University, in 1948, he defended his dissertation on the theory of stability under the supervision of a well-known expert in differential equations, Solomon Lefschetz; see the fundamental book *Differential Equations: Geometric Theory*, New York: Princeton University, 1957, by S.Lefschetz. In the summer of 1949, he was invited to RAND Corporation, where he studied in multistage decision processes. At that time in 1951, he developed the framework of dynamic programming for solving applied problems. Bellman's first book on this method, *Introduction to Dynamic Programming*, was published by RAND Corporation in 1953. He is author of 621 papers and 41 books, 21 of which were translated into different languages (in particular, more than a dozen into Russian). He applied dynamic programming in various fields of mathematics—calculus of variations, approximation theory, operations research, optimal control, to name a few. In 1965, he moved from RAND Corporation to the position of Professor of mathematics at the University of South Carolina, where he worked until his death. He passed away on March 19, 1984, in Los Angeles.

object or system from one state to another. Such a division of the process into separate sequential stages either naturally follows from the real properties of the system itself, or is artificially introduced into the problem statement from various considerations. The process stages are not necessarily determined by the passage of time and can be associated with changes in other parameters and characteristics of the system. Without going deep into details, we list some problems of this type, which stimulated the development of dynamic programming [188, p. 12]:

- the optimal allocation of resources (financial, material, energy, etc.) between several enterprises for gaining maximum profit in a certain period of time;
- optimal maintenance and update scheduling for technological equipment of enterprises;
- optimal inventory management (raw materials and finished products of enterprises) for smooth operation;
- minimum-cost road design under difficult terrain conditions;
- maximum-load goods transportation using vehicles.

Elementary resource allocation problem [15]. Consider an example of a resource allocation problem arising in mathematical economics and operations research. Suppose that we have some economic *resources*. This general term may cover people, money, machines, fissile material for nuclear power plants, water for agricultural and industrial purposes or for power generation, fuel for a spaceship, etc. Resources can be used in many different ways, which predetermines the conflict of interests. Each such way will be called a technological process or *production method*.

As the result of consuming all or part of resources by any separate process, we obtain some *gain*.

Let us make the following assumptions.

1)   The gains from different processes can be measured in common units.

2)   The gain from a given process does not depend on how many resources have been allocated to other processes.

3)   The total gain can be defined as the sum of gains from all separate processes.

The main problem is to distribute all available resources among processes in order to maximize the total gain.

Well, there are K different processes numbered by $1, 2, \ldots, K$. (Hereinafter, whenever two processes are considered, we will mean processes 1 and 2; if five processes are considered, then the matter concerns processes 1, 2, 3, 4 and 5, and so on.) The order of numbering of the processes actually makes no sense but, once adopted, it must be firmly observed.

Each process has a utility function expressing the dependence of gain on the amount of resources allocated to it. Denote by x_i the amount of resources allocated to process i and by $g_i(x_i)$ the corresponding gain from this process.

The utility function can be seen in Figure 3.3.1. The shape of this curve is determined by two important economic conditions as follows. First, small amounts of allocated resources, in essence, yield no significant gains; second, a further increase in these amounts finally causes the saturation effect (the law of decreasing gain). As it has been

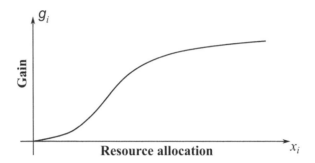

Figure 3.3.1 One example of utility function.

mentioned, x_i and $g_i(x_i)$ can be expressed in different units. The assumptions regarding the independence of processes and the additive property of their utility functions lead to the criterion (also called objective function)

$$F(x_1, \ldots, x_K) = g_1(x_1) + g_2(x_2) + \ldots + g_K(x_K),$$

which assesses the total utility of an allocation procedure.

The maximization problem arises due to a limited amount of available resources. Denoting this amount by a, we arrive at the constraint

$$x_1 + x_2 + \ldots + x_K = a, \quad a = const > 0,$$

where $x_i \geqslant 0$. Thus, the problem is to maximize the objective function $F(x_1, \ldots, x_K)$ in the variables x_i $(i = 1, \ldots, K)$ subject to this equality constraint.

Solution using method of Lagrange multipliers

According to this method, we introduce the auxiliary function

$$\varphi(x_1, \ldots, x_K) = g_1(x_1) + g_2(x_2) + \ldots + g_K(x_K) - \lambda [x_1 + \ldots + x_K - a]$$

and also the vector $\bar{x} = (x_1, \ldots, x_K)$.

The function $\varphi(x_1, \ldots, x_K)$ achieves maximum at a point $(x_1^*, \ldots, x_K^*) = \bar{x}^* \in \mathbf{R}^K$ under the sufficient conditions

$$\left. \frac{\partial \varphi(\bar{x})}{\partial x_i} \right|_{\bar{x}=(x_1,\ldots,x_K)=\bar{x}^*} = 0,$$

$$\left. \frac{\partial^2 \varphi(\bar{x})}{\partial x_i^2} \right|_{\bar{x}=\bar{x}^*} < 0 \qquad (i = 1, \ldots, K).$$

Hence,

$$\left. \frac{\partial \varphi(\bar{x})}{\partial x_i} \right|_{\bar{x}=\bar{x}^*} = \left. \frac{dg_i(x_i)}{dx_i} \right|_{x_i=x_i^*} - \lambda = 0,$$

$$\left. \frac{\partial^2 \varphi(\bar{x})}{\partial x_i^2} \right|_{\bar{x}=\bar{x}^*} = \left. \frac{d^2 g_i(x_i)}{dx_i^2} \right|_{x_i=x_i^*} < 0 \qquad (i = 1, \ldots, K).$$

Resolving the first K equalities with respect to x_i^* (i.e., constructing $x_i^* = h_i(\lambda)$), we find λ from the condition

$$\sum x_i^* = a,$$

which takes the form

$$h_1(\lambda) + h_2(\lambda) + \ldots + h_K(\lambda) = a.$$

These are standard operations for determining the maximum of $F(\bar{x})$ subject to the constraint $\sum x_i = a$.

Example 3.3.1 Let

$$-F(x_1, \ldots, x_K) = a_1 x_1^2 + a_2 x_2^2 + \ldots + a_K x_K^2, \qquad a_i = const > 0, \qquad (i = 1, \ldots, K)$$

where $x_i \geqslant 0$ satisfy the constraint $\sum_{i=1}^{K} x_i = a$. In this case,

$$\varphi(\bar{x}) = \sum_{i=1}^{K} \left(\lambda x_i - a_i x_i^2 \right),$$

and consequently

$$\left. \frac{\partial \varphi(\bar{x})}{\partial x_i} \right|_{\bar{x}=\bar{x}^*} = -2a_i x_i^* + \lambda = 0,$$

$$\left. \frac{\partial^2 \varphi(\bar{x})}{\partial x_i^2} \right|_{\bar{x}=\bar{x}^*} = -2a_i < 0 \qquad (i = 1, \ldots, K).$$

This gives

$$x_i^* = \frac{\lambda}{2a_i} \qquad (i = 1, \ldots, K),$$

but in view of the above constraint

$$\sum_{i=1}^{K} x_i = a,$$

we have

$$\sum_{i=1}^{K} \frac{\lambda}{2a_i} = a.$$

As a result,

$$\lambda = \frac{a}{\sum\limits_{i=1}^{K} \frac{1}{2a_i}},$$

and

$$x_i^* = \frac{\lambda}{2a_i} = \frac{\frac{1}{2a_i}a}{\sum\limits_{j=1}^{K}\frac{1}{2a_j}} = \frac{a}{a_i\sum\limits_{j=1}^{K}\frac{1}{a_j}} \quad (i = 1, \dots, K).$$

Thus, the maximum value of $-F(x_1, \dots, x_K)$ (which obviously coincides with the minimum value of $F(x_1, \dots, x_K)$) is

$$F(x_1^*, \dots, x_K^*) = \frac{a^2}{\sum\limits_{i=1}^{K}\frac{1}{a_i}}.$$

This example and its solution are typical for textbooks on mathematical analysis and calculus. However, the real problems arising in important applications are often least suitable for standard methods and require more subtle consideration. Let us clarify the features causing difficulties in this case.

Features

Consider in detail those fundamental obstacles occurring along the pathway of applying mathematical analysis methods to the general resource allocation problem.

Feature 1. (Local optimum)
As follows from the construction procedure above, the points $\bar{x}^*$ at which the function $F(\bar{x}) - \lambda \sum x_i$ achieves maximum for x_i^* satisfy the equation

$$\frac{dg_i(x_i)}{dx_i} - \lambda = 0 \quad (i = 1, \dots, K).$$

(Recall that $\bar{x}^* = \left(x_1^*, \dots, x_K^*\right)$.) For $g_i(x_i)$ shown in Figure 3.3.1, this equation may have two roots. There is no clarity which of the roots corresponds to the global maximum; hence, all combinations of values must be checked. This procedure requires assessing 2^K cases. If $K = 10$, then we have 1024 cases, which is not so large though; but for $K = 20$, the number of cases will exceed 10^6!

Moreover, utility functions can have a more complex nature (than in Figure 3.3.1), which creates a serious theoretical difficulty and also a considerable obstacle to numerical solution.

Note that the method of dynamic programming always yields a global (not local) optimum.

Feature 2. (Constraints)
With a traditional application of the classical calculus method, we actually neglect the fact that a maximum has to be found in some *a priori* bounded domain: the zero value of the first derivatives in combination with the definiteness of the second derivatives determine only *inner* optimum, in no way indicating of the maximum achieved on *the boundaries* of the domain of x. As an example, we take the function of one variable $g(x)$ in Figure 3.3.2. The derivative $\frac{dg(x)}{dx}$ vanishes at the points a_1 and a_2 of a local maximum of $g(x)$, yet being not equal to zero for $x = 0$, where $g(x)$ achieves global maximum on the interval $[0, x_0]$.

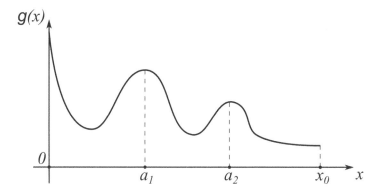

Figure 3.3.2 One example of Feature 2.

Sadly, this common situation occurs in controlled economic and technical processes, for which the conditions

$$c_i \leqslant x_i \leqslant b_i \quad (i = 1, \ldots, K)$$

are natural and reasonable. Note that in maximization problems with *many variables*, the number of all possible options (including the points of local maxima, boundary points) exceed all expectations and the very idea of enumeration has to be rejected.

In the optimization problems reducible to combinatorial problems that require evaluating each option separately, the total number of cases often increases exponentially (and even faster) with increasing the number of measurements. For example, if the total number of options is 2^K, then transition from K to $2K$ will even change the order of magnitude for computational time. Spending 100 minutes instead of 10 on computations usually does not cause problems, provided that the answer is of definite value. But the need to consume 100 hours instead of 10 for obtaining a numerical solution may raise the question: should this solution method still be preferred or rather withdrawn in favor of a faster counterpart?

With the method of dynamic programming being used, any condition of the form $a_i \leqslant x_i \leqslant b_i$ restricts the number of options at each stage. As a result, the search procedure becomes simpler and computational effort is reduced accordingly. In other words, the fewer actions are performed at each stage of dynamic programming, the faster the calculations are.

Feature 3. (Maximization on discrete sets)
The apparatus of classical analysis and calculus suits well optimization problems with a continuous change of independent variables. This type of change can be used as a rough approximation to real problems. However, in a series of cases, such smoothing crucially affects the accuracy of solution. One case corresponds to binary variables taking only two values, 0 and 1.

With one artificial technique or another, continuous change can be usually introduced. But as was claimed in 1962, "optimization on discrete sets basically requires new methods, and at present many important classes of problems are completely beyond our reach"; see the book [15, p. 31]. However, the active development of discrete programming—a branch of mathematics that studies and solves optimization problems on finite sets—began in the second half of the 20th century. As was emphasized in 1978, "universal effective methods for solving dynamic programming problems have not been created"; see [249, p. 204].

The application of dynamic programming to discrete sets (like in the example above) actually simplifies each stage, reducing computational efforts: the smaller the number of possible options is, the simpler the calculations are. The elementary problems are those in which each variable x_i can take only a few values, say, 0 or 1.

Feature 4. (Non-differentiable functions)
As is well known, there exist continuous functions that are defined on a certain interval and have no derivative at any point on this interval. To tell the truth, such a "pathology" would hardly occur in applications. Nevertheless, it is quite expectable to encounter objective functions suffering from other inconveniences such as discontinuity on a finite set of points and the existence of one-sided derivatives only. Step functions (Figure 3.3.3) are the simplest representatives of this class, serving as acceptable approximations of smoother but much more complex functions.

Another interesting and useful class of functions (Figure 3.3.4) consists of piecewise constant functions of the form

$$g(x) = \max\{a_1 x + b_1, a_2 x + b_2, \ldots, a_K x + b_K\}.$$

Here derivatives can be constructed on an *a priori* given interval of interest. If this such an interval is unknown, then again we have to deal with special annoying features of

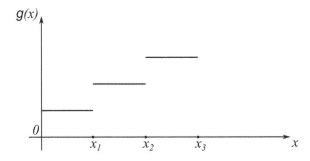

Figure 3.3.3 One example of Feature 4.

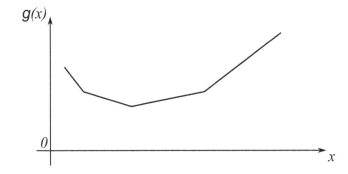

Figure 3.3.4 Another example of Feature 4.

combinatorial optimization problems. In this case, determining the domain of features often becomes an essential part of the solution.

The method of dynamic programming mainly operates with the tabular values of objective functions, and their analytical structure is of no sense.

Feature 5. (Linearity)

Difficulties of the opposite nature accompany optimization problems in which all the functions are linear. In this case, all derivatives exist but give little information, because the optima are surely reached at the boundary points of the domain of change. *Linear programming* is a special method for maximizing a linear objective function of the form

$$\sum_{i=1}^{K} c_i x_i,$$

where $x_i \geqslant 0$, subject to the constraints

$$\sum_{j=1}^{K} a_{ij} xj \leqslant b_i \quad (i = 1, \ldots, K).$$

This universal tool pays *almost no attention* to the structure of a process under study.

If some structural features (convexity, monotonicity, concavity, etc.) ate found, they can be effectively used in the method of discrete programming to simplify the search procedure. These issues were discussed in detail in the book [15].

Feature 6. (Stability)

As it has been mentioned, classical analysis and calculus methods are based on the continuous change of independent variables. Hence, the results obtained in the classical way turn out to be very sensitive to local changes and therefore to small errors. Consider the two functions $g(x)$ and $h(x)$ demonstrated in Figs. 3.3.5 and 3.3.6, respectively; the latter is an idealized "mathematical" function whereas the former a "physical" function.

Figure 3.3.6 has the following interpretation: if $h(x)$ is defined as the result of measurements or calculations, then its value at each point x is never equal to the number $h(x)$, since it has a spread of values. Therefore, if we determine the value of $h(x)$ with a high degree of accuracy, then the tangent direction at any given point is not set exactly. Thus, relying on optimization methods with differentiation becomes very problematic.

In fact, here it is advisable to use a method *ensuring that the errors in the answer will not exceed those in the initial data. This is a property called stability*.

The stability analysis of computational algorithms is a major challenge for modern theory of numerical analysis.

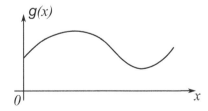

Figure 3.3.5 Idealized mathematical function.

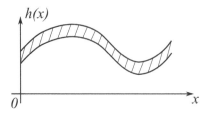

Figure 3.3.6 Physical function.

The concept of stability has close connection with the concept of *sensitivity*, which is due to the following circumstances. In most cases, determining the optimal behavior of a controlled system for single values of its parameters *is not enough*. Usually, it is required to find out whether the optimal behavior will remain the same or not if the parameters vary in some critical domain of values, i.e., whether the optimal behavior will be preserved or not with the indicated change in parameters.

For example, important parameters in the distribution procedures described in this section (the elementary resource allocation problem) are the amount of resources available and the number of processes. In the method of dynamic programming, the original problem is automatically included into a family of similar subproblems, in which the amount of resources varies from 0 to a and the number of processes from 1 to K. Therefore, after a numerical solution was obtained for the K-stage procedure, we can determine the relation between the parameters of the current state and the number of stages for a huge number of subproblems. In the resource allocation problem, the optimal gain is obtained as a function of the main parameters (the amount of current resources and the number of processes) and hence a sensitivity analysis automatically accompanies the construction of a numerical solution.

3.3.3   Bellman's principle of optimality

Three assumptions

As a matter of fact, dynamic programming has the following specifics: to find the optimal control, *a planned operation is divided into a number of successive stages (also called steps). As a result, the control process itself becomes multistage, evolving successively from step to step, and the control is optimized at each stage accordingly.*

Some operations naturally fall into stages; in others, such a partition has to be introduced artificially.

The method of dynamic programming proceeds from a series of *basic assumptions* regarding the character of problems to be solved. These assumptions are satisfied for a wide class of economic and technical systems and processes, and an attempt to reject them would significantly increase the complexity and cumbersomeness of solution methods.

Assumption I *No aftereffect*: the state $x(k)$ of a controlled system $\sum$ at stage k depends *only* on the state $x(k-1)$ and control $u(k-1, x(k-1))$ at the previous stage $k-1$.

This assumption can be written as

$$x(k) = f(k-1, x(k-1), u(k-1, x(k-1))), \qquad (3.3.5)$$

or in the developed form

$$x(1) = f(0, x(0), u(0, x(0))), \quad x(2) = f(1, x(1), u(1, x(1))), \ldots, x(K)$$
$$= f(K - 1, x(K - 1), u(K - 1, x(K - 1))).$$

The dependence of $x(k)$ on the earlier states $x(k - 2), x(k - 3), \ldots, x(0) = x_0$ (i.e., on the entire history of the process) has an implicit nature, being revealed through $x(k - 1)$ only.

The same assumption $F(k, x(k), u[k])$, $u[k] = u(k, x(k))$, holds for the partial objective functions $F(k, x(k), u(k))$ at each stage of the process, i.e.,

$$F(t_0, x(t_0), u[t_0]), F(t_0 + 1, x(t_0 + 1), u[t_0 + 1]), \ldots, F(K - 1, x(K - 1), u[K - 1]),$$

and the fact of achieving a terminal state $x(K)$ is stimulated by an additional payoff $\Phi(x(K))$.

Assumption II *The objective function is additive* with respect to the partition of a decision process into stages, i.e., the resulting objective function $\mathcal{J}(U, t_0, x_0)$ for the entire multistage process is represented as the *sum* of partial objective functions $\mathcal{J}_k$ at each stage:

$$\mathcal{J}(U, t_0, x_0) = \mathcal{J}_{t_0} + \mathcal{J}_{t_0+1} + \ldots + \mathcal{J}_{K-1} + \Phi(x(K)),$$

where

$$\mathcal{J}_k = F(k, x(k), u[k]) \quad (k = t_0, t_0+1, \ldots, K-1).$$

Hereinafter, an objective function $\mathcal{J}(U, t_0, x_0)$ of this type will be also called an *additive criterion*.

Assumption III The sequence of discrete time instants $0, 1, 2, \ldots, K-1, K$ of a multistage controlled process is *finite*, i.e., there exists an integer $L > 0$ such that $\max \{0, 1, \ldots, K\} < L < \infty$. The multistage controlled processes that satisfy Assumption III are often called *finite*.

From this point onwards, we consider the systems and controlled processes that satisfy Assumptions I–III above.

Remarks on optimization of multistage processes
Here are some preliminary remarks on optimization of multistage processes.

First of all, note that the optimization problem of a process on a set of stages *cannot be equivalently reduced to the independent optimization subproblems at separate stages*. Really, achieving a maximum gain at one of the stages may require too much resources, thereby decreasing the economic effect at subsequent stages and hence over the entire multistage process. For example, long-distance running is not the same as a sequence of sprints at maximum speed.[30] In this example, the objective function is the time to

---

[30]This was described well in the song *About a skate sprinter* by Vladimir S. Vysotsky, (1938–1980), Russian actor, poet, songwriter, and performer who was considered "the voice of the nation's heart":

For ten thousand I rushed like for a mile
And withered!
Gasping mouth, goggled eyes, this goddam race,
I cursed it!
Made three laps and then fell down in disgrace,
Exhausted!

See http://www.wysotsky.com/1033.htm?594

cover the entire distance; an admissible control (strategy) is a tactic that allows simply reaching the end of the distance; an optimal control is a tactic that allows covering the entire distance in minimum time.

Another example from industry is the enterprise modernization planning problem. Generally, modernization is performed under limited financial and other resources and is complicated by the requirement of uninterrupted production. As a result, modernization is carried out not instantly (at once), but in several stages. An optimal policy for the allocation and use of resources by stages is crucial for successful modernization of an enterprise.

This intuitively clear remark can be illustrated by the following mathematical property: the maximum (or minimum) of the sum of terms is generally not equal to the sum of their maxima (or minima). More precisely, for arbitrary functions $\varphi_1(t)$ and $\varphi_2(t)$, only the inequality

$$\max_t [\varphi_1(t) + \varphi_2(t)] \leqslant \max_t \varphi_1(t) + \max_t \varphi_2(t)$$

can be guaranteed; the equality may take no place. For example, let $\varphi_1(t) = t$ and $\varphi_2(t) = 1 - t$, where $0 \leqslant t \leqslant 1$; in this case,

$$1 = \max_{t \in [0,1]} [\varphi_1(t) + \varphi_2(t)] < \max_{t \in [0,1]} \varphi_1(t) + \max_{t \in [0,1]} \varphi_2(t) = 1 + 1 = 2.$$

However, if $\varphi_2(t) = A = const$, then

$$\max_t [\varphi_1(t) + A] = \max_t \varphi_1(t) + A,$$

i.e., a constant term can be taken outside the maximum operator. By analogy,

$$\min_t [\varphi_1(t) + \varphi_2(t)] \geqslant \min_t \varphi_1(t) + \min_t \varphi_2(t)$$

and

$$\min_t [\varphi_1(t) + A] = \min_t \varphi_1(t) + A.$$

Also, the following equality holds:

$$\min_t \varphi(t) = -\max_t \{-\varphi(t)\}.$$

Finally, using the difference equation

$$x(k+1) = f(k, x(k), u[k]), \quad x(0) = x_0 \quad (k = 0, \ldots, K-1),$$

the criterion

$$J(U, 0, x_0) = \Phi(x(K)) + \sum_{k=0}^{K-1} F(k, x(k), u[k])$$

can be transformed to

$$J(U, 0, x_0) = F(0, x_0, u[0]) + F(1, x(1), u[1]) + \ldots + F(K-1, x(K-1), u[K-1])$$
$$+ \Phi(x(K)) = \mathcal{F}(x_0, u[0], \ldots, u[K-1]),$$

where

$$
\begin{cases}
x(1) = f(0, x_0, u[0]), \\
x(2) = f(1, x(1), u[1]) = f\left(1, f(0, x_0, u[0]), u[1]\right), \\
x(3) = f(2, x(2), u[2]) = f\left(2, f\left(1, f(0, x_0, u[0]), u[1]\right), u[2]\right), \\
\quad \vdots \\
x(K) = f(K-1, x(K-1), u[K-1]) = f\left(K-1, f(K-2, x(K-2), u[K-2])\right), \\
\qquad u[K-1]) = \ldots = \varphi(0, x_0, u[0], \ldots, u[K-1]).
\end{cases}
$$

Hence, in final analysis the problem has been reduced to maximization of the function $\mathcal{F}(x_0, u[0], \ldots, u[K-1])$ in Km variables $(u[0], \ldots, u[K-1])$.

The difficulties to find

$$
\mathcal{J}^* = \max_U \mathcal{J}(U, 0, x_0) = \max \mathcal{F}(x_0, u[0], \ldots, u[K-1])
$$

have been discussed in detail at the end of Subsection 3.3.2. To summarize, an attempt to apply classical differential calculus methods for optimizing multistage processes leads to problems no less complicated than the original one, which is not an effective way. Therefore, the problem should be solved using other methods with due account for its specifics. Dynamic programming is the most famous and effective method of this class, and we proceed to its basic principle.

When planning a multistage operation, we have to choose control at each stage by considering not the narrow interests of this particular stage, but the wide interests of the entire operation. These two approaches do not always coincide!

How such a control strategy can be constructed? A general rule has already been formulated: in the process of multistage decision-making, control at each stage should be chosen with the future in mind. However, there exists an exception to this rule. Among all the stages, one can be planned simply, without regard to the future. What is this stage? Obviously the last one! Only this stage as such can be planned for obtaining the greatest gain, for example, by appointing a payoff for achieving the last stage.

With the last stage having been optimally planned, we may attach to it the previous stage (last but one), and so on, until the first stage is attached.

Therefore, the process of dynamic programming always evolves backwards in time: not from the beginning to the end, but *from the end to the beginning*. First of all, the last stage is planned. But how to plan it, if we do not know the outcome of the last but one stage? Obviously, different assumptions about the outcome of this stage have to be considered, and for each of them an appropriate control at the last stage has to be chosen.

Such an optimal control constructed under a certain condition (the outcome of a previous stage) is called *conditional optimal control*.

The method of dynamic programming requires finding at each stage a conditional optimal control for any of the possible outcomes of a previous stage.

First of all, we consider the elementary multistage process with two stages of scalar control, $K = 2$ and $\Phi(x(2)) = 0$. In this case, the vector of control variables (u_1, u_2) contains two components, and the objective function $\mathcal{J}$ is a function of two variables $u[0]$ and $u[1]$:

$$
\mathcal{J} = \mathcal{J}(u[0], u[1]) = F(0, x_0, u[0]) + F(1, x(1), u[1]),
$$

where $x(1) = f(0, x_0, u[0])$. For determining the optimal value $\mathcal{J}^*$, we have to calculate the maximum of the objective function with respect to both control variables $u[0], u[1]$:

$$
\mathcal{J}^* = \max_{u[0], u[1]} \{F(0, x_0, u[0]) + F(1, x(1), u[1])\}.
$$

The maximum in two variables *always* can be found by *successive* maximization with respect to each of the variables. Hence,

$$\mathcal{J}^* = \max_{u[0]} \left\{ \max_{u[1]} [F(0, x_0, u[0]) + F(1, x(1), u[1])] \right\}.$$

(Of course, this transformation is valid if the domains of the vectors $(u[0], u[1])$ used for the simultaneous and successive maximization of $\mathcal{J}^*$ coincide with each other.) The first term $F(0, x_0, u[0])$ of the above sum does not depend on the variable $u[1]$ with respect to which the inner maximum is calculated. Consequently, this term can be taken outside the inner maximum operator:

$$\max_{u[1]} \{F(0, x_0, u[0]) + F(1, x(1), u[1])\} = F(0, x_0, u[0]) + \max_{u[1]} [F(1, x(1), u[1])].$$

Thus, we obtain the equality

$$\mathcal{J}^* = \max_{u[0]} \left\{ F(0, x_0, u[0]) + \max_{u[1]} [F(1, x(1), u[1])] \right\}.$$

As the result of several transformations, simultaneous maximization with respect to the two variables $u[0]$, $u[1]$ has been reduced to two successive operations of maximization with respect to the separate variables $u[0]$ and $u[1]$. In other words, the two-dimensional maximization problem has been written as a system of two one-dimensional problems solved successively. As a rule, a one-dimensional problem is significantly simpler than a two-dimensional one (not to mention a multidimensional one). These transformations match a general rule: solving several problems of smaller dimension is easier than one problem of higher dimension.

The structure of the above equality suggests that calculation of $\mathcal{J}^*$ should be partitioned into 2 stages, which corresponds to a two-stage process and yields simpler formulas. For this purpose, we introduce a new auxiliary function $V^{(1)}(x(1))$ defined as the inner maximum:

$$V^{(1)}(x(1)) = \max_{u[1]} [F(1, x(1), u[1])].$$

The function $V^{(1)}(x(1))$ represents the maximum value of the partial objective function $F(1, x(1), u[1])$ at stage 2 of the process, provided that before this stage the controlled system was at the state $x(1)$. With the function $V^{(1)}(x(1))$, we may write

$$\mathcal{J}^* = \max_{u[0]} \left\{ F(0, x_0, u[0]) + V^{(1)}(x(1)) \mid x(1) = f(0, x_0, u[0]) \right\}.$$

Hereinafter, the symbol "—" means "subject to condition": the maximum of $F(0, x_0, u[0]) + V^{(1)}(x(1))$ with respect to the variable $u[0]$ is calculated *subject to the condition* that $x(1)$ is not arbitrary but given by $x(1) = f(0, x_0, u[0])$.

Let us reduce the two latter expressions to a homogeneous form. For uniform notations, we introduce two new functions: the function $V^{(2)}(x(2))$ that is identically zero ($V^{(2)}(x(2)) = 0$ for all $x(2)$) and the function $V^{(0)}(x_0)$ that is defined only for x_0 so that $V^{(0)}(x_0) = \mathcal{J}^*$ (the desired optimal value). With these notations, the expressions can be written as

$$V^{(1)}(x(1)) = \max_{u[1]} \left\{ F(1, x(1), u[1]) + V^{(2)}(x(2)) \mid x(2) = f(1, x(1), u[1]) \right\},$$

$$V^{(0)}(x_0) = \max_{u[0]} \left\{ F(0, x_0, u[0]) + V^{(1)}(x(1)) \mid x(1) = f(0, x_0, u[0]) \right\}.$$

We emphasize that the two equalities above has the same structure, differing only by the indexes (sub- and superscripts) of variables and functions. Thus, the transformations performed using the auxiliary functions $V^{(0)}(x_0)$, $V^{(1)}(x(1))$, and $V^{(2)}(x(2))$ have simplified the calculation procedure of the optimal value $\mathcal{J}^*$ by reducing the dimension and moreover, have led it to a *homogeneous* form corresponding to the structure of the process. The calculations are performed in *the descending order of* superscripts: $V^{(2)}(x(2)) = 0$ by definition; using this equality, we further calculate $V^{(1)}(x(1))$ and finally find $V^{(0)}(x_0) = \mathcal{J}^*$. Particular attention should be paid to the form of the last two equalities, since it expresses the general principle of optimality.

Bellman's principle of optimality is embodied in dynamic programming
According to R. Bellman, the basic principle of optimality of multistage decision processes can be expressed as follows: *"An optimal policy has the property that whatever the initial state and initial decision are, the remaining decisions must constitute an optimal policy with regard to the state resulting from the first decision."*; see the books [14, p. 82] and [15, pp. 36–37]. In other words, any part of an optimal trajectory, including the final one, is also optimal, and control errors causing deviations from the optimal trajectory cannot be subsequently corrected. Of course, such a general rule is not directly applicable to dynamic programming problems and needs further specification.

For complete presentation, we give the formalization of Bellman's principle of optimality from [49, p. 28]: "For a control to be optimal, *a necessary condition* is that for any duration of periods, the control at the second period is optimal with respect to the state where the process was after control at the first period."

Thus, Bellman's principle of optimality is implemented by "finding an optimal continuation of the process with respect to the state that has been achieved by a current time instant" [35, p. 14].

We proceed to the mathematical formalization of Bellman's principle of optimality subject to maximization problems.

For rigorously stating this principle, we again introduce a series of auxiliary functions $V^{(0)}(x_0)$, $V^{(1)}(x(1)), \ldots, V^{(K)}(x(K))$. The functions $V^{(k)}(x(k))$ ($k = 0, 1, 2, \ldots, K$) have an important economic interpretation, as they represent the maximum values of sums of the partial objective functions

$$F(k, x(k), u[k]) + \ldots + F(K - 1, x(K - 1), u[K - 1]) + \Phi(x(K))$$

that are calculated over all admissible truncated sets of controls $(u[k], \ldots, u[K - 1])$. In other words, $V^{(k)}(x(k))$ is **the conditionally optimal value of the objective function** in the case driving the system from the state $x(k)$ after stage k in the terminal state $x(K)$; its conditional optimality consists in referring to not the entire process but to its final part, and depends on the choice of the initial state $x(k)$ for the truncated process. Thus, the functions $V^{(k)}(x(k))$, often called **Bellman functions**, characterize the optimal properties of the controlled system Σ at *subsequent* stages of the process. Note that the simple and important relation

$$V^{(K)}(x(K)) = \Phi(x(K))$$

holds because the state $x(K)$ is terminal: the system Σ will no more change its state, and the corresponding economic effect (the payoff of the system Σ for reaching the desired state $x(K)$) is $\Phi(x(K))$.

Bellman's principle of optimality underlies the method of dynamic programming, and is expressed by the *functional equation*

$$V^{(j-1)}(x(j-1)) = \max_{u[j-1]} \left\{ F(j-1, x(j-1), u[j-1]) + V^{(j)}(x(j))|x(j) \right.$$

$$\left. = f(j-1, x(j-1), u[j-1]) \right\}, \tag{3.3.6}$$

in which index j runs over the numbers of all stages of the process *in backward direction*: $j = K, K-1, \ldots, 2, 1$. In addition, $V^{(K)}(x) = \Phi(x)$.

The functional Bellman equation is structurally a ***recursive*** equation. This means that in the sequence of functions $V^{(0)}(x(0))$, $V^{(1)}(x(1))$, $\ldots$, $V^{(K)}(x(K))$, each previous element is expressed through the next one.

What is important? In the functional Bellman equation, the maximum value $V^{(j-1)}(x(j-1))$ for each fixed $x(j-1)$ is calculated simultaneously with the corresponding value(s) of the variable $u[j-1]$ at which this maximum value is achieved. This value depends on the state $x(j-1)$ and therefore will be denoted by $\tilde{u}(j-1, x(j-1))$. In fact, $\tilde{u}(j-1, x(j-1))$ is a function (possibly, *multivalued*), which is called ***the conditionally optimal strategy***. Conditional character consists in its dependence on the choice of the state $x(j-1)$. Although the functions $\tilde{u}(j-1, x(j-1))$ are not implicitly participating in equation (3.3.6), they still are as important as the Bellman functions $V^{(j-1)}(x(j-1))$ and are used for designing the optimal strategy. In the course of calculations, the conditionally optimal strategies

$$\tilde{u}(K-1, x(K-1)), \ \tilde{u}(K-2, x(K-2)), \ \ldots, \ \tilde{u}(0, x(0))$$

are successively found and *memorized*.

With Bellman's principle of optimality, a specific problem with the optimal value $V^{(0)}(x(0))$ is treated not separately but as a representative of the *family* of similar problems with the optimal values $V^{(1)}(x(1))$, $\ldots$, $V^{(K)}(x(K))$ of smaller dimension, i.e., simpler ones. There exists a connection between the problems of this family, which is described by the functional equation (3.3.6). This equation can be used for successively calculating all functions $V^{(K-1)}(x(K-1))$, $\ldots$, $V^{(0)}(x(0))$, starting from the simplest function $V^{(K)}(x(K)) = \Phi(x(K))$. As a result, the solutions of the entire family of problems are constructed. Actually, the technique to replace an original problem with a family of homogeneous problems whose solutions determine the solution of the former problem is called ***the immersion and invariance (I&I) principle***.

The principle of optimality has almost the same statement for *minimization* problems: the only difference is that max in the functional equation (3.3.6) is obviously replaced by min.

Discussion of principle of optimality

For real applications, the recurrent-functional Bellman equations are often derived without using the principle of optimality. In each case, the corresponding Bellman equations are obtained by equivalent and very simple transformations that do not require additional facts. In essence, these transformations contain the principle of optimality or, more precisely put, they are a concrete implementation of the principle of optimality. In such cases, the validity of the principle of optimality is out of question, because strict mathematical transformations provide necessary grounds. However, the principle of optimality is a general statement, and its justification is of interest regardless of specific problems solved. Even for the processes described by equation (3.3.5), the principle of optimality as a necessary condition of optimality is not always true; see the discussion above. What is the connection between the principle of optimality and the sufficient conditions of optimality? The answer to this question plays a fundamental role for solving specific problems by the method of dynamic programming. Really, with the principle of optimality being applied as a necessary condition, we obtain relations and procedures to calculate controls that are not necessarily optimal. The effectiveness of dynamic programming consists in the development of algorithms leading to optimal controls. Hence, the principle of optimality in the form of a sufficient condition should be the basis of dynamic programming. From this viewpoint, the issue is scantily covered in the literature. For the optimization problems solved by dynamic programming, the principle of optimality is often declared but not explicitly used in its conventional form. When deriving the Bellman equations for such problems, the relations resulting from the principle of optimality are not written out, but the transformations leading to the Bellman equation are performed; their legitimacy follows not from the principle of optimality, but from common mathematical facts. In most cases, the transformations adopted for deriving the Bellman equations are such that the resulting relations usually express sufficient conditions of optimality. We will adhere to this approach in the next section when considering sufficient conditions of a Berge equilibrium in the multistage game $\Gamma(t_0, x_0)$ (3.3.1).

3.4   SUFFICIENT CONDITIONS OF BERGE EQUILIBRIUM

> I have had my results for a long time,
> but I do not yet know how to arrive at them.
> —Gauss[31]

In this section, sufficient conditions for the existence of Berge equilibria (Theorems 3.4.1 and 3.4.2) are established. These conditions are based on the method of dynamic programming and then used in Sections 3.5–3.9.

3.4.1   Problem statement

Like in Subsection 3.3.1, consider the K-stage noncooperative positional two-player game (3.3.1) with an initial position (t_0, x_0), which is described by an ordered quadruple

$$\Gamma(t_0, x_0) = \langle \{1, 2\}, \textstyle\sum_{\varnothing}, \{\mathfrak{A}_i\}_{i=1,2}, \{\mathcal{J}_i(U, t_0, x_0)\}_{i=1,2} \rangle, \tag{3.4.1}$$

The game (3.4.1) has the following notations:

1 and 2 as the numbers of players;

$\sum_{\varnothing}$ as a controlled system evolving over time in accordance with the vector difference equation

$$x(k + 1) = f(k, x(k), u_1, u_2), \quad x(t_0) = x_0 \quad (k = 0, 1, \ldots, K - 1), \tag{3.4.2}$$

with an integer $K > 0$ and an initial initial position $(t_0, x_0) \in \{0, 1, \ldots, K-1\} \times \mathbf{R}^n$;

$x \in \mathbf{R}^n$ and $u_i \in \mathbf{R}^{m_i}$ $(i = 1, 2)$;

$U_i(k)$ as a strategy of player i at a time instant $k \in \{0, 1, \ldots, K-1\}$ that is identified with an m_i-dimensional vector function $u_i(k, x)$:

---

[31]Carl Friedrich Gauss, original name Johann Friedrich Carl Gauss, (1777–1855), German mathematician, generally regarded as one of the greatest mathematicians of all time for his contributions to number theory, geometry, probability theory, the theory of functions, and potential theory.

$$U_i(k) \div u_i(k, x) \in \mathcal{U}_i(k, x) \quad \forall x \in \mathbf{R}^n,$$

where the subsets $\mathcal{U}_i(k, x) \subseteq \mathbf{R}^{m_i}$ are *a priori* given at all positions (k, x) $\forall x \in \mathbf{R}^n$. Denote

$$\mathfrak{A}_i(k) = \{U_i(k) \div u_i(k, x) \in \mathcal{U}_i(k, x) \quad \forall x \in \mathbf{R}^n\} \quad (i = 1, 2);$$

then the set of all strategy profiles at a time instant k has the form

$$U(k) = (U_1(k), U_2(k)) \in \mathfrak{A}_1(k) \times \mathfrak{A}_2(k) = \mathfrak{A}(k).$$

Assume that the players have chosen their strategies $U_i(k) \div u_i(k, x)$ for all $k = t_0$, $t_0+1, \ldots, K-1$,

$$U_i(k) \in \mathfrak{A}_i(k), \quad U(k) \div u(k, x) = (u_1(k, x), u_2(k, x)).$$

According to (3.3.3),

$$x(t_0 + 1) = f(t_0, x_0, u(t_0, x_0)),$$
$$x(t_0 + 2) = f(t_0 + 1, x(t_0 + 1), u(t_0 + 1, x(t_0 + 1))),$$
$$\vdots$$
$$x(K) = f(K - 1, x(K - 1), u(K - 1, x(K - 1))). \tag{3.4.3}$$

A strategy U_i of player i in the game (3.4.1) is identified with a collection

$$U_i \div \left(u_i(t_0, x), u_i(t_0 + 1, \tilde{x}(t_0 + 1)), \ldots, u_i(K - 1, \tilde{x}(K - 1))\right),$$

where $\tilde{x}(k)$ $(k = t_0+1, \ldots, K-1)$ are the solutions of (3.4.3) with $x_0 = x$. In other words, for each $x \in \mathbf{R}^n$ and any strategies $U_i(t_0) \div u_i(t_0, x)$, $U_i(t_0) \in \mathfrak{A}_i(t_0)$ $(i = 1, 2)$, chosen by the players, the corresponding solution

$$\tilde{x}(t_0 + 1) = f(t_0, x, u_1(t_0, x), u_2(t_0, x)),$$

for the strategies

$$U_i(t_0 + 1) \div u_i(t_0 + 1, x), \quad U_i(t_0 + 1) \in \mathfrak{A}_i(t_0 + 1) \quad (i = 1, 2)$$

chosen by the players at the time instant $t_0 + 1$ is given by

$$\tilde{x}(t_0 + 2) = f(t_0 + 1, \tilde{x}(t_0 + 1), u_1(t_0 + 1, \tilde{x}(t_0 + 1)), u_2(t_0 + 1, \tilde{x}(t_0 + 1))),$$

$$\vdots$$

for the strategies

$$U_i(K - 2) \div u_i(K - 2, x), \quad U_i(K - 2) \in \mathfrak{A}_i(K - 2) \quad (i = 1, 2)$$

chosen by the players at the time instant $K - 2t_0 + 1$, the state vector $\tilde{x}(K - 1)$ is given by

$$\tilde{x}(K - 1) = f(K - 2, \tilde{x}(K - 2), u_1(K - 2, \tilde{x}(K - 2)), u_2(K - 2, \tilde{x}(K - 2))).$$

Denote by $\mathfrak{A}_i$ the set of all strategies U_i of player i in the game (3.4.1).

For a strategy profile $U = (U_1, U_2) \in \mathfrak{A}_1 \times \mathfrak{A}_2 = \mathfrak{A}$ formed by specific strategies $U_i \in \mathfrak{A}_i$ ($i = 1, 2$) of the players, *a play of the game* evolves over time in accordance with the recursive relations (3.4.3). As a result, we obtain three sequences

$$\{x(k)\}_{k=0}^{K}, \quad \{u_i[k] = u_i(k, x(k))\}_{k=0}^{K-1} \quad (i = 1, 2),$$

on which the payoff function (3.3.4) of player i is defined:

$$\mathcal{J}_i(U, t_0, x_0) = \Phi_i(x(K)) + \sum_{k=t_0}^{K-1} F_i(k, x(k), u_1[k], u_2[k]). \tag{3.4.4}$$

A pair

$$\left(U^B = (U_1^B, U_2^B), \mathcal{J}(U^B, t_0, x_0) = \left(\mathcal{J}_1^B[t_0, x_0], \mathcal{J}_2^B[t_0, x_0] \right) \right) \in \mathfrak{A} \times \mathbf{R}^2$$

is a Berge equilibrium solution of game (3.4.1) if

1) U^B *is a Berge equilibrium for all initial positions* $(t_0, x_0) \in \{0, 1, \ldots, K-1\} \times \mathbf{R}^n$, *i.e.,*

$$\max_{U_2 \in \mathfrak{A}_2} \mathcal{J}_1(U_1^B, U_2, t_0, x_0) = \mathcal{J}_1(U^B, t_0, x_0),$$

$$\max_{U_1 \in \mathfrak{A}_1} \mathcal{J}_2(U_1, U_2^B, t_0, x_0) = \mathcal{J}_2(U^B, t_0, x_0) \tag{3.4.5}$$

$\forall \, (t_0, x_0) \in \{0, 1, \ldots, K-1\} \times \mathbf{R}^n$;

2)   *in this case,* $\mathcal{J}_i^B[t_0, x_0] = \mathcal{J}_i(U^B, t_0, x_0)$ ($i = 1, 2$).

In the next subsection, we derive sufficient conditions for the existence of a Berge equilibrium solution of the game (3.4.1). They are inspired by the method of dynamic programming; see Subsection 3.3.1.

3.4.2   Sufficient conditions

> The idea of dynamic programming consists in that
> maximization of a function of many variables
> is replaced by multiple maximization of
> a function of one or few variables.
> —Wentzel[32]

We define two sequences of scalar functions

$$V_i^{(K)}(x), \; V_i^{(K-1)}(x), \; \ldots, \; V_i^{(t_0)}(x) \quad (i = 1, 2).$$

First of all, let

$$V_i^{(K)}(x) = \Phi_i(x) \quad \forall x \in \mathbf{R}^n \quad (i = 1, 2) \tag{3.4.6}$$

---

[32] Elena S. Wentzel, (1907—2002), was a Soviet mathematician, author of textbooks on probability theory and operation research, Doctor of technical sciences (1954), Professor (1955); also known as Russian prose writer, under the pen name of I. Grekova. An English translation of a quote from [35, p. 10].

and also let a pair of functions $V_i^{(j)}(x)$ ($i = 1, 2$) with domains in $\mathbf{R}^n$ ($k = K$, $K-1, \ldots, j$; $i = 1, 2$) be constructed. Using the method of dynamic programming, we recursively define the functions $V_i^{(j-1)}(x)$ ($i = 1, 2$) in the following way: for all $x \in \mathbf{R}^n$,

$$V_1^{(j-1)}(x) = \max_{u_2(j-1,x)\in\mathcal{U}_2(j-1,x)} \left[V_1^{(j)}(f(j-1, x, u_1^B(j-1, x), u_2(j-1, x))) \right.$$

$$\left. + F_1(j-1, x, u_1^B(j-1, x), u_2(j-1, x)) \right],$$

$$V_2^{(j-1)}(x) = \max_{u_1(j-1,x)\in\mathcal{U}_1(j-1,x)} \left[V_2^{(j)}(f(j-1, x, u_1(j-1, x), u_2^B(j-1, x))) \right.$$

$$\left. + F_2(j-1, x, u_1(j-1, x), u_2^B(j-1, x)) \right] \quad (j = K-1, K-2, \ldots, 1, 0).$$

$$(3.4.7)$$

In formulas (3.4.6) and (3.4.7), the functions $F_i(j - 1, x, u_1, u_2)$ and $\Phi_i(x)$ are given by (3.4.4). As it will be demonstrated below, the recursive relations (3.4.6) and (3.4.7) can be used to design a Berge equilibrium solution $(U^B, \mathcal{J}^B[t_0, x_0])$ of the game (3.4.1). For proving a corresponding result, we suppose that *all the maxima involved are achieved*.

Theorem 3.4.1 *Consider game (3.4.1), assuming that there exist the following sequences of functions that are defined in $\mathbf{R}^n$ and are unique for each $i = 1, 2$ and $j = K-1, K-2, \ldots, 1, 0$:*

a)   *two sequences of m_i-dimensional vector functions*

$$\left\{ u_i^B(K-1, x), u_i^B(K-2, x), \ldots, u_i^B(1, x), u_i^B(0, x) \right\} \quad (i = 1, 2),$$

where $u_i^B(j, x) \in \mathcal{U}_i(j, x)$;

b)   *two sequences of scalar functions*

$$\left\{ V_i^{(K)}(x), V_i^{(K-1)}(x), \ldots, V_i^{(1)}(x), V_i^{(0)}(x) \right\} \quad (i = 1, 2),$$

such that for all $x \in \mathbf{R}^n$,

$$V_i^{(K)}(x) = \Phi_i(x) \tag{3.4.8}$$

and

$$V_1^{(j)}(x) = \max_{u_2(j,x)\in\mathcal{U}_2(j,x)} \left[F_1(j, x, u_1^B(j, x), u_2(j, x)) \right.$$

$$\left. + V_1^{(j+1)} \left(f(j, x, u_1^B(j, x), u_2(j, x)) \right) \right] = Idem \left[u_2 \to u_2^B \right], \tag{3.4.9}$$

$$V_2^{(j)}(x) = \max_{u_1(j,x)\in\mathcal{U}_1(j,x)} \left[F_2(j, x, u_1(j, x), u_2^B(j, x)) \right.$$

$$\left. + V_2^{(j+1)} \left(f(j, x, u_1(j, x), u_2^B(j, x)) \right) \right] = Idem \left[u_1 \to u_1^B \right] \tag{3.4.10}$$

$$(j = K-1, K-2, \ldots, 1, 0).$$

Then for any choice of the initial position $(t_0, x_0) \in \{0, 1, \ldots \ldots, K-1\} \times \mathbf{R}^n$, the strategy profile $U^B = (U_1^B, U_2^B)$ in which

$$U_i^B \div \left(u_i^B(t_0, x), u_i^B\left(t_0+1, \bar{x}^B(t_0+1)\right), \ldots, u_i^B\left(K-1, \bar{x}^B(K-1)\right) \right) \tag{3.4.11}$$

is a Berge equilibrium solution of game (3.4.1) *with the initial position* (t_0, x_0) *(i.e., conditions* (3.4.5) *hold), and the corresponding Berge equilibrium payoffs are*

$$\mathcal{J}_i\left(U^B, t_0, x_0\right) = \mathcal{J}_i^B[t_0, x_0] = V_i^{(t_0)}(x_0) \quad (i = 1, 2). \tag{3.4.12}$$

As before, Idem $\left[u_i \to u_i^B\right]$ *in* (3.4.9) *and* (3.4.10) *means that* u_i *in the bracketed expression* $[\dots]$ *is replaced by* u_i^B; *for each* $x \in \mathbf{R}^n$ *in* (3.4.11),

$$\bar{x}^B(t_0+1) = f\left(t_0, x, u_1^B(t_0, x), u_2^B(t_0, x)\right),$$

$$\bar{x}^B(t_0+2) = f\left(t_0+1, \bar{x}^B(t_0+1), u_1^B\left(t_0+1, \bar{x}^B(t_0+1)\right), u_2^B\left(t_0+1, \bar{x}^B(t_0+1)\right)\right),$$

$$\vdots$$

$$\bar{x}^B(K-1) = f\left(K-2, \bar{x}^B(K-2), u_1^B\left(K-2, \bar{x}^B(K-2)\right), u_2^B\left(K-2, \bar{x}^B(K-2)\right)\right),$$

$$\bar{x}^B(K) = f\left(K-1, \bar{x}^B(K-1), u_1^B\left(K-1, \bar{x}^B(K-1)\right), u_2^B\left(K-1, \bar{x}^B(K-1)\right)\right). \tag{3.4.13}$$

Proof Includes two stages as follows. In the first stage, equality (3.4.12) will be established; in the second stage, equalities (3.4.5).

First stage Let (t_0, x_0) be an arbitrary initial position from $\{0, 1, \dots, K-1\} \times \mathbf{R}^n$, and also let the requirements (3.4.8)–(3.4.10) be satisfied. We have to prove (3.4.12). According to (3.4.8)–(3.4.10) and (3.4.13) for $j = t_0, t_0+1, \dots, K$ and $x = x_0$, we write the equalities

$$\begin{cases} V_i^{(t_0)}(x_0) = F_i\left(t_0, x_0, u_1^B(t_0, x_0), u_2^B(t_0, x_0)\right) + V_i^{(t_0+1)}\left(x^B(t_0+1)\right), \\ V_i^{(t_0+1)}\left(x^B(t_0+1)\right) = F_i\left(t_0+1, x^B(t_0+1), u_1^B\left(t_0+1, x^B(t_0+1)\right), \\ \qquad u_2^B\left(t_0+1, x^B(t_0+1)\right)\right) + V_i^{(t_0+2)}\left(x^B(t_0+2)\right), \\ \vdots \\ V_i^{(K-1)}\left(x^B(K-1)\right) = F_i\left(K-1, x^B(K-1), u_1^B\left(K-1, x^B(K-1)\right), \\ \qquad u_2^B\left(K-1, x^B(K-1)\right)\right) + V_i^{(K)}\left(x^B(K)\right), \\ V_i^{(K)}(x^B(K)) = \Phi_i(x^B(K)) \quad (i = 1, 2), \end{cases} \tag{3.4.14}$$

where $x^B(t_0) = x_0$, $x^B(k)$ for $k = t_0 + 1, \dots, K$ are given by (3.4.13). (Note that $x = x_0$ and the line over $\bar{x}^B(k)$ are omitted.) In view of (3.4.4), summing up the left- and right-hand sides of (3.4.14) independently and collecting like terms yield

$$V_i^{(t_0)}(x_0) = \Phi_i\left(x^B(K)\right) + \sum_{k=t_0}^{K-1} F_i\left(k, x^B(k), u^B[k]\right) = \mathcal{J}_i\left(U^B, t_0, x_0\right)$$

$$= \mathcal{J}_i^B[t_0, x_0] \quad (i = 1, 2),$$

where $u^B[k] = \left(u_1^B\left(k, x^B(k)\right), u_2^B\left(k, x^B(k)\right)\right)$. Consequently, equalities (3.4.12) hold.

Second stage Let us establish the first equality of (3.4.5) only, since the second one can be demonstrated by analogy. We use the same (t_0, x_0) and $u_i^B(k, x)$ as in the first stage of this proof. Assume at a time instant $k \in \{t_0, t_0+1, \dots, K-1\}$ player 2 has chosen (and uses) his strategy $\tilde{U}_2(k) \div \tilde{u}_2(k, x) \in \mathcal{U}_2(k, x)$, $\tilde{U}_2(k) \in \mathfrak{A}_2(k)$, where

$\tilde{u}_2(k, x^B(k)) \neq u_2^B(k, x^B(k))$; at the other time instants $j = t_0, t_0 + 1, \ldots, k - 1, k + 1, \ldots, K - 1$ and $x \in \mathbf{R}^n$, player 2 uses the strategy $U_2^B(j) \div u_2^B(j, x)$ designed in the first stage of this proof. Thus, before the time instant k, player 2 uses his strategies $U_2^B(j) \div u_2^B(j, x)$, $U_2^B(j) \in \mathfrak{A}_2(j)$; at the time instant k, he uses a strategy $\tilde{U}_2(k) \div \tilde{u}_2(k, x)$, $\tilde{U}_2(k) \in \mathfrak{A}_2(k)$, such that in the position $(k, x^B(k))$ the values $u_2^B(k, x)$ and $\tilde{u}_2(k, x)$ differ from each other; starting from the time instant $k + 1$ till $K - 1$, player 2 again uses his strategy $U_2^B(j) \in \mathfrak{A}_2(j)$ $(j = k + 1, \ldots, K-1)$ designed in the first stage of this proof. We construct the corresponding discrete solution

$$x^B(t_0) = x_0, \ x^B(t_0 + 1), \ldots, x^B(k), \ \tilde{x}(k + 1), \ \tilde{x}(k + 2), \ldots, \tilde{x}(K) \tag{3.4.15}$$

to the system

$$\begin{cases} x^B(t_0+1) = f\left(t_0, x_0, u_1^B(t_0, x_0), u_2^B(t_0, x_0)\right), \\ x^B(t_0+2) = f\left(t_0+1, x^B(t_0+1), u_1^B\left(t_0+1, x^B(t_0+1)\right), u_2^B\left(t_0+1, x^B(t_0+1)\right), \\ \quad \vdots \\ x^B(k) = f\left(k-1, x^B(k-1), u_1^B\left(k-1, x^B(k-1)\right), u_2^B\left(k-1, x^B(k-1)\right), \\ \tilde{x}(k+1) = f(k, x^B(k), \tilde{u}_1(k, x^B(k)), u_2^B(k, x^B(k))), \\ \tilde{x}(k+2) = f\left(k+1, \tilde{x}(k+1), u_1^B\left(k+1, \tilde{x}(k+1)\right), u_2^B\left(k+1, \tilde{x}(k+1)\right), \\ \quad \vdots \\ \tilde{x}(K-1) = f\left(K-2, \tilde{x}(K-2), u_1^B\left(K-2, \tilde{x}(K-2)\right), u_2^B\left(K-2, \tilde{x}(K-2)\right), \\ \tilde{x}(K) = f\left(K-1, \tilde{x}(K-1), u_1^B\left(K-1, \tilde{x}(K-1)\right), u_2^B\left(K-1, \tilde{x}(K-1)\right). \end{cases} \tag{3.4.16}$$

This solution $\left(x_0, x^B(t_0 + 1), \ldots, x^B(k), \tilde{x}(k + 1), \ldots, \tilde{x}(K)\right)$ coincides with (3.4.13) till the time instant k; they differ merely starting from the time instant $k + 1$. Due to (3.4.9), along the solution (3.4.15) of the system (3.4.16) we obtain

$$\begin{cases} V_1^{(t_0)}(x_0) = F_1\left(t_0, x_0, u_1^B(t_0, x_0), u_2^B(t_0, x_0)\right) + V_1^{(t_0+1)}\left(x^B(t_0 + 1)\right), \\ V_1^{(t_0+1)}\left(x^B(t_0+1)\right) = F_1\left(t_0+1, x^B(t_0+1), u_1^B\left(t_0+1, x^B(t_0+1)\right), \\ \qquad u_2^B\left(t_0+1, x^B(t_0+1)\right)\right) + V_1^{(t_0+2)}\left(x^B(t_0+2)\right), \\ \quad \vdots \\ V_1^{(k-1)}\left(x^B(k-1)\right) = F_1\left(k-1, x^B(k-1), u_1^B\left(k-1, x^B(k-1)\right), \\ \qquad u_2^B\left(k-1, x^B(k-1)\right)\right) + V_1^{(k)}\left(x^B(k)\right), \\ V_1^{(k)}\left(x^B(k)\right) \geqslant F_1\left(k, x^B(k), \tilde{u}_1\left(k, x^B(k)\right), u_2^B\left(k, x^B(k)\right)\right) + V_1^{(k+1)}(\tilde{x}(k+1)), \\ V_1^{(k+1)}(\tilde{x}(k+1)) = F_1\left(k+1, \tilde{x}(k+1), u_1^B\left(k+1, \tilde{x}(k+1)\right), \\ \qquad u_2^B\left(k+1, \tilde{x}(k+1)\right)\right) + V_1^{(k+2)}(\tilde{x}(k+2)), \\ \quad \vdots \\ V_1^{(K-1)}(\tilde{x}(K-1)) = F_1\left(K-1, \tilde{x}(K-1), u_1^B\left(K-1, \tilde{x}(K-1)\right), \\ \qquad u_2^B\left(K-1, \tilde{x}(K-1)\right)\right) + V_1^{(K)}(\tilde{x}(K)), \\ V_1^{(K)}(\tilde{x}(K)) = \Phi_1(\tilde{x}(K)). \end{cases} \tag{3.4.17}$$

Note that inequality in (3.4.17) follows from (3.4.9) with $j = k$ and $x = x^B(k)$, i.e., from the chain of relations

$$V_1^{(k)}(x^B(k)) = F_1\left(k, x^B(k), u_1^B\left(k, x^B(k)\right), u_2^B\left(k, x^B(k)\right)\right) + V_1^{(k+1)}\left(x^B(k+1)\right)$$

$$= \max_{u_2(k, x^B(k)) \in \mathcal{U}_2(k, x^B(k))} \left\{ F_1\left(k, x^B(k), u_1^B\left(k, x^B(k)\right), u_2\left(k, x^B(k)\right)\right) \right.$$

$$\left. + V_1^{(k+1)}\left[f\left(k, x^B(k), u_1^B\left(k, x^B(k)\right), u_2\left(k, x^B(k)\right)\right)\right]\right\}$$

$$\geqslant F_1\left(k, x^B(k), u_1^B\left(k, x^B(k)\right), \tilde{u}_2\left(k, x^B(k)\right)\right) + V_1^{(k+1)}(\tilde{x}(k+1)).$$

Finally, summing up the left- and right-hand sides of (3.4.17) independently and collecting like terms yield

$$V_1^{(t_0)}(x_0) \geqslant F_1\left(t_0, x_0, u_1^B(t_0, x_0), u_2^B(t_0, x_0)\right)$$

$$+ F_1\left(t_0+1, x^B(t_0+1), u_1^B\left(t_0+1, x^B(t_0+1)\right), u_2^B\left(t_0+1, x^B(t_0+1)\right)\right) + \ldots$$

$$+ F_1\left(k-1, x^B(k-1), u_1^B\left(k-1, x^B(k-1)\right), u_2^B\left(k-1, x^B(k-1)\right)\right)$$

$$+ F_1\left(k, x^B(k), u_1^B\left(k, x^B(k)\right), \tilde{u}_2\left(k, x^B(k)\right)\right) + F_1\left(k+1, \tilde{x}(k+1),\right.$$

$$\left. u_1^B\left(k+1, \tilde{x}(k+1)\right), u_2^B\left(k+1, \tilde{x}(k+1)\right)\right) + \ldots + F_1\left(K-1, \tilde{x}(K-1),\right.$$

$$\left. u_1^B\left(K-1, \tilde{x}(K-1)\right), u_2^B\left(K-1, \tilde{x}(K-1)\right)\right) + \Phi_1(\tilde{x}(K)).$$

That is, due to (3.4.4), we have

$$V_1^{(t_0)}(x_0) \geqslant \mathcal{J}_1\left(U_1^B, \tilde{U}_2, t_0, x_0\right), \tag{3.4.18}$$

where $\tilde{U}_2 = (U_2^B(t_0), U_2^B(t_0+1), \ldots, U_2^B(k-1), \tilde{U}_2(k), U_2^B(k+1), \ldots, U_2^B(K-1))$, and $U_2^B(j) \div u_2^B(j, x)$, $U_2^B(j) \in \mathfrak{A}_2(j)$ $(j = t_0, \ldots, k-1)$, $\tilde{U}_2(k) \div \div \tilde{u}_2(k, x)$, $\tilde{U}_2(k) \in \mathfrak{A}_2(k)$. Next, $U_2^B(l) \div u_2^B(l, x)$, $U_2^B(l) \in \mathfrak{A}_2(l)$ $(l = k+1, \ldots, K-1)$, and then the strategy takes the form

$$\tilde{U}_2 \div \left(u_2^B(t_0, x), u_2^B\left(t_0+1, \bar{x}^B(t_0+1)\right), \ldots, u_2^B\left(k-1, \bar{x}^B(k-1)\right),\right.$$

$$\left. \tilde{u}_2\left(k, \bar{x}^B(k)\right), u_2^B\left(k+1, \tilde{\bar{x}}(k+1)\right), \ldots, u_2^B\left(K-1, \tilde{\bar{x}}(K-1)\right)\right),$$

where for each $x \in \mathbf{R}^n$,

$$\bar{x}^B(t_0+1) = f\left(t_0, x, u_1^B(t_0, x), u_2^B(t_0, x)\right),$$

$$\vdots$$

$$\bar{x}^B(k-1) = f\left(k-2, \bar{x}^B(k-2), u_1^B\left(k-2, \bar{x}^B(k-2)\right), u_2^B\left(k-2, \bar{x}^B(k-2)\right)\right),$$

$$\bar{x}^B(k) = f\left(k-1, \bar{x}^B(k-1), u_1^B\left(k-1, \bar{x}^B(k-1)\right), u_2^B\left(k-1, \bar{x}^B(k-1)\right)\right),$$

$$\tilde{\bar{x}}(k+1) = f\left(k, \bar{x}^B(k), u_1^B\left(k, \bar{x}^B(k)\right), \tilde{u}_2\left(k, \bar{x}^B(k)\right)\right),$$

$$\tilde{\bar{x}}(k+2) = f\left(k+1, \tilde{\bar{x}}(k+1), u_1^B\left(k+1, \tilde{\bar{x}}(k+1)\right), u_2^B\left(k+1, \tilde{\bar{x}}(k+1)\right)\right),$$

$$\vdots$$

$$\tilde{x}(K-1) = f\left(K-2, \tilde{x}(K-2), u_1^B\left(K-2, \tilde{x}(K-2)\right), u_2^B\left(K-2, \tilde{x}(K-2)\right)\right),$$
$$\tilde{x}(K) = f\left(K-1, \tilde{x}(K-1), u_1^B\left(K-1, \tilde{x}(K-1)\right), u_2^B\left(K-1, \tilde{x}(K-1)\right)\right).$$

As is easily seen, equalities (3.4.12) (for $i = 1$) in combination with (3.4.18) in the last analysis give the first equality of (3.4.5).

Remark 3.4.1 If player i deviates from the strategy U_i^B not only at the time instant $t = k$ but also at the other time instants $t = m$ ($m \neq k$), the same technique can be adopted for proving equality (3.4.5) for all $(t_0, x_0) \in \{0, \dots, K-1\} \times \mathbf{R}^n$.

Remark 3.4.2 In the course of this proof, we have succeeded in revealing the game-theoretic interpretation of the functions $V_i^{(t)}(x)$. As a matter of fact, they coincide with the Berge equilibrium payoffs $\mathcal{J}_i(U^B, t, x)$ ($i = 1, 2$) in the game (3.4.1) with any initial position $(t, x) \in \{0, \dots, K-1\} \times \mathbf{R}^n$ if both players use their strategies from the Berge equilibrium $U^B = (U_1^B, U_2^B)$, i.e.,

$$V_1^{(t)}(x) = \max_{U_2 \in \mathfrak{A}_2} \mathcal{J}_1\left(U_1^B, U_2, t, x\right) = \mathcal{J}_1\left(U^B, t, x\right),$$
$$V_2^{(t)}(x) = \max_{U_1 \in \mathfrak{A}_1} \mathcal{J}_2\left(U_1, U_2^B, t, x\right) = \mathcal{J}_2\left(U^B, t, x\right). \tag{3.4.19}$$

The functions $V_i^{(t)}(x)$ ($i = 1, 2$) satisfying (3.4.19) for all $(t, x) \in \{0, \dots, K-1\} \times \mathbf{R}^n$ can be naturally called *the Bellman functions* for the game (3.4.1).

Remark 3.4.3 For each integer $k = 0, 1, \dots, K$ in the state space $\mathbf{R}^n$, let a subset $X(k) \subseteq \mathbf{R}^n$ be some *a priori* defined *constraint on the state vector*. With such a system of restrictions, Theorem 3.4.1 takes the following form.

Theorem 3.4.2 *Consider game (3.4.1) with additional state-space constraints* $x(k) \in X(k)$ ($k = 0, 1, \dots, K$), *assuming that there exist the following sequences of functions that are unique for each* $i = 1, 2$ *and* $j = 0, \dots, K$:

a)   *sequence of scalar functions*

$$\left\{V_i^{(j)}(x)\right\}_{j=0}^{K} \quad (i = 1, 2),$$

where $V_i^{(j)}(x)$ *is defined for* $x \in X(j)$;

b)   *sequences of* m_i-*dimensional vector functions* $\left\{u_i^B(k, x)\right\}_{k=0}^{K-1}$, *where* $u_i^B(k, x)$ *are defined for* $x \in X(k)$, $u_i^B(k, x) \in \mathcal{U}_i(k, x)$ ($i = 1, 2$), *such that*

$$V_i^{(K)}(x) = \Phi_i(x) \quad \forall x \in X(K) \quad (i = 1, 2), \tag{3.4.20}$$

and for all $x \in X(j)$,

$$V_1^{(j)}(x) = \max_{u_2(j,x) \in \mathcal{U}_2(j,x)} \left[F_1\left(j, x, u_1^B(j, x), u_2(j, x)\right)\right.$$
$$\left. + V_1^{(j+1)}\left(f\left(j, x, u_1^B(j, x), u_2(j, x)\right)\right)\right] = Idem\left[u_2 \rightarrow u_2^B\right], \tag{3.4.21}$$

$$V_2^{(j)}(x) = \max_{u_1(j,x)\in\mathcal{U}_1(j,x)} \left[F_2\left(j, x, u_1(j,x), u_2^B(j,x)\right) \right.$$
$$\left. + V_2^{(j+1)}\left(f\left(j, x, u_1(j,x), u_2^B(j,x)\right)\right)\right] = Idem\left[u_1 \to u_1^B\right].$$

Then for any choice of the initial position $(t_0, x_0) \in \{0, 1, \ldots, K-1\} \times X(t_0)$, *the strategy profile* $U^B = \left(U_1^B, U_2^B\right)$ *in which*

$$U_i^B \doteq \left(u_i^B(t_0, x), u_i^B\left(t_0+1, \bar{x}^B(t_0+1)\right), \ldots, u_i^B\left(K-1, \bar{x}^B(K-1)\right)\right) \quad (i = 1, 2),$$

(3.4.22)

is a Berge equilibrium solution of game (3.4.1) *with the state-space constraints* $x(k) \in X(k)$, *and the corresponding Berge equilibrium payoffs are*

$$\mathcal{J}_i\left(U^B, t_0, x_0\right) = \mathcal{J}_i^B[t_0, x_0] = V_i^{(t_0)}(x_0) \quad (i = 1, 2).$$

In (3.4.22), $x \in X(t_0)$ *and for all* $x \in X(t_0)$,

$$x^B(t_0+1) = f\left(t_0, x, u_1^B(t_0, x), u_2^B(t_0, x)\right) \in X(t_0+1), \ \bar{x}^B(t_0+2) = f(t_0+1, \bar{x}(t_0+1),$$
$$u_1^B(t_0+1, \bar{x}(t_0+1)), u_2^B(t_0+1, \bar{x}(t_0+1))) \in X(t_0+2),$$

$$\vdots$$

$$\bar{x}^B(K-1) = f\left(K-2, \bar{x}(K-2), u_1^B(K-2, \bar{x}(K-2)), u_2^B(K-2, \bar{x}(K-2))\right) \in X(K-1).$$

Remark 3.4.4 Conditions (3.4.21) can be interpreted in game-theoretic terms. For this purpose, at each time instant $j = 0, 1, \ldots, K-1$ we construct the following single-stage noncooperative two-player game with the frozen state $x \in X(j)$:

$$\Gamma(j, x) = \Big\langle \{1, 2\}, \{\mathcal{U}_i(j, x)\}_{i=1,2}, \{F_i(j, x, u_1(j, x), u_2(j, x))$$
$$+ V_i^{(j+1)}(f(j, x, u_1(j, x), u_2(j, x)))\}_{i=1,2}\Big\rangle.$$

In this game, using an appropriate choice of his strategy $u_l(j, x)$, $x \in X(j)$, player 1 seeks to maximize the payoff function of the opponent (the other party to this conflict, $i \neq l$, $i, l = 1, 2$), i.e.,

$$F_i(j, x, u_1(j, x), u_2(j, x)) + V_i^{(j+1)}(f(j, x, u_1(j, x), u_2(j, x))).$$

According to conditions (3.4.21), the strategy profile $u^B(j, x) = (u_1^B(j, x), u_2^B(j, x))$ calculated above is a Berge equilibrium in the game $\Gamma(j, x)$ with a fixed position (j, x), $x \in X(j)$. In fact, by Theorem 3.4.2 the local Berge equilibrium conditions (3.4.21) in combination with the additional requirement of uniqueness and the boundary-value conditions (3.4.20) guarantee the global Berge equilibrium in the game (3.4.1) with the state-space constraints $x(k) \in X(k)$ $(k = 0, 1, \ldots, K)$.

Remark 3.4.5 Based on Theorem 3.4.2, we suggest the following *design procedure* of a Berge equilibrium and corresponding payoffs in the multistage positional noncooperative two-player game (3.4.1) with state-space constraints.

Well, the problem is to find a Berge equilibrium $U^B = (U_1^B, U_2^B)$ in the game (3.4.1) defined by (3.4.19) for all $(t_0, x_0) \in \{0, 1, \ldots, K-1\} \times X(t_0)$, in which the payoff functions of the players have the form

$$\mathcal{J}_i(U, t_0, x_0) = \Phi_i(x(K)) + \sum_{k=t_0}^{K-1} F_i(k, x(k), u_1[k], u_2[k]) \quad (i = 1, 2), \tag{3.4.23}$$

and the controlled discrete system evolves in accordance with the vector difference equation

$$x(k+1) = f(k, x(k), u_1(k, x(k)), u_2(k, x(k))) \quad (k = t_0, t_0+1, \ldots, K-1) \tag{3.4.24}$$

subject to the state-space constraints

$$u_i(k, x) \in \mathcal{U}_i(k, x), \quad x \in X(k). \tag{3.4.25}$$

The design procedure includes **three** stages as follows.
   Stage I (in the decreasing order of index $k = K, K-1, \ldots, t_0$).
In view of the requirement (3.4.20), for $k = K$ let

$$V^{(K)}(x) = \Phi_i(x) \quad \forall x \in X(k), \tag{3.4.26}$$

where $\Phi_i(x)$ are given by (3.4.23).
   For $k = K$ and each $x \in X(K-1)$, using the transition operator $f(\cdot)$ (3.4.24), find out whether there exist collections $u(K-1, x) = (u_1(K-1, x), u_2(K-1, x))$, $u_i(K-1, x) \in \mathcal{U}_i(K-1, x)$ $(i = 1, 2)$ such that

$$f(K-1, x, u(K-1, x)) \in X(K); \tag{3.4.27}$$

denote by $\overline{\mathcal{U}}(K-1, x)$ the set of collections $u(K-1, x)$ satisfying the inclusion (3.4.27); in parallel, find a subset $\bar{X}(K-1) \subseteq X(K-1)$ of all points $x = x(K-1) \in \bar{X}(K-1)$ that satisfy the inclusion (3.4.27). Next, at each point $x \in \bar{X}(K-1)$ consider the noncooperative single-stage two-player game

$$\Gamma_2^{(K-1)}(x) = \Big\langle \{1, 2\}, \{\mathcal{U}_i(K-1, x)\}_{i=1,2}, \{F_i(K-1, x, u_1, u_2)$$
$$+ V_i^{(K)}(f(K-1, x, u_1, u_2))\}_{i=1,2} \Big\rangle, \tag{3.4.28}$$

where $f(K-1, x, u_1(K-1, x), u_2(K-1, x)) = x(K)$. Due to (3.4.26),

$$V_i^{(K)} = \Phi_i(x(K)) \quad (i = 1, 2).$$

In the game (3.4.28), construct a Berge equilibrium $\left(u_1^B(K-1, x), u_2^B(K-1, x)\right)$, i.e.,

$$\max_{u_2(K-1,x)} \Big[F_1(K-1, x, u_1^B(K-1, x), u_2(K-1, x)) + \Phi_1(x^{(1)}(K)) \Big]$$
$$= \Big(\text{for } (u_1^B(K-1, x), u_2(K-1, x)) \in \overline{\mathcal{U}}(K-1, x) \Big)$$
$$= Idem \left[u_2 \to u_2^B \right] = F_1 \left(K-1, x, u^B(K-1, x) \right) + \Phi_1 \left(x^B(K) \right) = V_1^{(K-1)}(x), \tag{3.4.29}$$

$$\max_{u_1(K-1,x)} \left[F_2(K-1, x, u_1(K-1, x), u_2^B(K-1, x)) + \Phi_2(x^{(2)}(K)) \right]$$

$$= \left(\text{for } (u_1(K-1, x), u_2^B(K-1, x)) \in \overline{U}(K-1, x) \right)$$

$$= Idem \left[u_1 \to u_1^B \right] = F_2 \left(K-1, x, u^B(K-1, x) \right) + \Phi_2 \left(x^B(K) \right) = V_2^{(K-1)}(x),$$

where

$$x^B(K) = f(K-1, x, u^B(K-1, x)),$$
$$x^{(1)}(K) = f(K-1, x, u_1^B(K-1, x), u_2(K-1, x)),$$
$$x^{(2)}(K) = f(K-1, x, u_1(K-1, x), u_2^B(K-1, x))$$

for $x \in \bar{X}(K-1)$.

This gives $X^B(K-1)$—a subset $\bar{X}(K-1)$ of all points $x^B(K-1)$ ($\in X^B(K-1)$) with the property that the game $\Gamma_2^{(K-1)}(x^B(K-1))$ has a unique Berge equilibrium

$$U^B(K-1) = \left(U_1^B(K-1), U_2^B(K-1) \right) \in \mathfrak{A}(K-1),$$

$U_i^B(K-1) \div u_i^B(K-1, x) \in \mathcal{U}_i(K-1, x)$ $(i = 1, 2)$ at stage $K-1$. Construct the strategy profile $U^B(K-1)$ itself and in parallel find the two functions

$$V_i^{(K-1)}(x) = F_i \left(K-1, x, u^B(K-1, x) \right) + \Phi_i \left(x^B(K) \right) \quad (i = 1, 2),$$

defined for all $x \in X^B(K-1)$.

For $k = K - 2$, using (3.4.24) construct the set $\bar{X}(K-2) \subseteq \subseteq X(K-2)$ of points $x = x(K-2)$ such that

$$x(K-1) = f(K-2, x = x(K-2), u(K-2, x)) \in X^B(K-1) \tag{3.4.30}$$

(the set $X^B(K-1)$ has been already constructed for $k = K - 1$). In parallel, determine the subset $\overline{\mathcal{U}}(K-2, x) \subseteq \mathcal{U}(K-1, x)$ of the corresponding strategy profiles $u(K-2, x)$ whose elements $u(K-2, x) \in \overline{\mathcal{U}}(K-2, x)$ satisfy the inclusion (3.4.30). Next, for each $x \in \bar{X}(K-2, x)$ consider the specific noncooperative game

$$\Gamma_2^{(K-2)}(x) = \Big\langle \{1, 2\}, \{\overline{\mathcal{U}}_i(K-2, x)\}_{i=1,2}, \{F_i(K-2, x, u_1(K-2, x), u_2(K-2, x))$$
$$+ V_i^{(K-1)}(f(K-2, x, u_1(K-2, x), u_2(K-2, x)))\}_{i=1,2} \Big\rangle.$$

In the game $\Gamma_2^{(K-2)}(x)$, from the set $\bar{X}(K-2)$ we choose the points $x = = x^B(K-2) \in \bar{X}(K-2)$ for which there exists a unique Berge equilibrium

$$u^B(K-2, x) = \left(u_1^B(K-2, x), u_2^B(K-2, x) \right) \in \mathcal{U}(K-2, x),$$

given by the conditions

$$\max_{u_2(K-2,x) \in \mathcal{U}_2(K-2,x)} \Big[F_1(K-2, x, u_1^B(K-2, x), u_2(K-2, x))$$
$$+ V_1^{(K-1)}(f(K-2, x, u_1^B(K-2, x), u_2(K-2, x))) \Big]$$

$$= \Big(\text{for } (u_1^B(K-2, x), u_2(K-2, x)) \in \overline{U}(K-2, x)\Big) = Idem\Big[u_2 \to u_2^B\Big] = V_1^{(K-2)}(x),$$

$$\max_{u_1(K-2,x)\in\mathcal{U}_1(K-2,x)} \Big[F_2(K-2, x, u_1(K-2, x), u_2^B(K-2, x))$$

$$+ V_2^{(K-1)}(f(K-2, x, u_1(K-2, x), u_2^B(K-2, x)))\Big]$$

$$= \Big(\text{for } (u_1(K-2, x), u_2^B(K-2, x)) \in \overline{U}_2(K-2, x)\Big) = Idem[u_1 \to u_1^B] = V_2^{(K-2)}(x).$$

$$(3.4.31)$$

Denote by $X^B(K-2)$ the set of such states $x^B(K-2)$.

Thus, for stage $k = K-2$ the above-mentioned process yields

a)   the set $X^B(K-2)$ of all points $x^B(K-2)$ in each of which the game $\Gamma_2^{(K-2)}(x)$ with $x = x^B(K-2)$ has a unique Berge equilibrium $U^B(K-2) \div u^B(K-2, x) = \big(u_1^B(K-2, x), u_2^B(K-2, x)\big)$ given by equalities (3.4.31);

b)   the two scalar functions $V_i^{(K-2)}(x)$ $(i = 1, 2)$ defined for $x \in X^B(K-2)$;

c)   for each $x^* \in X^B(K-2)$, the strategy profiles $u^B(K-2, x) = = \big(u_1^B(K-2, x), u_2^B(K-2, x)\big)$ that satisfy (3.4.31).

Repeating this process in the decreasing order of index k finally gives the sequences

a)   of the sets $X^B(K-3), \ldots, X^B(t_0+1), X^B(t_0)$;

b)   of the scalar functions $V_i^{(K-3)}(x), \ldots, V_i^{(t_0+1)}(x), V_i^{(t_0)}(x)$ $(i = 1, 2)$, each $V_i^{(j)}(x)$ being defined on $X^B(j)$ $(j = t_0, t_0+1, \ldots, K)$;

c)   of the strategy profiles

$$U^B(K-1) \div \Big(u_1^B(K-1, x), u_2^B(K-1, x)\Big), \quad U^B(K-1) \in \mathfrak{A}(K-1),$$

$$U^B(K-2) \div \Big(u_1^B(K-2, x), u_2^B(K-2, x)\Big), \quad U^B(K-2) \in \mathfrak{A}(K-2),$$

$$\ldots$$

$$U^B(t_0+1) \div \Big(u_1^B(t_0+1, x), u_2^B(t_0+1, x)\Big), \quad U^B(t_0+1) \in \mathfrak{A}(t_0+1),$$

$$U^B(t_0) \div \Big(u_1^B(t_0, x), u_2^B(t_0, x)\Big), \quad U^B(t_0) \in \mathfrak{A}(t_0).$$

Stage II (for $k = t_0$). Calculate the Berge equilibrium payoffs

$$\mathcal{J}_i(U^B, t_0, x_0) = V^{(t_0)}(x_0) \quad (i = 1, 2). \tag{3.4.32}$$

Stage III (in the increasing order of index k). Successively apply the resulting Berge equilibria of the local games $\Gamma_2^{(k)}(x)$ and the transition operator $f(\cdot)$ (3.4.2) with the initial position (t_0, x_0) to find the realizations of the local Berge equilibria and the corresponding discrete trajectory x_0, $u^B(t_0, x_0) = u^B[t_0]$:

$$f\Big(t_0, x_0, u^B(t_0, x_0)\Big) = x^B(t_0 + 1), \quad u^B\Big(t_0 + 1, x^B(t_0 + 1)\Big) = u^B[t_0 + 1],$$

$$f\Big(t_0 + 1, x^B(t_0 + 1), u^B[t_0 + 1]\Big) = x^B(t_0 + 2), \quad u^B\Big(t_0 + 2, x^B(t_0 + 2)\Big) = u^B[t_0 + 2],$$

. . .

$$f\left(K-2, x^B(K-2), u^B[K-2]\right) = x^B(K-1), \quad u^B\left(K-1, x^B(K-1)\right) = u^B[K-1],$$

$$f\left(K-1, x^B(K-1), u^B[K-1]\right) = x^B(K).$$

In the end, the design procedure yields the discrete trajectory

$$\left(x_0, x^B(t_0+1), x^B(t_0+2), \ldots, x^B(K-1), x^B(K)\right),$$

the Berge equilibrium U^B (composed of the local ones) and the Berge equilibrium payoffs (3.4.32).

3.5   BERGE EQUILIBRIUM IN COURNOT DUOPOLY

Cést le sort le plus beau,
le plus digne d'envie.[33]

Prius[34]

One of the first research works devoted to the application of game theory in economics was the monograph *Recherches sur les principes mathématiques de la théorie de richesses* (Researches into the Mathematical Principles of the Theory of Wealth) [281] published by Cournot.[35] In Section 7 of that book (competition of firms), Cournot examined a special case of duopoly and solved the corresponding game using a concept representing a particular case of the well-known Nash equilibrium.

In Section 3.5 of this book, the static and dynamic cases of the Cournot model are considered. The most common solution of such games involves the concept of Nash equilibrium. However, the approach described below differs from the conventional one, because its main principle of behavior is based on the Golden Rule of ethics: "Do unto others as you would like them to do unto you." The concept of Berge equilibrium, which was first strictly formalized in the dissertation and the first papers K. Vaisman [25, 26, 342–346], matches well this approach. Berge equilibrium as a method of balancing conflicts has already become quite widespread among the researchers around the world; see [377].

This section is focused on a practical application of Berge equilibrium in the Cournot duopoly model.

3.5.1   Static case of the model

Consider an economic model of market competition known as *the Cournot duopoly model*.

---

[33] French "It is the most beautiful, most desirable fate." A fragment of *Le Chant des Girondins* (The Song of the Girondists), the national anthem of the French Second Republic, written for the drama *Le Chevalier de Maison-Rouge* by Alexandre Dumas with Auguste Maquet.

[34] Latin "previously."

[35] Antoine-Augustin Cournot, (1801–1877), was a French economist and mathematician. He was considered to be the first economist who, with competent knowledge of both subjects, endeavoured to apply mathematics to the treatment of economics.

Two firms supply a homogeneous product during a given time period. Let q_i be the quantity of products supplied by firm i $(i = 1, 2)$. *The production cost of firm i* is assumed to be a linearly function of the quantity q_i, i.e., can be written as $cq_i + d$, where the constants c and d specify the average variable and fixed cost, respectively. Variable cost include, wages, raw material purchases, and depreciation of equipment, whereas fixed cost the rent of premises, land, equipment, licences, and so on. The product price p is determined by the law of supply and demand depending on the total quantity $\tilde{q} = q_1 + q_2$ supplied by both firms. Let the price p be a linear function of the total supply, $p(\tilde{q}) = a - b\tilde{q}$, where $a = const > 0$ is initial price without product supply and a positive constant b (known as the coefficient of elasticity) shows the price drop in response to unit product supply. The revenue of firm 1 is given by

$$p(\tilde{q})q_1 = (a - b\tilde{q})q_1 = [a - b(q_1 + q_2)]q_1,$$

and its profit (revenue minus cost) can be calculated as

$$\psi_1(q_1, q_2) = [a - b(q_1 + q_2)]q_1 - (cq_1 + d) = aq_1 - bq_1^2 - bq_1q_2 - cq_1 - d;$$

by analogy, the profit of firm 2 has the form

$$\psi_2(q_1, q_2) = [a - b(q_1 + q_2)]q_2 - (cq_2 + d) = aq_2 - bq_2^2 - bq_1q_2 - cq_2 - d.$$

The functions $\psi_i(q_1, q_2)$ are strictly concave in q_i $(i = 1, 2)$, since $\frac{\partial^2 \psi_i}{\partial q_i^2} = -2b < 0$. As a result, the sufficient conditions for the existence of a quantity q_i^* that maximizes $\psi_i(q_1, q_2)$ in q_i $(i = 1, 2)$ are reduced to the system of two linear algebraic equations

$$\begin{cases} \frac{\partial \psi_1(q_1, q_2)}{\partial q_1}\big|_{q_1^*} = a - 2bq_1^* - bq_2 - c = 0, \\ \frac{\partial \psi_2(q_1, q_2)}{\partial q_2}\big|_{q_2^*} = a - bq_1 - 2bq_2^* - c = 0. \end{cases}$$

The solution (q_1^*, q_2^*) of this system has the form

$$q_1^* = \frac{a - c - bq_2}{2b}, \quad q_2^* = \frac{a - c - bq_1}{2b}.$$

Thus, in the static case of the Cournot duopoly model, the functional relation between the best responses of both competing firms is given by

$$q_1 = \frac{a - c - bq_2}{2b}, \quad q_2 = \frac{a - c - bq_1}{2b}.$$

This fact will be taken into account in the dynamic case of the model.

3.5.2 Dynamic case of the model

Let us introduce the following assumptions.

First, the products are supplied during a single period between two time instants (stages), i.e., $k = 0, 1$.

Second, there exists a time lag equal to one period; as a result, the best responses of firms to the competitor's actions have the form

$$q_1(k + 1) = \frac{a - c - bq_2(k)}{2b}, \quad q_2(k + 1) = \frac{a - c - bq_1(k)}{2b} \quad (k = 0).$$

Third, at the time instant $k = 0$ the initial quantities of products $q_i(0) = q_{i0}$ available from the stock of firm i $(i = 1, 2)$ are fixed.

Fourth, the management of each firm i (further referred to as *player i*) chooses and maintains *a rate of supply* $u_i[k]$ for each time instant k ($i = 1, 2$; $k = 0, 1$), e.g., by investing a portion ($\beta_i u_i$) into production and transferring the remaining portion $(1 - \beta_i)u_i$ to the other player ($i, j = 1, 2$; $j \neq i$); here $\beta_i \in [0, 1]$ are some constants. The same applies to the implementation of new technologies. Note that the control action u_i of player i at a time instant k depends on the quantities supplied by both firms at the time instant k. A strategy $U_i(k)$ of player i (the behavioral rule of firm i) at a time instant k, ($i = 1, 2$), ($k = 0, 1$), is identified with a scalar function $u_i(k, q_1, q_2)$. This fact will be denoted by $U_i \div u_i(k, q)$, where $q = (q_1, q_2)$. Then the mathematical model of the controlled dynamic system Σ that describes the supply of products at discrete time instants 0 and 1 can be written as

$$q_1(k+1) = \frac{a - c}{2b} - \frac{q_2(k)}{2} + \beta_1 u_1 + (1 - \beta_2)u_2, \quad q_1(0) = q_{10},$$

$$q_2(k+1) = \frac{a - c}{2b} - \frac{q_1(k)}{2} + (1 - \beta_1)u_1 + \beta_2 u_2, \quad q_2(0) = q_{20} \quad (k = 0). \qquad (3.5.1)$$

The system (3.5.1) consists of two difference (single-stage) linear equations. Denote by $\mathcal{U}_i(k)$ the set of all strategies $U_i(k)$ at stage k ($i = 1, 2$; $k = 0$). Then a strategy profile $U(0)$ is represented by an ordered pair

$$U(0) = (U_1(0) \div u_1(0, q), U_2(0) \div u_2(0, q)),$$

where $U_i(k) \in \mathcal{U}_i(k)$ ($k = 0$) and $q(0) = (q_{10}, q_{20})$.

Thus, a pair of strategies (U_1, U_2) forms *a strategy profile U*.

The game evolves over $k = 0, 1$ in the following way. Without making coalition with the opponent, each player i independently chooses his strategy

$$U_i(0) \in \mathcal{U}_i(0), \quad U_i(0) \div u_i(0, q) \quad (i = 1, 2);$$

as a result, two scalar functions $u_i(0, q_1, q_2)$ ($i = 1, 2$) are formed. Note that player i chooses an appropriate strategy $U_i(0) \in \mathcal{U}_i(0)$, $U_i(0) \div u_i(0, q_1, q_2)$, being guided by the concept of Berge equilibrium ("help everyone, forgetting about your individual interests"). This concept is used by him for the payoff functions of both players; see below. For obtaining these functions, we consider the controlled dynamic system (3.5.1) with the initial state $q(0) = q^0 = (q_1(0), q_2(0)) = (q_{10}, q_{20})$.

We construct

$$q_1(1) = \frac{a - c}{2b} - \frac{q_{20}}{2} + \beta_1 u_1(0, q_{10}, q_{20}) + (1 - \beta_2)u_2(0, q_{10}, q_{20}),$$

$$q_2(1) = \frac{a - c}{2b} - \frac{q_{10}}{2} + (1 - \beta_1)u_1(0, q_{10}, q_{20}) + \beta_2 u_2(0, q_{10}, q_{20}). \qquad (3.5.2)$$

This yields:

1) the two sequences

$$\{q_i(k)\}_{k=o}^{1} \ (i = 1, 2), \qquad (3.5.3)$$

that describe the trajectory of the system (3.5.1) with the specific strategies $U_i(0) \in \mathcal{U}_i(0)$, $U_i(0) \div u_i(0, q)$ ($i = 1, 2$) of the players;

2) the two realizations

$$\{u_i[k] = u_i(0, q(0))\} \ (i = 1, 2). \qquad (3.5.4)$$

Using (3.5.3) and (3.5.4), we design the payoff function of player i for assessing the latter's performance. Two circumstances have to be considered as follows.

First, each firm i ($i = 1, 2$) seeks to supply to the market as much products as possible, i.e., to maximize the sum

$$q_i^2(1) + q_i^2(0) \ (i = 1, 2)$$

with an appropriate choice of $U_i \in \mathcal{U}_i$.

Second, each firm i ($i = 1, 2$) simultaneously tries to consume as few resources of both firms as possible, i.e., to maximize the difference

$$-u_1^2[0] - u_2^2[0].$$

In view of these circumstances, the payoff function of player i can be written as

$$J_i(U, q_0) = q_i^2(1) + q_i^2(0) - u_1^2[0] - u_2^2[0] \ (i = 1, 2). \tag{3.5.5}$$

The ordered quadruple

$$\Gamma = \langle \{1, 2\}, \Sigma \div (3.5.1), \{\mathcal{U}_i\}_{i=1,2}, \{J_i(U, q_0) \div (3.5.5)\}_{i=1,2} \rangle,$$

forms a *single-stage noncooperative positional linear-quadratic two-player game*. The notations $\Sigma \div$ (3.5.1) and $J_i(U, q_0) \div$ (3.5.5) mean that the controlled system Σ is described by the system of difference equations (3.5.1) and the payoff function of player i is given by (3.5.5).

A strategy profile $U^B = (U_1^B, U_2^B) \in \mathcal{U}$ is a Berge equilibrium in the game Γ if

$$\max_{U_2 \in \mathcal{U}_2} J_1\left(U_1^B, U_2, q_0\right) = J_1\left(U^B, q_0\right),$$

$$\max_{U_1 \in \mathcal{U}_1} J_2\left(U_1, U_2^B, q_0\right) = J_2\left(U^B, q_0\right). \tag{3.5.6}$$

A triplet $(U^B, J_1^B = J_1(U^B, q_0), J_2^B = J_2(U^B, q_0))$ is called *a Berge equilibrium solution of the game* Γ.

According to the definition of *a Nash equilibrium* $U^e = (U_1^e, U_2^e)$ in the game Γ,

$$\max_{U_1 \in \mathcal{U}_1} J_1\left(U_1, U_2^e, q_0\right) = J_1\left(U^e, q_0\right),$$

$$\max_{U_2 \in \mathcal{U}_2} J_2\left(U_1^e, U_2, q_0\right) = J_2\left(U^e, q_0\right).$$

In any two-player game (in particular, Γ), a Nash equilibrium becomes a Berge equilibrium if the players exchange their payoff functions with each other. Therefore, the Berge equilibrium in the game Γ can be designed using the method of dynamic programming, which was adapted for Nash equilibrium in the Cournot duopoly model in the book [71, pp. 184–197]. Let us describe it in detail.

Remark 3.5.1 For obtaining the Berge equilibrium solution, we will employ the procedure dictated by Theorem 3.4.2, making some necessary modifications according to Definition (3.5.6). For the game Γ, this procedure can be reduced to two stages as follows.

Stage 1. For $k = 1$, construct the two functions

$$V_i^{(1)}(q) = q_i^2 \ (i = 1, 2).$$

Stage 2. For $k = 0$, find the four scalar functions $u_i(0, q)$ and $V_i^{(0)}(q)$ ($i = 1, 2$) from the conditions

$$V_1^{(0)}(q) = \max_{u_2} \left\{ q_1^2 - \left[u_1^B(0, q) \right]^2 - u_2^2 + \left[\frac{a-c}{2b} - \frac{q_2}{2} + \beta_1 u_1^B(0, q) + (1 - \beta_2)u_2 \right]^2 \right\}$$

$$= Idem \left\{ u_2 \to u_2^B(0, q) \right\},$$

$$V_2^{(0)}(q) = \max_{u_1} \left\{ q_2^2 - u_1^2 - \left[u_2^B(0, q) \right]^2 + \left[\frac{a-c}{2b} - \frac{q_1}{2} + (1 - \beta_1)u_1 + \beta_2 u_2^B(0, q) \right]^2 \right\}$$

$$= Idem \left\{ u_1 \to u_1^B(0, q) \right\}; \qquad (3.5.7)$$

like before, $Idem\{u_i \to u_i^B(0, q)\}$ means the bracketed expression with u_i replaced by $u_i^B(0, q)$ ($i = 1, 2$).

Then:

1) the Berge equilibrium in the game Γ is given by

$$U^B = (U_1^B, U_2^B), \ U_i^B = U_i^B(0) \ \ U_i^B(0) \div u_i^B(0, q_1, q_2) \quad (i = 1, 2);$$

2) the corresponding payoffs of the players are

$$J_i(U^B, q_0) = V_i^{(0)}(q_{10}, q_{20}) \quad (i = 1, 2);$$

the Berge equilibrium solution of the game Γ is formed by the triplet $(U^B, V_1^{(0)}(q_0), V_2^{(0)}(q_0))$.

3.5.3 Berge equilibrium design

Proposition 3.5.1. *The Berge equilibrium solution of the game Γ has the form*

$$\left(U^B = \left(U_1^B, U_2^B \right), J_1^B = J_1 \left(U^B, q_0 \right), J_2^B = J_2 \left(U^B, q_0 \right) \right),$$

where

$$U_1^B \div u_1^B(0, q) = \frac{1 - \beta_1}{2\beta_1(3 - \beta_1 - \beta_2)} \left[-\frac{a-c}{b}(3 - 2\beta_2) + q_1(2 - \beta_2) + q_2(1 - \beta_2) \right],$$

$$U_2^B \div u_2^B(0, q) = \frac{1 - \beta_2}{2\beta_2(3 - \beta_1 - \beta_2)} \left[-\frac{a-c}{b}(3 - 2\beta_1) + q_1(1 - \beta_1) + q_2(2 - \beta_1) \right],$$

$$J_1 \left(U^B, q_0 \right) = q_{10}^2 - \left[u_1^B(0, q_0) \right]^2 - \left[u_2^B(0, q_0) \right]^2$$

$$+ \left[\frac{a-c}{2b} - \frac{q_{20}}{2} + \beta_1 u_1^B(0, q_0) + (1 - \beta_2)u_2^B(0, q_0) \right]^2,$$

$$J_2 \left(U^B, q_0 \right) = q_{20}^2 - \left[u_1^B(0, q_0) \right]^2 - \left[u_2^B(0, q_0) \right]^2$$

$$+ \left[\frac{a-c}{2b} - \frac{q_{10}}{2} + (1 - \beta_1)u_1^B(0, q_0) + \beta_2 u_2^B(0, q_0) \right]^2.$$

Proof We will follow the procedure mentioned in Remark 3.5.1.

Stage 1 ($k = 1$). Construct the two scalar functions

$$V_i^{(1)}(q) = q_i^2 \ (i = 1, 2).$$

Stage 2 ($k = 0$). The first equality of (3.5.7) is satisfied for $u_2^B(0, q)$ if

$$\max_{u_2} \psi_1(u_2) = \psi_1(u_2^B(0, q)) \ \forall q \in \mathbb{R}^2, \tag{3.5.8}$$

where $\psi_1(u_2) = -u_2^2 + [\frac{a-c}{2b} - \frac{q_2}{2} + \beta_1 u_1^B(0, q) + (1 - \beta_2)u_2]^2$. In turn, condition (3.5.8) holds if

$$\frac{d\psi_1(u_2)}{du_2}\Big|_{u_2^B(0,q)} = -2u_2^B(0, q) + 2\left[\frac{a-c}{2b} - \frac{q_2}{2} + \beta_1 u_1^0(0, q)\right.$$

$$\left. + (1 - \beta_2)u_2^B(0, q)\right](1 - \beta_2) = 0,$$

$$\frac{d^2\psi_1(u_2)}{du_2^2} = -2 + 2(1 - \beta_2)^2 = 2[\beta_2 - 2]\beta_2 < 0.$$

(Here the second inequality is true, because $\beta \in [0, 1]$.) Due to the first equality and analogous second equalities of (3.5.6)–(3.5.7),

$$-2u_1^B(0, q) + 2\left[\frac{a-c}{2b} - \frac{q_1}{2} + (1 - \beta_1)u_1^B(0, q) + \beta_2 u_2^B(0, q)\right](1 - \beta_1) = 0,$$

and the desired strategies $u_i^B(0, q)$ ($i = 1, 2$) are obtained by solving the two linear inhomogeneous algebraic equations

$$\begin{cases} \beta_1(1-\beta_2)u_1^B(0, q) + \beta_2(\beta_2 - 2)u_2^B(0, q) = \left(-\frac{a-c}{2b} + \frac{q_2}{2}\right)(1 - \beta_2), \\ \beta_1(\beta_1 - 2)u_1^B(0, q) + \beta_2(1 - \beta_1)u_2^B(0, q) = \left(-\frac{a-c}{2b} + \frac{q_1}{2}\right)(1 - \beta_1). \end{cases} \tag{3.5.9}$$

The determinant of the system (3.5.9) is

$$\Delta = \begin{vmatrix} \beta_1(1 - \beta_2) & \beta_2(\beta_2 - 2) \\ \beta_1(\beta_1 - 2) & \beta_2(1 - \beta_1) \end{vmatrix} = \beta_1\beta_2(-3 + \beta_1 + \beta_2) \neq 0,$$

since $0 < \beta \leqslant 1$ ($i = 1, 2$). For calculating the solution $u_i^B(0, q)$ ($i = 1, 2$) of the system (3.5.9), we apply Cramer's rule:

$$\Delta_1 = \begin{vmatrix} \left(-\frac{a-c}{2b} + \frac{q_2}{2}\right)(1 - \beta_2) & \beta_2(\beta_2 - 2) \\ \left(-\frac{a-c}{2b} + \frac{q_1}{2}\right)(1 - \beta_1) & \beta_2(1 - \beta_1) \end{vmatrix}$$

$$= (1 - \beta_1)\beta_2 \begin{vmatrix} \left(-\frac{a-c}{2b} + \frac{q_2}{2}\right)(1 - \beta_2) & (\beta_2 - 2) \\ \left(-\frac{a-c}{2b} + \frac{q_1}{2}\right) & 1 \end{vmatrix}$$

$$= \frac{\beta_2(1 - \beta_1)}{2}\left[-\frac{a-c}{b}(3 - 2\beta_2) + q_1(2 - \beta_2) + q_2(1 - \beta_2)\right] > 0,$$

$$\Delta_2 = \begin{vmatrix} \beta_1(1 - \beta_2) & \left(-\frac{a-c}{2b} + \frac{q_2}{2}\right)(1 - \beta_2) \\ \beta_1(\beta_1 - 2) & \left(-\frac{a-c}{2b} + \frac{q_1}{2}\right)(1 - \beta_1) \end{vmatrix}$$

$$
= \beta_1(1-\beta_2) \begin{vmatrix} 1 & \left(-\frac{a-c}{2b}+\frac{q_2}{2}\right) \\ \beta_1 - 2 & \left(-\frac{a-c}{2b}+\frac{q_1}{2}\right)(1-\beta_1) \end{vmatrix}
$$

$$
= \frac{\beta_1(1-\beta_2)}{2}\left[-\frac{a-c}{b}(3-2\beta_1)+q_1(1-\beta_1)+q_2(2-\beta_1)\right].
$$

As a result,

$$
u_1^B(0,q) = \frac{\Delta_1}{\Delta} = \frac{1-\beta_1}{2\beta_1(-3+\beta_1+\beta_2)}\left[-\frac{a-c}{b}(3-2\beta_2)+q_1(2-\beta_2)+q_2(1-\beta_2)\right],
$$

$$
u_2^B(0,q) = \frac{\Delta_2}{\Delta} = \frac{1-\beta_2}{2\beta_2(-3+\beta_1+\beta_2)}\left[-\frac{a-c}{b}(3-2\beta_1)+q_1(1-\beta_1)+q_2(2-\beta_1)\right]
$$

$$(3.5.10)$$

and $U^B = \left(U_1^B, U_2^B\right) \div \left(u_1^B(0,q), u_2^B(0,q)\right)$.

Finally, according to (3.5.7),

$$
J_1^B = J_1\left(U^B, q_0\right) = V_1^{(0)}(q_0) = q_{10}^2 - \left[u_1^B(0,q_0)\right]^2 - \left[u_2^B(0,q_0)\right]^2
$$

$$
+ \left[\frac{a-c}{2b} - \frac{q_{20}}{2} + \beta_1 u_1^B(0,q_0) + (1-\beta_2)u_2^B(0,q_0)\right]^2,
$$

$$
J_2^B = J_2\left(U^B, q_0\right) = V_2^{(0)}(q_0) = q_{20}^2 - \left[u_1^B(0,q_0)\right]^2 - \left[u_2^B(0,q_0)\right]^2
$$

$$
+ \left[\frac{a-c}{2b} - \frac{q_{10}}{2} + (1-\beta_1)u_1^B(0,q_0) + \beta_2 u_2^B(0,q_0)\right]^2,
$$

where $u_i^B(0,q_0)$ are given by (3.5.10), $q_0 = \left(q_1^0, q_2^0\right)$.

This completes the proof of Proposition 3.5.1.

Let us summarize the outcomes of Subsection 3.5.3.

First, a dynamic single-stage case of the controlled Cournot duopoly has been considered, and the concepts of Berge equilibrium and Berge equilibrium solution of such a single-stage noncooperative positional two-player game have been formalized.

Second, an explicit form of the Berge equilibrium solution has been obtained.

3.5.4   Controlled Cournot duopoly with import

As it has been demonstrated in the previous subsection, dynamic programming can be adopted to find a Berge equilibrium in multistage conflicts. There arise natural questions (a) about a proper consideration of uncertain factors and (b) about different conceptual approaches to guarantees, that is, about the possible use of dynamic programming for multistage conflicts in these cases as well. Subsection 3.5.4 is intended to answer both questions within the Cournot duopoly model. Like in the previous subsection, the mathematical model under study is based on the principle of pricing in oligopolized markets proposed by Cournot [281]. The multistage statement of duopoly with possible imports is formalized as a noncooperative game with uncertainty. As a solution the Pareto-guaranteed equilibrium with the principle of vector maximin [365] is introduced. The design procedure of this solution involves a modification of dynamic programming proposed in the book [121] for multistage games.

Static case of the model

Two firms supply a homogeneous product during a given time period. Let q_i be the quantity of products supplied by firm i ($i = 1, 2$). Like in Subsections 3.5.1–3.5.3, *the production cost of firm i* is assumed to be a linearly function of the quantity q_i, i.e., can be written as $cq_i + d$, where the constants c and d specify the average variable and fixed cost, respectively. As before, the product price p is determined by the law of supply and demand depending on the total quantity $\tilde{q} = q_1 + q_2$ supplied by both firms. Let the price p be a linear function of the total supply, $p(\tilde{q}) = a - b\tilde{q}$, where $a = const > 0$ is initial price without product supply and a positive constant b (known as the coefficient of elasticity) shows the price drop in response to unit product supply. The revenue of firm 1 is given by

$$p(\tilde{q})q_1 = (a - b\tilde{q})q_1 = [a - b(q_1 + q_2)]q_1,$$

and its profit (revenue minus cost) can be calculated as

$$\psi_1(q_1, q_2) = [a - b(q_1 + q_2)]q_1 - (cq_1 + d) = aq_1 - bq_1^2 - bq_1q_2 - cq_1 - d;$$

by analogy, the profit of firm 2 has the form

$$\psi_2(q_1, q_2) = [a - b(q_1 + q_2)]q_2 - (cq_2 + d) = aq_2 - bq_2^2 - bq_1q_2 - cq_2 - d.$$

Due to the strict concavity of the functions $\psi_i(q_1, q_2)$ in q_i ($i = 1, 2$), the sufficient conditions for the existence of a quantity q_i^* that maximizes $\psi_i(q_1, q_2)$ in q_i ($i = 1, 2$) are reduced to the system of two linear algebraic equations

$$\begin{cases} \left.\frac{\partial \psi_1(q_1, q_2)}{\partial q_1}\right|_{q_1^*} = a - 2bq_1^* - bq_2 - c = 0, \\ \left.\frac{\partial \psi_2(q_1, q_2)}{\partial q_2}\right|_{q_2^*} = a - bq_1 - 2bq_2^* - c = 0. \end{cases}$$

As a result,

$$q_1^* = \frac{a - c - bq_2}{2b}, \quad q_2^* = \frac{a - c - bq_1}{2b}.$$

Thus, in the static case of the Cournot duopoly model with import, the functional relation between the best responses of both competing firms is given by

$$q_1 = \frac{a - c - bq_2}{2b}, \quad q_2 = \frac{a - c - bq_1}{2b}.$$

This fact will be taken into account in the dynamic case of the model.

Dynamic case of the model

For the model with import, we introduce the following assumptions.

First, the products are supplied during two periods between three time instants (stages), i.e., $k = 0, 1, 2$.

Second, there exists a time lag equal to one period; as a result, the best responses of firms to the competitor's actions have the form

$$q_1(k+1) = \frac{a - c - bq_2(k)}{2b}, \quad q_2(k+1) = \frac{a - c - bq_1(k)}{2b} \quad (k = 0, 1).$$

Third, at the time instant $k = 0$ the initial quantities of products $q_i(0) = q_{i0}$ available from the stock of firm i ($i = 1, 2$) are fixed.

Fourth, which is a new assumption, an importer is expected to enter the market. Both firms have no information about his intentions (maybe, he wants to take a certain market share, or get maximum profit, or simply show his presence in the market, or try to ruin the competitors, etc.). The planned quantities of imports $z(k)$ at time instants $k = 0, 1$ are also unknown. The only natural restriction of the uncertain factor has the form $z(k) \geqslant 0$, $k = 0, 1$. The uncertainty Z is identified with the two-component vector $(z(0), z(1))$, and this fact is indicated by $Z \div (z(0), z(1))$. The set of uncertainties Z is denoted by $\mathcal{Z}$.

Fifth, the management of each firm i (also called *player i*) chooses and maintains a *rate of supply* $u_i[k]$ for each time instant k ($i = 1, 2$; $k = 0, 1$), e.g., by investing into production or implementing new technologies. Note that the control action u_i of player i at a time instant k depends on the quantities supplied by both firms at the time instant k and also on a specific realization $z(k)$ of an uncertain factor z. Thus, a *strategy* $U_i(k)$ of player i (the behavioral rule of firm i) at a time instant k, ($i = 1, 2$), ($k = 0, 1$), is identified with a scalar function $u_i(k, q_1, q_2, z)$. This fact will be denoted by $U_i \div u_i(k, q, z)$, where $q = (q_1, q_2)$ and $u_i(k, q, z) \geqslant 0$ for $q_j \geqslant 0$ ($j = 1, 2$). Then the mathematical model of the controlled dynamic system $\sum$ that describes the supply of products at discrete time instants 0, 1, and 2 can be written as

$$\begin{cases} q_1(k+1) = \frac{a-c}{2b} - \frac{q_2(k)}{2} - \frac{z(k)}{2} + u_1(k, q(k), z(k)), \\ q_2(k+1) = \frac{a-c}{2b} - \frac{q_1(k)}{2} - \frac{z(k)}{2} + u_2(k, q(k), z(k)), \\ q_1(0) = q_{10}, \quad q_2(0) = q_{20} \quad (k = 0, 1). \end{cases} \quad (3.5.11)$$

Note that in the game (3.5.11), each player forms his strategy following the Golden Rule of ethics: "Behave to others as you would like them to behave to you"; also, see the epigraph to this subsection. Therefore, in the game (3.5.11), player 2 directs all the possibilities to help player 1; player 1 responds reciprocally.

The system (3.5.11) consists of two difference (two-stage) linear equations. Denote by $\mathfrak{A}_i(k)$ the set of all strategies $U_i(k)$ at stage k ($i = 1, 2$; $k = 0, 1$). Then a strategy U_i of player i is an ordered pair

$$(U_i(0) \div u_i(0, q, z), \ U_i(1) \div u_i(1, q(1), z(1))) \quad (i = 1, 2),$$

where $U_i(k) \in \mathfrak{A}_i(k)$ ($k = 0, 1$) and $q(1) = (q_1(1), q_2(1))$,

$$q_1(1) = \frac{a-c}{2b} - \frac{1}{2} q_2 - \frac{1}{2} z(0) + u_1(0, q_1, q_2, z(0)),$$

$$q_2(1) = \frac{a-c}{2b} - \frac{1}{2} q_1 - \frac{1}{2} z(0) + u_2(0, q_1, q_2, z(0)); \quad (3.5.12)$$

the set of all strategies U_i is denoted by $\mathfrak{A}_i$.

Thus, a pair of strategies (U_1, U_2) forms a *strategy profile* U of this game, i.e., $\{U\} = \mathfrak{A}$.

The game evolves over $k = 0, 1, 2$ in the following way. Without making coalition with the opponent, each player i independently chooses his strategy $U_i = (U_i(0), U_i(1)) \in \mathfrak{A}_i$; as a result, two scalar functions $u_i(0, q_1, q_2, z) \geqslant 0$ and $u_i(1, q_1, q_2, z) \geqslant 0$ are formed, where $q_1 \geqslant 0$, $q_2 \geqslant 0$, and $z \geqslant 0$. Note that player i chooses an appropriate strategy $U_i(0) \in \mathfrak{A}_i(0)$ and $U_i(1) \in \mathfrak{A}_i(1)$, seeking to maximize his payoff (the value of his payoff function $\mathcal{J}_i(U, q_0)$, $q_0 = (q_{10}, q_{20}, z)$, which will be explicitly defined below). In parallel, specific values $z(0), z(1)$ of the uncertainty are realized. Using (3.5.11) with

$k = 0$, $z = z(0)$ and $q_i(0) = q_{i0}$ $(i = 1, 2)$, we find the state vector $q(1) = (q_1(1), q_2(1))$. Then, again using (3.5.11) with $k = 1$, $z = z(1)$ and the scalar functions $u_i(1, q_1, q_2, z)$ $(i = 1, 2)$, we construct

$$\begin{cases} q_1(2) = \frac{a-c}{2b} - \frac{1}{2}q_2(1) - \frac{1}{2}z(1) + u_2(1, q_1(1), q_2(1), z(1)), \\ q_2(2) = \frac{a-c}{2b} - \frac{1}{2}q_1(1) - \frac{1}{2}z(1) + u_1(1, q_1(1), q_2(1), z(1)). \end{cases} \qquad (3.5.13)$$

This yields:

1) the two sequences

$$\{q_i(k)\}_{k=0}^{2} \qquad (i = 1, 2), \qquad (3.5.14)$$

that describe the trajectory of the system (3.5.11) with the specific strategies $U_i \div \{u_i(0, q, z), u_i(1, q, z)\}$, $U_i \in \mathfrak{A}_i$ $(i = 1, 2)$ of the players;

2) the two sequences of *realizations*

$$\{u_i[k] = u_i(k, q_1(k), q_2(k), z(k))\}_{k=0}^{1} \qquad (i = 1, 2), \qquad (3.5.15)$$

of the strategies $U_i \in \mathfrak{A}_i$ $(i = 1, 2)$ chosen by the players.

Using (3.5.14) and (3.5.15), we design the payoff function of player i for assessing the latter's performance. Three circumstances have to be considered as follows.

First, each firm i $(i = 1, 2)$ seeks to supply to the market as much products as possible, i.e., to maximize the sum

$$q_i^2(2) + \sum_{k=0}^{1} q_i^2(k)$$

with an appropriate choice of $U_i \in \mathfrak{A}_i$.

Second, each firm i $(i = 1, 2)$ simultaneously tries to consume as few internal resources as possible, i.e., to maximize the sum

$$\sum_{k=0}^{1} \left(-2u_i^2[k] \right).$$

Third, following…Germeier's principle of guaranteed result, player i $(i = 1, 2)$ chooses his strategy, expecting the maximum counteraction of the uncertainty, which can be achieved by incorporating the following term into the payoff function:

$$\sum_{k=0}^{1} \left(\frac{z^2(k)}{2} \right).$$

In view of these circumstances, *the payoff function* of player i can be written as

$$\mathcal{J}_i(U, q_0, z) = q_i^2(2) + \sum_{k=0}^{1} \left(q_i^2(k) - 2u_j^2[k] + \frac{z^2(k)}{2} \right) \qquad (i, j = 1, 2; i \neq j). \qquad (3.5.16)$$

The ordered quintuple

$$\Gamma = \langle \{1, 2\}, \textstyle\sum_o \div (3.5.11), \{\mathfrak{A}_i\}_{i=1,2}, \mathcal{Z}, \{\mathcal{J}_i(U, q_0, z) \div (3.5.16)\}_{i=1,2} \rangle,$$

forms a *two-stage noncooperative positional linear-quadratic two-player game under uncertainty*. The notations $\sum_o \div (3.5.11)$ and $\mathcal{J}_i(U, q_0, z) \div (3.5.16)$ mean that the controlled system $\sum_o$ is described by the system of difference equations (3.5.11) and the payoff function of player i is given by (3.5.16).

Definition 3.5.1 *A Pareto-guaranteed Berge equilibrium in the game Γ is a triplet $\left(U^B, \mathcal{J}_1^B[q_0], \mathcal{J}_2^B[q_0]\right) \in \mathfrak{A} \times \mathbb{R}^2$ for which there exists an uncertainty $Z^p \in \mathcal{Z}$ such that:*

1) *the uncertainty Z^p is Pareto-minimal in the bi-criteria choice problem*

$$\left\langle \sum_0, \mathcal{Z}, \{\mathcal{J}_i(U, Z, q_0)\}_{i=1,2} \right\rangle$$

for each $U \in \mathfrak{A}$;

2) *the strategy profile U^B is a Berge equilibrium in the noncooperative multistage two-player game*

$$\left\langle \{1, 2\}, \sum_0 (Z = Z^p), \{\mathfrak{A}_i\}_{i=1,2}, \{\mathcal{J}_i(U, Z^p, q_0)\}_{i=1,2} \right\rangle,$$

which is obtained from the game Γ by letting $Z = Z^p$. In this case, $\mathcal{J}_i^B[q_0] = \mathcal{J}_i(U^B, Z^p, q_0)$ is called the Pareto-guaranteed payoff of player i, and U^B the guaranteeing Berge equilibrium.

A pair $\left(\{q^B(k) \mid k = 0, 1, 2\}, U^B\right)$, where $\{q^B(0) = q_0, q^B(1), q^B(2)\}$ is the discrete trajectory of the system $\sum$ that is induced by the Berge equilibrium U^B, is called *a Berge equilibrium process of the game Γ.*

Remark 3.5.2 For obtaining the Berge equilibrium solution, we will employ the procedure of five stages as follows.

Stage 1 For $k = 2$, construct the three functions

$$V^{(2)}(q) = q_1^2 + q_2^2, \quad V_i^{(2)}(q) = q_i^2 \quad (i = 1, 2).$$

Stage 2 For $k = 1$, find the scalar function $z^P(1, u)$ from the condition

$$V^{(1)}(q) = \min_z \left\{ q_1^2 + q_2^2 - 2u_1^2 - 2u_2^2 + z^2 + \left[\frac{a - c}{2b} - \frac{q_2}{2} - \frac{z}{2} + u_1\right]^2 \right.$$
$$\left. + \left[\frac{a - c}{2b} - \frac{q_1}{2} - \frac{z}{2} + u_2\right]^2 \right\} = Idem\left\{z \rightarrow z^P(1, u)\right\}; \qquad (3.5.17)$$

like before, $Idem\left\{z \rightarrow z^P(1, u)\right\}$ means the bracketed expression with z replaced by $z^P(1, u)$.

Stage 3 For $k = 1$, find the four scalar functions $u_i^B(1, q)$ and $V_i^{(1)}(q)$ $(i = 1, 2)$ from the conditions

$$V_1^{(1)}(q) = \max_{u_2} \left\{ q_1^2 - 2u_2^2 + \frac{(z^P(1, u))^2}{2} + \left[\frac{a - c}{2b} - \frac{q_2}{2} - \frac{z^P(1, u)}{2} + u_2\right]^2 \right\}$$
$$= Idem\left\{u_2 \rightarrow u_2^B(1, q)\right\},$$

$$V_2^{(1)}(q) = \max_{u_1} \left\{ q_2^2 - 2u_1^2 + \frac{(z^P(1, u))^2}{2} + \left[\frac{a - c}{2b} - \frac{q_1}{2} - \frac{z^P(1, u)}{2} + u_1\right]^2 \right\}$$
$$= Idem\left\{u_1 \rightarrow u_1^B(1, q)\right\}. \qquad (3.5.18)$$

Stage 4 For $k = 0$, find the scalar function $z^P(0, q)$ from the condition

$$V^{(0)}(q) = \min_z \left\{ q_1^2 + q_2^2 - 2u_1^2 - 2u_2^2 + z^2 + V_1^{(1)}(q_1(1), q_2(1)) + V_2^{(1)}(q_1(1), q_2(1)) \right\}$$

$$= Idem \left\{ z \to z^P(0, u) \right\}. \tag{3.5.19}$$

Stage 5 For $k = 0$, find the four scalar functions $u_i^B(0, q)$ and $V_i^{(0)}(q)$ $(i = 1, 2)$ from the conditions

$$V_1^{(0)}(q) = \max_{u_2} \left\{ q_1^2 - 2u_2^2 + \frac{(z^P(0, u))^2}{2} + V_1^{(1)}(q_1(1), q_2(1)) \right\} = Idem \left\{ u_2 \to u_2^B(0, q) \right\},$$

$$V_2^{(0)}(q) = \max_{u_1} \left\{ q_2^2 - 2u_1^2 + \frac{(z^P(0, u))^2}{2} + V_2^{(1)}(q_1(1), q_2(1)) \right\} = Idem \left\{ u_1 \to u_1^B(0, q) \right\},$$

$$\tag{3.5.20}$$

where

$$q_1(1) = \frac{a - c}{2b} - \frac{1}{2}q_2 - \frac{1}{2}z^P(0, u) + u_2,$$

$$q_2(1) = \frac{a - c}{2b} - \frac{1}{2}q_1 - \frac{1}{2}z^P(0, u) + u_1. \tag{3.5.21}$$

Then:

1) the guaranteeing Berge equilibrium in the game Γ is given by $U^B = (U_1^B, U_2^B)$, $U_i^B = (U_i^B(0), U_i^B(1))$ and $U_i^B(0) \div u_i^B(0, q_1, q_2)$, $U_i^B(1) \div u_i^B(1, q_1^B(1), q_2^B(1))$ $(i = 1, 2)$, where

$$q_1^B(1) = \frac{a - c}{2b} - \frac{1}{2}q_2 - \frac{1}{2}z^P(0, u^B(0, q_1, q_2)) + u_2^B(0, q_1, q_2),$$

$$q_2^B(1) = \frac{a - c}{2b} - \frac{1}{2}q_1 - \frac{1}{2}z^P(0, u^B(0, q_1, q_2)) + u_1^B(0, q_1, q_2); \tag{3.5.22}$$

2) the corresponding Pareto-guaranteed payoffs of the players are $\mathcal{J}_i(U^B, q_0) = V_i^{(0)}(q_{10}, q_{20})$ $(i = 1, 2)$; the Berge equilibrium solution of the game Γ is formed by the triplet $(U^B, V_1^{(0)}(q_0), V_2^{(0)}(q_0))$.

Equilibrium solution design

For equilibrium design, we will follow the procedure discussed in Remark 3.5.2.

Stage 1 (for $k = 2$). We construct the three scalar functions

$$V^{(2)}(q) = q_1^2 + q_2^2, \quad V_i^{(2)}(q) = q_i^2 \quad (i = 1, 2).$$

Stage 2 (for $k = 1$). Equality (7) is satisfied for $z^P(1, u)$ if

$$\min_z W^{(1)}(z) = W^{(1)}(z^P(1, u)) \quad \forall q \in \mathbf{R}^2, \quad \forall u \in \mathfrak{A}, \tag{3.5.23}$$

where

$$W^{(1)}(z) = q_1^2 + q_2^2 - 2u_1^2 - 2u_2^2 + z^2 + \left[\frac{a - c}{2b} - \frac{q_2}{2} - \frac{z}{2} + u_2 \right]^2 + \left[\frac{a - c}{2b} - \frac{q_1}{2} - \frac{z}{2} + u_1 \right]^2.$$

In turn, condition (3.5.23) holds if

$$\left.\frac{\partial W^{(1)}(z)}{\partial z}\right|_{z^P(1,u)} = 3z^P(1,u) - \frac{a-c}{b} + \frac{q_1}{2} + \frac{q_2}{2} - u_1 - u_2 = 0 \tag{3.5.24}$$

and the function $W^{(1)}(z)$ is convex for any fixed $q \in \mathbf{R}^2$ and $u \in \mathfrak{A}$. Convexity follows from

$$\frac{\partial^2 W^{(1)}(z)}{\partial z^2} = 3 > 0.$$

From (3.5.24) we obtain

$$z^P(1,u) = \frac{a-c}{3b} - \frac{q_1}{6} - \frac{q_2}{6} + \frac{u_1}{3} + \frac{u_2}{3}. \tag{3.5.25}$$

Finally, we substitute (3.5.25) into the right-hand side of (3.5.11) to get the following system of difference equations for $k = 1$:

$$\begin{cases} q_1(2) = \frac{a-c}{3b} + \frac{q_1(1)}{12} - \frac{5}{12}q_2(1) + \frac{5}{6}u_2[1] - \frac{u_1[1]}{6}, \\ q_2(2) = \frac{a-c}{3b} - \frac{5}{12}q_1(1) + \frac{q_2(1)}{12} - \frac{u_2[1]}{6} + \frac{5}{6}u_1[1]. \end{cases} \tag{3.5.26}$$

Stage 3 (for $k = 1$). We take the functions

$$W_1^{(1)}(u) = q_1^2 - 2u_2^2 + \frac{1}{2}\left[\frac{a-c}{3b} - \frac{q_1}{6} - \frac{q_2}{6} + \frac{u_2}{3} + \frac{u_1}{3}\right]^2$$
$$+ \left[\frac{a-c}{3b} + \frac{q_1(1)}{12} - \frac{5}{12}q_2(1) + \frac{5}{6}u_2[1] - \frac{u_1[1]}{6}\right]^2,$$

$$W_2^{(1)}(u) = q_2^2 - 2u_1^2 + \frac{1}{2}\left[\frac{a-c}{3b} - \frac{q_1}{6} - \frac{q_2}{6} + \frac{u_2}{3} + \frac{u_1}{3}\right]^2$$
$$+ \left[\frac{a-c}{3b} - \frac{5}{12}q_1(1) + \frac{q_2(1)}{1} - \frac{u_2[1]}{6} + \frac{5}{6}u_1[1]\right]^2.$$

Equalities (3.5.18) are satisfied for $u^B(1,q) = (u_1^B(1,q), u_1^B(1,q))$ if

$$\max_{u_j} W_i^{(1)}(u) = W_i^{(1)}(u^B(1,q)) \quad (i,j = 1, 2; \ i \neq j) \ \forall q \in \mathbf{R}^2. \tag{3.5.27}$$

In turn, condition (3.5.27) holds if

$$\left.\frac{\partial W_1^{(1)}(u)}{\partial u_1}\right|_{u^B(1,q)} = -4u_2^B(1,q) + \frac{1}{3}\left[\frac{a-c}{3b} - \frac{q_1}{6} - \frac{q_2}{6} + \frac{u_2^B(1,q)}{3} + \frac{u_1^B(1,q)}{3}\right]$$
$$+ \frac{5}{3}\left[\frac{a-c}{3b} + \frac{q_1(1)}{12} - \frac{5}{12}q_2(1) + \frac{5}{6}u_2^B(1,q) - \frac{u_1^B(1,q)}{6}\right] = 0,$$

$$\left.\frac{\partial^2 \partial W_1^{(1)}(u)}{\partial u_1^2}\right|_{u^B(1,q)} = -\frac{5}{2} < 0,$$

$$\left.\frac{\partial W_2^{(1)}(u)}{\partial u_2}\right|_{u^B(1,q)} = -4u_1^B(1,q) + \frac{1}{3}\left[\frac{a-c}{3b} - \frac{q_1}{6} - \frac{q_2}{6} + \frac{u_2^B(1,q)}{3} + \frac{u_1^B(1,q)}{3}\right]$$

$$+ \frac{5}{3} \left[\frac{a-c}{3b} - \frac{5}{12} q_1(1) + \frac{1}{12} - \frac{u_2^B(1,q)}{6} + \frac{5}{6} u_1^B(1,q) \right] = 0,$$

$$\left. \frac{\partial^2 \partial W_2^{(1)}(u)}{\partial u_2^2} \right|_{u^B(1,q)} = -\frac{5}{2} < 0. \tag{3.5.28}$$

The second and fourth conditions of (3.5.28) are true; the first and third ones form a system of equations with the solution $u^B(1,q) = (u_1^B(1,q), u_2^B(1,q))$ given by

$$\begin{cases} u_1^B(1,q) = \frac{a-c}{4b} + \frac{3}{56} q_1 - \frac{17}{56} q_2, \\ u_2^B(1,q) = \frac{a-c}{4b} - \frac{17}{56} q_1 + \frac{3}{56} q_2. \end{cases} \tag{3.5.29}$$

Next, substituting (3.5.29) into (3.5.25) yields

$$z^P(1, u^B(1,q)) = \frac{1}{2} \left(\frac{a-c}{b} - \frac{q_1}{2} - \frac{q_2}{2} \right).$$

In view of (3.5.27) and (3.5.29), from (3.5.18) we arrive at

$$V_1^{(1)}(q) = \frac{1}{4} \left(\frac{a-c}{b} - q_2 \right)^2 + q_1^2 + \frac{45}{784} (q_1 - q_2)^2,$$

$$V_2^{(1)}(q) = \frac{1}{4} \left(\frac{a-c}{b} - q_1 \right)^2 + q_2^2 + \frac{45}{784} (q_1 - q_2)^2.$$

Thus, in Stage 3 we have found the four functions

$$u_1^B(1,q) = \frac{a-c}{4b} - \frac{17}{56} q_1 + \frac{3}{56} q_2,$$

$$u_2^B(1,q) = \frac{a-c}{4b} + \frac{3}{56} q_1 - \frac{17}{56} q_2,$$

$$V_1^{(1)}(q) = \frac{1}{4} \left(\frac{a-c}{b} - q_2 \right)^2 + q_1^2 + \frac{45}{784} (q_1 - q_2)^2,$$

$$V_2^{(1)}(q) = \frac{1}{4} \left(\frac{a-c}{b} - q_1 \right)^2 + q_2^2 + \frac{45}{784} (q_1 - q_2)^2.$$

Stage 4 (for $k = 0$). Equality (3.5.19) is satisfied for $z^P(0, u)$ if

$$\min_z W^{(0)}(z) = W^{(0)}(z^P(0,u)) \quad \forall q \in \mathbf{R}^2, \quad \forall u \in \mathfrak{A}, \tag{3.5.30}$$

where

$$W^{(0)}(z) = q_1^2 + q_2^2 - 2u_1^2 - 2u_2^2 + z^2 + V_1^{(1)}(q(1)) + V_2^{(1)}(q(1)) = q_1^2 + q_2^2 - 2u_1^2$$

$$- 2u_2^2 + z^2 + \frac{1}{4} \left(\frac{a-c}{b} - q_1(1) \right)^2 + \frac{1}{4} \left(\frac{a-c}{b} - q_2(1) \right)^2 + q_1^2(1) + q_2^2(1)$$

$$+ \frac{45}{392} (q_1(1) - q_2(1))^2,$$

and $q_1(1)$ and $q_2(1)$ are given by (3.5.12).

In turn, condition (3.5.30) holds if

$$\left. \frac{\partial W^{(1)}(z)}{\partial z} \right|_{z^P(0,u)} = 2z^P(1,u) - \frac{1}{2}\left(\frac{a-c}{b} - q_2(1)\right)\frac{\partial q_2(1)}{\partial z}$$

$$- \frac{1}{2}\left(\frac{a-c}{b} - q_1(1)\right)\frac{\partial q_1(1)}{\partial z} + 2q_1(1)\frac{\partial q_1(1)}{\partial z} + 2q_2(1)\frac{\partial q_2(1)}{\partial z}$$

$$= \frac{13}{4}z^P(0,u) - \frac{3}{4}\frac{a-c}{b} + \frac{5q_1}{8} + \frac{5q_2}{8} - \frac{5u_1}{4} - \frac{5u_2}{4} \qquad (3.5.31)$$

and

$$\left. \frac{\partial^2 W^{(1)}(z)}{\partial z^2} \right|_{z^P(0,u)} = \frac{13}{4} > 0. \qquad (3.5.32)$$

Inequality (3.5.32) is true for all z, in particular, for $z = z^P(0,u)$; from (3.5.31) it follows that

$$z^P(0,u) = \frac{1}{13}\left(\frac{3(a-c)}{b} - \frac{5q_1}{2} - \frac{5q_1}{2} + 5u_1 + 5u_2\right). \qquad (3.5.33)$$

Substituting (3.5.33) into (3.5.11), we obtain the following system of two difference equations for $k = 0$:

$$\begin{cases} q_1(1) = \frac{5(a-c)}{13b} + \frac{5}{52}q_1(0) - \frac{21}{52}q_2(0) + \frac{21}{26}u_2[0] - \frac{5}{26}u_1[0], \\ q_2(1) = \frac{5(a-c)}{13b} - \frac{21}{52}q_1(0) + \frac{5}{52}q_2(0) - \frac{5}{26}u_2[0] + \frac{21}{26}u_1[0]. \end{cases} \qquad (3.5.34)$$

Stage 5 (for $k = 0$). According to (3.5.21), (3.5.33) and (3.5.34),

$$V_1^{(0)}(q) = \max_{u_2}\left\{ q_1^2 - 2u_2^2 + \frac{1}{2}\frac{1}{13^2}\left(\frac{3(a-c)}{b} - \frac{5q_1}{2} - \frac{5q_2}{2} + 5u_2 + 5u_1\right)^2 \right.$$

$$+ \frac{1}{4}\left(\frac{1}{13}\right)^2\left(\frac{8(a-c)}{b} + \frac{21}{4}q_1 - \frac{5}{4}q_2 + \frac{5}{2}u_2 - \frac{21}{2}u_1\right)^2$$

$$+ \left(\frac{1}{13}\right)^2\left(\frac{5(a-c)}{b} + \frac{5}{4}q_1 - \frac{21}{4}q_2 + \frac{21}{2}u_2 - \frac{5}{2}u_1\right)^2$$

$$\left. + \frac{45}{784}\left(\frac{q_1}{2} - \frac{q_2}{2} + u_2 - u_1\right) = Idem\left\{u_2 \rightarrow u_2^B(0,q)\right\}; \qquad (3.5.35)$$

by analogy,

$$V_2^{(0)}(q) = \max_{u_1}\left\{ q_2^2 - 2u_1^2 + \frac{1}{2}\frac{1}{13^2}\left(\frac{3(a-c)}{b} - \frac{5q_2}{2} - \frac{5q_1}{2} + 5u_1 + 5u_2\right)^2 \right.$$

$$+ \frac{1}{4}\left(\frac{1}{13}\right)^2\left(\frac{8(a-c)}{b} + \frac{21}{4}q_2 - \frac{5}{4}q_1 + \frac{5}{2}u_1 - \frac{21}{2}u_2\right)^2$$

$$+ \left(\frac{1}{13}\right)^2\left(\frac{5(a-c)}{b} + \frac{5}{4}q_2 - \frac{21}{4}q_1 + \frac{21}{2}u_1 - \frac{5}{2}u_2\right)^2$$

$$\left. + \frac{45}{784}\left(\frac{q_2}{2} - \frac{q_1}{2} - u_2 + u_1\right) = Idem\left\{u_1 \rightarrow u_1^B(0,q)\right\}, \qquad (3.5.36)$$

Using the results of Stage 3, we write

$$u_1^e(0, q) = \frac{1}{18}\left(\frac{5(a-c)}{b} + \frac{11765}{5246}q_1 - \frac{32749}{5246}q_2\right),$$

$$u_2^e(0, q) = \frac{1}{18}\left(\frac{5(a-c)}{b} - \frac{32749}{5246}q_1 + \frac{11765}{5246}q_2\right). \tag{3.5.37}$$

Substituting (3.5.37) into (3.5.35) and (3.5.36), and performing routine calculations in *Maple 2017*, we finally arrive at

$$V_1^{(0)}(q) \approx -0.5839\frac{a-c}{b}q_{20} + 0.3707q_{20}^2 - 0.3246q_{10}q_{20} + 0.1882\frac{a-c}{b}q_{10}$$

$$+ 0.3025\left(\frac{a-c}{b}\right)^2 + 1.2564q_{10}^2,$$

$$V_2^{(0)}(q) \approx -0.5839\frac{a-c}{b}q_{10} + 0.3707q_{10}^2 - 0.3246q_{10}q_{20} + 0.1882\frac{a-c}{b}q_{20}$$

$$+ 0.3025\left(\frac{a-c}{b}\right)^2 + 1.2564q_{20}^2.$$

In fact, the following result has been established.

Proposition 3.5.2 *The Pareto-guaranteed Berge equilibrium in the game* Γ *is the triplet* $(U^B, \mathcal{J}_1^B[q_0], \mathcal{J}_2^B[q_0])$, *where the guaranteed payoffs of the players are*

$$\mathcal{J}_1^B[q_0](q) \approx -0.5839\frac{a-c}{b}q_{20} + 0.3707q_{20}^2 - 0.3246q_{10}q_{20} + 0.1882\frac{a-c}{b}q_{10}$$

$$+ 0.3025\left(\frac{a-c}{b}\right)^2 + 1.2564q_{10}^2,$$

$$\mathcal{J}_2^B[q_0](q) \approx -0.5839\frac{a-c}{b}q_{10} + 0.3707q_{10}^2 - 0.3246q_{10}q_{20} + 0.1882\frac{a-c}{b}q_{20}$$

$$+ 0.3025\left(\frac{a-c}{b}\right)^2 + 1.2564q_{20}^2,$$

and the guaranteeing Berge equilibrium is

$$U^B = (U_1^B, U_2^B), \quad U_i^B = (U_i^B(0), U_i^B(1)) \quad (i = 1, 2),$$

where

$$U_1^B(0) \div u_1^B(0, q) = \frac{1}{18}\left(\frac{5(a-c)}{b} + \frac{11765}{5246}q_1 - \frac{32749}{5246}q_2\right),$$

$$U_1^B(1) \div u_1^B(1, q) = \frac{a-c}{4b} + \frac{3}{56}q_1(1) - \frac{17}{56}q_2(1),$$

$$U_2^B(0) \div u_2^B(0, q) = \frac{1}{18}\left(\frac{5(a-c)}{b} - \frac{32749}{5246}q_1 + \frac{11765}{5246}q_2\right),$$

$$U_2^B(1) \div u_2^B(1, q) = \frac{a-c}{4b} - \frac{17}{56}q_1(1) + \frac{3}{56}q_2(1),$$

and $q_1(1)$ *and* $q_2(1)$ *are calculated by substituting (3.5.37) into (3.5.34).*

Postscript

Uncertainty is chasing us throughout life! How should we struggle with uncertainty? The economic applications of Subsection 3.5.4 have indicated two possible approaches as follows.

Consider the multicriteria problem under uncertainty in which the DM has to maximize a set of criteria $f_i(x, y)$, where $x \in \mathbf{X}$ is an alternative and $y \in \mathbf{Y}$ is an uncertain factor. The first approach is to associate with each criterion $f_i(x, y)$ *its strong guarantee*

$$f_i[x] = \min_{y \in \mathbf{Y}} f_i(x, y) \leqslant f_i(x, y) \quad \forall \, y \in \mathbf{Y}.$$

In this case, the original mathematical model is replaced by the model of guarantees, where the role of criteria is played by the guarantees $f_i[x]$. (Of course, this makes sense only for the class of *interval uncertainties* about which the DM knows only the lower and upper limits of possible values). Such guarantees $f_i[x]$ do exist and are continuous if, e.g., the sets $\mathbf{X}$ and $\mathbf{Y}$ are compact and the functions $f_i(x, y)$ are continuous on $\mathbf{X} \times \mathbf{Y}$. However, this approach suffers from an obvious drawback: the strong guarantees $f_i[x]$ are the smallest among all possible ones, but we are interested in maximization of the criteria.

The second approach is to perform transition to the vector guarantees $f_i[x]$, $i \in \mathbb{N} = \{1, \ldots, N \geqslant 2\}$. The core of such an approach consists in using *vector guarantees*, which represent some vector minimum (in the sense of Slater, Pareto, Borwein, or Joffrion—any of the vector minima) with respect to $y \in \mathbf{Y}$ from the theory of multicriteria choice problems $\Gamma_v = \langle \mathbf{X}, \mathbf{Y}, \{f(x, y) = (f_1(x, y), ..., f_N(x, y))\} \rangle$. The vector minimal alternatives $x \in \mathbf{X}$ have the following guaranteeing property in Γ_v: with the alternative $x \neq x^*$ used in the problem Γ_v, the values of all components $f_i[x^*]$ cannot be reduced simultaneously, i.e., the inequalities

$$f_i(x, y) < f_i[x^*] \quad \forall \, y \in \mathbf{Y}, \; i \in \mathbb{N},$$

are inconsistent.

No doubt, the second approach (vector guarantees) is preferable to the DM than its counterpart based on the strong guarantees $\min_{y \in \mathbf{Y}} f_i(x, y) = f_i[x]$: a strong guarantee never exceeds a vector guarantee, and there exist vector guarantees under the same requirements to the elements of Γ_v as strong guarantees (i.e., in the case of compact sets $\mathbf{X}$ and $\mathbf{Y}$ and $f_i(x, y) \in C(\mathbf{X} \times \mathbf{Y})$).

Exercises, problems, and solution tips

> When studying science, the examples are more useful than the rules.
> —Newton[36]

Appendix 1 presents exercises in the form of propositions and statements as well as problems, most of which were actively used for writing this book. Some solution tips are given.

1) *Prove* the following relations, in which a set $X \subset \mathbf{R}^n$ is compact and scalar functions $F_i(x)$, $i \in \{\varnothing, 1, 2, \ldots, N\}$, with the domain X are continuous:

(a)
$$
\begin{cases}
\max\limits_{x \in X} F(x) = -\min\limits_{x \in X}[-F(x)], \\
\min\limits_{x \in X} F(x) = -\max\limits_{x \in X}[-F(x)];
\end{cases}
$$

(b)
$$
\begin{cases}
\left|\max\limits_{x \in X} F(x)\right| \leqslant \max\limits_{x \in X} |F(x)|, \\
\left|\min\limits_{x \in X} F(x)\right| \leqslant \max\limits_{x \in X} |F(x)|;
\end{cases}
$$

(c)
$$
\begin{cases}
\max\limits_{x \in X}[F_1(x) + F_2(x)] \leqslant \max\limits_{x \in X} F_1(x) + \max\limits_{x \in X} F_2(x), \\
\max\limits_{x \in X}[F_1(x) + F_2(x)] \geqslant \max\limits_{x \in X} F_1(x) + \max\limits_{x \in X_1} F_2(x),
\end{cases}
$$
where
$$
X_1 = \{x \in X \mid F_1(x) = \max\limits_{z \in X} F_1(z)\};
$$

(d)
$$
\begin{cases}
\left|\max\limits_{x \in X} F_1(x) - \max\limits_{x \in X} F_2(x)\right| \leqslant \max\limits_{x \in X} |F_1(x) - F_2(x)|, \\
\left|\min\limits_{x \in X} F_1(x) - \min\limits_{x \in X} F_2(x)\right| \leqslant \max\limits_{x \in X} |F_1(x) - F_2(x)|;
\end{cases}
$$

(e) $\left(\max\limits_{x \in X} |F(x)|\right)^2 = \max\limits_{x \in X} (F(x))^2;$

(f) $\min\limits_{x \in X}(\alpha F(x) + \beta) = \alpha \min\limits_{x \in X} F(x) + \beta \ \ \forall \alpha, \beta \in \mathbf{R} \ \wedge \alpha > 0;$

(g) $\min\limits_{x \in X} \sum\limits_{i=1}^{N} F_i(x) \geqslant \sum\limits_{i=1}^{N} \min\limits_{x \in X} F_i(x);$

(h) $\max\limits_{x \in X} \sum\limits_{i=1}^{N} F_i(x) \leqslant \sum\limits_{i=1}^{N} \max\limits_{x \in X} F_i(x);$

(i) if $\min\limits_{x \in X} F_i(x) \geqslant 0$ $(i = 1, 2, \ldots, N)$, then

---

[36]Isaac Newton, in full Sir Isaac Newton, (1643–1727), English physicist and mathematician, who was the culminating figure of the Scientific Revolution of the 17th century.

1) $\quad\min\limits_{x\in X}\prod\limits_{i\in\mathbb{N}}F_i(x) \;=\; \prod\limits_{i\in\mathbb{N}}\min\limits_{x\in X}F_i(x),$

2) $\quad\min\limits_{x\in X}\min\limits_{i\in\mathbb{N}}F_i(x) \;=\; \min\limits_{i\in\mathbb{N}}\min\limits_{x\in X}F_i(x);$

(j) $\quad\min\limits_{x\in X}\max\limits_{i\in\mathbb{N}}F_i(x) \;\geqslant\; \max\limits_{i\in\mathbb{N}}\min\limits_{x\in X}F_i(x);$

(k) $\quad$ if $\min\limits_{x\in X}F_j(x)\geqslant-\infty$ $(j=1,2)$, then

$$F_1(x)F_2(x)+\left[\min_{x\in X}F_1(x)\right]\cdot\left[\min_{x\in X}F_2(x)\right] \geqslant F_1(x)\min_{x\in X}F_2(x)+\left[\min_{x\in X}F_1(x)\right]F_2(x).$$

Solution tips (a) Let $\max\limits_{x\in X}F(x)=F(\bar{x})$ and $\min\limits_{x\in X}[-F(x)]=-F(\bar{\bar{x}})$.
Then

$$\max_{x\in X}F(x)=F(\bar{x})=-[-F(\bar{x})]\leqslant-\min_{x\in X}[-F(x)]=-[-F(\bar{\bar{x}})],$$

$$-\min_{x\in X}[-F(x)]=-[-F(\bar{\bar{x}})]=F(\bar{\bar{x}})\leqslant\max_{x\in X}F(x)=F(\bar{x}).$$

Combining these two chains of relations yields

$$\max_{x\in X}F(x)=-\min_{x\in X}[-F(x)].$$

(b) $\quad\left|\max\limits_{x\in X}F(x)\right|=|F(\bar{x})|\leqslant\max\limits_{x\in X}|F(x)|,$

$\quad\left|\min\limits_{x\in X}F(x)\right|=\left|-\max\limits_{x\in X}[-F(x)]\right|\leqslant\max\limits_{x\in X}|-F(x)|=\max\limits_{x\in X}|F(x)|.$

(c) $\quad$ The function $F_1(x)=C=const.$ on the set X_1; therefore,

$$\max_{x\in X}[F_1(x)+F_2(x)]\geqslant\max_{x\in X_1}[C+F_2(x)]=C+\max_{x\in X_1}F_2(x)=\max_{x\in X}F_1(x)+\max_{x\in X_1}F_2(x),$$

since $X_1\subset X$.

(d) $\quad$ Due to (c),

$$\max_{x\in X}F_1(x)\leqslant\max_{x\in X}F_2(x)+\max_{x\in X}|F_1(x)-F_2(x)|,$$

$$\max_{x\in X}F_2(x)\leqslant\max_{x\in X}F_1(x)+\max_{x\in X}|F_1(x)-F_2(x)|,$$

which directly implies (d).

(e) $\quad$ First of all,

$$\left(\max_{x\in X}|F(x)|\right)^2=(|F(\bar{x})|)^2\leqslant\max_{x\in X}(F(x))^2;$$

on the other hand, for all $x\in X$,

$$[F(x)]^2=(|F(x)|)^2\leqslant\left(\max_{x\in X}|F(x)|\right)^2.$$

Combining these two inequalities finally gives the requisite equality (e).

(g) For all $x \in X$ and $i = 1, 2$, $\min_{x \in X}[F_1(x) + F_2(x)] \geqslant \min_{x \in X} F_1(x) + \min_{x \in X} F_2(x)$. For obtaining the desired result, apply mathematical induction.

(j) Let $x \in X$ and $i = 1, 2$. Then for all $x \in X$,

$$\max_{x \in X}\{F_1(x), F_2(x)\} \geqslant \max\{\min_{x \in X} F_1(x), \min_{x \in X} F_2(x)\},$$

and consequently

$$\min_{i=1,2}\left[\max_{x \in X}\{F_1(x), F_2(x)\}\right] \geqslant \max\left\{\min_{x \in X} F_1(x), \min_{x \in X} F_2(x)\right\}.$$

(k) Expand the inequality

$$\left[F_1(x) - \min_{x \in X} F_1(x)\right] \cdot \left[F_2(x) - \min_{x \in X} F_2(x)\right] \geqslant 0.$$

2) Consider the noncooperative two-player game

$$\Gamma_2 = \langle\{1, 2\}, \{X_i\}_{i=1,2}, \{f_i(x_1, x_2)\}_{i=1,2}\rangle,$$

in which each player $i = 1, 2$ chooses his strategy $x_i \in X_i \subset \mathbf{R}^{n_i}$, thereby forming a strategy profile $x = (x_1, x_2) \in X = X_1 \times X_2$. The payoff function $f_i(x) = f_i(x_1, x_2)$ of each player $i = 1, 2$ is defined on X. A strategy profile $x^e = (x_1^e, x_2^e)$ *is a Nash equilibrium* (NE) in the game Γ_2 if, for all $x_i \in X_i$ and $i = 1, 2$,

$$f_1\left(x_1, x_2^e\right) \leqslant f_1(x^e), \; f_2\left(x_1^e, x_2\right) \leqslant f_2\left(x^e\right) \; \forall x_i \in X_i \; (i = 1, 2).$$

Strategy profiles $x^{(1)}$ and $x^{(2)}$ are called *equivalent* in the game Γ_2 if $f_i(x^{(1)}) = f_i(x^{(2)})$ $(i = 1, 2)$. An alternative $y^P = (y_1^P, \ldots, y_N^P) \in Y$ is called Pareto-maximal in an N-criteria choice problem $\Gamma_v = \langle Y, \{y = (y_1, \ldots, y_N)\}\rangle$ if, for all $y \in Y \subseteq \mathbf{R}^N$, the system of inequalities $y_j \geqslant y_j^P$ $(j = 1, \ldots, N)$, with at least one strict inequality, is inconsistent. Denote by $P(Y)$ the set of Pareto-maximal alternatives in the problem Γ_v.

Proposition A.1.1 If in the game

$$\Gamma_2^{**} = \left\langle\{1, 2\}, \left\{Y, Z = Y \subset \mathbf{R}_>^N\right\}, \{f_i(y, z)\}_{i=1,2}\right\rangle$$

the payoff functions of the players are given by

$$f_1(y, z) = \min_{j \in \mathbb{N}} \frac{y_j}{z_j}, \; f_2(y, z) = \min_{j \in \mathbb{N}} \frac{z_j}{y_j},$$

then $\{(y^0, y^0) \mid y^0 \in P(Y)\}$ are the Nash equilibria in the game Γ_2^{**}, and they are equivalent.

This result was proved in [213, pp. 94–96].

3) For a scalar function $f(x, y)$, where $x \in \mathbf{R}^n$ and $y \in \mathbf{R}^m$, denote by $\frac{\partial f(x, y)}{\partial x} = \mathrm{grad}_x f(x, y)$ an n-dimensional column vector composed of the partial derivatives of $f(x, y)$ with respect to the coordinates of x.

Prove that, for constant matrices A and B of dimensions $n \times n$ and $n \times m$, respectively, and for constant vectors a and b of dimensions n and m, respectively,

$$\frac{\partial}{\partial x}(x'Ax) = (A + A')x, \quad \frac{\partial}{\partial y}(x'By) = B'x,$$

$$\frac{\partial}{\partial x}(x'By) = By, \quad \frac{\partial^2}{\partial x^2}(x'Ax) = A + A', \quad \frac{\partial}{\partial x}(a'x) = a,$$

and hence

$$\frac{\partial}{\partial x}\left[x'Ay + a'x\right] = Ay + a, \quad \frac{\partial}{\partial y}\left[x'Ay + b'y\right] = A'x + b,$$

$$\frac{\partial}{\partial x}\left[x'Ax + a'x\right] = (A + A')x + a,$$

for any $x \in \mathbf{R}^n$ and $y \in \mathbf{R}^m$.

Solution tips Write the left- and right-hand sides of the equalities in the coordinate-wise form and then check them.

4)   *Consider* the function

$$f(x, y) = x'Ax + 2x'By + y'Cy + 2a'x + 2c'y + d, \quad x \in \mathbf{R}^n, y \in \mathbf{R}^m,$$

with constant matrices A, B, and C (where $A = A'$ and $C = C'$), constant vectors a and c of compatible dimensions, and a scalar d. Prove that for $A < 0$ and $C > 0$, this function has a saddle point (x^o, y^o):

$$\max_{x \in \mathbf{R}^n} f\left(x, y^o\right) = f\left(x^o, y^o\right) = \min_{y \in \mathbf{R}^m} f\left(x^o, y\right).$$

Find x^o, y^o, and $f(x^o, y^o)$ in explicit form.

Solution tips A saddle point (x^o, y^o) exists under the sufficient conditions

$$\left.\frac{\partial f(x, y^o)}{\partial x}\right|_{x=x^o} = 2Ax^o + 2By^o + 2a = 0_n,$$

$$\left.\frac{\partial f(x^o, y)}{\partial y}\right|_{y=y^o} = 2B'x^o + 2Cy^o + 2c = 0_m,$$

$$\frac{\partial^2 f(x, y^o)}{\partial x^2} = 2A < 0, \quad \frac{\partial^2 f(x^o, y)}{\partial y^2} = 2C > 0.$$

Solve the system of the first and second linear vector equations in the two vector variables x^o and y^o to obtain

$$x^o = A^{-1}\left[B\left(C - B'A^{-1}B\right)^{-1}\left(c - B'A^{-1}a\right) - a\right],$$

$$y^o = \left(C - B'A^{-1}B\right)^{-1}\left(B'A^{-1}a - c\right),$$

since $A < 0$ and $C > 0$. Then substitute the resulting vectors $x = x^o$, $y = y^o$ into $f(x, y)$:

$$f(x^o, y^o) = -\left(c' - a'A^{-1}B\right)\left(C - B'A^{-1}B\right)^{-1}\left(c - B'A^{-1}a\right) + d - a'A^{-1}a.$$

5) *Find* the saddle points (min and max) of the following functions:
$$\underset{y}{}\qquad\underset{x}{}$$

(a) $f(x, y) = \begin{cases} -(x - \frac{1}{3})^2 + y^2 + \frac{1}{3}y - \frac{1}{3} & \text{for } 0 \leqslant x \leqslant y \leqslant 1, \\ -(x - \frac{2}{3})^2 + y^2 - \frac{1}{3}y & \text{for } 0 \leqslant y \leqslant x \leqslant 1; \end{cases}$

(b) $f(x, y) = -x_1^2 - x_2^2 - 4x_1y_1 - x_2y_2 - 8y_2 + 6x_1 + 2y_1^2 + 4y_2^2, \quad 0 \leqslant x_i \leqslant 1, \ 0 \leqslant y_i \leqslant 1$
$(i = 1, 2)$;

(c) $f(x, y) = \begin{cases} \frac{1}{x} - \frac{1}{y} & \text{for } x \neq 0, \ y \neq 0, \\ 0 & \text{for } x = 0 \ \text{and/or} \ y = 0; \end{cases}$

(d) Find constraints on the constants α, β, and γ under which the function
$f(x, y) = -xy + \alpha x + \beta y - \gamma$ has a saddle point within the unit square.
Answers: a) $(\frac{2}{3}, \frac{1}{6})$; b) $[(1, 0); (1, 1)]$; c) $(0, 0)$;
d) $-xy + \alpha x + \beta y - \gamma = -(x - \beta)(y - \alpha) - \gamma + \beta\alpha$; if $\alpha > 0$ and $\beta < 1$,
then (β, α) is a saddle point.

6) *Calculate* $\max_x \min_y f(x, y)$ and $\min_y \max_x f(x, y)$ for the following functions:

(a) $f(x, y) = (x - y)^2, \quad 0 \leqslant x, \ y \leqslant 1$;

(b) $f(x, y) = (x - y)^2 - 0.5x^2, \quad -1 \leqslant y \leqslant 1, \ -0.5 \leqslant x \leqslant 0.5$;

(c) $f(x, y) = -[x - y(1 - y^2)]^2, \quad -1 \leqslant x, \ y \leqslant 1$.
Answers: (a) 0 and $\frac{1}{4}$; (b) 0 and 0.5; (c) $-\frac{4}{27}$ and 0.

7) *Establish* the minimax inequality

$$\max_{y \in Y} \min_{x \in X} f(x, y) \leqslant \min_{x \in X} \max_{y \in Y} f(x, y).$$

This result was proved in [43, p. 31].
Due to the minimax inequality,

$$\max_{v \in Y} f(x, v) \geqslant \min_{z \in X} f(z, y) \quad \forall x \in X, \ y \in Y.$$

8) Let $f(x, y) = \varphi(x) + \psi(y)$, $Y(x) = \{y \mid g(x, y) \geqslant 0\}$,
$X(y) = \{x \mid g(x, y) \geqslant 0\}$, $X = \{x \mid Y(x) \neq \varnothing\}$, and $Y = \{y \mid X(y) \neq \varnothing\}$.
Prove the inequality

$$\sup_{y \in Y} \inf_{x \in X(y)} f(x, y) \geqslant \inf_{x \in X} \sup_{y \in Y(y)} f(x, y).$$

Solution tips For $x \in X$ and $y \in Y$, it follows that $X(y) \subset X$ and $Y(x) \subset Y$.
Therefore,

$$\sup_{y \in Y} \inf_{x \in X(y)} f(x, y) = \sup_{y \in Y} \left[\psi(y) + \inf_{x \in X(y)} \varphi(x) \right] \geqslant \sup_{y \in Y} \psi(y) + \inf_{x \in X} \varphi(x)$$

$$\geqslant \inf_{x \in X} \left[\sup_{y \in Y(x)} \psi(y) + \varphi(x) \right] = \inf_{x \in X} \sup_{y \in Y(x)} f(x, y).$$

An example of the strict inequality is given by $f(x, y) = y - x$ and $g(x, y) = 1 - x - y$.

9)   *Prove* the equality

$$\left[\lambda x^{(1)} + (1 - \lambda)x^{(2)}\right]' A \left[\lambda x^{(1)} + (1 - \lambda)x^{(2)}\right] = \left[x^{(1)}\right]' A x^{(1)}$$
$$+ 2(1 - \lambda)\left[x^{(2)} - x^{(1)}\right]' A x^{(1)} + (1 - \lambda)^2 \left[x^{(2)} - x^{(1)}\right]' A \left[x^{(2)} - x^{(1)}\right],$$

(A.1.1)

where A is a symmetric constant matrix of dimensions $n \times n$.

Solution tips The following chain of equalities holds:

$$\lambda x^{(1)} + (1 - \lambda)x^{(2)} = \lambda x^{(1)} + (1 - \lambda)x^{(2)} - (1 - \lambda)x^{(1)} + (1 - \lambda)x^{(1)}$$
$$= x^{(1)} + (1 - \lambda)\left(x^{(2)} - x^{(1)}\right).$$

Substitute this equality into the right-hand side of (A.1.1) and perform trivial multiplications to derive (A.1.1).

10)   *Check* the identity

$$\sum_{i,j=1}^{n} a_{ij} x_i x_j = \sum_{i,j=1}^{n} \frac{a_{ij} + a_{ji}}{2} x_i x_j.$$

As a matter of fact, this identity can be used to symmetrize any quadratic form $x'Ax$.

11)   *Prove* the following equalities for nonsingular matrices A and A_i ($i = 1, .., N$) of the same order:

(a) $(A_1 \cdot A_2 \cdot ... \cdot A_{N-1} \cdot A_N)^{-1} = A_N^{-1} \cdot A_{N-1}^{-1} \cdot ... \cdot A_2^{-1} \cdot A_1^{-1}$;

(b) $[A']^{-1} = [A^{-1}]'$;

(c) $(\alpha A)^{-1} = \frac{1}{\alpha} A^{-1}$ for any value $\alpha \neq 0$; in particular, $(-A)^{-1} = -A^{-1}$;

(d) $\det A^{-1} = [\det A]^{-1}$.

Solution tips (a) Post-multiply both sides of the equality $(A_1 A_2)^{-1}(A_1 A_2) = E$, first by A_2^{-1} and then by A_1^{-1}, to obtain $(A_1 A_2)^{-1} = A_2^{-1} A_1^{-1}$. Next, apply mathematical induction.

(a)   Take advantage of the formula

$$A^{-1} = \frac{1}{\det A}\hat{A} = \frac{1}{\det A}\begin{bmatrix} A_{11} \ A_{21}...A_{n1} \\ A_{12} \ A_{22}...A_{n2} \\ .... \\ A_{1n} \ A_{2n}...A_{nn} \end{bmatrix},$$

(A.1.2)

where A_{ij} is the algebraic adjunct of an element a_{ij} of the matrix A, and $\hat{A}$ denotes the adjoint matrix. Construct $[A^{-1}]'$ and $[A']^{-1}$ using (A.1.2).

(b)   Utilize the facts that $\det(\alpha A) = \alpha^n \det A$ and, for a matrix αA, the adjoint matrix is given by $\widehat{\alpha A} = \alpha^{n-1}\hat{A}$, where $\hat{A}$ has form (A.1.2).

(c)   Since $AA^{-1} = E$, $\det(A \cdot A^{-1}) = \det A \cdot \det A^{-1} = 1$.

12) *Prove* the following results:

   (a) Adding a quadratic form with sufficiently small coefficients to a (positive or negative) definite quadratic form will not affect the latter's definiteness.

   (b) Adding a positive (negative) definite matrix to a positive (negative, respectively) definite quadratic form will not affect the latter's definiteness.

   (c) A linear combination of positive (negative) definite quadratic forms with positive coefficients is also a positive (negative, respectively) definite quadratic form.

   Solution tips (a) Let $x'A_1x$ be positive definite and $\varepsilon x'A_2x$ be indefinite, where A_i ($i = 1, 2$) are symmetric matrices of the same order and $\varepsilon > 0$ is a small parameter. According to [38, p. 109], for any $x \in \mathbf{R}^n$ the inequality $\lambda_i x'x \leqslant x'A_i x \leqslant \Lambda_i x'x$ holds, where λ_i and Λ_i are the smallest and greatest roots, respectively, of the characteristic equation $\det[A_i - \lambda E] = 0$ ($i = 1, 2$); note that $\lambda_1 > 0$ because $A_1 > 0$, whereas $\lambda_2 = -|\lambda_2| < 0$ because A_2 is an indefinite matrix. Therefore, $x'A_1x + \varepsilon x'A_2x \geqslant (\lambda_1 - \varepsilon|\lambda_2|)x'x > 0$ for $x'x \neq 0$ and any $\varepsilon \in \left(0, \frac{\lambda_1}{|\lambda_2|}\right)$.
   (b) In the case $A_1 > 0$ and $A_2 \geqslant 0$, the root $\lambda_2 = 0$ and consequently $x'A_1x + x'A_2x \geqslant \lambda_1 x'x > 0$ for $x'x \neq 0$.

13) *Prove* the following implications for a symmetric matrix A of dimensions $n \times n$:

   (a) $A < 0 \Rightarrow A^{-1} < 0$;

   (b) $A < 0 \Rightarrow \det A^{-1} \neq 0 \wedge \det A \neq 0$;

   (c) $A < 0 \Rightarrow B'AB \leqslant 0$ ($B'AB < 0$ if $\det B \neq 0$).

   Solution tips (a) If $\{\lambda_i\}$ are all roots of the equation $\det[A - \lambda E] = 0$, then $\{\frac{1}{\lambda_i}\}$ are all roots of the equation $\det[A^{-1} - \lambda E] = 0$; see [38, p. 54]. In addition, $\lambda_i < 0$ due to $A < 0$. As a result, $\frac{1}{\lambda_i} < 0$ and hence $A^{-1} < 0$.
   (b) This implication is a corollary of the necessary and sufficient conditions of $A < 0$.
   (c) $A < 0 \Leftrightarrow x'Ax < 0$ for $x'x \neq 0$; if $x = By$, where $y \in \mathbf{R}^m$, then the quadratic form $y'B'ABy$ is negative definite and is equal to 0 only for $By = 0_n$; if $\det B \neq 0$, then the equation $By = 0_n$ has the trivial solution $y = 0_m$ only.

14) *Prove* the following result: a linear combination of concave (convex) functions with positive coefficients is also a concave (convex, respectively) function.
   This fact can be directly established using inductive reasoning for the definition of a concave function. Also, see [43, p. 118].

15) A circular city is divided into two equal parts by a narrow river running along the diameter. *Choose* a best site for constructing a bridge in order to minimize the path from any point of one part of the city to any point of the other.
   Solution tips The construction site of the bridge represents some point x on the diameter AB, as is illustrated in Figure A.1.1. The choice of an appropriate point $x \in AB$ is the strategy of the civil engineering firm; the points y_1 and y_2 in the two opposite parts of the city act as uncertainties. The sets of admissible values of y_1 and y_2 are the semicircles Π_1 and Π_2, respectively. Then the payoff function (criterion) of the civil engineering firm is the sum of the distances

   $$f(x, y_1, y_2) = \rho(y_1, x) + \rho(y_2, x).$$

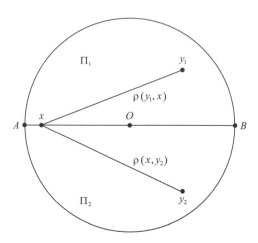

Figure A.l.l Exercise 15.

The goal is to minimize $f(x, y_1, y_2)$ under possible realization of any uncertainties $y_i \in \Pi_i$ $(i = 1, 2)$. For any point $x \in AB$, $x \neq 0$, it is possible to specify points $y_i^* \in \Pi_i$ $(i = 1, 2)$ such that $f(x, y_1^*, y_2^*) > 2R$. Therefore,

$$\min_{x \in AB} \max_{(y_1, y_2) \in \Pi_1 \times \Pi_2} f(x, y_1, y_2) = \min_{(y_1, y_2) \in \Pi_1 \times \Pi_2} f(0, y_1, y_2) = 2R.$$

16)   Recall the definition of a metrical projection. *The metrical projection* of a point $x \in \mathbf{R}^n$ onto a set $X \subseteq \mathbf{R}^n$ is the nearest to x point of the set X, i.e., a point p such that

$$p \in X, \quad \|p - x\| = \inf_{u \in X} \|u - x\|.$$

Denote by $\mathcal{P}_X x$ the projection of a point x onto X. If $x \in X$, then obviously $\mathcal{P}_X x = x$. If a point x does not belong to X, then its projection onto X may be nonunique or even may not exist. Illustrate these facts by simple examples as follows.

Consider the set $X = (-1, 1)$ in the space $\mathbf{R}$. Show that there exists no projection of the point $x = 2$ onto the set X. Study the points $u_k = 1 - \frac{1}{k} \in X$, $k = 1, 2, \ldots$, to observe that $|u_k - x| = 1 + \frac{1}{k}$. Also, it is obvious that $|u - x| > 1$ for any u from X. Therefore, $\inf_{u \in X} \|u - x\| = 1$, but the set X does not contain any points u such that $|u - x| = 1$.

Let $X = \{x \in E^n \mid \|x\| = 1\}$ (the unit sphere); clearly, the projection of the origin $0 = (0, 0, \ldots, 0)$ onto the set X is any element of the set X, since $\|u - 0\| = 1$ for all $u \in X$.

The following propositions were established in convex analysis.

1)   On the existence and uniqueness of a metrical projection: *Let X be a closed set in the space $\mathbf{R}^n$. Then for any point $x \in \mathbf{R}^n$ there exists a metrical projection onto the set X. Moreover, if the set X is convex, then this projection is unique.*

2)   On the characteristic property of a metrical projection: *Let $X \subseteq \mathbf{R}^n$ be a closed convex set. A point p is the metrical projection of a point x onto the set X if and only if*

$$\langle p - x, u - p \rangle \geqslant 0, \quad \forall u \in X.$$

Really, p is the solution of the minimization problem of the strongly convex continuous function $g(u) = \|u - x\|^2$ on a closed convex set X. This is equivalent to the inequality

$$\langle g'(p), u - p \rangle \geqslant 0, \quad \forall u \in X.$$

It remains to observe that $g'(u) = 2(u - x)$ and $g'(p) = 2(p - x)$.

Proposition 2) has a simple geometrical meaning (see Figure A.1.2): the vector $(p - x)$ must form an acute or right angle with any vector of the form $(u - p)$, where $u \in X$.

3) On the nonstrict contractability of a projective mapping: *Let $X \subseteq \mathbf{R}^n$ be a closed convex set. Then for any $x_1, x_2 \in \mathbf{R}^n$,*

$$\|\mathcal{P}_X x_1 - \mathcal{P}_X x_2\| \leqslant \|x_1 - x_2\|,$$

i.e., the distance between the projections of two arbitrary points onto X does not exceed the distance between these points.

Using Propositions 1)–3), *derive* explicit formulas for the projections onto some sets in the space $\mathbf{R}^n$.

(a) *Find* the projections onto the n-dimensional parallelepiped

$$X = \{x = (x_1, \ldots, x_n) \in \mathbf{R}^n \mid \alpha_i \leqslant x_i \leqslant \beta_i, \ i = 1, \ldots, n\},$$

where $\alpha_i, \beta_i, \alpha_i < \beta_i$, are given numbers, $i = 1, \ldots, n$.
Solution tips The convexity and closedness of this parallelepiped is elementary checked using the corresponding definitions. Let $x = (x_1, \ldots, x_N) \notin X$. Take $p = (p_1, \ldots, p_N)$, where

$$p_i = \begin{cases} \alpha_i, & x_i < \alpha_i, \\ \beta_i, & x_i > \beta_i, \\ x_i, & \alpha_i \leqslant x_i \leqslant \beta_i, \end{cases} \quad i = 1, \ldots, n.$$

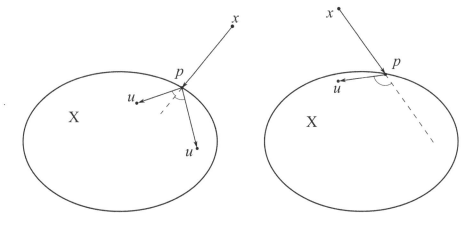

Figure A.l.2 Geometrical interpretation of Proposition 2).

Then $(p_i - x_i)(u_i - p_i) \geqslant 0$ for all u_i, $\alpha_i \leqslant u_i \leqslant \beta_i$, $i = 1, \ldots, n$. Hence, summing over i from 1 to n yields $\langle p - x, u - p \rangle \geqslant 0$ for all $u \in X$. By Proposition 2) the resulting point p is the projection of x onto the set X. In the case $x \in X$, $\mathcal{P}_X x = x$; see the discussion above.

(b) *Find* the projection onto the hyperplane

$$H = \left\{ x \in \mathbf{R}^n \mid \langle c, x \rangle = \alpha \right\}, \quad c \in \mathbf{R}^n, \ c \neq 0, \ \alpha \in \mathbf{R}.$$

Solution tips As is well known, a hyperplane is a closed convex set. Therefore, the projection of any point onto this hyperplane exists and is unique. An explicit formula for such a projection can be derived using geometrical considerations: in fact, it suffices to drop a perpendicular to the hyperplane under consideration. Since the vector c is the normal to the hyperplane, the projection p of an arbitrary point x should be found in the form $p = x + tc$, where t is some number; see Figure A.1.3.

Due to $p \in H$,

$$\langle c, p \rangle = \alpha \Leftrightarrow \langle c, x + tc \rangle = \alpha \Leftrightarrow t\|c\|^2 = \alpha - \langle c, x \rangle \Leftrightarrow t = \frac{\alpha - \langle c, x \rangle}{\|c\|^2}.$$

Thus,

$$p = \mathcal{P}_H x = x + \frac{\alpha - \langle c, x \rangle}{\|c\|^2} c. \tag{A.1.3}$$

Now, justify rigorously this result using Proposition 2). Substituting formula (A.1.3) into Proposition 2) gives

$$
\begin{aligned}
\langle p - x, u - p \rangle &= \left\langle x + \frac{\alpha - \langle c, x \rangle}{\|c\|^2} c - x, \ u - x - \frac{\alpha - \langle c, x \rangle}{\|c\|^2} c \right\rangle \\
&= \left(\frac{\alpha - \langle c, x \rangle}{\|c\|^2} \right) \cdot \left(\langle c, u \rangle - \langle c, x \rangle - \frac{\alpha - \langle c, x \rangle}{\|c\|^2} \langle c, c \rangle \right) \\
&= \left(\frac{\alpha - \langle c, x \rangle}{\|c\|^2} \right) \cdot (\langle c, u \rangle - \alpha).
\end{aligned}
$$

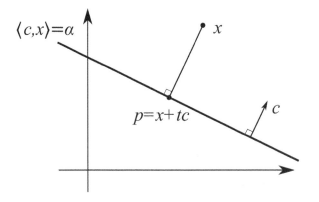

Figure A.1.3 Exercise 16b.

However, if $u \in H$, then $\langle c, u \rangle = \alpha$, and consequently $\langle p - x, u - p \rangle \equiv 0$ for all $u \in H$. The inequality from Proposition 2) holds, and the requisite formula is valid.

Similar considerations can be used to obtain formulas for the metrical projection into the half-space $HS = \{x \in E^n \mid \langle c, x \rangle \leqslant \alpha\}$:

$$
\mathcal{P}_{HS}x = \begin{cases} x + \dfrac{\alpha - \langle c, x \rangle}{\|c\|^2} c & \text{if } \langle c, x \rangle > \alpha, \\ x & \text{if } \langle c, x \rangle \leqslant \alpha \end{cases}
$$

and the layer $F = \{x \in E^n \mid \alpha \leqslant \langle c, x \rangle \leqslant \beta\}$:

$$
\mathcal{P}_F x = \begin{cases} x + \dfrac{\alpha - \langle c, x \rangle}{\|c\|^2} c & \text{if } \langle c, x \rangle < \alpha, \\ x + \dfrac{\beta - \langle c, x \rangle}{\|c\|^2} c & \text{if } \langle c, x \rangle > \beta, \\ x & \text{if } \alpha \leqslant \langle c, x \rangle \leqslant \beta. \end{cases}
$$

Their validity is established by analogy with the case of a hyperplane.

(c) *Find* the projection onto the ball $B = \{x \in \mathbf{R}^n \mid \|x - x_0\| \leqslant R\}$, $x_0 \in \mathbf{R}^n$, $R \geqslant 0$.

Solution tips Like a hyperplane, a ball is a closed convex set. Hence, the projection of any point onto a ball exists and is unique. An explicit formula for such a projection can be derived using rather obvious geometrical considerations: a point belonging to the ball will coincide with its projection; a point x outside the ball will be projected into the intersection p (see Figure A.1.4) of the ball's boundary with the line containing x and the ball's center. Show that this intersection is the nearest to x point.

Obviously, the projection of a point x lying outside a ball onto this ball can be only a point on its boundary. Consider an arbitrary point q on the ball's boundary; due to the triangle inequality,

$$\|x-q\| \geqslant \|x-x_0\| - \|q-x_0\| = \|x-p\| + \|p-x_0\| - \|q-x_0\| = \|x-p\| + R - R = \|x-p\|,$$

i.e., p is the ball's nearest boundary point to x. Finally, the vector p represents the sum of the vector x_0 and a vector of length R that is collinear

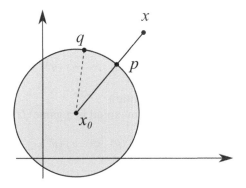

Figure A.1.4 Exercise 16c.

to the vector $(x - x_0)$. Therefore, the requisite projection onto the ball can be calculated by the formula

$$p = P_B x = \begin{cases} x_0 + \dfrac{R}{\|x - x_0\|}(x - x_0) & \text{if } \|x - x_0\| > R. \\ x & \text{if } \|x - x_0\| \leqslant R. \end{cases} \tag{A.1.4}$$

For a rigorous substantiation of the considerations above, use Proposition 2) on the characteristic property of a metrical projection. Consider an arbitrary point x lying outside the ball B, i.e., $\|x - x_0\| > R$. Application of the ball projection formula (A.1.4) gives

$$\langle p - x, u - p \rangle = \left\langle (x_0 - x) + \frac{R}{\|x - x_0\|}(x - x_0), u - x_0 - \frac{R}{\|x - x_0\|}(x - x_0) \right\rangle$$

$$= \frac{\|x - x_0\| - R}{\|x - x_0\|}\left(\langle x_0 - x, u - x_0 \rangle - \frac{R}{\|x - x_0\|}\langle x_0 - x, x - x_0 \rangle \right)$$

$$= \frac{\|x - x_0\| - R}{\|x - x_0\|}\left(\langle x_0 - x, u - x_0 \rangle + R\|x - x_0\| \right).$$

Note that if $u \in B$, then $\|u - x_0\| \leqslant R$, and by the Cauchy–Bunyakovsky inequality

$$\langle x_0 - x, u - x_0 \rangle \geqslant -\|u - x_0\| \cdot \|x - x_0\| \geqslant -R\|x - x_0\|.$$

Hence, $\langle p - x, u - p \rangle \geqslant 0$ for all $u \in B$, and the validity of the ball projection formula (A.1.4) is established.

17)   *Find* the sets of Slater-maximal and Pareto-maximal strategy profiles and also the set of Nash equilibria in the following games:

(a) $\langle\{1, 2\}, \{\mathcal{U}_1 = \mathcal{U}_2 = [0, 1]\}, \{\mathcal{J}_1(u_1, u_2) = u_1 + u_2, \mathcal{J}_2(u_1, u_2) = -(u_1 + u_2)\}\rangle$;

(b) $\langle\{1, 2\}, \{\mathcal{U}_1 = \mathcal{U}_2 = \{1; 2\}\}, \{\mathcal{J}_1(1, 1) = \mathcal{J}_1(2, 2) = -1,$
$\mathcal{J}_1(1, 2) = \mathcal{J}_1(2, 1) = 0, \; \mathcal{J}_2(u_1, u_2) = -\mathcal{J}_1(u_1, u_2)\}\rangle.$

Solution tips For an N-dimensional vector function $f(u) = (f_1(u), \dots, f_N(u))$ and strategy profiles $u \in \mathcal{U} \subset \mathbf{R}^m$, introduce the notations $f(u_1) > f(u_2) \Leftrightarrow f_i(u_1) > f_i(u_2) \; \forall i \in \mathbf{N} = \{1, \dots, N\}$ and

$$f(u_1) \geq f(u_2) \Leftrightarrow f_i(u_1) \geqslant f_i(u_2) \; (i \in \mathbf{N}) \; \wedge \; f(u_1) \neq f(u_2).$$

A strategy profile $u^S \in \mathcal{U}$ is Slater-maximal in a game $\langle \mathcal{U}, f(u) \rangle$ if

$$f(u^S) \not< f(u) \quad \forall u \in \mathcal{U} \quad \left(f(u^S) \not< f(u) \Leftrightarrow \neg f(u^S) < f(u) \right);$$

in this case, $f(u^S)$ is the Slater maximum.
A strategy profile $u^P \in \mathcal{U}$ is Pareto-maximal in a game $\langle \mathcal{U}, f(u) \rangle$ if

$$f(u^P) \not\leq f(u) \quad \forall u \in \mathcal{U} \quad \left(f(u^P) \not\leq f(u) \Leftrightarrow \neg f(u^P) \leq f(u) \right);$$

in this case, $f(u^P)$ is the Pareto maximum.
   A vector maximum for $\langle \mathcal{U}, f(u) \rangle$ is simultaneously a vector minimum for $\langle \mathcal{U}, -f(u) \rangle$.

Game (a). Since $\mathcal{J}_1 + \mathcal{J}_2 = 0$, the sets of Pareto-maximal and Slater-maximal strategy profiles coincide with one another, and any strategy from $[0, 1] \times [0, 1]$ is such. Demonstrate that the strategy profile $(1, 0)$ is a Nash equilibrium in the game (a), i.e.,

$$\mathcal{J}_1(1, 0) \geqslant \mathcal{J}_1(u_1, 0) \quad \forall u_1 \in [0, 1],$$
$$\mathcal{J}_2(1, 0) \geqslant \mathcal{J}_2(1, u_2) \quad \forall u_2 \in [0, 1].$$

These inequalities can be rewritten as $1 \geqslant u_1$ and $-1 \geqslant -1 - u_2 \ \forall u_i \in [0, 1]$ $(i = 1, 2)$, or in the equivalent forms $u_1 \leqslant 1$ and $u_2 \geqslant 0$. Both inequalities hold due to $u_i \in [0, 1]$ $(i = 1, 2)$. Note that the strategy profile $(1, 0)$ is a unique Nash equilibrium. Assume on the contrary that there exists another Nash equilibrium $(u_1^e, u_2^e) = u^e \neq (1, 0)$. Then by definition,

$$\mathcal{J}_1(u^e) \geqslant \mathcal{J}_1(u_1, u_2^e), \quad \mathcal{J}_2(u^e) \geqslant \mathcal{J}_2(u_1^e, u_2) \quad \forall u_i \in [0, 1] \quad (i = 1, 2)$$

or

$$u_1^e \geqslant u_1, \quad u_2^e \leqslant u_2 \quad \forall u_i \in [0, 1] \quad (i = 1, 2),$$

which is impossible, because $u^e \neq (1, 0)$.

Game (b). The payoff matrix of this game has the form $\begin{pmatrix} 1,-1 & 0,0 \\ 0,0 & 1,-1 \end{pmatrix}$. Hence, due to $\mathcal{J}_1 + \mathcal{J}_2 = 0$, any admissible strategy profile is both Pareto- and Slater-maximal, but is not a Nash equilibrium. In other words, the set of Nash equilibria is empty.

18)   Each of the participants can allocate his own money resources for personal and social needs. Assume that the investment effect is linear and the efficiency of allocating money resources for personal needs is higher than for social needs. This leads to the N-player game

$$\Gamma_{(18)} = \left\langle \mathbb{N} = \{1, \ldots, N\}, \{U_i = [0, 1]\}_{i \in \mathbb{N}}, \{\mathcal{J}_i(u) = \lambda u_i + \sum_{j=1}^{N}(1 - u_j), \ \lambda = const > 1\}_{i \in \mathbb{N}} \right\rangle.$$

Prove that a unique Nash equilibrium in the game $\Gamma_{(18)}$ has the form $u^e = \underbrace{(1, 1, \ldots, 1)}_{N \text{ times}}$.

Solution tips According to the definition, $\mathcal{J}_i(u^e) = \max\limits_{u_i \in [0,1]} \mathcal{J}_i(u^e \| u_i)$, $i \in \mathbb{N}$, or

$$\lambda \geqslant \lambda u_i + (1 - u_i) = (\lambda - 1)u_i + 1,$$

which always holds, since $\lambda > 1$ and $u_i \in [0, 1]$. Assume on the contrary that this Nash equilibrium is not unique: let $u^* \neq (1, \ldots, 1) = u^e$ be another Nash equilibrium in the game $\Gamma_{(18)}$. Then

$$\mathcal{J}_i(u^*) \geqslant \mathcal{J}_i(u^* \| u_i) \quad \forall u_i \in [0, 1] \quad (i \in \mathbb{N}),$$

which gives

$$\lambda u_i^* + \sum_{j=1}^{N}(1 - u_j^*) \geqslant \lambda u_i + \sum_{j=1}^{N}(1 - u_j^*) + (1 - u_i)$$

or equivalently

$$\lambda u_i^* + (1 - u_i^*) \geqslant \lambda u_i + (1 - u_i) \quad \forall u_i \in [0, 1] \quad (i \in \mathbb{N}).$$

If $u_i^* \neq 1$, then these inequalities cannot be satisfied for any $u_i \in [0, 1]$.

19) *Find* Nash equilibria in the following games:

(a) $\mathbb{N} = \{1, 2\}, \quad \mathcal{U}_1 = \mathcal{U}_2 = [0, 1],$
   $\mathcal{J}_1(u_1, u_2) = -u_1^2 + 5u_1 u_2 + u_2^2,$ and
   $\mathcal{J}_2(u_1, u_2) = -(u_1 - u_2)^2 - \alpha u_2,$
   where α is a real number;

(b) $\mathbb{N} = \{1, \ldots, N\}, \quad \mathcal{U}_i = [0, 1] \; (i \in \mathbb{N}), \quad u = (u_1, \ldots, u_N),$

$$\mathcal{J}_i(u) = \min\left\{ c_i u_i^p, \; \prod_{l=1}^{N} \left(\frac{1 - u_i}{\sqrt[q]{c_l} - 1} \right)^q \prod_{l=1}^{N} c_l^{q/p} \right\},$$

   where $p, q > 0$, and $c_i > 1$ are real numbers $(i \in \mathbb{N})$.

Solution tips Utilize *the principle of equilibration* suggested by Yu. Germeier [53]: if $\phi_i(t)$ are continuous increasing functions on the closed interval $[0, 1]$ such that $\phi_i(0) = 0 \; (i \in \mathbb{N})$, then an alternative u^0 that maximizes the criterion

$$\min_{i \in \mathbb{N}} \phi_i(u_i)$$

on the set $\{u \in \mathbf{R}_>^N \mid \sum_{i \in \mathbb{N}} u_i = 1\}$ is unique and satisfies the equalities

$$\phi_i(u^0) = const \quad (i \in \mathbb{N}).$$

Answers: (a) $u_i^e = (1, \max\{1 - \frac{\alpha}{2}, 0\})$ if $\alpha > \frac{6}{5}$; (b) $u^e = \left(\frac{1}{\sqrt[p]{c_1}}, \ldots, \frac{1}{\sqrt[p]{c_N}} \right).$

20) *Prove* that there exist no Slater and Pareto minima in the bi-criteria choice problem

$$\Gamma_{(20)} = \langle \mathcal{U} = \mathbf{R}^m, \{\mathcal{J}_i(u) = u' C_i u + 2c_i' u + d_i\}_{i=1,2} \rangle$$

with constant symmetric matrices $C_i < 0 \; (i = 1, 2)$ of dimensions $m \times m$, any constant m-dimensional vectors c_i, and any values d_i.

   Solution tips First, note that the absence of Slater minima implies the absence of Pareto minima as well. Second, the problem $\Gamma_{(20)}$ has no Slater minimum if, for each alternative $u \in \mathbf{R}^m$, there exists $\bar{u} = \bar{u}(u) \in \mathbf{R}^m$ such that

$$-\lambda_i \bar{u}' \bar{u} + 2c_i' \bar{u} + d_i < u' C_i u + 2c_i' u + d_i \quad (i = 1, 2), \tag{A.1.5}$$

where $-\lambda_i = const < 0$ is the greatest root of the equation $\det[C_i - \lambda E_m] = 0$. (This is because $\bar{u}' C_i \bar{u} \leqslant -\lambda_i \bar{u}' u$.) Introduce the notations $\gamma_i = -u' C_i u - 2c_i' u$ and $\bar{u} = \beta e_m$, where e_m is the vector of ones in the space $\mathbf{R}^m$. The inequality in

$$-\lambda_i \bar{u}' \bar{u} + 2c_i' \bar{u} + \gamma_i = -\lambda_i m \beta^2 + 2(c_i', e_m)\beta + \gamma_i < 0$$

holds for any

$$\beta > \beta_i = \frac{|c_i' e_m| + \sqrt{(c_i', e_m)^2 + |\gamma_i m \lambda_i|}}{\lambda_i m}.$$

Then conditions (A.1.5) are satisfied with $\bar{u} = \beta^* e_m$ for all constants $\beta^* > \max_{i=1,2} \beta_i$.

21) *Find* a Nash equilibrium for the game
$\Gamma_{(21)} = \langle\{1, 2\}, \{\mathcal{U}_1 = \mathcal{U}_2 = [0, 1]\}, \{\mathcal{J}_1(u) = -u_1^2 + f_1(u_2),$
$\mathcal{J}_2(u) = f_2(u_1) - e^{-u_2}\}\rangle$, where $f_1(u_2)$ and $f_2(u_1)$ are continuous functions on the closed interval $[0, 1]$.

Solution tips The sets $\mathcal{U}_1$ and $\mathcal{U}_2$ are convex compact sets, whereas the payoff functions $\mathcal{J}_i(u_1, u_2)$ are concave in u_i $(i = 1, 2)$. According to [41], there exists a Nash equilibrium in the game $\Gamma_{(21)}$. This is the strategy profile $u^e = (u_1^e, u_2^e) = (0, 1)$, which directly follows from the definition.
Note that the game

$$\langle\{1, 2\}, \{\mathcal{U}_1 = [0, 1], \mathcal{U}_2 = [0, 1)\}, \{\mathcal{J}_1(u) = u_2 - u_1, \mathcal{J}_2(u) = u_1 + u_2\}\rangle$$

has no Nash equilibrium, despite that the functions $\mathcal{J}_i(u_1, u_2)$ are concave in u_i $(i = 1, 2)$, the set $\mathcal{U}_1$ is a convex compact set, and the set $\mathcal{U}_2$ is convex and bounded. (The latter set is not closed and hence not compact.)

22) For the zero-sum two-player game with $\mathbb{N} = \{1, 2\}$ and $\mathcal{J}_1(u) = -\mathcal{J}_2(u) = \mathcal{J}(u)$, the Nash equilibrium coincides with the saddle point $u^e = (u_1^e, u_2^e)$, i.e.,

$$\max_{u_1} \mathcal{J}(u_1, u_2^e) = \mathcal{J}(u^e) = \min_{u_2} \mathcal{J}(u_1^e, u_2);$$

the saddle point exists if and only if

$$\max_{u_1} \min_{u_2} \mathcal{J}(u_1, u_2) = \min_{u_2} \max_{u_1} \mathcal{J}(u_1, u_2). \tag{A.1.6}$$

Prove that the game

$$\left\langle\{1, 2\}, \{\mathcal{U}_i = [0, 1]\}_{i=1, 2}, \left\{\mathcal{J}(u) = \begin{cases} k_1(u_1 - u_2) & \text{for } u_1 \geqslant u_2 \\ k_2(u_2 - u_1) & \text{for } u_2 \geqslant u_1 \end{cases}\right\}\right\rangle$$

has no saddle point.

Clearly, $\min_{u_2} \mathcal{J}(u_1, u_2) = 0$ $\forall u_1 \in \mathcal{U}_1$, and consequently $\max_{u_1} \min_{u_2} \mathcal{J}(u_1, u_2) = 0$.
On the other hand, for a given u_2 the function $\mathcal{J}(u_1, u_2)$ has the graph presented in Figure A.1.5.
Then

$$\max_{u_1} \mathcal{J}(u_1, u_2) = \begin{cases} k_1(1 - u_2) & \text{for } u_2 \leqslant \frac{k_1}{k_1 + k_2}, \\ k_2 u_2 & \text{for } u_2 \geqslant \frac{k_1}{k_1 + k_2}, \end{cases}$$

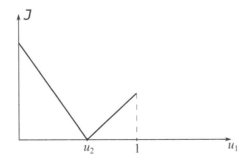

Figure A.I.5 Exercise 22.

which gives

$$\min_{u_2} \max_{u_1} \mathcal{J}(u_1, u_2) = \frac{k_1 k_2}{k_1 + k_2} > 0.$$

As a result,

$$0 = \max_{u_1} \min_{u_2} \mathcal{J}(u_1, u_2) < \frac{k_1 k_2}{k_1 + k_2} = \min_{u_2} \max_{u_1} \mathcal{J}(u_1, u_2),$$

i.e., the game under consideration has no saddle point and hence no Nash equilibrium.

Interestingly, this game describes the following economic situation. One of the firms (player 1) is trying to supplant the other (player 2) from one of two markets. Player 1 gains income on both markets, in the amounts of k_1 (the first market) and k_2 (the second market) per one monetary unit of the goods sold. Each of the firms allocates unit capital for operations. The strategy of each player is the share of capital allocated to a corresponding market. Denote by u_1 the share of capital allocated by player 1 to the first market; then $(1 - u_1)$ is the residual share intended for the second market. The strategy of player 2 is the share u_2 of its capital allocated to the first market; hence, the residual share $(1 - u_2)$ is invested in the second market. The firm investing more in a given market sweeps it and receives a payoff proportional to the surplus of its capital.

If $u_1 \geqslant u_2$, then the surplus of capital in the first market is $u_1 - u_2$ and the income of firm 1 will be $k_1(u_1 - u_2)$. (In the case $u_1 \geqslant u_2$, $1 - u_1 \leqslant 1 - u_2$ and firm 1 cannot sweep the second market.) If $u_1 \leqslant u_2$, then $1 - u_1 \geqslant 1 - u_2$, i.e., firm 1 allocates to the second market not less than the opponent. In this case, the surplus of capital is $(1 - u_1) - (1 - u_2) = u_2 - u_1$, and the income of firm 1 will be $k_2(u_2 - u_1)$. In fact, these features are reflected by the payoff functions in Exercise 22.

23) *Find* the sets of all Slater and Pareto minima, $S(D)$ and $P(D)$, for the following sets:

(a) $D = \left\{ \mathcal{J} = (\mathcal{J}_1, \ldots, \mathcal{J}_N) \mid \sum_{i=1}^{N} a_i \mathcal{J}_i^2 = a, \text{ with constants } a_i > 0, a > 0 \right\}$;

(b) $D = \{ \mathcal{J} = (\mathcal{J}_1, \ldots, \mathcal{J}_N) \mid a_i \leqslant \mathcal{J}_i \leqslant b_i \ (i = 1, \ldots, N) \}$ (a parallelepiped in the Euclidean space $\mathbf{R}^N$);

(c) $D = \{ \mathcal{J} = (\mathcal{J}_1, \ldots, \mathcal{J}_N) \mid \mathcal{J}_i \leqslant 0 \ (i = 1, \ldots, N) \wedge \sum_{i \neq j} \mathcal{J}_i \geqslant 1 - N$

$(j = 1, \ldots, N) \}$.

Answers: (a) $P(D) = S(D) = D \cap \mathbf{R}^N_\leqslant$, where $\mathbf{R}^N_\leqslant = \{ (\mathcal{J}_1, \ldots, \mathcal{J}_N) \mid \mathcal{J}_i \leqslant 0 \ (i \in \mathbb{N}) \}$.
(b) $P(D) = \{ a = (a_1, \ldots, a_N) \}$ and $S(D) = \{ \mathcal{J} \in D \mid \exists j \in \mathbb{N} \text{ for which } \mathcal{J}_j = a_j \}$.
(c) $P(D) = \{ \mathcal{J} \in D \mid \exists k, j \in \mathbb{N} \ (k \neq j) \text{ for which } \sum_{i \neq k} \mathcal{J}_i = 1 - N, \ \sum_{i \neq j} \mathcal{J}_i = 1 - N \}$
and $S(D) = \{ \mathcal{J} \in D \mid \exists j \in \mathbb{N} \text{ for which } \sum_{i \neq j} \mathcal{J}_i = 1 - N \}$.

24) *Find* the sets of all Slater-minimal and Pareto-minimal alternatives, $S(\mathcal{U})$ and $P(\mathcal{U})$, in the following bi-criteria choice problems $\langle \mathcal{U}, \{ \mathcal{J}_1(u), \mathcal{J}_2(u) \} \rangle$:

(a) $\mathcal{U} = \left[0, \frac{3+\sqrt{13}}{4} \right]$, $\mathcal{J}_1(u) = -u$, $\mathcal{J}_2(u) = -u^3 + 3u^2 - 2u$;

(b) $\mathcal{U} = [0, 1]$, $\mathcal{J}_1(u) = -au - b(1 - u)$, $\mathcal{J}_2(u) = -u^\alpha(1 - u)^\beta$, where a, b, α, and β are positive constants.

Answers: (a) $P(\mathcal{U}) = \left[1 - \frac{1}{\sqrt{3}}, \frac{3}{2}\right) \cup \left\{\frac{3 + \sqrt{13}}{4}\right\}$ and $S(\mathcal{U}) = P(\mathcal{U}) \cup \left\{\frac{3}{2}\right\}$;

(b) If $a = b$, then $P(\mathcal{U}) = \left\{\frac{\alpha}{\alpha + \beta}\right\}$ and $S(\mathcal{U}) = [0, 1]$;

if $a > b$, then $P(\mathcal{U}) = S(\mathcal{U}) = \left[\frac{\alpha}{\alpha + \beta}, 1\right]$;

if $a < b$, then $P(\mathcal{U}) = S(\mathcal{U}) = \left[0, \frac{\alpha}{\alpha + \beta}\right]$.

25)   (a) *Prove* that if a set $\mathcal{F} \in \mathbf{R}^2$ is closed and convex, then the set of Pareto minima $P(\mathcal{F})$ for $\mathcal{F}$ is closed as well.
   This result was proved in [213, pp. 143–144].
   (b) *Give* a supportive example to illustrate that in the space $\mathbf{R}^3$, the convexity and closedness of $\mathcal{F}$ not necessarily imply the closedness of the set of Pareto minima $P(\mathcal{F})$ for $\mathcal{F}$.
   The set $\mathcal{F}$ in Figure A.1.6 (a top cone) is convex and closed. However, the sequence of Pareto minima $\{\mathcal{J}^{(k)}\}_1^\infty$ converges to $\mathcal{J}^*$, which is not a Pareto minimum for $\mathcal{F}$. (In fact, here the Pareto minimum is $P(\mathcal{F})$.)

26)   *Prove* that for $C > 0 \wedge D > 0$ or $C < 0 \wedge D < 0$, the solution $\Theta(\cdot) \in C_{n \times n}[0, \vartheta]$ of the matrix system

$$\dot{\Theta} + \Theta A(t) + A'(t)\Theta - \Theta D^{-1}\Theta = 0_{n \times n}, \quad \Theta(\vartheta) = C,$$

has the form

$$\Theta(t) = \left[X^{-1}(t)\right]' \left\{C^{-1} + \int_t^\vartheta X^{-1}(\tau)D^{-1}\left[X^{-1}(\tau)\right]' d\tau\right\}^{-1} X^{-1}(t),$$

where $X(t)$ is the fundamental matrix of the equation $\dot{x} = Ax$:

$$\frac{dX}{dt} = A(t)X, \quad X(\vartheta) = E_n \ \Rightarrow \ \dot{X}^{-1}(t) = -X^{-1}(t)A(t) \ \wedge \ \left[\dot{X}^{-1}(t)\right]'$$

$$= -A(t)\left[X^{-1}(t)\right]'.$$

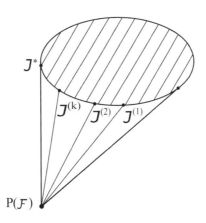

Figure A.I.6 Exercise 25.

27)   *Prove* that the solution $\Theta(\cdot) \in C_{n \times n}[0, \vartheta]$ of the system

$$\dot{\Theta} + \Theta A(t) + A'(t)\Theta + G(t) = 0, \quad \Theta(\vartheta) = C,$$

can be represented as

$$\Theta(t) = \left[X^{-1}(t)\right]' \left\{ C^{-1} + \int_t^{\vartheta} X'(\tau)G(\tau)X(\tau)d\tau \right\} X^{-1}(t),$$

where $X(t)$ has been determined in Exercise 26.

28)   *Prove* that the solution $\xi(\cdot) \in C_n[0, \vartheta]$ of the system with corresponding boundary conditions

$$\dot{\xi} = A(t)\xi + f(t), \quad a)\, \xi(t_0) = \xi_0 \in \mathbf{R}^n, \quad b)\, \xi(\vartheta) = \xi^0 \in \mathbf{R}^n,$$

has the following form:

(a) $\xi(t) = X(t)\left\{ X^{-1}(t_0)\xi_0 + \int_t^{\vartheta} X^{-1}(\tau)f(\tau)d\tau \right\}, \quad X(t_0) = E_n,$

(b) $\xi(t) = \tilde{X}(t)\left\{ \tilde{X}^{-1}(\vartheta)\xi^0 - \int_t^{\vartheta} \tilde{X}^{-1}(\tau)f(\tau)d\tau \right\}, \quad \dot{\tilde{X}}(t) = A(t)\tilde{X}(t), \quad \tilde{X}(\vartheta) = E_n.$

29)   *Prove* that the solution $\Xi(\cdot) \in C_{n \times n}[t_0, \vartheta]$ of the matrix system with corresponding boundary conditions

$$\dot{\Xi} = A(t)\Xi + B(t), \quad a)\, \Xi(t_0) = \Xi_0, \quad b)\, \Xi(\vartheta) = \Xi^0,$$

has the following form:

(a) $\Xi(t) = X(t)\left\{ X^{-1}(t_0)\Xi_0 + \int_t^{\vartheta} X^{-1}(\tau)B(\tau)d\tau \right\}, \quad X(t_0) = E_n,$

(b) $\Xi(t) = \tilde{X}(t)\left\{ \tilde{X}^{-1}(\vartheta)\Xi^0 - \int_t^{\vartheta} \tilde{X}^{-1}(\tau)B(\tau)d\tau \right\}, \quad \dot{\tilde{X}}(t) = A(t)\tilde{X}(t), \quad \tilde{X}(\vartheta) = E_n.$

30)   The model of a controlled system has the form

$$\dot{x} = u$$

where $x, u \in \mathbf{R}$, and the performance criterion is given by

$$J(U, t_0, x_0) = -\frac{1}{2}x^2(1) - \frac{1}{2}\int_0^1 u^2[t]dt \to \max_u.$$

Find the maximizing strategy $U^* \div u^*(t, x)$.
    Solution According to the method of dynamic programming,

$$\max_u \left\{ \frac{\partial V_1}{\partial t} + \frac{\partial V_1}{\partial x}u - \frac{1}{2}u^2 \right\} = 0, \quad V_1(1, x) = -\frac{1}{2}x^2.$$

Obtain the structure of the requisite strategy by maximizing the expression in curly brackets:

$$\left.\frac{\partial\{\ldots\}}{\partial u}\right|_{u(t,x,V_1)} = 0 \;\Rightarrow\; u(t, x, V_1) = \frac{\partial V_1}{\partial x}.$$

Substituting $u(t, x, V_1)$ into $\{\ldots\}_{u(t,x,V_1)} = 0$ yields $\frac{\partial V_1}{\partial t} + \frac{1}{2}\left[\frac{\partial V_1}{\partial x}\right]^2 = 0$, where (as before) $V_1(1, x) = -\frac{1}{2}x^2$. Next, find the solution $V_1(t, x)$ of the resulting equation in the form $V_1(t, x) = \frac{1}{2}K(t)x^2$, where $K(t)$ is an unknown function. Obviously, the latter satisfies the differential equation

$$\dot{K} = -K^2, \quad K(1) = -1.$$

The solution is $K(t) = \frac{1}{t-2}$, which finally gives the maximizing strategy

$$U^* \div u^*(t, x) = \frac{1}{t-2}x, \quad 0 \leqslant t \leqslant 1,$$

and the maximum value of the performance criterion

$$\mathcal{J}(U^*, t_0 = 0, x_0) = -\frac{1}{4}x_0^2 \quad \forall x_0 \neq 0.$$

31) The model of a controlled system has the form

$$\dot{x}_1 = u, \quad \dot{x}_2 = x_1,$$

and the quadratic performance criterion is given by

$$\mathcal{J}(U, t_0, x_0) = -\frac{1}{2}\int_0^2 u^2[t]dt - \frac{1}{2}\left[x_1^2(2) + x_2^2(2)\right] \;\to\; \max_u,$$

where $x = (x_1, x_2) \in \mathbf{R}^2$, $u \in \mathbf{R}$, and $t \in [0, 2]$. *Find* the maximizing strategy $U^* \div u^*(t, x)$.

Solution Like in Exercise 30, write

$$\max_u \left\{\frac{\partial V}{\partial t} + \frac{\partial V}{\partial x_1}u + \frac{\partial V}{\partial x_2}x_1 - \frac{1}{2}u^2\right\} = 0, \quad V(2, x) = -\frac{1}{2}x_1^2 - \frac{1}{2}x_2^2.$$

Then

$$\frac{\partial}{\partial u}\{\ldots\}_{u(t,x,V)} = 0 \quad V(2, x) = -\frac{1}{2}x_1^2 - \frac{1}{2}x_2^2,$$

and consequently $u(t, x, V) = \frac{\partial V}{\partial x_1}$. As a result, $V(t, x)$ satisfies the equation

$$\frac{\partial V}{\partial t} + \frac{1}{2}\left\{\frac{\partial V}{\partial x_1}\right\}^2 + \frac{\partial V}{\partial x_2}x_1 = 0, \quad V(2, x) = -\frac{1}{2}x_1^2 - \frac{1}{2}x_2^2.$$

Find its solution in the form

$$V(t, x) = \frac{1}{2}K_{11}(t)x_1^2 + K_{12}(t)x_1x_2 + \frac{1}{2}K_{22}(t)x_2^2,$$

where $K_{11}(t)$, $K_{12}(t)$, and $K_{22}(t)$ are unknown functions. First, substitute $V(t, x)$ into the equation and its boundary condition; then, collect like terms and set the factors at the identical powers of x_1 and x_2 equal to zero. These manipulations will finally yield

$$\dot{K}_{11} = -2K_{12} - K_{11}^2, \quad \dot{K}_{12} = -K_{22} - K_{12}K_{11}, \quad \dot{K}_{22} = -K_{12}^2,$$
$$K_{11}(2) = -1, \quad K_{12}(2) = 0, \quad K_{22}(2) = -1.$$

The solution of this system has the form

$$K_{11}(t) = \frac{-[12 + 4(2-t)^2(5-t)]}{12(3-t) + (2-t)^3(6-t)}, \quad K_{12}(t) = \frac{-6(2-t)(4-t)}{12(3-t) + (2-t)^3(6-t)},$$
$$K_{22}(t) = \frac{-12(3-t)}{12(3-t) + (2-t)^3(6-t)}.$$

The maximizing strategy and the corresponding maximum value of the quadratic performance criterion are

$$U^* \div u^*(t, x) = K_{11}(t)x_1 + K_{12}(t)x_2 = -\frac{[12 + 4(2-t)^2(5-t)]x_1 + 6(2-t)(4-t)x_2}{12(3-t) + (2-t)^3(6-t)}$$

and

$$J(U^*, t_0 = 0, x_0) = -\frac{23}{42}x_{10} - \frac{3}{7}x_{10}x_{20} - \frac{3}{8}x_{20}^2 \quad \forall x_0 = (x_{10}, x_{20}) \in \mathbf{R}^2,$$

respectively.

32)   The model of a controlled system has the form

$$\dot{x} = ax + u, \quad x(t_0) = x_0.$$

Find a positional strategy $U^* \div u^*(t, x)$ maximizing the performance criterion

$$J(U, t_0, x_0) = -\int_{t_0}^{\vartheta}(x^2(t) + u^2[t])dt \rightarrow \max_u.$$

Solution For solving this problem, it suffices to find a function $V(t, x)$ from the equation

$$\frac{\partial}{\partial u}\left\{\frac{\partial V}{\partial t} + (ax + u)\frac{\partial V}{\partial x} + x^2 = u^2\right\}_{u(t,x,V)} = 0$$

with the boundary condition

$$V(\vartheta, x) = 0.$$

The first equation yields

$$u(t, x, V) = -\frac{1}{2}\frac{\partial V}{\partial x}.$$

Substitute this result into $\frac{\partial}{\partial u}\{\ldots\}_{u(t,x,V)} = 0$ to obtain

$$\frac{\partial V}{\partial t} + ax\frac{\partial V}{\partial x} - \frac{1}{4}\left(\frac{\partial V}{\partial x}\right)^2 + x^2 = 0, \quad V(\vartheta, x) = 0.$$

Find the function $V(t, x)$ in the form

$$V(t, x) = \gamma(t)x^2.$$

Then the requisite function $\gamma(t)$ satisfies the differential equation with separable variables

$$\frac{d\gamma}{dt} + 2a\gamma - \gamma^2 + 1 = 0$$

with the boundary condition

$$\gamma(\vartheta) = 0.$$

As a result,

$$\frac{d\gamma}{\gamma^2 - 2a\gamma - 1} = dt.$$

The equation $\gamma^2 - 2a\gamma - 1 = 0$ has the two real roots $\gamma_{1,2} = a \pm \sqrt{a^2 + 1}$, and hence

$$\frac{1}{\gamma^2 - 2a\gamma - 1} = \frac{1}{2\sqrt{a^2 + 1}} \left(\frac{1}{\gamma - \gamma_1} - \frac{1}{\gamma - \gamma_2} \right),$$

$$\frac{1}{2\sqrt{a^2 + 1}} \ln \frac{\gamma - \gamma_1}{\gamma - \gamma_2} = t + C,$$

$$\frac{\gamma - \gamma_1}{\gamma - \gamma_2} = Ce^{2\sqrt{a^2 + 1}\,t}.$$

From the boundary condition $\gamma(\vartheta) = 0$ it follows that

$$C = \frac{\gamma_1}{\gamma_2} e^{-2\vartheta\sqrt{a^2 + 1}},$$

and finally

$$\gamma(t) = \frac{a + \sqrt{a^2 + 1} - (a + \sqrt{a^2 + 1})e^{-2\vartheta\sqrt{a^2 + 1}}}{1 - \frac{a + \sqrt{a^2 + 1}}{a - \sqrt{a^2 + 1}} e^{-2(\vartheta - t)\sqrt{a^2 + 1}}}.$$

Thus, the maximizing strategy has the form

$$U^* \div u^*(t, x) = -\frac{a + \sqrt{a^2 + 1} - (a + \sqrt{a^2 + 1})e^{-2(\vartheta - t)\sqrt{a^2 + 1}}}{1 - \frac{a + \sqrt{a^2 + 1}}{a - \sqrt{a^2 + 1}} e^{-2(\vartheta - t)\sqrt{a^2 + 1}}}\, x.$$

Now, calculate the maximum value of the performance criterion:

$$\mathcal{J}(U^*, o, x_0) = V(0, x_0) = \frac{a + \sqrt{a^2 + 1} - (a + \sqrt{a^2 + 1})e^{-2\vartheta\sqrt{a^2 + 1}}}{1 - \frac{a + \sqrt{a^2 + 1}}{a - \sqrt{a^2 + 1}} e^{-2(\vartheta - t)\sqrt{a^2 + 1}}}\, x_0^2.$$

33) Consider a differential three-player game in which the controlled system has the dynamics

$$\dot{x}_i = u_i, \quad x_i(0) = 1, \quad (i = 1, 2, 3), \tag{A.1.7}$$

where $x = \{x_1, x_2, x_3\}$ and $x_i, u_i \in \mathbf{R}$; $t \in [0, 1]$ denotes the time variable; the set of admissible strategies of player i is $\mathcal{U}_i = \{U_i \div \div u_i(t, x) \mid u_i(t, x) = Q_i(t)x, \ Q_i(t) \in C_{3 \times 3}[0, 1]\}$. The initial position is $(t_*, x_*) = (0, (1, 1, 1))$, and the game ends at the terminal time instant $\vartheta = 1$. The terminal payoff functions of the players have the form

$$\mathcal{J}_1(U) = -x_1^2(1) + 2x_1(1)x_2(1) + 2x_2(1)x_3(1),$$
$$\mathcal{J}_2(U) = -x_2^2(1) + 2x_1(1)x_2(1) - 2x_2(1)x_3(1), \tag{A.1.8}$$
$$\mathcal{J}_3(U) = -x_3^2(1) + 2x_1(1)x_3(1) - 2x_2(1)x_3(1),$$

where $U = (U_1, U_2, U_3) \in \mathcal{U} = \mathcal{U}_1 \times \mathcal{U}_2 \times \mathcal{U}_3$.

Denote this game by $\Gamma_{(33)}$.

Due to $-(x_1 - x_2 - x_3)^2 \leqslant 0$ and the special forms (A.1.7) and (A.1.8) of the controlled system and payoff functions, respectively, the equalities $\mathcal{J}_i(U^e) = \max_{U_i} \mathcal{J}_i(U^e \| U_i)$ $(i = 1, 2, 3)$ hold if

$$x_1^e = x_2^e + x_3^e, \tag{A.1.9}$$

where $x_i^e(t)$ $(i = 1, 2, 3)$, $0 \leqslant t \leqslant 1$, is the solution of (A.1.7) with $u_i = u_i^e(t, x)$, $U_i^e \div u_i^e(t, x)$, and $U^e = (U_1^e, U_2^e, U_3^e)$ is the Nash equilibrium of the game $\Gamma_{(33)}$.

In contrast to the conventional definition adopted in this book, for the game $\Gamma_{(33)}$ the concept of Nash equilibrium is considered under the *fixed* initial position $(t_*, x_*) = (0, (1, 1, 1))$.

Let $\alpha > 0$ be a parameter. Then, for any $\alpha \in (0, +\infty)$ the strategy profile $U^e(\alpha) = (U_1^e(\alpha), U_2^e(\alpha), U_3^e(\alpha))$ in which

$$U_1^e(\alpha) \div \alpha x_1, \quad U_j^e(\alpha) \div (\alpha - \ln 2)x_j \quad (j = 2, 3) \tag{A.1.10}$$

implements equality (A.1.9). Therefore, $U^e(\alpha)$ is a Nash equilibrium for the game $\Gamma_{(33)}$ with the initial position $(0, (1, 1, 1))$, and also

$$x_1^e(1) = e^\alpha, \quad x_2^e(1) = x_3^e(1) = \frac{1}{2}e^\alpha.$$

Substituting these value into (A.1.8) gives

$$\mathcal{J}_1(U^e(\alpha)) = e^{2\alpha}, \quad \mathcal{J}_2(U^e(\alpha)) = \mathcal{J}_3(U^e(\alpha)) = \frac{1}{4}e^{2\alpha}. \tag{A.1.11}$$

Note a pair of interesting facts as follows.

First, the game $\Gamma_{(33)}$ has *a continuum of* Nash equilibria: these are the strategy profiles $U^e(\alpha) = (U_1^e(\alpha), U_2^e(\alpha), U_3^e(\alpha))$ (A.1.10) for all admissible values $\alpha \in (0, +\infty)$. Different values α correspond to different payoffs (A.1.11) of the players.

Second, the strategy profiles (A.1.10) are not *the complete set* of Nash equilibria. For example, equality (A.1.9) also holds for the strategy profiles $U^e(\alpha) = (U_1^e(\alpha, \beta), U_2^e(\alpha, \beta), U_3^e(\alpha))$, where

$$U_1^e(\alpha) \div \alpha x_1, \quad U_2^e(\alpha, \beta) \div (\alpha - \ln \beta) x_2, \quad U_3^e(\alpha, \beta) \div \left(\alpha - \ln \frac{\beta}{\beta - 1} \right) x_3, \quad \text{(A.1.12)}$$

for any $\alpha > 0$ and $\beta > 1$. Hence, the strategy profiles (A.1.12) are Nash equilibria too.

Now, discuss *some negative features* of Nash equilibrium.

Feature 1 When choosing the concept of Nash equilibrium as a solution of a differential game, before the start of the game the players have to *agree what specific Nash equilibrium* (out of many possible!) they will use.

The matter is that in the absence of such an agreement, each of the players can use his own strategy from different Nash equilibria corresponding to different α in (A.1.10), and their combination

$$U(\alpha_1, \alpha_2, \alpha_3) = (U_1^e(\alpha_1), U_2^e(\alpha_2), U_3^e(\alpha_3))$$

will not form a Nash equilibrium (*no interchangeability*). For example, for $0 < \alpha_1 < \alpha_2 < \alpha_3$ it follows that

$$x_1^e(1) = e^{\alpha_1} < x_2^e(1) + x_3^e(1) = \frac{1}{2} \left(e^{\alpha_2} + e^{\alpha_3} \right),$$

and hence condition (A.1.9) is not satisfied. In other words, the strategy profile $U(\alpha_1, \alpha_2, \alpha_3)$ is not a Nash equilibrium.

As a result, each player can no longer make independent decisions: negotiations with the others are necessary. This fact reduces the noncooperative character of the game, actually moving it to the class of cooperative games, where negotiations between the parties on a coordinated choice of strategies are allowed. Therefore, new solution concepts without this drawback (the need for preliminary negotiations) are required for noncooperative games. In such solutions, each player can independently make a decision (choose his own strategy), regardless of the actions of the others. This book is devoted to the Berge equilibrium, one of such solutions.

Feature 2 When implementing a selected Nash equilibrium, each player must be absolutely sure that the others will adhere to the preliminary agreement during the game, i.e., that they will use their strategies from a pre-selected Nash equilibrium (otherwise they will not gain the maximum possible payoff). Each player needs confidence in the actions of the others! However, in real applications (especially, in economics), such a hope seems very problematic. It would be more natural for each player to choose and use his own strategy (based on his own interests), and if the choice of the others matches the interests of this player, he will gain a "good" payoff. Such a practical approach is also implemented by the concept of Berge equilibrium.

Feature 3 The set of Nash equilibria *is internally unstable*: there may exist two Nash equilibria such that the payoffs of all players in one of them are strictly greater than in the other. For example, according to (A.1.11), for $\alpha_2 > \alpha_1 > 0$ it follows that

$$\mathcal{J}_i \left(U^e(\alpha_2) \right) > \mathcal{J}_i \left(U^e(\alpha_1) \right) \quad (i = 1, 2, 3). \tag{A.1.13}$$

Here the players' payoffs in two different Nash equilibria from (A.1.10) are compared with each other:

$$U^e(\alpha_2) \div (\alpha_2 x_1, (\alpha_2 - \ln 2) x_2, (\alpha_2 - \ln 2) x_3),$$
$$U^e(\alpha_1) \div (\alpha_1 x_1, (\alpha_1 - \ln 2) x_2, (\alpha_1 - \ln 2) x_3).$$

The drawback (A.1.13) can be eliminated using new solution concepts in which the set $\mathfrak{U}^*$ of equilibria would have *internal stability*, i.e., for any $U^{(1)} \in \mathfrak{U}^*$ and $U^{(2)} \in \mathfrak{U}^*$ the system of inequalities

$$\mathcal{J}_i\left(U^{(1)}, t_*, x_*\right) > \mathcal{J}_i\left(U^{(2)}, t_*, x_*\right) \quad (i = 1, 2, 3),$$

would be inconsistent. Such solutions are the active equilibrium and its special cases—the equilibrium in threats and counter-threats and the absolute equilibrium. In addition to other advantages for noncooperative games, these solutions have internal stability.

Feature 4 Generally speaking, Nash equilibria are *improvable*, i.e., for a Nash equilibrium U^e there may exist another strategy profile $U^* \in \mathfrak{U}$ (not necessarily a Nash equilibrium) in which

$$\mathcal{J}_i\left(U^*, t_*, x_*\right) > \mathcal{J}_i\left(U^e, t_*, x_*\right) \quad (i = 1, 2, 3). \tag{A.1.14}$$

For example, see inequalities (A.1.13). This feature actually expresses an analog of the Prisoner's Dilemma.

The drawback (A.1.14) can be avoided by using Pareto-maximal strategy profiles as new solutions of the game.

The negative properties (internal stability and improveability) of Nash equilibrium as a solution of noncooperative games can be removed by considering only those Nash equilibria that are simultaneously Pareto-maximal. However, the classes of differential games in which there exist Pareto-maximal Nash equilibria are very narrow. Some examples of such differential games are available in [195, 213, 155]; also, see Exercise 2.

In conclusion, note that the negative features of the Nash equilibrium identified here do not reduce its merits. After all, there are spots in the sun and, as J. Hopkinson said, "We cannot get more out of the mathematical mill than we put into it, though we may get it in a form infinitely more useful for our purpose."[37]

34) Two geological associations are engaged in the exploration and mining of mineral resources at n fields. The financial resources for the operating activity of the first and second associations are allocated with the rates a and b, respectively. Exploration and production can only occur during a given period from the time instant $t = 0$ to $t = \vartheta$. The profit obtained from mining at field j is c_j; at the time instant ϑ, this profit is distributed between the associations in proportion to the amounts spent by that time on the exploration and production at this field. Moreover, if these amounts are equal to zero, the profits obtained by both associations are also zero.

Construct a corresponding noncooperative differential game, considering the total profit at the time ϑ obtained from mining in all fields as the payoff of each association. Also, find a Nash equilibrium in this game.

Solution The dynamics of this game are described by

$$\dot{x}_j = u_1^{(j)}, \quad \dot{y}_j = u_2^{(j)},$$
$$x_j[0] = y_j[0] = 0, \quad j \in \{1, \ldots, n\}. \tag{A.1.15}$$

---

[37]John Hopkinson, (1849–1898), was a British engineer and physicist who invented the three-wire system for electricity distribution and improved the design and efficiency of electric generators.

Here, $x_j[t]$ and $y_j[t]$ are the amounts of financial resources spent by associations 1 and 2, respectively, on the exploration and production at field j by a time instant t; $u_k^{(j)}$ specifies the rate with which these financial resources are allocated to field j from association k ($k = 1, 2$) at a time instant $t \in [0, \vartheta]$, where $\vartheta = const > 0$; the sets of all admissible strategies of the associations have the following form:

$$\mathfrak{U}_1 = \left\{ U_1 \div u_1(t, x, y) \mid \sum_{j=1}^{n} u_1^{(j)} = a, \ u_1^{(j)} \geqslant 0, \ j = 1, \ldots, n \right\}$$

for association 1
and

$$\mathfrak{U}_2 = \left\{ U_2 \div u_2(t, x, y) \mid \sum_{j=1}^{n} u_2^{(j)} = b, \ u_2^{(j)} \geqslant 0, \ j = 1, \ldots, n \right\}$$

for association 2. Construct the vectors $u_k = (u_k^{(1)}, \ldots, u_k^{(n)})$, $x = (x_1, \ldots, x_n)$, and $y = (y_1, \ldots, y_n)$. Assume that the vector functions $u_k(t, x, y)$ in $\mathfrak{U}_k$ are such that, for $u_k = u_k(t, x, y)$, the system of differential equations (A.1.15) has a unique solution that is extendable to $[0, \vartheta]$. (For example, they are continuously differentiable and satisfy the Lipschitz condition in x and y.) From the system of differential equations (A.1.15) and constraints in $\mathfrak{U}_k$ it follows that

$$\sum_{j=1}^{n} x_j[\vartheta] = a\vartheta, \quad \sum_{j=1}^{n} y_j[\vartheta] = b\vartheta.$$

The payoff functions of the associations are terminal:

$$\mathcal{J}_1(U_1, U_2) = \sum_{j=1}^{n} \frac{c_j x_j[\vartheta]}{x_j[\vartheta] + y_j[\vartheta]}, \quad \mathcal{J}_1(U_1, U_2) = \sum_{j=1}^{n} \frac{c_j y_j[\vartheta]}{x_j[\vartheta] + y_j[\vartheta]}.$$

$\left(\text{If } x_j[\vartheta] = y_j[\vartheta] = 0, \text{ then let } \dfrac{x_j[\vartheta]}{x_j[\vartheta] + y_j[\vartheta]} = \dfrac{y_j[\vartheta]}{x_j[\vartheta] + y_j[\vartheta]} = 0. \right)$

Consider the solution $(x^e[t], y^e[t])$, $0 \leqslant t \leqslant \vartheta$, of the system (A.1.15) that is induced by $u_k = u_k^e(t, x, y)$, $k = 1, 2$, where $U^e = (U_1^e, U_2^e) \div (u_1^e(t, x, y), u_2^e(t, x, y))$. Show that for this solution, any Nash equilibrium satisfies the conditions

$$x_j^e[\vartheta] > 0, \quad y_j^e[\vartheta] > 0 \quad (j = 1, \ldots, n).$$

Really, assume on the contrary that $x_i[\vartheta] = y_i[\vartheta] = 0$. In this case, under a fixed strategy of the opponent, each association will benefit from allocating a sufficiently small amount $\varepsilon = const > 0$ to field i, reducing by ε its amount spent on another field. This obviously contradicts the definition of Nash equilibrium. Next, assume that $x_i[\vartheta] > 0$ and $y_i[\vartheta] = 0$. In this case, association 1 will increase its profit, reallocating an amount $\varepsilon = const > 0$, $0 < \varepsilon < x_i[\vartheta]$, from field i to any other field.

All Nash equilibria of the static game

$$\Gamma = \left\langle \{1, 2\}, \{X, Y\}, \{F_1(x, y) = \sum_{j=1}^{n} \frac{c_j x_j}{x_j + y_j}, F_2(x, y) = \sum_{j=1}^{n} \frac{c_j y_j}{x_j + y_j} \} \right\rangle,$$

where

$$X = \left\{ x \in \mathbf{R}^n \left| \sum_{j=1}^n x_j = a\vartheta \right. \right\},$$

$$Y = \left\{ y \in \mathbf{R}^n \left| \sum_{j=1}^n y_j = b\vartheta \right. \right\},$$

can be found using Lagrange's method of multipliers:

$$\frac{\partial}{\partial x_i} \left(F_1(x, y) - \lambda \sum_{j=1}^n x_j \right) = 0, \quad \sum_{j=1}^n x_j = a\vartheta,$$

$$\frac{\partial^2}{\partial x_i^2} \left(F_1(x, y) - \lambda \sum_{j=1}^n x_j \right) = -\frac{2c_i y_i}{(x_i + y_i)^3} < 0,$$

$$\frac{\partial}{\partial y_i} \left(F_2(x, y) - \mu \sum_{j=1}^n y_j \right) = 0, \quad \sum_{j=1}^n y_j = b\vartheta,$$

$$\frac{\partial^2}{\partial y_i^2} \left(F_2(x, y) - \mu \sum_{j=1}^n y_j \right) = -\frac{2c_i x_i}{(x_i + y_i)^3} < 0 \quad (i = 1, \ldots, n).$$

Hence,

$$\frac{y_i}{x_i} = \frac{\lambda}{\mu} = \frac{b}{a}, \quad y_i = \frac{b}{a} x_i \quad (i = 1, \ldots, n),$$

which gives

$$\mu = \frac{a \sum\limits_{j=1}^n c_j}{\vartheta (a^2 + b^2)}, \quad \lambda = \frac{b \sum\limits_{j=1}^n c_j}{\vartheta (a^2 + b^2)},$$

and the Nash equilibria of the static game Γ are

$$x_i^e = \frac{a c_i \vartheta}{\sum\limits_{j=1}^n c_j}, \quad y_i^e = \frac{b c_i \vartheta}{\sum\limits_{j=1}^n c_j} \quad i = (1, \ldots, n).$$

In view of the system of differential equations (A.1.15), the corresponding Nash equilibrium has the form

$$U^e = (U_1^e, U_2^e) \div (u_1^e(t, x, y), u_2^e(t, x, y)),$$

where

$$u_1^e(t, x, y) = \left(\frac{a}{\sum\limits_{j=1}^n c_j} c_1, \frac{a}{\sum\limits_{j=1}^n c_j} c_2, \ldots, \frac{a}{\sum\limits_{j=1}^n c_j} c_n \right),$$

$$u_2^e(t, x, y) = \left(\dfrac{b}{\displaystyle\sum_{j=1}^{n} c_j} c_1, \dfrac{b}{\displaystyle\sum_{j=1}^{n} c_j} c_2, \ldots, \dfrac{b}{\displaystyle\sum_{j=1}^{n} c_j} c_n \right).$$

The Nash equilibrium U^e is Pareto-maximal, since

$$\mathcal{J}_1(U_1, U_2) + \mathcal{J}_2(U_1, U_2) = 2 \sum_{j=1}^{n} c_j = const.$$

(According to this equality, an increase of one payoff function inevitably leads to a decrease of the other.)

35) *The three-player pulling game.* A point M is moving in the three-dimensional space $\mathbf{R}^3$, and its velocity has the form

$$x = u_1 + u_2 + u_3, \tag{A.1.16}$$

where the control vectors are subjected to the inequality constraints

$$\|u_i\| \leqslant \alpha_i \quad (\alpha_i = const, \ i = 1, 2, 3). \tag{A.1.17}$$

At the initial time instant $t = 0$, the point is at the origin $x(0) = 0_3 \in \mathbf{R}^3$. The set of all admissible strategies of player i has the form $\mathfrak{U}_i = \{ U_i \div u_i(t, x) \mid u_i(t, x)$ are continuous and satisfy the Lipschitz condition in $x \}$.

Three points x_i^* $(i = 1, 2, 3)$ are given in the space $\mathbf{R}^3$, and the terminal time instant $\vartheta = const > 0$ of motion satisfies the condition

$$\vartheta \sum_{j=1}^{3} \alpha_j < \min_{i=1,2,3} \|x_i^*\|.$$

The goal of player i is to pull by the time instant ϑ the point M as close to x_i^* as possible using an appropriately chosen strategy $U_i \div u_i(t, x)$, $U_i \in \mathfrak{U}_i$. Then the payoff function of player i can be written as

$$\mathcal{J}_i(U_1, U_2, U_3) = -\|x(\vartheta) - x_i^*\| \quad (i = 1, 2, 3),$$

where $x(t)$, $0 \leqslant t \leqslant \vartheta$, is the solution of (A.1.16) for the players' strategies $u_i = u_i(t, x)$ $(i = 1, 2, 3)$.

Find a Nash equilibrium in this game.

Solution This game with the control constraints (A.1.16), (A.1.17) can be solved using the following sufficient conditions of Nash equilibrium. The corresponding Lyapunov–Bellman function $V_i(t, x)$ and the strategies $U_i^e \div u_i^e(t, x)$, $U_i \in \mathfrak{U}_i$ $(i = 1, 2, 3)$, must satisfy the relations

$$\max_{\|u_i\| \leqslant \alpha_i} \left\{ \frac{\partial V_i}{\partial t} + \left[\frac{\partial V_i}{\partial x} \right]' \left[u_i + \sum_{k=1, k \neq i}^{3} u_k^e(t, x) \right] \right\} = 0,$$

$$V_i(\vartheta, x) = -\|x_i^* - x_i\| \quad (i = 1, 2, 3). \tag{A.1.18}$$

Hence,

$$u_i^e = \alpha_i \left\| \frac{\partial V_i}{\partial x} \right\|^{-1} \frac{\partial V_i}{\partial x} \quad (i = 1, 2, 3). \tag{A.1.19}$$

Substituting (A.1.19) into (A.1.18) gives

$$\left[\frac{\partial V_i}{\partial x}\right]' \sum_{j=1}^{3} \alpha_j \left\|\frac{\partial V_j}{\partial x}\right\|^{-1} \frac{\partial V_j}{\partial x} = 0,$$

$$V_i(\vartheta, x) = -\|x_i^* - x\| \quad (i = 1, 2, 3). \tag{A.1.20}$$

The solution of (A.1.20) has the form

$$V_i(t, x) = -\|x_i^* - x\| \quad (i = 1, 2, 3) \tag{A.1.21}$$

if

$$\sum_{j=1}^{3} \alpha_j \frac{x_j^* - x}{\|x_j^* - x\|} = 0. \tag{A.1.22}$$

In view of (A.1.21), from (A.1.19) it follows that

$$U_i^e \div u_i^e(t, x) = \alpha_i \frac{x_i^* - x}{\|x_i^* - x\|} \quad (i = 1, 2, 3). \tag{A.1.23}$$

Condition (A.1.22) holds for the velocity (A.1.16) if

$$\sum_{j=1}^{3} \alpha_j x_j^* \|x_j^*\|^{-1} = 0_3. \tag{A.1.24}$$

Thus, under conditions (A.1.24) the Nash equilibrium $U^e = (U_1^e, U_2^e, U_3^e)$ is given by (A.1.23). A similar situation was described in Krylov's[38] fable *Swan, Pike and Crawfish*: "...But anyway the cart's still there today."

36) *Find* a Berge equilibrium in the game

$$\langle\{1, 2\}, \{X_i\}_{i=1,2}, \{\mathcal{J}_i(x_1, x_2)\}_{i=1,2}\rangle,$$

where $X_1 = X_2 = [0, 1]$, $\mathcal{J}_1(x_1, x_2) = -(x_1 - x_2)^2 - \alpha x_2$, $\mathcal{J}_2(x_1, x_2) = -x_1^2 + 5x_1 x_2 + x_2^2$, and $\alpha > \frac{5}{6}$.
   Answer: $x^B = (x_1^B, x_2^B) = \left(1, \max\{1 - \frac{\alpha}{2}, 0\}\right)$.

37) *Competition problem* The performance of two competing production enterprises is assessed using the same continuous-time criteria $F_1(z), \dots, F_N(z)$ (e.g., profit, product quality, environmental damage, product cost, etc.). Denote by $\mathbb{N} = \{1, \dots, N\}$ the set of performance criteria. The strategies of enterprises 1 and 2 are $x \in X \in comp\ \mathbf{R}^n$ and $y \in Y$, respectively, where $X = Y$. Their strategies can be bonuses, penalties, wage funds, etc. The payoff functions have the form

$$\mathcal{J}_1(x, y) = \min_{i \in \mathbb{N}}[F_i(x) - F_i(y)]$$

for enterprise 1 and

$$(\mathcal{J}_2(x, y) = \min_{i \in \mathbb{N}}[F_i(y) - F_i(x)])$$

---

[38] Ivan A. Krylov, (1768/69–1844), was a Russian writer of innocent-sounding fables that satirized contemporary social types in the guise of beasts. His command of colloquial idiom brought a note of realism to Russian classical literature. Many of his aphorisms have become part of everyday Russian speech.

for enterprise 2. Here $\mathcal{J}_1(x, y) \neq -\mathcal{J}_2(x, y)$ and hence the game is nonzero-sum. *Find* a Nash equilibrium in this game.

Solution The following result was demonstrated in [155, p. 69–73]: a strategy profile $(x^\alpha, y^\alpha) \in X \times Y$ such that

1) x^α is a Pareto-maximal alternative in the multicriteria choice problem

$$\langle X, \{F_i(x)\}_{i \in \mathbb{N}} \rangle$$

2) $x^\alpha = y^\alpha$,

is a Pareto-maximal Nash equilibrium in the game

$$\langle \{1, 2\}, \{X, Y\}, \{\mathcal{J}_i(x, y)\}_{i=1,2} \rangle.$$

38)   Consider the multicriteria dynamic problem

$$\dot{x} = A(t)x + B(t)u, \qquad x(t_*) = x_*, \tag{A.1.25}$$

$$\mathcal{J}_i(U, t_*, x_*) = x'(\vartheta)C_i x(\vartheta) + \int_{t_*}^{\vartheta} (x'(t)G_i x(t) + u'[t]D_i u[t])dt,$$

$$i \in \mathbb{N} = \{1, 2, \dots, N\} \tag{A.1.26}$$

with the set of strategies

$$\mathfrak{U} = \{U \div u(t, x) \mid u(t, x) = Q(t)x \ \forall Q(\cdot) \in C_{m \times n}[0, \vartheta]\},$$

where $x \in \mathbf{R}^n$ and $u \in \mathbf{R}^m$; $A(t) \in C_{n \times n}[0, \vartheta]$, $B(t) \in C_{n \times m}[0, \vartheta]$, and $Q(t) \in C_{m \times n}[0, \vartheta]$ are time-varying matrices of appropriate dimensions; C_i, G_i, and D_i are symmetric constant matrices of dimensions $n \times n$, $n \times n$, and $m \times m$, respectively; ϑ and t_*, $\vartheta > t_* \geqslant 0$, are constants; $x(t)$, $t_* \leqslant t \leqslant \vartheta$, is the solution of (A.1.25) for $u = u(t, x)$, $u[t] = u(t, x(t))$. For the N-dimensional vectors $\mathcal{J}^{(1)} = (\mathcal{J}_1^{(1)}, \dots, \mathcal{J}_N^{(1)})$ and $\mathcal{J}^{(2)} = (\mathcal{J}_1^{(2)}, \dots, \mathcal{J}_N^{(2)})$, introduce the notations:

$$\mathcal{J}^{(1)} = \mathcal{J}^{(2)} \Leftrightarrow \mathcal{J}_i^{(1)} = \mathcal{J}_i^{(2)}, \quad i \in \mathbb{N};$$

$$\mathcal{J}^{(1)} \neq \mathcal{J}^{(2)} \Leftrightarrow \neg \mathcal{J}^{(1)} = \mathcal{J}^{(2)};$$

$$\mathcal{J}^{(1)} \geqslant \mathcal{J}^{(2)} \Leftrightarrow \mathcal{J}_i^{(1)} \geqslant \mathcal{J}_i^{(2)}, \quad i \in \mathbb{N};$$

$$\mathcal{J}^{(1)} \geq \mathcal{J}^{(2)} \Leftrightarrow (\mathcal{J}^{(1)} \geqslant \mathcal{J}^{(2)}) \wedge (\mathcal{J}^{(1)} \neq \mathcal{J}^{(2)});$$

$$\mathcal{J}^{(1)} > \mathcal{J}^{(2)} \Leftrightarrow \mathcal{J}_i^{(1)} > \mathcal{J}_i^{(2)}, \quad i \in \mathbb{N};$$

$$\mathcal{J}^{(1)} \not\geq \mathcal{J}^{(2)} \Leftrightarrow \neg \mathcal{J}^{(1)} \geq \mathcal{J}^{(2)};$$

$$\mathcal{J}^{(1)} \not> \mathcal{J}^{(2)} \Leftrightarrow \neg \mathcal{J}^{(1)} > \mathcal{J}^{(2)}.$$

Construct the N-dimensional vector

$$\mathcal{J}(U, t_*, x_*) = (\mathcal{J}_1(U, t_*, x_*), \dots, \mathcal{J}_N(U, t_*, x_*)).$$

A strategy $U^S \in \mathfrak{U}$ is *Slater-maximal* if for all $U \in \mathfrak{U}$,

$$\mathcal{J}(U, t_*, x_*) \not> \mathcal{J}(U^S, t_*, x_*);$$

a strategy $U^P \in \mathfrak{U}$ is *Pareto-maximal* if for all $(t_*, x_*) \in [0, \vartheta) \times \mathbf{R}^n$, $\mathcal{J}(U, t_*, x_*) \not\geq \mathcal{J}(U^P, t_*, x_*)$.

According to the definitions, any Pareto-maximal strategy is Slater-maximal; generally speaking, the converse does not hold.

Problem For (A.1.25), (A.1.26), *establish* the existence conditions of a Pareto-maximal strategy and *find* it in explicit form.

Solution For the set of constants α_i ($i \in \mathbb{N}$), introduce the matrices

$$C(\alpha) = \sum_{i \in \mathbb{N}} \alpha_i C_i, \quad G(\alpha) = \sum_{i \in \mathbb{N}} \alpha_i G_i, \quad D(\alpha) = \sum_{i \in \mathbb{N}} \alpha_i D_i.$$

The following **assertion** is true: *if there exist constants $\alpha_i > 0$, $i \in \mathbb{N}$, such that $D(\alpha) < 0$ and the Riccati matrix equation*

$$\dot{\Theta} + A'(t)\Theta + \Theta A(t) + G(\alpha) - \Theta B(t) D^{-1}(\alpha) B'(t)\Theta = 0_{n \times n} \tag{A.1.27}$$

has a solution $\Theta^*(t)$ that is extendable to $[0, \vartheta]$, then a Pareto-maximal strategy is given by

$$U^P \div u^P(t, x) = -D^{-1}(\alpha) B'(t)\Theta^*(t)x. \tag{A.1.28}$$

In other words, for all $t \in [t_*, \vartheta]$ the strategy U^P remains Pareto-maximal in problem (A.1.25), (A.1.26), where (t_*, x_*) are replaced by $(t, x^P(t))$ and $x^P(t)$ ($t_* \leqslant t \leqslant \vartheta$) is the solution of (A.1.25) for $u = u^P[t] = u^P(t, x^P(t))$.

This fact can be established using the following chain of implications.

1) If $\exists\, \alpha_i = const > 0$ such that

$$\left[\sum_{i \in \mathbb{N}} \alpha_i \mathcal{J}_i(U^P, t_*, x_*) = \max_{U \in \mathfrak{U}} \alpha_i \mathcal{J}_i(U, t_*, x_*) \right] \Rightarrow \left[U^P \text{ is Pareto-maximal} \right]. \tag{A.1.29}$$

The proof is by contradiction.

2) If $D(\alpha) < 0$, then there exists a Lyapunov–Bellman function $V(t, x)$ such that the function

$$W(t, x, u, V) = \frac{\partial V}{\partial t} + \left[\frac{\partial V}{\partial x} \right]' [A(t)x + B(t)u] + x'G(\alpha)x + u'D(\alpha)u$$

satisfies for all $(t, x) \in [0, \vartheta] \times \mathbf{R}^n$ the relations

$$W(t, x, u^P(t, x), V(t, x)) = 0,$$
$$W(t, x, u(t, x), V(t, x)) \leqslant 0, \quad \forall\, U \div u(t, x), \quad U \in \mathfrak{U},$$
$$V(\vartheta, x) = x'C(\alpha)x, \quad \forall\, x \in \mathbf{R}^n.$$

In this case, the function $V = x'\Theta(t)x$ is adopted to prove (A.1.28) and compile equation (A.1.27).

Therefore, the Pareto-maximal strategy (A.1.28) is obtained from the solution $\Theta^*(t)$ of equation (A.1.27). In this context, let us make an important *remark* as follows: if $C(\alpha) < 0$, $D(\alpha) < 0$, and $G(\alpha) = 0_{n \times n}$, then the solution $\Theta^*(t)$ of equation (A.1.27) has the form

$$\Theta^*(t) = \left[\mathbf{X}^{-1}(t) \right]' \left\{ C^{-1}(t) + \int_t^\vartheta \mathbf{X}^{-1}(\tau) B(\tau) D^{-1}(\alpha) B'(\tau) [\mathbf{X}^{-1}(\tau)]' d\tau \right\}^{-1} \mathbf{X}^{-1}(t),$$

where $\mathbf{X}(t)$ is the fundamental matrix of the system $\dot{x} = Ax$, $\mathbf{X}(\vartheta) = E_n$.

39)   Consider the differential linear-quadratic noncooperative four-player game

$$\langle \mathbb{N} = \{1, 2, 3, 4\}, \; \Sigma, \; \{\mathfrak{U}_i\}_{i \in \mathbb{N}}, \; \{\mathcal{J}_i(U, t_*, x_*)\}_{i \in \mathbb{N}} \rangle, \tag{A.1.30}$$

where the controlled system Σ has the linear dynamics

$$\dot{x} = A(t)x + \sum_{i=1}^{4} B_i(t) u_i, \quad x(t_*) = x_*; \tag{A.1.31}$$

the strategy set of player i is

$$\mathfrak{U}_i = \{U_i \div u_i(t, x) \mid u_i(t, x) = Q_i(t)x, \; Q_i(t) \in C_{n \times n}[0, \vartheta]\};$$

the payoff function of player i is

$$\mathcal{J}_i(U, t_*, x_*) = x'(\vartheta) C_i x(\vartheta) + \int_{t_*}^{\vartheta} \sum_{j=1}^{4} u_j'[t] D_{ij} u_j[t] dt \quad (i = 1, 2, 3, 4). \tag{A.1.32}$$

In formulas (A.1.30)–(A.1.32), the vectors $x, u_i \in \mathbf{R}^n$; the matrices $A(t), B_i(t) \in C_{n \times n}[0, \vartheta]$; C_i and D_{ij} are symmetric constant matrices of dimensions $n \times n$; $\vartheta = const > 0$; an initial position $(t_*, x_*) \in [0, \vartheta] \times \mathbf{R}^n$; finally, $U = (U_1, U_2, U_3, U_4) \in \mathfrak{U} = \prod_{i=1}^{4} \mathfrak{U}_i$.

Let us formalize the concept of equilibrium in threats and counter-threats in this game.

First, introduce the matrices

$$C(\alpha) = \sum_{j=1}^{4} \alpha_j C_j, \quad D_i(\alpha) = \sum_{j=1}^{4} \alpha_j D_{ji} \quad (i = 1, 2, 3, 4)$$

for positive scalars α_j, which will be determined from the following **assertion**.

If there exists a set of positive numbers $(\alpha_1^*, \dots, \alpha_4^*) = \alpha^*$ *such that*

$$C(\alpha^*) < 0, \quad D_i(\alpha^*) < 0 \quad (i \in \mathbb{N} = \{1, 2, 3, 4\}), \tag{A.1.33}$$

then a Pareto-maximal strategy profile in game (A.1.30) *has the form*

$$U_i^{\alpha} \div u_i^{\alpha}(t, x) = Q_i^{\alpha}(t)x = -D_i^{-1}(\alpha)B_i'(t)\Theta(t)x, \quad i \in \mathbb{N}, \tag{A.1.34}$$

where the symmetric matrix

$$\Theta(t) = \left[X^{-1}(t) \right]' \left\{ C_{\alpha^*}^{-1} + \int_t^{\vartheta} X^{-1}(\tau) \sum_{i=1}^{4} B_i(\tau) D_i^{-1}(\alpha^*) B_i'(\tau) [X^{-1}(\tau)]' d\tau \right\}^{-1} X^{-1}(t),$$

and $X(t)$ *has been calculated in Exercise 38.*

Now, consider the equilibrium in threats and counter-threats for the four-player game (A.1.30).

Player i ($i \in \mathbb{N}$) is said to impose *a threat* to a strategy profile U^{α} if there exists a strategy $U_i^t \in \mathfrak{U}_i$ such that

$$\mathcal{J}_i(U_i^t, U_{\mathbb{N} \setminus i}^{\alpha}, t_*, x_*) > \mathcal{J}_i(U^{\alpha}, t_*, x_*). \tag{A.1.35}$$

Here $U_{\mathbb{N}\setminus i}$ denotes the set of strategies of all players except for U_i, and

$$\left(U_i^t, U_{\mathbb{N}\setminus i}^\alpha\right) = \left(U_1^\alpha, \ldots, U_{i-1}^\alpha, U_i^t, U_{i+1}^\alpha, \ldots, U_N^\alpha\right).$$

Player j $(j \neq i)$ is said to have *a counter-threat* in response to a threat U_i^t of player i if there exists a strategy $U_j^k \in \mathfrak{U}_j$ such that

$$\mathcal{J}_i\left(U_i^t, U_j^k, U_{\mathbb{N}\setminus\{i\cup j\}}^\alpha, t_*, x_*\right) < \mathcal{J}_i\left(U^\alpha, t_*, x_*\right), \tag{A.1.36}$$

$$\mathcal{J}_j\left(U_i^t, U_j^k, U_{\mathbb{N}\setminus\{i\cup j\}}^\alpha, t_*, x_*\right) > \mathcal{J}_j\left(U_i^t, U_{\mathbb{N}\setminus i}^\alpha, t_*, x_*\right); \tag{A.1.37}$$

here $\left(U_i^t, U_j^k, U_{\mathbb{N}\setminus\{i\cup j\}}^\alpha\right) = \left(U_1^\alpha, \ldots, U_{i-1}^\alpha, U_i^t, U_{i+1}^\alpha, \ldots, U_{j-1}^\alpha, U_j^k, U_{j+1}^\alpha, \ldots, U_N^\alpha\right)$.

Implementing his counter-threat in the game (A.1.30), player j will "punish" player i for his deviation, reducing the latter's payoff in comparison with the initial one in the strategy profile U^α (see (A.1.36)). Moreover, player j benefits from such a counter-threat, because his own payoff will increase in comparison with the strategy profile $(U_i^t, U_{\mathbb{N}\setminus i}^\alpha)$ obtained by implementing the threat (see inequality (A.1.37)).

The concepts of threats and counter-threats applied to the noncooperative four-player game (A.1.30) lead to the following **definition**.

A strategy profile $U^\alpha \in \mathfrak{U}$ is called *an equilibrium in threats and counter-threats for the differential game* (A.1.30) *with an initial position* $(t_*, x_*) \in [0, \vartheta) \times \mathbf{R}^n$, $x_* \neq 0_n$, if

(a)   the strategy profile U^α is Pareto-maximal, i.e., for any $U \in \mathfrak{U}$ the system of inequalities

$$\mathcal{J}_i(U, t_*, x_*) \geqslant \mathcal{J}_i(U^\alpha, t_*, x_*), \quad i \in \mathbb{N},$$

with at least one strict inequality, is inconsistent;

(b)   either all players impose no threats to U^α, i.e., ·

$$\mathcal{J}_i(U_i, U_{\mathbb{N}\setminus i}^\alpha, t_*, x_*) \leqslant \mathcal{J}_i(U^\alpha, t_*, x_*), \quad \forall U_i \in \mathfrak{U}_i, \quad i \in \mathbb{N},$$

or at least one of the other players $j \neq i$ has a counter-threat in response to each threat of any player i.

Prove the following **assertion.** *Assume for game* (A.1.30) *there exists a set of positive numbers* $(\alpha_1^*, \ldots, \alpha_4^*) = \alpha^*$ *that satisfies conditions* (A.1.33); *in addition,*

$$D_{ii} > 0, \quad i \in \mathbb{N}, \tag{A.1.38}$$

and, at least for a single number $j \in \mathbb{N}$ $(j \neq i)$,

$$D_{ij} < 0, \quad i \in \mathbb{N}. \tag{A.1.39}$$

Then for any initial position $(t_*, x_*) \in [0, \vartheta) \times \mathbf{R}^n$, $\|x_*\| \neq 0$, *game* (A.1.30)

1)   *has no Nash equilibria;*
2)   *has an equilibrium in threats and counter-threats* U^α *given by* (A.1.34).

Proof Using $D_{ii} > 0$ and $\|x_*\| \neq 0$, establish that there are no Nash equilibria in the game (A.1.30) with the initial positions specified above.

Next, demonstrate that in this game each player imposes a threat to U^α. For this purpose, write the value of the payoff function (A.1.32) for $U = U^\alpha$ as

$$
\mathcal{J}_i(U^\alpha, t_*, x_*) = x'(\vartheta) C_i x(\vartheta) + \int_{t_*}^{\vartheta} \left\{ u_i'[t] D_{ii} u_i[t] + \sum_{j=1, j\neq i}^{4} x'(t) (Q_j^\alpha(t))' D_{ij} Q_j^\alpha(t) x(t) \right\} dt,
$$

(A.1.40)

and the system (A.1.31) as

$$
\dot{x} = \left[A(t) + \sum_{j=1, j\neq i}^{4} B_j(t) Q_j^\alpha(t) \right] x + B_i(t) u_i, \quad x(t_*) = x_*,
$$

(A.1.41)

where, due to (A.1.34),

$$
Q_j^\alpha(t) = -D_j^{-1}(\alpha^*) B_j'(t) \Theta(t);
$$

$x(t)$, $t_* \leqslant t \leqslant \vartheta$, is the solution of (A.1.31) for $u_i = Q_i^\alpha(t) x$, $u_i[t] = Q_i^\alpha(t) x(t)$.

In view of $D_{ii} > 0$ and $\|x_*\| \neq 0$, there exists a constant $\beta_i(U^\alpha) > 0$ such that inequality (A.1.35) is satisfied for the strategy $U_i^t \doteq \beta_i(U^\alpha) e_n' x$, where e_n denotes an n-dimensional vector composed of ones. Hence, each player i imposes a threat.

Now, using $D_{ii} > 0$, $D_{ij} < 0$, and $\|x_*\| \neq 0$, prove that player j has a counter-threat in response to the threat $U_i^t \doteq Q_i^t(t) x$ ($U_i^t \in \mathfrak{U}_i$) of player i.

Really, if $U \neq (U_i^t, U_{\mathbb{N} \setminus i}^\alpha)$, the payoff function of player j can be represented as

$$
\mathcal{J}_j(U_i^t, U_{\mathbb{N} \setminus i}^\alpha, t_*, x_*) = x'(\vartheta) C_j x(\vartheta) + \int_{t_*}^{\vartheta} \left\{ u_j'[t] D_{jj} u_j[t] \right.
$$

$$
\left. + x'(t) \left[\sum_{k=1, k\neq i, k\neq j}^{4} (Q_k^\alpha(t))' D_{jk} Q_k^\alpha(t) + (Q_i^t(t))' D_{ji} Q_i^t(t) \right] x(t) \right\} dt, \quad \text{(A.1.42)}
$$

and the system (A.1.18) as

$$
\dot{x} = \left[A(t) + \sum_{k=1, k\neq i, k\neq j}^{4} B_k(t) Q_k^\alpha(t) + B_i(t) Q_i^t(t) \right] x + B_j(t) u_j,
$$

$$
x(t_*) = x_*.
$$

(A.1.43)

In (A.1.42), the vector function $x(t)$, $t_* \leqslant t \leqslant \vartheta$, is the solution of (A.1.31) for $u_j = Q_j^\alpha(t) x, j \in \mathbb{N}, u_i = Q_i^t(t) x$ and $u_j[t] = Q_j^\alpha(t) x(t)$. Since $D_{jj} > 0$ and $\|x_*\| \neq 0$, there exists a constant $\beta_j^* = const > 0$ such that inequality (A.1.37) is satisfied for the strategy

$$
U_j^k \doteq \beta x, \quad \forall \beta \geqslant \beta_j^*.
$$

(A.1.44)

Thus, the second inequality of (A.1.37) holds.

Finally, construct a strategy U_j^k of player j for which inequality (A.1.36) will be satisfied. Introduce the function

$$W_i(t, x, u, V_i) = \frac{\partial V_i}{\partial t} + \left[\frac{\partial V_i}{\partial x}\right]' \left[A(t)x + \sum_{j=1}^{4} B_j(t)u_j\right] + \sum_{j=1}^{4} u_j' D_{ij} u_j. \quad \text{(A.1.45)}$$

Using the same technique as described above, find a Lyapunov–Bellman function $V_i(t, x)$ that satisfies the conditions

$$W_i(t, x, u^\alpha(t, x), V_i) = 0, \quad V_i(\vartheta, x) = x' C_i x, \quad \text{(A.1.46)}$$

where $u^\alpha(t, x) = (u_1^\alpha(t, x), \ldots, u_4^\alpha(t, x))$ and $u_i^\alpha(t, x) = Q_i^\alpha(t)x$. Substitute $u_j^\alpha(t, x) = Q_j^\alpha(t)x$ into (A.1.46) and take into account (A.1.45) and (A.1.44) to obtain the following partial differential equation for $V_i(t, x)$:

$$\frac{\partial V_i}{\partial t} + \left[\frac{\partial V_i}{\partial x}\right]' \left[A(t)x + \sum_{j=1}^{4} B_j(t)Q_j^\alpha(t)\right]x + x' \sum_{j=1}^{4} (Q_j^\alpha(t))' D_{ij} Q_j^\alpha(t)x = 0,$$

$$V_i(\vartheta, x) = x' C_i x. \quad \text{(A.1.47)}$$

In (A.1.40), let

$$V_i(t, x) = x' \Theta_i x, \quad \Theta_i'(t) = \Theta_i(t)$$

and then equate the corresponding coefficients of the resulting quadratic forms of the vector x with each other to get the following matrix linear inhomogeneous equation with continuous coefficients:

$$\dot{\Theta}_i + \Theta_i \left[A(t) + \sum_{j=1}^{4} B_j(t)Q_j^\alpha(t)\right] + \left[A'(t) + \sum_{j=1}^{4} (Q_j^\alpha(t))' B_j'(t)\right]\Theta_i$$

$$+ \sum_{j=1}^{4} (Q_j^\alpha(t))' D_{ij} Q_j^\alpha(t) = 0_{n \times n}, \quad \Theta_i(\vartheta) = C_i.$$

This equation has a unique continuous solution $\Theta_i^*(t)$ that is extendable to $[0, \vartheta]$. Then the function

$$V_i^*(t, x) = x' \Theta_i^*(t)x$$

satisfies equalities (A.1.47) and hence (A.1.46) as well, for all $t \in [0, \vartheta]$ and $x \in \mathbf{R}^n$:

$$W_i(t, x, u^\alpha(t, x), V_i^*(t, x)) = 0, \quad V_i^*(\vartheta, x) = x' C_i x, \quad \forall t \in [0, \vartheta], \quad x \in \mathbf{R}^n. \quad \text{(A.1.48)}$$

Let $x^\alpha(t)$, $t_* \leqslant t \leqslant \vartheta$, be the solution of (A.1.31) for $u_i = u_i^\alpha(t, x)$ $i \in \mathbb{N}$. Then from (A.1.48) it follows that

$$W_i(t, x^\alpha(t), u^\alpha(t, x^\alpha(t)), V_i^*(t, x^\alpha(t))) = 0, \quad \forall t \in [t_*, \vartheta],$$

$$V_i^*(\vartheta, x^\alpha(\vartheta)) = [x^\alpha(\vartheta)]' C_i x^\alpha(\vartheta). \quad \text{(A.1.49)}$$

Integrate both sides of the first equality in (A.1.49) from t_* to ϑ and apply the second equality from (A.1.49) to obtain

$$\mathcal{J}_i\left(U^\alpha, t_*, x_*\right) = V_i^*(t_*, x_*). \tag{A.1.50}$$

Next, using the functions (A.1.45) and $V_i^*(t, x)$, construct $(u_{\mathbb{N}\setminus\{i \cup j\}} = (u_1, \ldots, u_{i-1}, u_{i+1}, \ldots, u_{j-1}, u_{j+1}, \ldots, u_4))$

$$W_i(t, x, u_i = Q_i^t(t)x, u_j, u_{\mathbb{N}\setminus\{i\cup j\}}^\alpha(t, x), V_i^*(t, x)) = \frac{\partial V_i^*(t, x)}{\partial t}$$

$$+ \left[\frac{\partial V_i^*(t, x)}{\partial x}\right]' \left\{\left[A(t) + \sum_{k\in\mathbb{N}\setminus\{i\cup j\}} B_k(t)Q_k^\alpha(t) + B_i(t)Q_i^t(t)\right] x\right.$$

$$+ B_j(t)u_j\Big\} + u_j'D_{ij}u_j + x\left[(Q_i^t(t))'D_{ii}Q_i^t(t) + \sum_{k\in\mathbb{N}\setminus\{i\cup j\}} (Q_k^\alpha(t))'D_{ik}Q_k^\alpha(t)\right]x$$

$$= x'\widetilde{P}_i(t)x + x'\widetilde{M}_i(t)u_j + u_j'D_{ij}u_j, \tag{A.1.51}$$

where the matrices

$$\widetilde{P}_i(t) = \dot{\Theta}_i^*(t) + \Theta_i^*(t)\left[A(t) + B_i(t)Q_i^t(t) + \sum_{k\in\mathbb{N}\setminus\{i\cup j\}} B_k(t)Q_k^\alpha(t)\right]$$

$$+ \left[A'(t) + (Q_i^t(t))'B_i'(t) + \sum_{k\in\mathbb{N}\setminus\{i\cup j\}} (Q_k^\alpha(t))'B_k'(t)\right]\Theta_i^*(t)$$

$$+ (Q_i^t(t))'D_{ii}Q_i^t(t) + \sum_{k\in\mathbb{N}\setminus\{i\cup j\}} (Q_k^\alpha(t))'D_{ik}Q_k^\alpha(t),$$

and

$$\widetilde{M}_i(t) = \Theta_i^*(t)B_j(t) + B_j'(t)\Theta_i^*(t)$$

are continuous on $[0, \vartheta]$ and do not depend on u_j.

Denote by $(-\lambda)$ the greatest root of the equation $\det[D_{ij} - \lambda E_n] = 0$. Due to $D_{ij} = D_{ij}' < 0$, $(-\lambda^*)$ is a negative real number for which

$$u_j'D_{ij}u_j \leqslant -\lambda^* u_j'u_j \quad \forall\, u_j \in \mathbf{R}^n.$$

In view of this inequality, for (A.1.51) it follows that

$$W_i(t, x, u_i = Q_i^t(t)x, u_j, u_{\mathbb{N}\setminus\{i\cup j\}}^\alpha(t, x), V_i^*(t, x)) \leqslant (-\lambda^*\|u_j\|^2 + x'\widetilde{M}_i(t)u_j + x'\widetilde{P}_i(t)x).$$

Since the elements of the matrices $\widetilde{M}_i(t)$ and $\widetilde{P}_i(t)$ are uniformly bounded on the compact set $[0, \vartheta]$, there exists a sufficiently great number $\bar{\beta}_j > 0$ for which the quadratic form

$$x'\left(-\lambda^*\bar{\beta}_j^2 E_n + \widetilde{M}_i(t)\bar{\beta}_j + \widetilde{P}_i(t)\right)x \quad \forall\, \beta_j \geqslant \bar{\beta}_j$$

is negative definite for all $t \in [0, \vartheta]$. (Hence, for each $t \in [0, \vartheta]$ the conditions of Sylvester's criterion are satisfied.) As a result,

$$\overline{W}_i[t, x] = W_i(t, x, u_i = Q_i^t(t)x, u_j = \beta_j x, u_{\mathbb{N}\setminus\{i\cup j\}}^\alpha(t, x), V_i^*(t, x)) < 0 \quad \text{(A.1.52)}$$

for all $\beta_j \geqslant \bar{\beta}_j$ and $x \in \mathbf{R}^n$, $x \neq 0_n$.

Now, let $\bar{x}(t)$, $t_* \leqslant t \leqslant \vartheta$, be the solution of (A.1.31) for $u_i = Q_i^t(t)x$, $u_j = \beta x$, and $u_{\mathbb{N}\setminus\{i\cup j\}} = u_{\mathbb{N}\setminus\{i\cup j\}}^\alpha(t, x)$. Due to the implication

$$[\|x_*\| \neq 0] \Rightarrow [\|\bar{x}(t)\| \neq 0 \ \forall t \in [t_*, \vartheta]],$$

the substitution of $x = \bar{x}(t)$ into (A.1.52) gives

$$\overline{W}_i[t, \bar{x}(t)] < 0 \quad \forall t \in [t_*, \vartheta]. \tag{A.1.53}$$

Finally, integrate both sides of (A.1.53) from t_* to ϑ to obtain

$$\mathcal{J}_i(U_i^t, U_j^k, U_{\mathbb{N}\setminus\{i\cup j\}}^\alpha, t_*, x_*) < V_i^*(t_*, x_*),$$

where $U_j^k \doteq \beta_j x$ for all $\beta_j \geqslant \bar{\beta}_j$. In combination with (A.1.50), this inequality yields

$$\mathcal{J}_i(U_i^t, U_j^k, U_{\mathbb{N}\setminus\{i\cup j\}}^\alpha, t_*, x_*) < \mathcal{J}_i(U^\alpha, t_*, x_*)$$

for any strategies $U_j^k \doteq \beta_j x$ and all constants $\beta_j \geqslant \widetilde{\beta}_j$. Then the strict inequalities (A.1.36) and (A.1.37) both hold for the strategy $U_j^k \doteq \widetilde{\beta}_j x$, where $\widetilde{\beta}_j = \max\{\beta_j^*, \bar{\beta}_j\}$. This fact finally proves the existence of a counter-threat of player j ($j \neq i$) in response to the threat U_i^t imposed by player i.

40)   Consider the differential game (A.1.30) with the coalition structure

$$\mathcal{K}_1 = [K_1 = \{1, 2\}, \quad K_2 = \{3\}, \quad K_3 = \{4\}].$$

Define the equilibrium in threats and counter-threats for the game

$$\Gamma(\mathcal{K}_1) = \langle (A.1.30), \mathcal{K}_1 \rangle$$

with an initial position (t_*, x_*), and *derive* sufficient conditions of its existence.

Solution The coalition K_1 imposes a threat U^α (A.1.41) if $\exists (U_1^t, U_2^t) \in \mathfrak{U}_1 \times \mathfrak{U}_2$:

$$\mathcal{J}_j\left(U_1^t, U_2^t, U_3^\alpha, U_4^\alpha, t_*, x_*\right) > \mathcal{J}_j(U^\alpha, t_*, x_*) \quad (j = 1, 2).$$

The coalition K_2 has a counter-threat in response to a threat (U_1^t, U_2^t) if $\exists U_3^k \in \mathfrak{U}_3$:

$$\mathcal{J}_j\left(U_1^t, U_2^t, U_3^k, U_4^\alpha, t_*, x_*\right) < \mathcal{J}_j\left(U^\alpha, t_*, x_*\right) \quad (j = 1, 2),$$

$$\mathcal{J}_3\left(U_1^t, U_2^t, U_3^k, U_4^\alpha, t_*, x_*\right) > \mathcal{J}_3\left(U_1^t, U_2^t, U_3^\alpha, U_4^\alpha, t_*, x_*\right).$$

A counter-threat U_4^k of the coalition K_3 in response to a threat (U_1^t, U_2^t) is defined by analogy.

The coalition K_2 imposes a threat U^α if $\exists U_3^t \in \mathfrak{U}_3$:

$$\mathcal{J}_3\left(U_1^\alpha, U_2^\alpha, U_3^t, U_4^\alpha, t_*, x_*\right) > \mathcal{J}_3\left(U^\alpha, t_*, x_*\right).$$

The coalition K_1 has a counter-threat in response to a threat U_3^t if $\exists\ (U_1^t, U_2^t) \in \mathfrak{U}_1 \times \mathfrak{U}_2$:

$$\mathcal{J}_j\left(U_1^k, U_2^k, U_3^t, U_4^\alpha, t_*, x_*\right) > \mathcal{J}_j\left(U_1^\alpha, U_2^\alpha, U_3^t, U_4^\alpha, t_*, x_*\right) \quad (j = 1, 2),$$

$$\mathcal{J}_3\left(U_1^k, U_2^k, U_3^t, U_4^\alpha, t_*, x_*\right) < \mathcal{J}_3\left(U^\alpha, t_*, x_*\right).$$

The coalition K_3 has a counter-threat in response to a threat U_3^t if $\exists\ U_4^k \in \mathfrak{U}_4$:

$$\mathcal{J}_3\left(U_1^\alpha, U_2^\alpha, U_3^t, U_4^k, t_*, x_*\right) < \mathcal{J}_3\left(U^\alpha, t_*, x_*\right),$$

$$\mathcal{J}_4\left(U_1^\alpha, U_2^\alpha, U_3^t, U_4^k, t_*, x_*\right) > \mathcal{J}_4\left(U_1^\alpha, U_2^\alpha, U_3^t, U_4^\alpha, t_*, x_*\right).$$

A threat posed by the coalition K_3 as well as counter-threats of the coalitions K_1 and K_2 in response to a such threat are defined by analogy.

A Pareto-maximal strategy profile U^α is called *an equilibrium in threats and counter-threats in the game* $\Gamma(\mathcal{K}_1)$ with an initial position (t_*, x_*) if each of the other coalitions $K_j(j \neq i)$ has a counter-threat in response to any threat posed by any coalition K_i.

The following **assertion** is true. Assume that for the game $\Gamma(\mathcal{K}_1)$:

1) there exists a set of positive numbers $\alpha_1^*, \ldots, \alpha_4^*$ such that conditions (A.1.33) are satisfied and

2)
$$\left\{\begin{array}{llll} D_{11} > 0 & & D_{13} < 0 & D_{14} < 0 \\ & D_{22} > 0 & D_{23} < 0 & D_{24} < 0 \\ D_{31} < 0 & D_{32} < 0 & D_{33} > 0 & D_{34} < 0 \\ D_{41} < 0 & D_{42} < 0 & D_{43} < 0 & D_{44} < 0 \end{array}\right\}.$$

Then for any initial position $(t_*, x_*) \in [0, \vartheta) \times \mathbf{R}^n$, $\|x_*\| \neq 0$, an equilibrium in threats and counter-threats is the strategy profile U^α given by (A.1.34).

Proof Rests on the following fact, which has been actually established in Exercise 39. Assume that player j in the game (A.1.30) knows the set of strategies $U_{\mathbb{N}\backslash j}^* \in \mathfrak{U}_{\mathbb{N}\backslash j}$ and the strategy profile $U^\alpha \in \mathfrak{U}$, and also $D_{ij} > 0\ (< 0)$. Then there exists a constant $\beta^* = \beta^*(U_{\mathbb{N}\backslash j}^*, U^\alpha) > 0$ such that, for any $\beta \geqslant \beta^*$ and initial positions $(t_*, x_*) \in [0, \vartheta) \times \mathbf{R}^n$, $\|x_*\| \neq 0$, the inequality

$$\mathcal{J}_i(U_{\mathbb{N}\backslash j}^*, U_j^\beta, t_*, x_*) > \mathcal{J}_i(U^\alpha, t_*, x_*)$$

(the inequality $\mathcal{J}_i(U_{\mathbb{N}\backslash j}^*, U_j^\beta, t_*, x_*) < \mathcal{J}_i(U^\alpha, t_*, x_*)$, respectively)

holds for the strategy $U_j^\beta \div \beta x$, where $\beta \geqslant \beta^*$.

41) Consider the differential game $\Gamma(\mathcal{K}_2) = \langle (\text{A.1.30}), \mathcal{K}_2 \rangle$ with the coalition structure

$$\mathcal{K}_2 = [\, K_1 = \{1, 2, 3\}, \quad K_2 = \{4\}\,].$$

Define the equilibrium in threats and counter-threats for the game $\Gamma(\mathcal{K}_2)$ and *derive* sufficient conditions of its existence.

Solution tips By analogy with Exercise 40, establish the following **result**.

Assume that for the game $\Gamma(\mathcal{K}_2)$ there exists a set of positive numbers $\alpha_1^*, \ldots, \alpha_4^*$ such that conditions (A.1.33) are satisfied and

$$D_{ii} > 0, \quad D_{i4} < 0, \quad D_{4i} < 0, \quad (i = 1, 2, 3, 4).$$

Then for any initial position $(t_*, x_*) \in [0, \vartheta) \times \mathbf{R}^n$, $\|x_*\| \neq 0$, an equilibrium in threats and counter-threats is the strategy profile U^α given by (A.1.34).

42)   Consider the noncooperative N-player game

$$\langle \, \mathbb{N}, \, \{X_i\}_{i \in \mathbb{N}}, \, \{F_i(x)\}_{i \in \mathbb{N}} \, \rangle,$$

where $\mathbb{N} = \{1, \ldots, N\}$, $x_i \in X_i \in comp \, \mathbf{R}^{n_i}$, $x = (x_1, \ldots, x_N) \in X = \prod\limits_{i \in \mathbb{N}} X_i$, and $F_i(x) \in C[X, \mathbf{R}^1]$.

A strategy profile $x^B \in X$ is called *a strong Berge equilibrium* in this game if

$$F_i(x^B) > F_i(x_i^B, x_{\mathbb{N} \setminus i}), \quad i \in \mathbb{N},$$

for all $x_{\mathbb{N} \setminus i} \in X_{\mathbb{N} \setminus i} = \prod\limits_{j \in \mathbb{N}, j \neq i} X_j$ and $x_j \neq x_j^B, j \in \mathbb{N} \setminus i$.

For $y_i \in X_i$ $(i \in \mathbb{N})$ and $y = (y_1, \ldots, y_N)$, introduce the function

$$\Phi(x, y) = \sum_{i \in \mathbb{N}} \left\{ [F_i(x_i, y_{\mathbb{N} \setminus i}) - F_i(x)] \prod_{j \in \mathbb{N}, j \neq i} \|y_j - x_j\|^2 \right\}.$$

Prove that if (x^0, y^0) is a strong saddle point of $\Phi(x, y)$, i.e.,

$$\Phi\left(x^0, y\right) < \Phi\left(x^0, y^0\right) < \Phi\left(x, y^0\right), \quad \forall \, x \in X, \, y \in X, \, x \neq x^0, \, y \neq y^0,$$

then x^0 is a strong Berge equilibrium.

Solution tips $[\, \Phi(x, y^0)|_{x=y^0} = 0 \,] \; \Rightarrow \; [\, \Phi(x^0, y) < 0, \, \forall y \in X, \, y \neq y^0 \,] \; \Rightarrow$
(for $y_i = x_i^0$) $[\, \{F_i(x_i^0, y_{\mathbb{N} \setminus i}) - F_i(x^0)\} \prod\limits_{j \in \mathbb{N} \setminus i} \|y_j - x_i^0\|^2 < 0 \, \forall \, y_{\mathbb{N} \setminus i} \in X_{\mathbb{N} \setminus i}, i \in \mathbb{N} \,]$
$\Rightarrow [\, F_i(x_i^0, x_{\mathbb{N} \setminus i}) < F_i(x^0) \, \forall x_{\mathbb{N} \setminus i} \in X_{\mathbb{N} \setminus i} \, (i \in \mathbb{N}), \, x_j \neq x_j^0 \, (j \in \mathbb{N} \setminus i) \,]$.

Pareto equilibrium in threats and counter-threats for a differential three-player game

In order to govern, the question is not to follow out
a more or less valid theory but to build
with whatever materials are at hand.
The inevitable must be accepted and turned to advantage.
—Napoleon I

In this appendix, a linear-quadratic positional differential three-player game is considered. Coefficient criteria under which the game has no Nash equilibrium but simultaneously has an equilibrium in threats and counter-threats are established.

A.2.1 INTRODUCTION

According to the opinion of leading figures in the mathematical theory of games, an equilibrium as an acceptable solution of a differential game should have the property of *stability*: a player who is unilaterally deviating from an equilibrium cannot increase his payoff. The solution [310, 311] proposed by 21 years-old Princeton University postgraduate John F. Nash, Jr., which was later called the *Nash equilibrium* (NE), fully meets this requirement. NE has gradually become a "reigning" concept in economics, sociology, and military sciences. In 1994 J. Nash, together with R. Selten and J. Harsanyi, was awarded the Nobel Prize in Economic Sciences "for their pioneering analysis of equilibria in the theory of non-cooperative games." In fact, Nash laid the foundations of a scientific method that made a huge impact on the development of the world economy. Nowadays, opening almost any periodical on economics, operations research, systems analysis, or game theory, we are likely to come across publications dealing with the concept of Nash equilibrium. However, there are spots on the sun: the set of Nash equilibria can be internally and externally unstable. As an example, consider the simplest noncooperative two-player game in normal form:

$$\left\langle \{1, 2\}, \{X_i = [-1, 1]\}_{i=1,2}, \left\{f_i(x_1, x_2) = 2x_1x_2 - x_i^2\right\}_{i=1,2}\right\rangle.$$

The set of Nash equilibria has the form

$$X^e = \left\{x^e = (x_1^e, x_2^e) = (\alpha, \alpha) | \forall \alpha = const \in [-1, 1]\right\}, \ f_i(x^e) = \alpha^2 \ (i = 1, 2).$$

The elements of this set, actually representing a segment of the bisecting line of the first and third quadrants, are remarkable for the following properties. *First*, for

$x^{(1)} = (0,0) \in X^e$ and $x^{(2)} = (1,1) \in X^e$, $f_i(x^{(1)}) = 0 < f_i(x^{(2)}) = 1$ $(i = 1, 2)$ and hence the set X^e is *internally unstable*. Second, $f_i(x^{(1)}) = 0 < f_i(\frac{1}{4}, \frac{1}{3})$ $(i = 1, 2)$, meaning that the set X^e is also *externally unstable*.

The external and internal instability of the set of Nash equilibria is a negative feature for its practical use. In the former case, there exists a strategy profile dominating a given NE (for all players); in the latter case, such a strategy profile is also a Nash equilibrium. The consequences of external and internal instability could be avoided using the Pareto maximality of a Nash equilibrium. However, such a property is a rather exotic phenomenon; at least we only know three cases in which a Nash equilibrium is simultaneously Pareto-maximal [213, pp. 92–93; 195, 365, 366, 389]. Thus, for eliminating the difficulties associated with external and internal instability, Pareto maximality will be introduced as an additional requirement for the concept of equilibrium in threats and counter-threats (ETC) [36, 37], which is discussed below [39–43].

Let us proceed to the analysis of ETC. For the first time in the literature, it appeared in the book [189]; also, see [387]. A perennial stream of publications has been devoted to the study of the positive and negative properties of Nash equilibrium as a key solution of noncooperative games in economic applications. They are mainly associated with nonuniqueness (and hence the absence of equivalence, interchangeability, external instability) as well as with instability against the simultaneous deviation of two or more players. The well-known Prisoner's Dilemma also revealed the property of improvability. These negative properties for positional differential games were considered in the book [155] by V. Zhukovskiy and N. Tynyanskii. The two options finally suggested by the authors were either to use the Nash equilibria that are simultaneously free from some of the drawbacks mentioned above, or to introduce new solutions of noncooperative games that possess the advantages of Nash equilibria and would allow eliminating its individual drawbacks. For the class of differential games, one of such opportunities is related to the equilibrium in threats and counter-threats, the subject of Appendix 2. It is based on the concept of threats and counter-threats, known in the classical theory of games. The theoretical foundations of this concept go back to the work of E. Vilkas [383, 384]. The term "active equilibrium" was proposed by E. Smol'yakov in 1983; for details, refer to [231]. Note that for the first time in differential games, the equilibrium in threats and counter-threats was used in 1974 by E. Vaisbord [381, 382]. Later on, it was considered by V. Zhukovskiy in [155]. In our opinion, this concept is still underused in differential games.

A threat is a promise to bring any evil or trouble; see [21]. A threat is not necessarily a real action, as it may consist in reporting the possibility of such an action (intimidation!). For mitigating an aggressive character, in some publications the term "threat" is used as a synonym of "objection." A message about the player's action that nullifies a threat is called a counter-threat (counter-objection). As it has been mentioned, the concept of threats and counter-threats figured in the early research on matrix games [189]. However, the analysis was limited either to the static version of such games or to differential two-player games [51, 354, 386, 388, 390–393]. The differential games of three or more players were not considered, which is also a motivation for writing this appendix.

A.2.2 PROBLEM STATEMENT

Consider a noncooperative linear-quadratic differential three-player game in normal form described by an ordered quadruple

$$\Gamma = \langle \{1, 2, 3\}, \Sigma, \{\mathcal{U}_i\}_{i=1,2,3}, \{J_i(U, t_0, x_0)\}_{i=1,2,3} \rangle.$$

In the game Γ, the set of players is $\{1, 2, 3\}$, and the dynamics of the controlled system Σ satisfy the vector linear differential equation

$$\Sigma \div \dot{x} = A(t)x + u_1 + u_2 + u_3, \ x(t_0) = x_0, \tag{A.2.1}$$

with the following notations: $x \in \mathbb{R}^n$ as the n-dimensional state vector of the system Σ; $[t_0, \vartheta]$ as a finite time interval of the game with a fixed terminal time instant $\vartheta = const > 0$; $u_i \in \mathbb{R}^n$ as the control action of player i ($i = 1, 2, 3$); $(t, x) \in [0, \vartheta] \times \mathbb{R}^n$ as a pair determining a current position in the game Γ; finally, (t_0, x_0) as an initial position, where $0 \leq t_0 < \vartheta$. The matrix $A(t)$ of dimensions $n \times n$ is assumed to have continuous elements on $[0, \vartheta]$, which is denoted by $A(\cdot) \in C_{n \times n}[0, \vartheta]$.

A *strategy* U_i of player i will be identified with an n-dimensional vector function $u_i(t, x)$, and this fact will be denoted by $U_i \div u_i(t, x)$. Then *the set of strategies of player i* can be written as

$$\mathcal{U}_i = \{U_i \div u_i(t, x), u_i(t, x) = Q_i(t)x | \forall Q_i(\cdot) \in C_{n \times n}[0, \vartheta]\}.$$

Thus, player i chooses his strategy by specifying a matrix $Q_i(t)$ of dimensions $n \times n$ ($i = 1, 2, 3$) from the space $C_{n \times n}[t_0, \vartheta]$.

The game evolves over time in the following way. Without forming coalitions with other players, each player i chooses his strategy $U_i \div Q_i(t)x$, which yields a *strategy profile* $U = (U_1, U_2, U_3) \in \mathcal{U} = \mathcal{U}_1 \times \mathcal{U}_2 \times \mathcal{U}_3$ of the game. Next, each player finds the solution $x(t), t_0 \leq t \leq \vartheta$, of the system (A.2.1) with $u_i = Q_i(t)x$ ($i = 1, 2, 3$), i.e.,

$$\dot{x} = [A(t) + Q_1(t) + Q_2(t) + Q_3(t)]x(t), \ x(t_0) = x_0. \tag{A.2.2}$$

The system of linear homogeneous differential equations (A.2.2) with continuous coefficients on $[t_0, \vartheta]$ has a continuous solution $x(t)$ that is extendable to $[t_0, \vartheta]$. Then each player constructs the *realization* of his strategy $u_i[t] = Q_i(t)x(t)$ ($i = 1, 2, 3$) and also the corresponding realization of the strategy profile $u[t] = (u_1[t], u_2[t], u_3[t])$, which consists of the three n-dimensional continuous vectors $u_1[t], u_2[t]$, and $u_3[t]$ on $[t_0, \vartheta]$.

The payoff function of player i is a quadratic functional

$$J_i(U_1, U_2, U_3, t_0, x_0) = x'(\vartheta)C_i x(\vartheta) + \int_{t_0}^{\vartheta} (u_1'[t]D_{i1}u_1[t] + u_2'[t]D_{i2}u_2[t]$$

$$+ u_3'[t]D_{i3}u_3[t])dt \quad (i = 1, 2, 3), \tag{A.2.3}$$

defined on the continuous quadruples $(x(t), u_1[t], u_2[t], u_3[t] | t \in [t_0, \vartheta])$. Without loss of generality, let the constant matrices C_i and D_{ij} ($i, j = 1, 2, 3$) of dimensions $n \times n$ be symmetric. As before, the prime denotes transposition, i.e., x' is a row vector. The value of the functional (A.2.3) is called *the payoff* of player i. Assume that in the game Γ each player seeks to maximize his payoff with an appropriate choice of his strategy.

The aim of this research is to reveal a rather general class of noncooperative linear-quadratic differential three-player games in normal form Γ that have no Nash equilibrium but simultaneously have an equilibrium in threats and counter-threats.

To this effect, we will associate with the game Γ the tri-criteria dynamic choice problem

$$\Gamma_\nu = \langle \Sigma, \mathcal{U}, \{J_i(U, t_0, x_0)\}_{i=1,2,3} \rangle.$$

Here the controlled dynamic system Σ coincides with (A.2.1); the set of alternatives $\mathcal{U}$ coincides with the set of strategy profiles $\mathcal{U} = \prod_{i=1}^{3} \mathcal{U}_i$ of the game Γ; the three criteria $J_i(U, t_0, x_0)$ ($i = 1, 2, 3$) are given by (A.2.3).

The decision-maker's goal in the problem Γ_ν is to choose an alternative $U^P \in \mathcal{U}$ for which the three criteria (A.2.3) will take the *maximum* possible values. A conventional approach to such problems is based on Pareto maximality.

Definition A.2.1 *An alternative* $U^P = (U_1^P, U_2^P, U_3^P) \in \mathcal{U}$ *is called Pareto-maximal in the problem* Γ_ν *if* $\forall U \in \mathcal{U}$ *and* $\forall (t_0, x_0) \in [0, \vartheta) \times \mathbb{R}^n$, $x_0 \neq 0_n$, *the system of inequalities*

$$J_i(U, t_0, x_0) \geq J_i(U^P, t_0, x_0) \ (i = 1, 2, 3),$$

with at least one strict inequality, is inconsistent. In this case, the vector $J^P = J^P[t_0, x_0] = (J_1(U^P, t_0, x_0), J_2(U^P, t_0, x_0), J_3(U^P, t_0, x_0))$ *is called a Pareto maximum in the problem* Γ_ν *[317].*

Note two circumstances, which are immediate from Definition A.2.1.

Property A.2.1

$$J_i\left(\hat{U}, t_0, x_0\right) > J_i\left(U^P, t_0, x_0\right) \Rightarrow J_j(\hat{U}, t_0, x_0) < J_j(U^P, t_0, x_0)$$

for at least one number $j = 1, 2, 3$; $j \neq i$.

Property A.2.2 If the condition

$$\max_{U \in \mathcal{U}} \{J_1(U, t_0, x_0) + \beta J_2(U, t_0, x_0) + \gamma J_3(U, t_0, x_0)\} = Idem \left\{U \to U^P\right\}$$

holds for constants $\beta > 0$ and $\gamma > 0$, then the alternative U^P is Pareto-maximal in Γ_ν. Recall that $Idem\{U \to U^P\}$ denotes the bracketed expression from (A.2.4) with U replaced by U^P.

Now, consider the equilibrium solutions of the game Γ, where $J = (J_1, J_2, J_3) \in \mathbb{R}^3$.

Definition A.2.2 *A pair* $(U^e, J^e = J(U^e, t_0, x_0)) \in \mathcal{U} \times \mathbb{R}^3$ *is called a Nash equilibrium in the game* Γ *if*

$$\begin{cases} \max_{U_1 \in \mathcal{U}} J_1(U_1, U_2^e, U_3^e, t_0, x_0) = J_1(U_1^e, U_2^e, U_3^e, t_0, x_0) = J_1^e, \\ \max_{U_2 \in \mathcal{U}} J_2(U_1^e, U_2, U_3^e, t_0, x_0) = J_2(U_1^e, U_2^e, U_3^e, t_0, x_0) = J_2^e, \\ \max_{U_3 \in \mathcal{U}} J_3(U_1^e, U_2^e, U_3, t_0, x_0) = J_3(U_1^e, U_2^e, U_3^e, t_0, x_0) = J_3^e \end{cases} \quad (A.2.4)$$

for any $(t_0, x_0) \in [0, \vartheta) \times \mathbb{R}^n$, $x_0 \neq 0_n$. (Here 0_n denotes a zero vector of dimension n.)

In fact, the concept of equilibrium in threats and counter-threats looks more sophisticated.

Let $U = (U_1, U_2, U_3)$ be some fixed strategy profile of the game Γ. Player 1 is said to pose *a threat to the strategy profile U* if there exists his strategy $U_1^T \in \mathcal{U}_1$ such that

$$J_1\left(U_1^T, U_2, U_3, t_0, x_0\right) > J_1\left(U_1, U_2, U_3, t_0, x_0\right). \quad (A.2.5)$$

The presence of a threat does not mean its mandatory implementation, but only *animus denuntiandi*.[39] The implementation of the threat defined above is beneficial to player 1: according to (A.2.5), his payoff will be increased in comparison with the one in the strategy profile U.

In response to a threat U_1^T of player 1, player 2 is said to pose *an incomplete counter-threat* if there exists his strategy $U_2^C \in \mathcal{U}_2$ such that

$$J_1\left(U_1^T, U_2^C, U_3, t_0, x_0\right) \leq J_1\left(U_1, U_2, U_3, t_0, x_0\right); \tag{A.2.6}$$

player 2 is said to pose *a complete counter-threat* if there exists his strategy $U_2^C \in \mathcal{U}_2$ such that inequality (A.2.6) is satisfied simultaneously with

$$J_2\left(U_1^T, U_2^C, U_3, t_0, x_0\right) > J_2\left(U_1^T, U_2, U_3, t_0, x_0\right). \tag{A.2.7}$$

A complete counter-threat of player 3 in response to a threat U_1^T is defined by analogy.

In the presence of an incomplete counter-threat, player 2 can choose his strategy U_2^C for making the payoff of player 1 (who poses an original threat) equal to a value not exceeding his original payoff in the strategy profile U; see (A.2.6). (Note that he may even reduce the payoff of player 1!) Everything happens in accordance with the well-known motto of Napoleon I: ordre, contre-ordre, désordre.[40] Therefore, the presence of an incomplete counter-threat negates the implementation of a threat. In addition, a complete counter-threat motivates player 2 to choose U_2^C, because his payoff in the resulting strategy profile $\left(U_1^T, U_2^C, U_3\right)$ yielded by implementing the threat and counter-threat will increase in comparison with the one in the strategy profile $\left(U_1^T, U_2, U_3\right)$ yielded by implementing the threat U_1^T.

A threat of player 2 (or player 3) to a strategy profile U and a (complete) counter-threat of one of the two other players are defined by analogy.

Naturally enough, if in response to each threat posed by any player to U, at least one of the other players has a counter-threat, then it makes no sense for the former to implement his threat: as the result of the counter-threat of another player, his payoff will not increase (but it may even decrease!).

Definition A.2.3 *A strategy profile* $U^P = (U_1^P, U_2^P, U_3^P) \in \mathcal{U}$ *is called an active equilibrium in the game* Γ *if, for any initial position* $(t_0, x_0) \in [0, \vartheta) \times \mathbb{R}^n$, $x_0 \neq 0_n$:

1) *the alternative* U^P *is Pareto-maximal in the tri-criteria dynamic choice problem* Γ_ν;
2) *in response to each threat* $U_i^T \in \mathcal{U}_i$ *of any player, at least one of the other players has an incomplete counter-threat.*

Definition A.2.4 *A pair* $(U^P, J^P) \in \mathcal{U} \times \mathbb{R}^3$ *is called an equilibrium in threats and counter-threats in the differential game* Γ *if, for any initial position* $(t_0, x_0) \in [0, \vartheta) \times \mathbb{R}^n$, $x_0 \neq 0_n$,

1) *the alternative* U^P *is Pareto-maximal in the tri-criteria dynamic choice problem* Γ_ν;
2) *in response to each threat of any player, at least one of the other players has a complete counter-threat.*

As before, $J^P = (J_1^P, J_2^P, J_3^P)$ *and* $J_i^P = J_i(U^P, t_0, x_0)$ $(i = 1, 2, 3)$.

---

[39]Latin "Intention to menace."
[40]French "Order, counter-order, disorder."

From Definitions A.2.3 and A.2.4, it follows that any equilibrium in threats and counter-threats is at the same time an active equilibrium. According to Definition A.2.2, a Nash equilibrium allows for no threats, and only the "best" Nash equilibrium (in the sense of Pareto maximality) will be an active equilibrium.

As it has been already mentioned, the concept of threats and counter-threats presented here proceed from the well-known concept of threats and counter-threats from the classical game theory [387]. It was used in [387] to define stable coalition structures, which were apparently first considered for cooperative differential games in [155]. The concept of threats and counter-threats for differential games was adopted by E. Vaisbord in 1974 (see [181, 381]) and then developed by V. Zhukovskiy [365, 382]. The corresponding theoretical aspects were studied by E. Vilkas [383, 384]. An original classification of solutions of noncooperative games, including the equilibrium in threats and counter-threats, was proposed by E. Smol'yakov [231]. He also introduced the term "active equilibrium" based on the above definition of an incomplete counter-threat. The concept of active equilibrium for noncooperative positional differential games was also employed in [354]. A method for proving the existence of an active equilibrium was suggested by V. Zhukovskiy in [354] and then successfully used by Bulgarian mathematicians to establish the existence of such a solution in positional differential two-player games described by partial differential equations [388, 393], stochastic differential equations [391], equations with a constant delay [390], and also in a Banach space [392].

Active equilibria and equilibria in threats and counter-threats have all the positive properties of Nash equilibria [89, p. 49]. More specifically:

1) They are stable against the deviations of an individual player.

2) They satisfy individual rationality.

3) They coincide with the saddle point in the case of zero-sum two-player games.

At the same time, unimprovable equilibria are free from the following disadvantages [89, p. 58]:

– They exist in a number of cases when there is no Nash equilibrium (for example, in the game Γ).

– Unlike a Nash equilibrium, they are unimprovable and internally stable (due to Pareto maximality).

– The presence of a Nash equilibrium in the game implies the existence of certain types of unimprovable equilibria in which the payoffs of all players are no smaller than in the Nash equilibrium.

– Only the "best" Nash equilibria (in the sense of Pareto maximality) are equilibria in threats and counter-threats. However, such best equilibria exist only in particular types of games [195, 213, 389].

Note that these properties hold for noncooperative positional differential games as well. In [89] the mathematical formalization of players' strategies and the motions of the dynamic system induced by them was adopted for analysis. This formalization was pioneered by N. Krasovskii [179] for a positional differential zero-sum two-player game.

A.2.3   PARETO-MAXIMAL STRATEGY PROFILES AND PAYOFFS

Hereinafter, the notation $D < 0$ (> 0) means that a quadratic form $x'Dx$ is negative definite (positive definite, respectively).

First of all, let us present an auxiliary result—Lemma A.2.1.

Consider the tri-criteria *static* problem

$$\Gamma_3 = \left\langle X = \mathbb{R}^{3n}, \{f_i(u) = u_1' D_{i1} u_1 + u_2' D_{i2} u_2 + u_3' D_{i3} u_3\}_{i=1,2,3} \right\rangle,$$

in which the decision-maker chooses an alternative (strategy profile) $u = (u_1, u_2, u_3) \in \mathbb{R}^{3n}$ for maximizing simultaneously the three components of a vector criterion $f(u) = (f_1(u), f_2(u), f_3(u))$. For this problem, Definition A.2.1 can be reformulated as follows: an alternative u^P is *Pareto-maximal* in Γ_3 if $\forall u \in \mathbb{R}^{3n}$ the system of inequalities $f_i(u) \geq f_i(u^P)$ $(i = 1, 2, 3)$, with at least one strict inequality, is inconsistent.

An analog of Property A.2.2 is used below.

Lemma A.2.1 *Assume that in the problem Γ_3 the symmetric matrices D_{ij} of dimensions $n \times n$ and positive numbers Λ_{ii} and λ_{ij} $(i, j = 1, 2, 3; j \neq i)$ are such that*

$$D_{ii} > 0, \quad D_{ij} < 0 \text{ (for } j \neq i), \quad \Lambda_{11}\Lambda_{22} < \lambda_{12}\lambda_{21}. \tag{A.2.8}$$

Then there exist numbers $\beta > 0$ and $\gamma > 0$ for which the quadratic forms $x'D_i x$ $(i = 1, 2, 3)$ in the expression

$$f(u) = f_1(u) + \beta f_2(u) + \gamma f_3(u) = u_1' D_1 u_1 + u_2' D_2 u_2 + u_3' D_3 u_3$$

become negative definite.

Here

$$D_i = D_{1i} + \beta D_{2i} + \gamma D_{3i} \ (i = 1, 2, 3); \tag{A.2.9}$$

in addition, Λ_{ii} is the greatest root of the characteristic equation

$$\Delta_{ii}(\Lambda) = \det[D_{ii} - \Lambda E_n] = 0 \ (i = 1, 2, 3),$$

whereas $-\lambda_{ij}$ $(i, j = 1, 2, 3; j \neq i)$ is the greatest (smallest by magnitude) root of the equation

$$\delta_{ij}(\lambda) = \det[D_{ij} - \lambda E_n] = 0;$$

finally, E_n denotes an identity matrix of dimensions $n \times n$.

Proof Due to the symmetry of all nine matrices D_{ii}, D_{ij} $(i, j = 1, 2, 3; j \neq i)$ figuring in Γ_3, the roots of the characteristic equations $\Delta_{ii}(\Lambda) = 0$ and $\delta_{ij}(\lambda) = 0$ are real; in addition, the roots of $\Delta_{ii}(\Lambda) = 0$ are positive, whereas the roots of $\delta_{ij}(\lambda) = 0$ are negative. Denote by Λ_{ii} the greatest of n roots of the equation $\det[D_{ii} - \Lambda E_n] = 0$, and also denote by $-\lambda_{ij}$ the greatest root of the equation $\delta_{ij}(\lambda) = 0$. From [51, p. 281] it follows that, for all $u_i \in \mathbb{R}^n$,

$$u_i' D_{ii} u_i \leq \Lambda_{ii} u_i' u_i \ (i = 1, 2, 3),$$
$$u_j' D_{ij} u_j \leq -\lambda_{ij} u_j' u_j \ (i, j = 1, 2, 3; \ j \neq i).$$

Hence,

$$
\begin{aligned}
f(u) = f_1(u) + \beta f_2(u) + \gamma f_3(u) &= u'_1[D_{11} + \beta D_{21} + \gamma D_{31}]u_1 \\
&+ u'_2[D_{12} + \beta D_{22} + \gamma D_{32}]u_2 + u'_3[D_{13} + \beta D_{23} + \gamma D_{33}]u_3 \\
&\leq [\Lambda_{11} - \beta\lambda_{21} - \gamma\lambda_{31}]u'_1 u_1 + [-\lambda_{12} + \beta\Lambda_{22} - \gamma\lambda_{32}]u'_2 u_2 \\
&+ [-\lambda_{13} - \beta\lambda_{23} + \gamma\Lambda_{33}]u'_3 u_3.
\end{aligned}
$$

Thus, $f(u) < 0 \ \forall u \in \{\mathbb{R}^{3n} \setminus \{0_{3n}\}\}$ if

$$
\begin{cases}
\Lambda_{11} - \beta\lambda_{21} - \gamma\lambda_{31} < 0, \\
-\lambda_{12} + \beta\Lambda_{22} - \gamma\lambda_{32} < 0, \\
-\lambda_{13} - \beta\lambda_{23} + \gamma\Lambda_{33} < 0.
\end{cases}
\tag{A.2.10}
$$

Consequently, the first two inequalities in (A.2.10) hold if

$$
\frac{\Lambda_{11}}{\lambda_{21}} < \beta < \frac{\lambda_{12}}{\Lambda_{22}} \Rightarrow f(u) < 0 \ \forall u \in \left\{ \mathbb{R}^{3n} \setminus \{0_{3n}\} \right\},
$$

i.e., in the case $\Lambda_{11}\Lambda_{22} < \lambda_{12}\lambda_{21}$ $\left(\text{for example, for } \beta = \frac{1}{2}\left(\frac{\Lambda_{11}}{\lambda_{21}} + \frac{\lambda_{12}}{\Lambda_{22}} \right) \right)$.

Similar considerations can be used to establish that the third inequality in (A.2.10) holds under the condition

$$
0 < \gamma < \frac{\lambda_{13}}{\Lambda_{33}} + \frac{1}{2}\left(\frac{\Lambda_{11}}{\lambda_{21}} + \frac{\lambda_{12}}{\Lambda_{22}} \right) \frac{\lambda_{23}}{\Lambda_{33}},
$$

e.g., for $\gamma = \frac{1}{2}\left[\frac{\lambda_{13}}{\Lambda_{33}} + \frac{1}{2}\left(\frac{\Lambda_{11}}{\lambda_{21}} + \frac{\lambda_{12}}{\Lambda_{22}} \right) \frac{\lambda_{23}}{\Lambda_{33}} \right]$.

Remark A.2.1 By analogy with Lemma A.2.1, the following result is the case. Assume that in the problem Γ_3 the symmetric matrices D_{ij} of dimensions $n \times n$ and positive numbers Λ_{ii} and λ_{ij} $(i, j = 1, 2, 3; j \neq i)$ defined in Lemma A.2.1 are such that

$$
D_{ii} > 0, \ D_{ij} < 0 \ (\text{for } j \neq i), \ \Lambda_{11}\Lambda_{33} < \lambda_{13}\lambda_{31}.
$$

Then for

$$
\beta = \frac{1}{2}\left[\frac{\lambda_{12}}{\Lambda_{22}} + \frac{1}{2}\left(\frac{\lambda_{13}}{\Lambda_{33}} + \frac{\Lambda_{11}}{\lambda_{31}} \right) \frac{\lambda_{32}}{\Lambda_{22}} \right] \quad \text{and}
$$

$$
\gamma = \frac{1}{2}\left(\frac{\lambda_{13}}{\Lambda_{33}} + \frac{\Lambda_{11}}{\lambda_{31}} \right),
$$

the quadratic form

$$
f(u) = f_1(u) + \beta f_2(u) + \gamma f_3(u) = u'_1 D_1 u_1 + u'_2 D_2 u_2 + u'_3 D_3 u_3
$$

becomes negative definite.

Really, the new numbers β and γ in this expression also satisfy the strict inequalities (A.2.10).

Note that in addition to the two solutions (β, γ) (see Lemma A.2.1 and Remark A.2.1), the strict inequalities (A.2.10) may have a continuum of other solutions. As it will be demonstrated below, each of them induces a specific equilibrium in threats and counter-threats in the differential game Γ (of course, under the conditions $D_{ii} > 0$ and $D_{ij} < 0$ $(i, j = 1, 2, 3; j \neq i)$).

Lemma A.2.2 *The solutions $x(t)$ of the system $\dot{x} = K(t)x$, $x(t_0) = x_0$, where $K(\cdot) \in C_{n \times n}[0, \vartheta]$, satisfy the nontrivial property*

$$x_0 \neq 0_n \Rightarrow x(t) \neq 0_n \; \forall t \in [t_0, \vartheta];$$

here 0_n denotes a zero vector from the space $\mathbb{R}^n$.

Proof Assume on the contrary that $\exists t_1 \in (t_0, \vartheta]$ such that $x(t_1) = 0_n$. In other words, at the time instant t_1 two different solutions of the system $\dot{x} = K(t)x$ are passing through the position $(t_1, 0_n)$, namely, the trivial one $x^{(1)}(t) = 0_n \; \forall t \in [0, \vartheta]$ and the nontrivial one $x^{(2)}(t_1)$ induced by the nonzero initial condition $x_0 \neq 0_n$. This obviously contradicts the existence and uniqueness theorem for first-order linear differential equations.

Proposition A.2.1 *Assume that in the differential game Γ,*

$$D_{ii} > 0, \; D_{ij} < 0, \; C_i < 0 \; (i, j = 1, 2, 3; \; j \neq i), \; \Lambda_{11}\Lambda_{22} < \lambda_{12}\lambda_{21}. \qquad (A.2.11)$$

Then a Pareto-maximal alternative U^P in the tri-criteria choice problem Γ_v has the form

$$U^P = \left(U_1^P, U_2^P, U_3^P \right) \div \left(u_1^P(t, x), u_2^P(t, x), u_3^P(t, x) \right) = u^P(t, x)$$
$$= \left(Q_1^P(t)x, Q_2^P(t)x, Q_3^P(t)x \right) = \left(-D_1^{-1}\Theta^P(t)x, -D_2^{-1}\Theta^P(t)x, -D_3^{-1}\Theta^P(t)x \right),$$
$$(A.2.12)$$

where

$$\Theta^P(t) = [X^{-1}(t)]' \left\{ C^{-1} + \int_t^\vartheta X^{-1}(\tau)[D_1^{-1} + D_2^{-1} + D_3^{-1}]X^{-1}(\tau)d\tau \right\}^{-1} X^{-1}(t),$$
$$(A.2.13)$$

is a continuous symmetric matrix of dimensions $n \times n$ on the time interval $[0, \vartheta]$;

$$D_i = D_{1i} + \beta D_{2i} + \gamma D_{3i} \; (i = 1, 2, 3), \qquad (A.2.14)$$

are constant symmetric matrices of dimensions $n \times n$;

$$\beta = \frac{1}{2}\left[\frac{\Lambda_{11}}{\lambda_{21}} + \frac{\lambda_{12}}{\Lambda_{22}} \right], \; \gamma = \frac{1}{2}\left[\frac{\lambda_{13}}{\Lambda_{33}} + \frac{1}{2}\left(\frac{\Lambda_{11}}{\lambda_{21}} + \frac{\lambda_{12}}{\Lambda_{22}} \right)\frac{\lambda_{23}}{\Lambda_{33}} \right]$$

are scalars, which can be also written as

$$\beta = \frac{1}{2}\left[\frac{\lambda_{12}}{\Lambda_{22}} + \frac{1}{2}\left(\frac{\lambda_{13}}{\Lambda_{33}} + \frac{\Lambda_{11}}{\lambda_{31}} \right)\frac{\lambda_{32}}{\Lambda_{22}} \right], \; \gamma = \frac{1}{2}\left[\frac{\lambda_{13}}{\Lambda_{33}} + \frac{\Lambda_{11}}{\lambda_{31}} \right];$$

Λ_{ii} is the greatest root of the characteristic equation $\det[D_{ii} - \Lambda E_n] = 0$ $(i = 1, 2, 3)$; $-\lambda_{ij}$ is the greatest root of the characteristic equation $\det[D_{ij} - \lambda E_n] = 0$ $(i, j = 1, 2, 3; j \neq i)$; E_n denotes an identity matrix of dimensions $n \times n$; finally, $X(t)$ means the fundamental matrix of the system $\dot{x} = A(t)x$, $X(\vartheta) = E_n$.

Proof We construct a Pareto-maximal strategy profile U^P using Lemma A.2.1 (more specifically, formula (A.2.4)) and the method of dynamic programming from [156, p. 112]. In view of Property A.1.2, here the application of dynamic programming reduces to two stages as follows. In the *first* stage, for the problem Γ_3 find two positive numbers β and γ, a continuously differentiable scalar function $V(t, x) = x'\Theta(t)x$, $\Theta(t) = \Theta'(t)$ $\forall t \in [0, \vartheta]$, and also three n-dimensional vector functions $u_i(t, x, V)$ $(i = 1, 2, 3)$ such that $\forall x \in \mathbb{R}^n$

$$V(\vartheta, x) = x'Cx, \quad C = C_1 + \beta C_2 + \gamma C_3; \tag{A.2.15}$$

then, using the scalar function

$$W(t, x, u_1, u_2, u_3, V) = \frac{\partial V}{\partial t} + \left[\frac{\partial V}{\partial x}\right]'(A(t)x + u_1 + u_2 + u_3) + u_1'D_1u_1 + u_2'D_2u_2$$
$$+ u_3'D_3u_3,$$

find three n-dimensional vector functions $u_i(t, x, V)$ $(i = 1, 2, 3)$ from

$$\max_{u_1 u_2 u_3} W(t, x, u_1, u_2, u_3, V) = Idem\{u_i \to u_i(t, x, V)\} \ (i = 1, 2, 3) \tag{A.2.16}$$

for any $t \in [0, \vartheta]$, $x \in \mathbb{R}^n$, and $V \in \mathbb{R}$. $\left(\frac{\partial V}{\partial x} = \text{grad}_x V.\right)$ The functions $u_i(t, x, V)$ in (A.2.16) exist under the following sufficient conditions: for all $(t, x) \in [0, \vartheta) \times \mathbb{R}^n$,

$$\frac{\partial W}{\partial u_i}\Big|_{u(t,x,V)} = \frac{\partial V}{\partial x} + 2D_iu_i(t, x, V) = 0_n \ (i = 1, 2, 3),$$

$$\frac{\partial^2 W}{\partial u_i^2} = 2D_i < 0 \ (i = 1, 2, 3), \tag{A.2.17}$$

where (as before) 0_n denotes an n-dimensional zero vector from the space $\mathbb{R}^n$ and $D_i < 0$ by Lemma A.2.1.

From (A.2.17) it follows that

$$u_i(t, x, V) = -\frac{1}{2}D_i^{-1}\frac{\partial V}{\partial x} \ (i = 1, 2, 3). \tag{A.2.18}$$

Then

$$W(t, x, u(t, x, V), V) = W[t, x, V] = \frac{\partial V}{\partial t} + \left[\frac{\partial V}{\partial x}\right]'A(t)x$$
$$- \frac{1}{4}\left(\frac{\partial V}{\partial x}\right)'\left[D_1^{-1} + D_2^{-1} + D_3^{-1}\right]\frac{\partial V}{\partial x}.$$

The *second* stage. Find the solution $V = V^P(t, x) = x'\Theta^P x$, $\Theta^P = \left[\Theta^P(t)\right]'$, of the partial differential equation

$$W[t, x, V] = 0$$

with the boundary condition

$$V(\vartheta, x) = x'Cx \ \forall x \in \mathbb{R}^n,$$

where $C = C_1 + \beta C_2 + \gamma C_3$. In other words, for all $t \in [0, \vartheta]$ and for all $x \in \mathbb{R}^n$,

$$W\left[t, x, V(t, x) = x'\Theta^P x\right] = 0, \quad V(\vartheta, x) = x'Cx \ \forall x \in \mathbb{R}^n.$$

Consequently, the symmetric matrix $\Theta^P(t)$ of dimensions $n \times n$ satisfies the Riccati matrix differential equation

$$\dot{\Theta}^P(t) + \Theta^P(t)A(t) + A(t)\Theta^P(t) - \Theta^P(t)\left[D_1^{-1} + D_2^{-1} + D_3^{-1}\right]\Theta^P(t) = 0_{n \times n},$$

$$\Theta^P(\vartheta) = C = C_1 + \beta C_2 + \gamma C_3,$$

where $0_{n \times n}$ denotes a zero matrix of dimensions $n \times n$.

As is well known, the solution $\Theta^P(t)$ of the resulting Riccati matrix differential equation has the form (A.2.13). (Here the implication

$$C_i < 0 \ (i = 1, 2, 3) \Rightarrow C_1 + \beta C_2 + \gamma C_3 < 0$$

has been taken into account.) Formula (A.2.18) in combination with the implication

$$\left[V(t, x) = x'\Theta^P(t)x\right] \Rightarrow \left[\frac{\partial V(t, x)}{\partial x} = 2\Theta^P(t)x\right]$$

finally yields (A.2.12). Thus, a Pareto-maximal alternative U^P in the tri-criteria choice problem Γ_v is given by (A.2.12)–(A.2.14).

Now, we construct the Pareto-maximal payoffs $J^P = \left(J_1\left(U^P, t_0, x_0\right), J_2\left(U^P, t_0, x_0\right),\right.$ $\left. J_3\left(U^P, t_0, x_0\right)\right) = \left(J_1^P, J_2^P, J_3^P\right)$, again using the method of dynamic programming.

Proposition A.2.2 *Let conditions (A.2.11) of Proposition A.2.1 be valid. Assume that for the differential game Γ, there are scalar continuously differentiable functions $V_i(t, x) = x'\Theta_i(t)x \ (i = 1, 2, 3)$ such that*

1) $V_i(\vartheta, x) = x'C_i x \ \forall x \in \mathbb{R}^n$;

2) *the system of three partial differential equations*

$$\frac{\partial V_i}{\partial t} + \left(\frac{\partial V_i}{\partial x}\right)'\left(N(t)x + x'\Theta^P(t)M_i(t)\Theta^P(t)x\right) = 0,$$

$$V_i(\vartheta, x) = x'C_i x \ \forall x \in \mathbb{R}^n \ (i = 1, 2, 3), \tag{A.2.19}$$

has the solution $V_i(t, x) = x'\Theta_i(t)x$, $[\Theta_i(t)]' = \Theta_i(t) \ (i = 1, 2, 3)$.

Then for any initial position $(t_0, x_0) \in [0, \vartheta] \times \mathbb{R}^n$, $x_0 \neq 0_n$,

$$J_i^P = J_i\left(U^P, t_0, x_0\right) = x_0'\Theta_i^P(t_0)x_0 \ (i = 1, 2, 3).$$

In (A.2.19),

$$N(t) = A(t) - \left(D_1^{-1} + D_2^{-1} + D_3^{-1}\right)\Theta^P(t),$$

$$M_i(t) = \Theta^P(t)\left[D_1^{-1}D_{i1}D_1^{-1} + D_2^{-1}D_{i2}D_2^{-1} + D_3^{-1}D_{i3}D_3^{-1}\right]\Theta^P(t) \ (i = 1, 2, 3),$$

are continuous matrices of dimensions $n \times n$; the matrices D_i of dimensions $n \times n$ and $\Theta^P(t)$ are given by (A.2.13), (A.2.14);

$$\Theta_i(t) = \left[Y^{-1}(t)\right]'\left\{C_i - \int_t^\vartheta Y'(\tau)\Theta^P(\tau)M_i(\tau)\Theta^P(\tau)Y(\tau)d\tau\right\}Y^{-1}(t) \ (i = 1, 2, 3)$$

$$\tag{A.2.20}$$

are symmetric matrices of dimensions $n \times n$; finally, $Y(t)$ denotes the fundamental matrix of the homogeneous system $\dot{y} = N(t)y$, $Y(\vartheta) = E_n$.

Proof We construct three scalar functions

$$W_i[t, x, V_i] = \frac{\partial V_i}{\partial t} + \left[\frac{\partial V_i}{\partial x}\right]' \left(N(t)x + \left[u_1^P(t, x)\right]' D_{i1} u_1^P(t, x)\right.$$

$$+ \left[u_2^P(t, x)\right]' D_{i2} u_2^P(t, x) + \left[u_3^P(t, x)\right]' D_{i3} u_3^P(t, x) \ (i = 1, 2, 3),$$

$$(A.2.21)$$

where $u_i^P(t, x)$ are the n-dimensional vector functions determined by (A.2.12).

Find the solution $V_i(t, x)$ $(i = 1, 2, 3)$ of the system of three partial differential equations

$$W_i[t, x, V_i] = 0, \ V_i(\vartheta, x) = x' C_i x \ \forall x \in \mathbb{R}^n \ (i = 1, 2, 3) \tag{A.2.22}$$

as the quadratic form $V_i(t, x) = x' \Theta_i(t) x$, $[\Theta_i(t)]' = \Theta_i(t)$ $(i = 1, 2, 3)$.

Let us establish two facts as follows.

First, the solution of the system (A.2.21), (A.2.22) has the property

$$V_i(t_0, x_0) = J_i(U^P, t_0, x_0) \ (i = 1, 2, 3), \tag{A.2.23}$$

where the strategy profile $U^P = (U_1^P, U_2^P, U_3^P)$ has the form (A.2.12). Really, if U^P is a strategy profile from (A.2.12)–(A.2.14), then by (A.2.21) and (A.2.22) the solution $x^P(t)$ of the system $\dot{x} = N(t)x$, $x(t_0) = x_0 \neq 0_n$, for $x = x^P(t)$ will be

$$0 = W_i[t, x^P(t), V_i(t, x^P(t))] = \frac{\partial V_i(t, x^P(t))}{\partial t} + \left[\frac{\partial V_i(t, x^P(t))}{\partial x}\right]' N(t) x^P(t)$$

$$+ \sum_{j=1}^{3} \left[u_j^P(t, x^P(t))\right]' D_{ij} u_j^P(t, x^P(t)) = \bar{W}_i[t] \ \forall t \in [t_0, \vartheta] \ (i = 1, 2, 3).$$

Integrating both sides of this identity from t_0 to ϑ subject to the boundary conditions from (A.2.22) yields

$$0 = \int_{t_0}^{\vartheta} \bar{W}_i[t] dt = \int_{t_0}^{\vartheta} \frac{dV_i^P(t, x^P(t))}{dt} dt + \int_{t_0}^{\vartheta} \sum_{j=1}^{3} \left[u_j^P(t, x^P(t))\right]' D_{ij} u_j^P\left(t, x^P(t)\right) dt$$

$$= V_i^P(\vartheta, x^P(\vartheta)) - V_i^P(t_0, x^P(t_0)) + \int_{t_0}^{\vartheta} \sum_{j=1}^{3} \left[u_j^P(t, x^P(t))\right]' D_{ij} u_j^P\left(t, x^P(t)\right) dt$$

$$= x'(\vartheta) C_i x(\vartheta) + \int_{t_0}^{\vartheta} \sum_{j=1}^{3} \left[u_j^P(t, x^P(t))\right]' D_{ij} u_j^P\left(t, x^P(t)\right) dt - V_i^P\left(t_0, x^P(t_0)\right)$$

$$= J_i(U^P, t_0, x_0) - V_i^P(t_0, x^P(t_0)) \ (i = 1, 2, 3),$$

and the requisite equality (A.2.23) is immediate.

Second, the solution $V_i(t, x)$ $(i = 1, 2, 3)$ of the system (A.2.22) has the form $V_i(t, x) = x' \Theta_i(t) x$, where a symmetric matrix $\Theta_i(t)$ of dimensions $n \times n$ is given by (A.2.20). Really, substituting $V_i(t, x) = x' \Theta_i(t) x$ into (A.2.22) leads to (A.2.23) if $\Theta_i(t)$ $(i = 1, 2, 3)$ is the solution of the matrix linear inhomogeneous differential equation

$$\dot{\Theta}_i + \Theta_i N + N \Theta_i + \Theta^P(t) M_i \Theta^P(t) = 0_{n \times n}, \ \Theta_i(\vartheta) = C_i \ (i = 1, 2, 3). \tag{A.2.24}$$

A direct substitution of (A.2.20) into equation (A.2.24) shows that this symmetric matrix $\Theta_i(t)$ of dimensions $n \times n$ is really the requisite solution. The proof of Proposition A.2.2 is complete.

Remark A.2.2 Propositions $A.2.1$ and $A.2.2$ considered together finally yield the following result concerning the Pareto-maximal solution $(U^P, J^P) \in \mathcal{U} \times \mathbb{R}^3$ of the game Γ.

Assume that in the differential game Γ:

1) the symmetric constant matrices D_{ij} and C_i of dimensions $n \times n$ are such that

$$D_{ii} > 0, \ D_{ij} < 0, \ C_i < 0 \ (i, j = 1, 2, 3; j \neq i);$$

2) $[\Lambda_{11} \Lambda_{22} < \lambda_{12} \lambda_{21}]$.

Then for all $(t_0, x_0) \in [0, \vartheta) \times \mathbb{R}^n$, $x_0 \neq 0_n$,

$$U^P \div u^P(t, x) = \left(-D_1^{-1} \Theta^P(t)x, -D_2^{-1} \Theta^P(t)x, -D_3^{-1} \Theta^P(t)x \right),$$

$$J^P = \left(J_1^P, J_2^P, J_3^P \right), \ J_i^P = x_0' \Theta_i(t_0) x_0 \ (i = 1, 2, 3),$$

where

$$\Theta^P(t) = \left[X^{-1}(t) \right]' \left\{ C^{-1} + \int_t^\vartheta X^{-1}(\tau) \left[D_1^{-1} + D_2^{-1} + D_3^{-1} \right] X^{-1}(\tau) d\tau \right\}^{-1} X^{-1}(t),$$

$$\Theta_i(t) = \left[Y^{-1}(t) \right]' \left\{ C_i - \int_t^\vartheta Y'(\tau) \Theta^P(\tau) M_i(\tau) \Theta^P(\tau) Y(\tau) d\tau \right\} Y^{-1}(t),$$

are symmetric matrices of dimensions $n \times n$; the matrices $X(t)$ and $Y(t)$ of dimensions $n \times n$ are the fundamental matrices of the systems $\dot{x} = A(t)x$, $X(\vartheta) = E_n$, and $\dot{y} = N(t)y$, $Y(\vartheta) = E_n$, respectively;

$$C = C_1 + \beta C_2 + \gamma C_3, \ D_i = D_{i1} + \beta D_{i2} + \gamma D_{i3},$$

$$N(t) = A(t) - \left(D_1^{-1} + D_2^{-1} + D_3^{-1} \right) \Theta^P(t),$$

$$M_i(t) = \Theta^P(t) \left[D_1^{-1} D_{i1} D_1^{-1} + D_2^{-1} D_{i2} D_2^{-1} + D_3^{-1} D_{i3} D_3^{-1} \right] \Theta^P(t),$$

$$\beta = \frac{1}{2} \left[\frac{\Lambda_{11}}{\lambda_{21}} + \frac{\lambda_{12}}{\Lambda_{22}} \right], \ \gamma = \frac{1}{2} \left[\frac{\lambda_{13}}{\Lambda_{33}} + \frac{1}{2} \left(\frac{\Lambda_{11}}{\lambda_{21}} + \frac{\lambda_{12}}{\Lambda_{22}} \right) \frac{\lambda_{23}}{\Lambda_{33}} \right],$$

finally, Λ_{ii} and $-\lambda_{ij}$ are the greatest roots of the characteristic equations $\det[D_{ii} - \Lambda E_n] = 0$ and $\det[D_{ij} - \lambda E_n] = 0$, respectively $(i, j = 1, 2, 3; j \neq i.)$

A.2.4 LEMMAS OF MAJORANTS

Now, consider some results that:

1) can be used to establish directly the absence of Nash equilibria in the differential game Γ (of course, under conditions (A.2.11));

2) implement the concept of equilibrium in threats and counter-threats for the differential game Γ.

What is important? These results involve the definiteness of quadratic forms figuring in the integral terms of the payoff functions (A.2.3).

Without special mention, from this point onwards assume that conditions (A.2.11) are satisfied. Hence, there exists a Pareto-maximal alternative

$$U^P = \left(U_1^P, U_2^P, U_3^P \right) \div \left(u_1^P(t, x), u_2^P(t, x), u_3^P(t, x) \right) = u^P(t, x)$$

$$= \left(Q_1^P(t)x, Q_2^P(t)x, Q_3^P(t)x \right) = \left(-D_1^{-1}\Theta^P(t)x, -D_2^{-1}\Theta^P(t)x, -D_3^{-1}\Theta^P(t)x \right)$$

in the tri-criteria dynamic choice problem Γ_v.

Lemma A.2.3 *Let the payoff function (A.2.3) be such that $D_{11} > 0$. Then for a Pareto-maximal strategy profile U^P in the game Γ there exists a constant $\alpha^{(1)}(U^P, t_0, x_0) > 0$ such that, for all $\alpha \geq \alpha^{(1)}(U^P, t_0, x_0) > 0$ and the strategy $U_1^T \div \alpha x$ of player 1, the inequality*

$$J_1 \left(U_1^T, U_2^P, U_3^P, t_0, x_0 \right) > J_1 \left(U_1^P, U_2^P, U_3^P, t_0, x_0 \right) \tag{A.2.25}$$

holds for any initial positions $(t_0, x_0) \in [0, \vartheta) \times [\mathbb{R}^n \setminus \{0_n\}]$.

Proof According to Proposition A.2.2, there exists a Bellman function $V_1(t, x) = x'\Theta_1(t)x$ such that

$$J_1 \left(U^P, t_0, x_0 \right) = V_1 (t_0, x_0) = x_0'\Theta_1(t_0)x_0,$$

where the symmetric matrix $\Theta_1(t)$ of dimensions $n \times n$ is continuous on $[0, \vartheta)$ and has the form (A.2.20) $(i = 1)$.

Consider the strategy $U_1^T \div u_1^T(t, x) = \alpha x$ of player 1, in which the numerical parameter $\alpha > 0$ will be determined below. Due to the symmetry of the matrix D_{11} and $D_{11} > 0$,

$$u_1'D_{11}u_1 \geq \lambda_1\|u_1\|^2 = \lambda_1 u_1'u_1 \quad \forall u_1 \in \mathbb{R}^n, \tag{A.2.26}$$

where $\|\cdot\|$ denotes the Euclidean norm and $\lambda_1 > 0$ is the smallest root of the characteristic equation $\det[D_{11} - \lambda E_n] = 0$; see [38, p. 89].

We take the symmetric matrix $\Theta^P(t)$ of dimensions $n \times n$ from (A.2.13)–(A.2.15) and also the strategies $U_2^P \div Q_2^P(t)x$ and $U_3^P \div Q_3^P(t)x$ of players 2 and 3, respectively, from (A.2.12). We introduce the scalar function

$$W_1[t, x] = W_1 \left(t, x, u_1^T(t, x) = \alpha x, u_2^P(t, x) = Q_2^P(t)x, u_3^P(t, x) \right.$$

$$= Q_3^P(t)x, V_1(t, x) = x'\Theta_1(t)x \right) = \frac{\partial V_1(t, x)}{\partial t} + \left[\frac{\partial V_1(t, x)}{\partial x} \right]' \left(A(t)x + u_1^T(t, x) \right.$$

$$+ u_2^P(t, x) + u_3^P(t, x) \right) + \left[u_1^T(t, x) \right]' D_1 u_1^T(t, x) + \left[u_2^P(t, x) \right]' D_2 u_2^P(t, x)$$

$$+ \left[u_3^P(t, x) \right]' D_3 u_3^P(t, x) \geq x' \frac{d\Theta_1(t)}{dt}x + 2x'\Theta_1(t) \left[A(t) + \alpha E_n + Q_2^P(t) + Q_3^P(t) \right] x$$

$$+ x' \left(\lambda_1 \alpha^2 E_n \right) x + x' \left[Q_2^P(t) \right]' D_{12}Q_2^P(t)x + x' \left[Q_3^P(t) \right]' D_{13}Q_3^P(t)x$$

$$= x' \left\{ \frac{d\Theta_1(t)}{dt} + \Theta_1(t) \left[A(t) + \alpha E_n + Q_2^P(t) + Q_3^P(t) \right] + \left[A'(t) + \alpha E_n + \left(Q_2^P(t) \right)' \right. \right.$$

$$\left. + \left(Q_3^P(t) \right)' \right] \Theta_1(t) + \lambda_1 \alpha^2 E_n + \left[Q_2^P(t) \right]' D_{12} Q_2^P(t) + \left[Q_3^P(t) \right]' D_{13} Q_3^P(t) \right\} x$$

$$= x' M_1(t, \alpha) x.$$

The matrix $M_1(t, \alpha)$ in curly brackets is symmetric and has the form

$$M_1(t, \alpha) = \lambda_1 \alpha^2 E_n + 2\alpha \Theta_1(t) + K_1(t),$$

where

$$K_1(t) = \dot{\Theta}_1(t) + \Theta_1(t) \left[A(t) + Q_2^P(t) + Q_3^P(t) \right] + \left[Q_2^P(t) \right]' D_{12} Q_2^P(t)$$

$$+ \left[Q_3^P(t) \right]' D_{13} Q_3^P(t) + \left[A'(t) + \left(Q_2^P(t) \right)' + \left(Q_3^P(t) \right)' \right] \Theta_1(t)$$

is a symmetric and continuous matrix of dimensions $n \times n$.

The elements of the matrices $\Theta_1(t)$ and $K_1(t)$ are continuous on $[0, \vartheta]$ and hence uniformly bounded on the compact set $[0, \vartheta]$. The factor α^2 enters only the diagonal elements of the matrix $M_1(t, \alpha)$. Recall that $\lambda_1 > 0$ is the smallest root of the characteristic equation $\det[D_{11} - \lambda E_n] = 0$, and E_n denotes an identity matrix of dimensions $n \times n$. Therefore, the constant $\alpha = \alpha^{(1)}(U^P, t_0, x_0) > 0$ can be chosen sufficiently great for making all principal minors of the matrix $M_1(t, \alpha)$ positive for all $t \in [0, \vartheta]$ and for all $\alpha \geq \alpha^{(1)}(U^P, t_0, x_0)$. (For a complete presentation, this fact will be proved at the end of this section, just prior to Remark A.2.3.) Then, according to Lemma A.2.2 and [38, p. 88], the quadratic form $x' M_1(t, \alpha) x$ is positive definite for all $t \in [0, \vartheta]$ and constants $\alpha \geq \alpha^{(1)}(U^P, t_0, x_0)$.

Now, we show the existence of a constant $\alpha^{(1)}(U^P, t_0, x_0) > 0$ such that, for all $\alpha \geq \alpha^{(1)}(U^P, t_0, x_0)$, the quadratic form $x' M_1(t, \alpha) x$ is positive definite for all $t \in [0, \vartheta]$ and $x \in \mathbb{R}^n$. Note that the matrix $M_1(t, \alpha)$ of dimensions $n \times n$ is symmetric. By Sylvester's criterion the quadratic form $x' M_1(t, \alpha) x$ is positive definite if all principal minors Δ_r $(r = 1, \ldots, n)$ of the matrix $M_1(t, \alpha)$ are positive. The minors Δ_r are located in the first r rows and first r columns of the matrix $M_1(t, \alpha)$ $(r = 1, \ldots, n)$:

$$\Delta_r(t, \alpha) = \begin{vmatrix} \lambda_1 \alpha^2 n + \alpha l_{11}(t) + k_{11}(t) & \ldots & \alpha l_{1r}(t) + k_{1r}(t) \\ \vdots & \ddots & \vdots \\ \alpha l_{r1}(t) + k_{r1}(t) & \ldots & \lambda_1 \alpha^2 n + \alpha l_{rr}(t) + k_{rr}(t) \end{vmatrix}.$$

They must be positive for all $t \in [0, \vartheta]$ and for all $\alpha \geq \alpha^{(1)}(U^P, t_0, x_0) > 0$. Expanding the determinants $\Delta_r(t, \alpha)$ and rearranging the terms in the descending order of the power of the parameter α give

$$\Delta_r(t, \alpha) = a_0 \alpha^{2r} + a_1(t) \alpha^{2r-1} + \ldots + a_{2r-1}(t) \alpha + a_{2r}(t),$$

where $a_0 = \lambda_1^r n^r = const > 0$ and the other coefficients $a_1(t), \ldots, a_{2r}(t)$ are continuous on the compact set $[0, \vartheta]$, hence being uniformly bounded. This uniform boundedness guarantees the existence of a constant $\Omega_r > 0$ such that

$$\max_{0 \leq t \leq \vartheta} \left\{ a_p(t) | \, p = 0, 1, \ldots, 2r \right\} < \Omega_r.$$

Let us demonstrate that if

$$\alpha > \frac{\Omega_r}{|a_0|} + 1 = \alpha^{(1)}(U^P, t_0, x_0),$$

then

$$|a_1(t)\alpha^{2r-1} + a_2(t)\alpha^{2r-2} + \ldots + a_{2r-1}(t)\alpha + a_{2r}(t)| < |a_0\alpha^{2r}|.$$

In other words, for a sufficiently large $|\alpha|$ the sign of this polynomial is determined by the sign of its leading coefficient. Really,

$$\left|a_1(t)\alpha^{2r-1} + a_2(t)\alpha^{2r-2} + \ldots + a_{2r-1}(t)\alpha + a_{2r}(t)\right| \leq |a_1(t)|\,\alpha^{2r-1} + |a_2(t)|\,\alpha^{2r-2}$$

$$+ \ldots + |a_{2r-1}(t)|\,\alpha + |a_{2r}(t)| \leq \Omega_r\left(\alpha^{2r-1} + \alpha^{2r-2} + \ldots + \alpha + 1\right) = \Omega_r\frac{\alpha^{2r} - 1}{\alpha - 1}.$$

In addition,

$$\left[\alpha > \frac{\Omega_r}{a_0} + 1\right] \Rightarrow [\Omega_r < a_0(\alpha - 1)].$$

Replacing Ω_r in this inequality by a greater value $a_0(\alpha - 1)$ yields

$$\left|a_1(t)\alpha^{2r-1} + a_2(t)\alpha^{2r-2} + \ldots + a_{2r}(t)\right| < a_0(\alpha - 1)\frac{\alpha^{2r} - 1}{\alpha - 1} = a_0(\alpha^{2r} - 1) < a_0\alpha^{2r}.$$

Thus, for all $\alpha \geq \Omega_r = \alpha^{(r)}(U, t_0, x_0) > 0$ and for all $t \in [0, \vartheta]$,

$$\left|a_1(t)\alpha^{2r-1} + a_2(t)\alpha^{2r-2} + \ldots + a_{2r}(t)\right| < a_0\alpha^{2r},$$

meaning that for a sufficiently large α the sign of the polynomial $\Delta_r(t, \alpha)$ is determined by the sign of its leading coefficient. Finally, for each $r = 1, \ldots, n$ calculate $\Omega_r > 0$ and let $\alpha^{(1)}(U^P, t_0, x_0) = \max_{r=1,\ldots,n} \Omega_r$.

Then, for $\alpha^{(1)} = \alpha^{(1)}(U^P, t_0, x_0)$ it follows that

$$\tilde{W}_1[t, x] = x'M_1(t, \alpha^{(1)})x > 0 \quad \forall t \in [0, \vartheta], \ \forall x \in \mathbb{R}^n \setminus \{0_n\}. \tag{A.2.27}$$

Denote by $\tilde{x}(t)$, $t \in [0, \vartheta]$, the solution of the vector differential equation

$$\dot{x} = A(t)x + \alpha^{(1)}x + Q_2^P(t)x + Q_3^P(t)x, \ x(t_0) = x_0 \neq 0_n.$$

Since $[x_0 \neq 0_n] \Rightarrow (\tilde{x}(t) \neq 0_n \ \forall t \in [0, \vartheta])$ (see Lemma A.2.2),

$$\tilde{W}_1[t, \tilde{x}(t)] > 0 \quad \forall t \in [0, \vartheta].$$

Hence, integrating both sides of this inequality from t_0 to ϑ subject to the boundary condition $\Theta_1(\vartheta) = C_1$ and taking into account $u_1^T[t] = \alpha^{(1)}\tilde{x}(t)$ give

$$0 = \int_{t_0}^{\vartheta} \tilde{W}_1[t, \tilde{x}(t)]dt = \int_{t_0}^{\vartheta} \left\{ \frac{\partial V_1(t, x)}{\partial t} + \left[\frac{\partial V_1(t, x)}{\partial x} \right]' [A(t)x + \alpha^{(1)} E_n x \right.$$

$$+ \left. Q_2^P(t)x + Q_3^P(t)x] \right\}_{x=\tilde{x}(t)} dt + \int_{t_0}^{\vartheta} \left\{ (\alpha^{(1)})^2 x' D_{11} x + x' [Q_2^P(t)]' D_{12} Q_2^P(t)x \right.$$

$$+ \left. x'[Q_3^P(t)]' D_{13} Q_3^P(t)x \right\}_{x=\tilde{x}(t)} dt = \int_{t_0}^{\vartheta} \frac{dV_1(t, \tilde{x}(t))}{dt} dt + \int_{t_0}^{\vartheta} \sum_{j=1}^{3} [u_j^T[t]]' D_{1j} u_j^T[t]dt$$

$$= [\tilde{x}(\vartheta)]' C_1 \tilde{x}(\vartheta) + \int_{t_0}^{\vartheta} \sum_{j=1}^{3} [u_j^T[t]]' D_{1j} u_j^T[t]dt - V_1(t_0, x_0)$$

$$= J_1(U_1^T, U_2^P, U_3^P, t_0, x_0) - V_1(t_0, x_0).$$

This result in combination with the equality $J_1(U_1^P, U_2^P, U_3^P, t_0, x_0) = V_1(t_0, x_0)$ finally proves Lemma A.2.3.

Remark A.2.3 Consider the inner optimization problem in the game Γ: for fixed strategies $U_2^P \in \mathcal{U}_2$ and $U_3^P \in \mathcal{U}_3$ of players 2 and 3, respectively, and for any $(t_0, x_0) \in [0, \vartheta) \times [\mathbb{R}^n \setminus \{0_n\}]$, find $\max_{U_1 \in \mathcal{U}_1} J_1(U_1, U_2^P, U_3^P, t_0, x_0)$ subject to the constraint (A.2.1). As a matter of fact, Lemma A.2.3 claims that, for $D_{11} > 0$ and $x_0 \neq 0_n$, this maximization problem has no solution. Really, whatever strategy $U_1 \in \mathcal{U}_1$ is chosen by player 1, there always exists another strategy $\tilde{U}_1 \in \mathcal{U}_1$ of this player such that

$$J_1\left(\tilde{U}_1, U_2^P, U_3^P, t_0, x_0 \right) > J_1\left(U_1, U_2^P, U_3^P, t_0, x_0 \right) \quad \forall (t_0, x_0) \in [0, \vartheta) \times [\mathbb{R}^n \setminus \{0_n\}].$$

When choosing an appropriate solution of the game Γ, this result allows eliminating directly those concepts of equilibrium that involve the maximization of the payoff function of player 1. (For example, if $D_{11} > 0$, the concept of Nash equilibrium should not be used as the solution of the game Γ.)

Thus, under conditions (A.2.11) the differential game Γ *has no Nash equilibrium* $U^e \in \mathcal{U}$. At the same time, the strategy $U_1^T \div \alpha x$, $\forall \alpha \geq \alpha^{(1)} = \alpha^{(1)}(U^P, t_0, x_0)$, implements the threat of player 1 to the Pareto-maximal strategy profile U^P; see (A.2.5). In the lemmas below, the initial position (t_0, x_0) is fixed and coincides with the one from Lemma A.2.3; in addition, the threatening strategy $U_1^T \div \alpha x$ of player 1 has a constant scalar $\alpha = \alpha^{(1)}$. Recall that conditions (A.2.11) are assumed to hold without special mention.

In fact, Lemma A.2.3 establishes the following result.

Proposition A.2.3 *Assume that in the game Γ at least one of the constant symmetric matrices $D_{ii}(i = 1, 2, 3)$ of dimensions $n \times n$ is positive definite. Then this game has no Nash equilibrium, i.e., there does not exist a strategy $U_i^e \in \mathcal{U}_i$ satisfying the corresponding requirement from Definition A.2.2.*

Note that, first, the condition $D_{ii} > 0$ with a frozen number $i \in \{1, 2, 3\}$ breaks only the ith equality from Definition A.2.2. This *is enough for the absence of a Nash equilibrium* U^e in the game Γ. If $D_{ii} > 0$ for all $i = 1, 2, 3$, then *the three* equalities from Definition A.2.2 will be violated.

Second, the equivalence

$$D > 0 \Leftrightarrow -D < 0$$

is obvious. (Here $-D$ means that all elements of the matrix D are multiplied by -1.)

Then Lemma A.2.3 also implies the following.

Lemma A.2.4 *Let the payoff function (A.2.3) be such that $D_{12} < 0$. Then there exists a constant $\alpha^{(2)} = \alpha^{(2)}(U^P, U_1^T, t_0, x_0) > 0$ such that, for all $\alpha \geq \alpha^{(2)}$ and the strategy $U_2^C \div \alpha x$ of player 2,*

$$J_1\left(U_1^T, U_2^C, U_3^P, t_0, x_0\right) < J_1\left(U^P, t_0, x_0\right). \tag{A.2.28}$$

In other words, in the game Γ the strategy $U_2^C \div \alpha x \; \forall \alpha \geq \alpha^{(2)}$ implements an incomplete counter-threat in response to the threat U_1^T of player 1.

Proof With some obvious modifications, the proof is immediate from Lemma A.2.3. For complete presentation, it will be given below, following the same stages as in the proof of Lemma A.2.3. More specifically, in the *first* stage, we construct a Bellman function $\bar{V}_1(t, x) = x'\bar{\Theta}_1(t)x$, $\bar{\Theta}_1(t) = \bar{\Theta}'_1(t)$, such that

$$J_1\left(U_1^T, U_2^P, U_3^P, t_0, x_0\right) = \bar{V}_1(t_0, x_0). \tag{A.2.29}$$

In the *second* stage, we verify (A.2.28) for the strategy $U_2^C \div \alpha x$ of player 2 for all $\alpha \geq \alpha^{(2)}$.

The first stage Consider the scalar Bellman function (dictated by the method of dynamic programming)

$$\begin{aligned}
\bar{W}_1[t, x, \bar{V}_1] &= W_1(t, x, u_1^T(t, x) = \alpha^{(1)}x, u_2^P(t, x) \\
&= -D_2^{-1}\Theta^P(t)x, u_3^P(t, x) = -D_3^{-1}\Theta^P(t)x, \bar{V}_1) \\
&= \frac{\partial \bar{V}_1}{\partial t} + \left[\frac{\partial \bar{V}_1}{\partial x}\right]'\left[A(t) + \alpha^{(1)}E_n - \left(D_2^{-1} + D_3^{-1}\right)\Theta^P(t)\right]x \\
&\quad + \left(\alpha^{(1)}\right)^2 x'D_{11}x + (u_2^P(t, x))'D_{12}u_2^P(t, x) + (u_3^P(t, x))'D_{13}u_3^P(t, x).
\end{aligned} \tag{A.2.30}$$

Then $\bar{W}_1[t, x, \bar{V}_1 = x'\Theta_1^C(t)x] = 0$ if $\Theta_1^C(t)$ is the solution of the matrix linear inhomogeneous differential equation

$$\begin{aligned}
&\dot{\Theta}_1^C(t) + \Theta_1^C\left[A(t) + \alpha^{(1)}E_n - \left(D_2^{-1} + D_3^{-1}\right)\Theta^P(t)\right] + \left[A'(t) + \alpha^{(1)}E_n\right. \\
&\quad \left. - \Theta^P(t)\left(D_2^{-1} + D_3^{-1}\right)\right]\Theta_1^C + \left(\alpha^{(1)}\right)^2 x'D_{11} + \Theta^P(t)\left[D_2^{-1}D_{12}D_2^{-1}\right. \\
&\quad \left. + D_3^{-1}D_{13}D_3^{-1}\right]\Theta^P(t) = 0_{n \times n}, \left[\Theta_1^C(\vartheta) = C_1\right] \Leftarrow \left[x'\Theta_1^C(\vartheta)x = x'C_1x \; \forall x \in \mathbb{R}^n\right].
\end{aligned}$$

The solution $\Theta_1^C(t)$ of this equation exists, is unique and extendable leftwards to $[t_0, \vartheta]$. Now, let the n-dimensional vector function $x^C(t)$ be the solution of the homogeneous linear differential equation

$$\dot{x} = \left[A(t) + \alpha^{(1)}E_n - \left(D_2^{-1} + D_3^{-1}\right)\Theta^P(t)\right]x, \; x(t_0) = x_0.$$

In view of (A.2.30), for all $t \in [t_0, \vartheta]$ it follows that

$$\bar{W}_1\left[t, x = x^C(t), \bar{V}_1(t, x^C(t))\right] = \left(x^C(t)\right)' \Theta_1^C(t) x^C(t))] = \frac{d\bar{V}_1(t, x^C(t))}{dt}$$

$$+ \left[u_1^C(t, x^C(t))\right]' D_{11} u_1^C(t, x^C(t)) + \left[u_2^P(t, x^C(t))\right]' D_{12} u_2^P(t, x^C(t))$$

$$+ \left[u_3^P(t, x^C(t))\right]' D_{13} u_3^P(t, x^C(t)) = \bar{W}_1^C[t] = 0.$$

Integrating both parts of this equality from t_0 to ϑ gives

$$u_1^C[t] = \alpha^{(1)} x^C(t), \ u_j^P[t] = -D_j^{-1} \Theta^P(t) x^C(t) \ (j = 2, 3),$$

$$\bar{V}_1\left(\vartheta, x^C(\vartheta)\right) - \bar{V}_1(t_0, x_0) + \int_{t_0}^{\vartheta} \left\{\left(u_1^C[t]\right)' D_{11} u_1^C[t] + (u_2^P[t])' D_{12} u_2^P[t]\right.$$

$$+ \left(u_3^P[t]\right)' D_{13} u_3^P[t]\bigg\} dt = J_1\left(U_1^C, U_2^P, U_3^P, t_0, x_0\right) - \bar{V}_1(t_0, x_0) = 0,$$

which finally proves (A.2.29).

The second stage Due to the symmetry of the negative definite constant matrix D_{12} of dimensions $n \times n$, there exists $\lambda_{12} = const > 0$ such that

$$u_2' D_{12} u_2 \le -\lambda_{12} u_2' u_2 = -\lambda_{12} \|u_2\|^2, \tag{A.2.31}$$

where $-\lambda_{12}$ is the greatest root of the characteristic equation $\det[D_{12} - \lambda E_n] = 0$. In view of (A.2.30), get back to the scalar function $\bar{W}_1[t, x, \bar{V}_1 = x' \Theta_1^C(t) x] = \overline{\overline{W}}_1[t, x]$, where by (A.2.33)

$$\overline{\overline{W}}_1[t, x] = \frac{\partial \bar{V}_1(t, x)}{\partial t} + \left(\frac{\partial \bar{V}_1(t, x)}{\partial x}\right)' \left[A(t) + \alpha^{(1)} E_n + \alpha E_n - D_3^{-1} \Theta^P(t)\right] x$$

$$+ \left(\alpha^{(1)}\right)^2 x' D_{11} x + \alpha^2 x' D_{12} x + x' \Theta^P(t) D_3^{-1} D_{13} D_3^{-1} \Theta^P(t) x$$

$$\ge -\lambda_{12} \alpha^2 x' x + 2x' \Theta^C(t) \left[A(t) + \alpha^{(1)} E_n + \alpha E_n - D_3^{-1} \Theta^P(t)\right] x$$

$$+ x' \Theta^P(t) D_3^{-1} D_{13} D_3^{-1} \Theta^P(t) x + \left(\alpha^{(1)}\right)^2 x' D_{11} x = x' L(t, \alpha) x, \ L(t, \alpha)$$

$$= L'(t, \alpha),$$

and $L(t, \alpha) = -\lambda_{12} \alpha^2 E_n + 2\alpha \Theta_1^C(t) + \Xi_1(t)$.

The elements of the matrix $L(t, \alpha)$ are continuous on the compact set $[0, \vartheta]$ and hence uniformly bounded on it. The factor α^2 enters only the diagonal elements of the matrix $L(t, \alpha)$. Then there exists a sufficiently great positive number $\alpha^{(2)}$ such that, for all $\alpha \ge \alpha^{(2)}$, all principal odd-order (even-order) minors of the matrix $L(t, \alpha)$ are negative (positive, respectively). According to [38, p. 88], the quadratic form $x'L(t, \alpha)x$ is negative definite for all $t \in [t_0, \vartheta]$ and constants $\alpha \ge \alpha^{(2)}$.

We take again the scalar function $\overline{\overline{W}}_1[t, x]$, replacing x by the solution $x^C(t)$ of the homogeneous vector differential equation

$$\dot{x}^C = \left[A(t) + \alpha^{(1)} E_n + \alpha E_n - D_3^{-1} \Theta^P(t)\right] x^C, \ x^C(t_0) = x_0. \tag{A.2.32}$$

According to Lemma A.2.2, $x^C(t) \neq 0_n$ for all $t \in [t_0, \vartheta]$. Hence, for $\alpha \geq \alpha^{(2)}$,

$$\overline{W}_1\left[t, x^C(t)\right] = \left[x^C(t)\right]' L(t, \alpha) x^C(t) < 0, \quad t \in [t_0, \vartheta].$$

Integrating both parts of this strict inequality from t_0 to ϑ and using $u_1^T(t, x) = \alpha^{(1)} x$, $\bar{V}_1(\vartheta, x) = x' C_1 x$, and $x = x^C(t)$ yield

$$0 > \int_{t_0}^{\vartheta} \overline{W}_1[t, x^C(t)] dt = \int_{t_0}^{\vartheta} \left\{ \frac{\partial \bar{V}_1(t, x^C(t))}{\partial t} + \left[\frac{\partial \bar{V}_1(t, x^C(t))}{\partial x} \right]' \left[A(t) + (\alpha^{(1)} + \alpha) E_n \right. \right.$$

$$\left. - D_3^{-1} \Theta^P(t) \right] x^C(t) + \left[u_1^T(t, x^C(t)) \right]' D_{11} u_1^T \left(t, x^C(t) \right) + \alpha^2 \left(x^C(t) \right)' D_{12} x^C(t)$$

$$+ \left(x^C(t) \right)' \Theta^P(t) D_3^{-1} D_{13} D_3^{-1} \Theta^P(t) x^C(t) \Big\} dt = \int_{t_0}^{\vartheta} \frac{d\bar{V}_1}{dt} dt$$

$$+ \int_{t_0}^{\vartheta} \left\{ \left[u_1^T(t, x^C(t)) \right]' D_{11} u_1^T \left(t, x^C(t) \right) + \alpha^2 \left(x^C(t) \right)' D_{12} x^C(t) \right.$$

$$+ \left(x^C(t) \right)' \Theta^P(t) D_3^{-1} D_{13} D_3^{-1} \Theta^P(t) x^C(t) \Big\} dt = \bar{V}_1 \left(\vartheta, x^C(\vartheta) \right) - \bar{V}_1(t_0, x_0)$$

$$+ \int_{t_0}^{\vartheta} \left\{ u_1^T[t] D_{11} u_1^T[t] + \alpha^2 \left(x^C(t) \right)' D_{12} x^C(t) \right.$$

$$+ \left(x^C(t) \right)' \Theta^P(t) D_3^{-1} D_{13} D_3^{-1} \Theta^P(t) x^C(t) \Big\} dt = J_1 \left(U_1^T, U_2^C, U_3^P, t_0, x_0 \right)$$

$$- \bar{V}_1(t_0, x_0).$$

Due to (A.2.29), this directly leads to (A.2.28).

The two assertions below (Lemmas A.2.5 and A.2.6) can be established by analogy with Lemmas A.2.3 and A.2.4. Recall that an initial position (t_0, x_0), a continuous matrix $\Theta^P(t)$ of dimensions $n \times n$, and an incomplete threat strategy $U_2^C \div \alpha^{(2)} x$ figuring in Lemmas A.2.3 and A.2.4 are assumed to be frozen and conditions (A.2.29) are assumed to hold.

Lemma A.2.5 *The condition* $D_{22} > 0$ *implies the existence of a value* $\alpha^{(3)} = \alpha^{(2)}(U^P, U_1^T, t_0, x_0) = const > 0$ *such that, for all* $\alpha \geq \alpha^{(3)}$ *and the strategy* $U_2^C \div \alpha x$,

$$J_2 \left(U_1^T, U_2^C, U_3^P, t_0, x_0 \right) > J_2 \left(U_1^T, U_2^P, U_3^P, t_0, x_0 \right). \tag{A.2.33}$$

In other words, the strategy $U_2^C \div (\max\{\alpha^{(2)}, \alpha^{(3)}\}) x$ *of player 2 implements a complete counter-threat (jointly with* $U_2^C \div \alpha^{(2)} x$*) in response to the threat of player 1 to* U^P.

Proof First, consider an incomplete counter-threat U_1^T of player 1 to U^P such that

$$J_1 \left(U_1^T, U_2^P, U_3^P, t_0, x_0 \right) > J_1 \left(U_1^P, U_2^P, U_3^P, t_0, x_0 \right), \tag{A.2.34}$$

Due to the Pareto maximality of U^P and Property A.2.1,

$$J_2 \left(U_1^T, U_2^P, U_3^P, t_0, x_0 \right) < J_2 \left(U_1^P, U_2^P, U_3^P, t_0, x_0 \right). \tag{A.2.35}$$

Next, by analogy with the proof of Lemma A.2.3, first we demonstrate the existence of a Bellman function $V_2(t, x) = x'\Theta_2(t)x$ for which

$$J_2\left(U_1^C, U_2^P, U_3^P, t_0, x_0\right) = V_2(t_0, x_0) = x_0'\Theta_2(t_0)x_0; \tag{A.2.36}$$

second, using $D_{22} = D_{22}'$, $D_{22} > 0$, and the inequality

$$u_2 D_{22} u_2 \geq \lambda_{22}\|u_2\|^2 = \lambda_{22} u_2' u_2 \;\forall u_2 \in \mathbb{R}^n, \tag{A.2.37}$$

where $\lambda_{22} > 0$ is the smallest root of the characteristic equation $\det[D_{22} - \lambda E_n] = 0$, we construct

$$\bar{W}_2[t, x] = W_2(t, x, u_1^T(t, x) = \alpha^{(1)}x, u_2(t, x) = \alpha x,$$

$$u_3^P(t, x) = -D_3^{-1}\Theta^P(t)x, \; V_2 = x'\Theta_2(t)x) \geq \frac{\partial V_2(t, x)}{\partial t} + \left(\frac{\partial V_2}{\partial x}\right)'\left[A(t)x + u_1^T(t, x)\right.$$

$$\left. + u_2(t, x) - D_3^{-1}\Theta^P(t)x\right] + \left[u_1^T(t, x)\right]' D_{21} u_1^T(t, x) + \alpha^2 \lambda_{22} x' x$$

$$+ x'\Theta^P(t)D_3^{-1}D_{23}D_3^{-1}\Theta^P(t)x = x'\left\{\frac{d\Theta_2(t)}{dt} + \Theta_2(t)\left[A(t) + \left(\alpha^{(1)}\right)^2 E_n\right.\right.$$

$$\left. + \alpha^2\lambda_{22}E_n - D_3^{-1}\Theta^P(t)\right] + \left[A'(t) + \left(\alpha^{(1)}\right)^2 E_n + \alpha^2\lambda_{22}E_n - \Theta^P(t)D_3^{-1}\right]\Theta_2(t)$$

$$\left. + \left(\alpha^{(1)}\right)^2 E_n + \alpha^2\lambda_{22}E_n + \Theta^P(t)D_3^{-1}D_{23}D_3^{-1}\Theta^P(t)\right\}x = x'M_2(t, \alpha)x.$$

For sufficiently large $\alpha \geq \alpha^{(3)}$, the quadratic form $x'M_2(t, \alpha)x > 0$ by Sylvester's criterion; moreover, $x(t)$ is the solution of the system

$$\dot{x} = \left[A(t) + \alpha^{(1)}E_n + \alpha E_n - D_3^{-1}\Theta^P(t)\right]x, \; x(t_0) = x_0. \tag{A.2.38}$$

If $x_0 \neq 0$, then $x(t) \neq 0_n$ (Lemma A.2.2); therefore, $x'(t)M_2(t, \alpha)x(t) > 0$ for all $t \in [t_0, \vartheta]$. Integrating both sides of this inequality gives

$$0 < \int_{t_0}^{\vartheta} x'(t)M_2(t, \alpha)x(t)dt = \int_{t_0}^{\vartheta} \frac{dV_2(t, x(t))}{dt}dt + \int_{t_0}^{\vartheta} \left\{u_1^T[t]D_{21}u_1^T[t]\right.$$

$$\left. + u_2^C[t]D_{22}u_2^C[t] + u_3^P[t]D_{23}u_3^P[t]\right\}dt = J_2\left(U_1^T, U_2^C, U_3^P, t_0, x_0\right) - V_2(t_0, x_0).$$

This result in combination with (A.2.36) finally establishes (A.2.33).

The next assertion is proved by analogy with Lemma A.2.4.

Lemma A.2.6 *Assume that U_2^T is a threat of player 2 to a Pareto-maximal alternative $U^P = (U_1^P, U_2^P, U_3^P)$ in the problem Γ_v, i.e., there exists a strategy $U_2^T \div \alpha x$ such that, for $\alpha \geq \alpha^{(2)}$,*

$$J_2\left(U_1^P, U_2^T, U_3^P, t_0, x_0\right) < J_2\left(U^P, t_0, x_0\right). \tag{A.2.39}$$

(Such a strategy U_2^T exists due to $D_{22} > 0$.)

Then

$$D_{21} < 0 \Rightarrow \exists \alpha^{(4)} = const > 0 : \forall \alpha = const \geq \alpha^{(4)}$$

$$J_2(U_1^C, U_2^T, U_3^P, t_0, x_0) < J_2(U^P, t_0, x_0)$$

for the strategy $U_1^C \div \alpha x$. In other words, U_1^C implements an incomplete counter-threat to the strategy profile U^P in the game Γ.

A.2.5 PROOF OF EXISTENCE

Theorem A.2.1 *Assume that the game Γ satisfies conditions $(A.2.11)$. Then the quadruple*

$$\left(U^P, J_1^P, J_2^P, J_3^P \right) = \left(\left(U_1^P, U_2^P, U_3^P \right), J_1 \left(U^P, t_0, x_0 \right), J_2 \left(U^P, t_0, x_0 \right), J_3 \left(U^P, t_0, x_0 \right) \right)$$

$$= \left(\left(-D_1^{-1} \Theta^P(t)x, -D_2^{-1} \Theta^P(t)x, -D_3^{-1} \Theta^P(t)x \right), \right.$$

$$\left. x_0' \Theta_1(t_0)x_0, x_0' \Theta_2(t_0)x_0, x_0' \Theta_3(t_0)x_0 \right)$$

is an equilibrium in threats and counter-threats for the differential game

$$\Gamma = \langle \{1, 2, 3\}, \Sigma \div (A.2.2), \{\mathcal{U}_i\}_{i=1,2,3}, \{J_i(U, t_0, x_0) \div (A.2.3)\}_{i=1,2,3} \rangle;$$

here

$$\Theta^P(t) = \left[X^{-1}(t) \right]' \left\{ C^{-1} + \int_{t_0}^{\vartheta} X^{-1}(\tau) \left[D_1^{-1} + D_2^{-1} + D_3^{-1} \right] \left[X^{-1}(\tau) \right]' d\tau \right\}^{-1} X^{-1}(t),$$

$$D_i = D_{1i} + \beta D_{2i} + \gamma D_{3i}, \quad C = C_1 + \beta C_2 + \gamma C_3,$$

$$\beta = \frac{1}{2} \left[\frac{\Lambda_{11}}{\lambda_{21}} + \frac{\lambda_{12}}{\Lambda_{22}} \right], \quad \gamma = \frac{1}{2} \left[\frac{\lambda_{13}}{\Lambda_{33}} + \frac{1}{2} \left(\frac{\Lambda_{11}}{\lambda_{21}} + \frac{\lambda_{12}}{\Lambda_{22}} \right) \frac{\lambda_{23}}{\Lambda_{33}} \right],$$

where Λ_{ii} is the smallest root of the equation $\det[D_{ii} - \Lambda E_n] = 0$; $-\lambda_{ij}$ is the greatest root of the equation $\det[D_{ij} - \lambda E_n] = 0$; $X(t)$ denotes the fundamental matrix of the system $\dot{x} = A(t)x$, $X(\vartheta) = E_n$ $(i, j = 1, 2, 3; j \neq i)$; finally, the symmetric matrices $\Theta_i(t)$ $(i = 1, 2, 3)$ are given by $(A.2.20)$.

Proof Note that two conclusions, the absence of a Nash equilibrium in the game Γ and the presence of a threat U_1^T posed by player 1 to a Pareto-maximal alternative U^P in the tri-criteria choice problem Γ_v, immediately follow from $D_{11} > 0$; see Remark A.2.3. The existence of a Pareto-maximal alternative and Pareto-maximal payoffs in the problem Γ_v (including their explicit form in this case) has been established in Propositions A.2.1 and A.2.2, respectively. The condition $D_{21} < 0$ allows constructing an incomplete counter-threat U_2^C of player 2 in response to the threat of player 1 (Lemma A.2.4), and the condition $D_{22} > 0$ and Lemma A.2.5 enable transforming the incomplete counter-threat U_2^C of player 2 into the complete one $\bar{U}_2^C$. The requirement $D_{22} > 0$ simultaneously implies the absence of a Nash equilibrium (for all $U_1 \in \mathbf{U}_1$, $\max_{U_1} J(U_1, U_2^e, U_3^e, t_0, x_0)$ is not achieved) as well as the ability of player 2 to design analytically a threat $U_2^T \in \mathbf{U}_2$ to U^P in the game Γ:

$$J_2 \left(U_1^C, U_2^T, U_3^P, t_0, x_0 \right) \leq J_2 \left(U^P, t_0, x_0 \right). \tag{A.2.40}$$

The condition $D_{21} < 0$ and Lemma A.2.6 guarantee the existence of an incomplete counter-threat $U_1^C \in \mathbf{U}_1$ of player 1 to the threat U_2^T of player 2:

$$J_2 \left(U_1^C, U_2^T, U_3^P, t_0, x_0 \right) < J_2 \left(U^P, t_0, x_0 \right). \tag{A.2.41}$$

Finally, the Pareto maximality of U^P and Property 2.1 lead to

$$J_1 \left(U_1^P, U_2^T, U_3^P, t_0, x_0 \right) < J_1 \left(U^P, t_0, x_0 \right). \tag{A.2.42}$$

Due to $D_{11} > 0$ and Lemma A.2.3, there exists a $\bar{U}_1^C \in \mathbf{U}_1$ such that

$$J_1 \left(\bar{U}_1^C, U_2^T, U_3^P, t_0, x_0 \right) > J_1 \left(U_1^P, U_2^T, U_3^P, t_0, x_0 \right). \tag{A.2.43}$$

The counter-threat in response to the threat posed by player 3 to U^P is designed by analogy.

Thus, in the game Γ, one of the other players always has a complete counter-threat in response to a threat posed by any player to the Pareto-maximal strategy profile U^P. This concludes the proof of Theorem A.2.1.

A.2.6   CONCLUSIONS

Let us summarize the outcomes of Appendix 2, in which the linear-quadratic positional differential game Γ has been considered. It has been established that under conditions (A.2.11), this game has no Nash equilibrium and simultaneously has an equilibrium in threats and counter-threats. In view of this fact, it is topical to investigate further the properties of this equilibrium, including the issues of existence and identification of other classes of games (in particular, non-differential ones) in which there exist no Nash equilibrium and simultaneously there exists an equilibrium in threats and counter-threats; see Proposition A.2.3. Of certain interest is the stability analysis of coalition structures [387], which will be the subject of future research.

Also, let us mention some questions that have arisen while writing Appendix 2.

1) The design procedure of a Pareto-maximal strategy profile here has been reduced to obtaining the constants $\beta > 0$ and $\gamma > 0$ that satisfy the system of three strict inequalities (A.2.10). The explicit form of the possible values of β and γ can be found prior to Remark A.2.1; however, it would be desirable to derive a general form of the solutions (A.2.10) as well as to establish its connection with the equilibrium in threats and counter-threats of the game Γ (excluding the form of β and γ from Remark A.2.1).

2) It is interesting to study the equilibrium in threats and counter-threats for differential games of four and more players under interval uncertainty.

Guaranteed solution for risk-neutral decision-maker: an analog of maximin in single-criterion choice problems

> Only those who will risk going too far
> can possibly find out how far one can go.
> —Eliot[41]

In this appendix, single-criterion choice problems under uncertainty (SCPUs) are considered. The principle of minimax regret and the Savage–Niehans risk function are introduced. A possible approach to solving an SCPU for a decision-maker who simultaneously seeks to increase his outcome and reduce his risk ("to kill two birds with one stone") is proposed. The explicit form of such a solution for the linear-quadratic setup of the SCPU is obtained.

A.3.1 INTRODUCTION

In the middle of the twentieth century, American mathematician and statistician, professor Leonard Savage (the University of Michigan) and Swiss economist, professor Jürg Niehans (the University of Zurich) independently proposed an approach to solving a single-criterion choice problem under uncertainty (SCPU), later called the principle of minimax regret or the Savage–Niehans principle. Along with Wald's principle of guaranteed outcome (maximin), the principle of minimax regret is crucial for guaranteed decision-making in SCPUs. The main role in this principle is played by the regret function, which determines the Savage–Niehans risk in SCPUs. In recent years, such a risk has been widely used in microeconomic analysis and applications. Appendix 3 of this book proposes a possible approach to solving SCPUs for a risk-neutral decision-maker, who simultaneously seeks to increase his outcome and reduce his risk ("to kill two birds with one stone"). The explicit form of such a solution for the linear-quadratic statement of the SCPU of a fairly general form is obtained.

A.3.1.1 Interval uncertainties

The mathematical model of decision-making under conflict considered below is described by the single-criterion choice problem under uncertainty (SCPU). Note that

---

[41]Thomas Stearns Eliot, (1888–1965), was an American-English poet, playwright, literary critic, and editor, a leader of the Modernist movement in poetry.

the case of interval uncertainty will be studied: the decision-maker knows only the ranges of admissible values of uncertain factors, and their probabilistic characteristics are absent, for one reason or another. The uncertainties occur due to the incomplete (inaccurate) information about the practical use of any strategies chosen by the decision-maker. For example, an economic system is often subject to unexpected, difficult-to-predict disturbances, both of *exogenous* origin (the disruption and variation of the quantity (range) of supply, demand fluctuations for the products supplied by a given enterprise, etc.) and *endogenous* origin (the emergence of new technologies, breakdowns and replacement of equipment, etc.). The question naturally arises: how to take into account the presence of uncertainties when choosing strategies?

The following aspects are described in the economic literature.

First, modern economic systems are characterized by a large number of elements and functional relations between them, a high degree of dynamism, the presence of non-functional relations between the elements, and the action of subjective factors due to the participation of individuals or their groups in the operation of such systems; in other words, an economic system usually operates under the uncertainty of its external and internal environment.

Second, as it has been already mentioned, the sources of uncertainties in economic systems are the incomplete or insufficient information about economic processes and their conditions; random or deliberate opposition from other economic agents; random factors that cannot be predicted due to the unexpectedness of their occurrence.

Third, the uncertainties are estimated using deterministic and probabilistic-statistical approaches as well as the approaches based on fuzzy logic.

For a detailed treatment of uncertainty in economic systems, see Section 1.3 of this book. Interval uncertainties were surveyed in the books [88–92,243] and other publications.

Each type of uncertainty requires its own approach for proper consideration. In this appendix, the analysis will be restricted to the class of interval uncertainties: only the ranges of admissible values of uncertain factors are known, without any probabilistic characteristics. The uncertainties will be taken into account using the method proposed by V. Zhukovskiy in [119–123]. This method allows passing from the original single-criterion choice problem under uncertainty (SCPU) to an equivalent single-criterion choice problem without uncertainty.

A.3.1.2   Principle of minimax regret

Traditionally, one of the most important challenges in the mathematical theory of SCPUs is the development of optimality principles, i.e., the answer to the following questions: What behavior of the decision-maker should be considered optimal (reasonable, appropriate)? Does an optimal solution exist and how can it be constructed? This appendix gives a possible answer to both questions for SCPUs.

The mathematical theory of games recommends making *the concept of stability* the cornerstone of optimality: a player's deviation from the optimal strategy introduced below cannot improve but at the same time can worsen his payoff (as well as the associated risk).

Let us proceed to the formal statement. Consider a single-criterion choice problem under uncertainty $\Gamma^{(1)} = \langle X, Y, f(x, y) \rangle$. In $\Gamma^{(1)}$, the decision-maker chooses his alternative $x \in X \subseteq \mathbf{R}^n$, seeking to maximize the value of a scalar criterion $f(x, y)$

for all possible realizations of the uncertainty $y \in Y \subseteq \mathbf{R}^m$. Recall that only the range of admissible values of the uncertainty is known.

The presence of uncertainties leads to the set of outcomes

$$f(x, Y) = \{f(x, y) \mid \forall y \in Y\},$$

that is induced by $x \in X$. The set $f(x, Y)$ can be reduced using risks.

Risk management is a topical problem of economics: in 1990, H. Markowitz [306] was awarded the Nobel Prize in Economic Sciences "for having developed the theory of portfolio choice." What is a proper comprehension of risk? A well-known Russian expert in optimization, T. Sirazetdinov, claims that today there is no rigorous mathematical definition of risk [227, p. 31]. The monograph [245, p. 15] even suggested sixteen possible concepts of risk. Most of them require statistical data on uncertainty. However, in many cases the decision-maker does not possess such information for objective reasons. Precisely these situations will be studied in Appendix 3.

Thus, here *risks* will be understood as *possible deviations of realized values from the desired ones*. Note that this definition (in particular, the Savage–Niehans risk) is in good agreement with the conventional notion of microeconomic risk; for example, see [241, pp. 40–50].

In 1939 A. Wald, a Romanian mathematician who emigrated to the USA in 1938, introduced the maximin principle, also known as the principle of guaranteed outcome [351, 352]. This principle allows finding a guaranteed outcome in a single-criterion choice problem under uncertainty (SCPU). Almost a decade later, Swiss economist J. Niehans (1948) and American mathematician, economist, and statistician L. Savage (1951) suggested the principle of minimax regret (PMR) for building guaranteed risks in the SCPUs [316, 322]. In the modern literature, this principle is also referred to as the Savage risk or the Savage–Niehans criterion. Interestingly, during World War II Savage worked as an assistant of J. von Neumann, which surely contributed to the appearance of the PMR. Note that the authors of two most remarkable dissertations in economics and statistics are annually awarded the Savage Prize, which was established in the USA as early as 1971.

For the single-criterion choice problem $\Gamma^{(1)} = \langle X, Y, f(x, y) \rangle$, the principle of minimax regret is to construct a pair $(x^r, R^r_f) \in X \times \mathbf{R}$ that satisfies the chain of equalities

$$R^r_f = \max_{y \in Y} R_f(x^r, y) = \min_{x \in X} \max_{y \in Y} R_f(x, y), \tag{A.3.1}$$

where *the Savage–Niehans risk function* has the form

$$R_f(x, y) = \max_{z \in X} f(z, y) - f(x, y). \tag{A.3.1}$$

The value R^r_f given by (A.3.1) is called *the Savage–Niehans risk* in the problem $\Gamma^{(1)}$. The risk function $R_f(x, y)$ assesses the difference between the realized value of the criterion $f(x, y)$ and its best-case value $\max_{z \in X} f(z, y)$ from the DM's view. Obviously, the DM strives for reducing $R_f(x, y)$ as much as possible with an appropriately chosen alternative $x \in X$, naturally expecting the strongest opposition from the uncertainty in accordance with the principle of guaranteed outcome; see formula (A.3.1). Therefore, following (A.3.1) and (A.3.2), the DM is an *optimist* who seeks for the best-case value $\max_{x \in X} f(x, y)$. In contrast, the pessimistic DM is oriented towards the worst-case outcome—the Wald maximin solution $(x^0, f^0 = \max_{x \in X} \min_{y \in Y} f(x, y) = \min_{y \in Y} f(x^0, y))$.

In the sequel, assume that the DM in the problem $\Gamma^{(1)}$ is optimistic: he constructs the Savage–Niehans risk function (A.3.2) for $f(x, y)$. Note two important aspects as follows. *First*, the criterion $f(x, y)$ from $\Gamma^{(1)}$ has its own risk $R_f(x, y)$; see (A.3.2). *Second*, the DM tries to choose alternatives $x \in X$ in order to reduce the risk $R_f(x, y)$, expecting any realization of *the strategic uncertainty* $y(\cdot) \in Y^X$, $y(x) : X \to Y$.

Remark A.3.1 The models $\Gamma^{(1)}$ naturally arise, e.g., in economics: a seller in a market is interested to maximize his profits under import uncertainty.

In many publications on macroeconomics [241, 243], all decision-makers are divided into three categories: *risk-averse*, *risk-neutral*, and *risk-seeking*. In this appendix, the DM is assumed to be a risk-neutral person and, as it has been mentioned above, an optimist.

A.3.I.3   Hierarchical interpretation of principle of minimax regret

Consider two hierarchical interpretations as follows. The first arises when the Savage–Niehans risk function $R_f(x, y) = \max_{z \in X} f(z, y) - f(x, y)$ is constructed, whereas the second when the solution $(x^r, R_f^r) \in X \times \mathbf{R}$ of the problem $\Gamma^{(1)}$ for the risk-seeking DM is obtained.

Hierarchical interpretation of Savage–Niehans risk function design

Hierarchical games represent a mathematical model of a conflict with a fixed sequence of moves and information exchange between its parties [395, p. 477]. In Russia, the intensive research of hierarchical games was initiated in the second half of the 20th century by Yu. Germeier [53, 54] (the founder of the Department of Operations Research at the Faculty of Computational Mathematics and Cybernetics, Moscow State University) and then continued by his scholars. Hierarchical two-player games describe the interaction between the upper (Leader) and lower (Follower) levels of the hierarchy. Such games have a given sequence of moves, i.e., an order in which each player chooses his strategies and (possibly) reports them to the partner.

An important element of hierarchical games is to choose the class of admissible strategies depending on the information available to the players. In the theory of hierarchical games, *the informational extension of the game* was rigorously formulated in [396]. In a particular case, this extension leads the so-called *strategic uncertainties*, i.e., m-dimensional vector functions $y(x) : X \to Y$, $y(\cdot) \in Y^X$, which are used along with pure uncertainties $y \in Y$ in the game $\Gamma^{(1)}$.

Now, let us discuss the hierarchical interpretation of risk function design for the SCPU $\Gamma^{(1)}$. Assume that the lower-level player (Follower) can apply only his pure strategy $y \in Y$, whereas the upper-level player (Leader) can adopt "any conceivable information" [179, p. 353]. Thus, further analysis will be confined to the Follower's pure strategies $y \in Y$ and the Leader's counterstrategies $x(y) : Y \to X$, $x(\cdot) \in X^Y$, i.e., the set of functions $x(y)$ with Y as the domain of definition and X as the codomain. For risk function design, consider the two-level two-stage hierarchical game

$$\Gamma_R = \left\langle X^Y, Y, f(x, y) \right\rangle.$$

In this game, *the first move* is made by Follower (the lower-level player), who reports his admissible pure strategies to the upper level.

The second move belongs to Leader (the upper-level player), who performs the following actions. *First*, he analytically constructs the counterstrategy

$$x(y) \in Y(x) = Arg \max_{x \in X} f(x, y) \ \forall y \in Y,$$

i.e., finds the scalar function $f[y] = f(x(y), y) = \max_{x \in X} f(x, y)$; *second*, he designs the Savage–Niehans risk function

$$R_f(x, y) = f[y] - f(x, y).$$

Solution of choice problem $\Gamma^{(1)}$ for risk-seeking DM

Assume that the Savage–Niehans risk function has the explicit form $R_f(x, y) = \max_{z \in X} f(z, y) - f(x, y)$, and the problem is to construct a pair $(x^r, R_f^r) \in X \times \mathbf{R}$ defined as the solution of the SCPU $\Gamma^{(1)}$ for the risk-seeking DM:

$$R_f^r = \min_{x \in X} \max_{y \in Y} R_f(x, y) = \max_{y \in Y} R_f(x^r, y).$$

In the problem $\Gamma^{(1)}$, suppose that Leader applies only a pure alternative (strategy) $x \in X$, whereas the other player (Follower) can adopt any conceivable information [178; 179, p. 353], including his knowledge of the strategy $x \in X$, to form his strategy (uncertainty) as a function $y(x) : X \to Y$, $y(\cdot) \in Y^X$. (This hypothesis is well known as *the informational discrimination of Leader*.) As a result, the criterion in the choice problem $\Gamma^{(1)}$ is defined as the scalar function $f(x, y(x))$.

Recall that in the theory of differential games, the functions $y(\cdot) \in Y^X$ (the set of m-dimensional vector functions with the domain of definition X and the codomain Y) are called *counterstrategies*. The problem $\Gamma^{(1)}$ in which counterstrategies describe the behavior of uncertain factors is called *the minimax game* [178, 179].

Thus, consider the hierarchical two-level three-stage game of two players (Leader and Follower) in which, in contrast to Γ_R, Leader and Follower use a pure strategy $x \in X$ and a counterstrategy $y(x) : X \to Y$, $y(\cdot) \in Y^X$, respectively.

The first move is made by Leader, who reports his admissible strategies $x \in X$ to the lower level.

The second move is made by Follower, who analytically constructs $y(x)$ in accordance with

$$\max_{y(\cdot) \in Y^X} R_f(x, y) = R_f(x, y(x)) = R_f[x] \ \forall x \in X,$$

assuming that the vector function $y(x)$ is unique (e.g., for a scalar function $R_f(x, y)$ that is strictly concave in y for each $x \in X$), and then reports $R_f[x]$ to the upper level.

The third move is made by Leader, who constructs a strategy $x^r \in X$ such that $\min_{x \in X} R_f[x] = R_f[x^r] = R_f^r$.

This three-move game-theoretic framework *completely* matches the concept of the Leader's guaranteed outcome in the problem $\Gamma^{(1)}$ (in the Germeier sense) if the Follower's payoff function considered in [122–140] is replaced by $-R_f(x, y)$. Moreover, in the game Γ_R, Leader can calculate the Follower's response and immediately implement the third move *if he knows the behavioral rule of the opponent*. Once again, note that the analog and modification of this three-move framework is convenient to design the guaranteed solution in outcomes and risks for the risk-seeking DM, both in noncooperative and cooperative conflicts.

Remark A.3.2 The minimax solution for the risk-seeking DM is determined by the pair $\left(x^r, R_f^r = \min_{x \in X} \max_{y(\cdot) \in Y^X} R_f(x, y) = \max_{y \in Y} R_f(x^r, y) \right)$ for two solutions as follows:

a)   For each alternative $x \in X$, the inner maximum $\max_{y \in Y} R_f(x, y) = R_f(x, y(x)) = R_f[x]$ (see move 2) gives the greatest Savage–Niehans risk of the form

$$R_f[x] = \max_{y \in Y} R_f(x, y) \geq R_f(x, y) \; \forall y \in Y.$$

In other words, $R_f(x, y)$ cannot exceed $R_f[x]$ for all $y \in Y$, and hence $R_f[x]$ can be considered the DM's guarantee obtained by choosing the alternative x. Note that due to (A.3.2), $R_f(x, y) \geq 0$; therefore, the Savage–Niehans risk function takes the values $R_f(x, y) \in [0, R_f[x]]$ for all $(x, y) \in X \times Y$.

b)   Like any DM, the risk-seeking one would like to implement his decisions (the choice of $x \in X$) with the smallest risk (ideally, zero!). This aspect explains his third move.

Therefore, in the problem $\Gamma^{(1)}$ the risk-seeking DM is suggested to use the alternative x^r to obtain the smallest (minimum) guarantee $R_f[x^r] = R_f(x^r, y(x^r)) \geq R(x^r, y) \; \forall y \in Y$. The same technique can be applied to formalize the strongly-guaranteed solution in outcomes and risks of the problem $\Gamma^{(1)}$.

Here is an important result from operations research that concerns informed uncertainties and strategies.

Lemma A.3.1 *If in the choice problem $\Gamma^{(1)} = \langle X, Y, f(x, y) \rangle$ the sets X and Y are compact and the criterion $f(x, y)$ is continuous on $X \times Y$, then the maximum (minimum) function $\max_{x \in X} f(x, y) \left(\min_{y \in Y} f(x, y) \right)$ is continuous on $Y(X)$.*

Lemma A.3.1 is a well-known fact that can be found in almost any textbook on operations research; for example, see [204, p. 54].

Remark A.3.3 Lemma A.3.1 implies the continuity of the risk function (A.3.2) on $X \times Y$ (of course, only if in the problem $\Gamma^{(1)}$ the sets X and Y are compact and the criterion $f(x, y)$ is continuous on $X \times Y$.)

Remark A.3.4 Assume that in the problem $\Gamma^{(1)}$, $X \in comp \; \mathbf{R}^n$, $Y \in comp \; \mathbf{R}^m$, and $f(\cdot) \in C(X \times Y)$. Then there exists the guaranteed solution in risks (x^r, R_f^r) of this problem.

Really, the Savage–Niehans risk function $R_f(x, y)$ (A.3.2) is continuous on $X \times Y$ (see Remark A.3.2). In this case, by Lemma A.3.1 the function $\max_{y \in Y} R_f(x, y) = R_f[x]$ is also continuous on X. (There exists a Borel measurable counterstrategy (selector) $y(x) : X \to Y$ such that

$$\max_{y \in Y} R_f(x, y) = R_f(x, y(x)) = R_f[x] \; \forall x \in X,$$

and $R_f[x]$ is continuous on X). According to the Weierstrass extreme-value theorem, on a compact set X a continuous function $R_f[x]$ achieves minimum at the point $x^r \in X$. If both sets X and Y are compact and the function $f(x, y)$ is continuous, then the guaranteed solution in risks (x^r, R_f^r) defined by (A.3.1) exists.

Thus, using x^r, the risk-seeking DM obtains a guarantee in risks $R_f^r \geq R_f(x^r, y) \; \forall y \in Y$, and for all $x \in X$ this guarantee will be smallest among all other guarantees $R_f[x] \geq R_f(x, y)$ for all alternatives $x \in X$. Such a procedure is characteristic of the risk-seeking DM. In this appendix, we will consider a similar procedure for the risk-neutral DM.

A.3.2   NEW APPROACH TO SCPU FOR RISK-NEUTRAL DM

A.3.2.I   Preliminaries

Let us utilize the approach proposed for noncooperative games in [397, 398]. For this purpose, from the SCPU $\Gamma^{(1)}$ we will pass to the problem of guarantees without any uncertainties.

At conceptual level, the DM's goal so far has been to choose an appropriate alternative maximizing his outcome. But this is not enough for the risk-neutral DM! He seeks for an alternative that would not only *increase his outcome but also reduce his risk, as much as possible*. Recall that the DM forms the Savage–Niehans risk function $R_f(x, y)$ (A.3.2), the value of which is called the DM's risk, and the Savage–Niehans risk R_f^r itself is determined by the chain of equalities (A.3.1). The pair (x^r, R_f^r) is the solution of the choice problem $\Gamma^{(1)}$ for the risk-seeking DM: the value $R_f(x, y)$ characterizes his risk when choosing and implementing the alternative $x \in X$, which he strives to minimize simultaneously with outcome improvement. In this context, two questions arise naturally:

1)   How can we combine the two objectives of the decision-maker (outcome increase with simultaneous risk reduction) using *only one criterion*?

2)   How can we implement these objectives *in a single alternative*, in such a way that uncertainty is also accounted for?

A.3.2.2   How to combine DM's desire to increase outcome and reduce risks?

Construction of Savage–Niehans risk function. Recall that, according to the principle of minimax regret, the DM's risk is defined by the value of the Savage–Niehans risk function $R_f(x, y) = \max_{z \in X} f(z, y) - f(x, y)$, where $f(x, y)$ denotes the DM's criterion in the choice problem $\Gamma^{(1)}$. Thus, to construct the risk function $R_f(x, y)$ for the DM, first the dependent maximum $f[y] = \max_{x \in X} f(x, y) \forall y \in Y$ needs to be found. To calculate $f[y]$, following the theory of two-level hierarchical games (see Subsection A.3.1.3), assume the *discrimination* of the lower-level player, who forms the uncertainty $y \in Y$ and sends this information to the upper level for constructing a counterstrategy $x(y) : Y \to X$ such that

$$\max_{x \in X} f(x, y) = f(x(y), y) = f[y] \ \forall y \in Y.$$

The set of such strategies is denoted by X^Y. (Actually, this set consists of n-dimensional vector functions $x(y) : Y \to X$ with the domain of definition Y and the codomain X.) Thus, to construct the first term in (A.3.2) at the upper level of the hierarchy, we have to solve the single-criterion choice problem $\langle X^Y, Y, f(x, y) \rangle$ for each uncertainty $y \in Y$; here X^Y is the set of counterstrategies $x(y) : Y \to X$. The problem itself consists in determining the scalar function $f[y]$ defined by

$$f[y] = \max_{x(\cdot) \in X^Y} f(x, y) \ \forall y \in Y. \tag{A.3.3}$$

Then, the Savage–Niehans risk functions are constructed by formula (A.3.2).

Continuity of risk function, strongly-guaranteed outcomes and risks

Hereinafter, the collection of all compact sets of Euclidean space $\mathbf{R}^k$ is denoted by *comp* $\mathbf{R}^k$, and if a scalar function $\psi(x)$ on the set X is continuous, we write $\psi(\cdot) \in C(X)$.

The main role below will be played by the following result.

Proposition A.3.1 *If* $X \in comp$ $\mathbf{R}^n$, $Y \in comp$ $\mathbf{R}^m$, *and* $f(\cdot) \in C(X \times Y)$, *then*:

(a)   *the maximum function* $\max_{x \in X} f(x, y)$ *is continuous on* Y;

(b)   *the minimum function* $\min_{y \in Y} f(x, y)$ *is continuous on* X.

Corollary A.3.1 *If in the choice problem* $\Gamma^{(1)}$ *the sets* $X \in comp$ $\mathbf{R}^n$ *and* $Y \in comp$ $\mathbf{R}^m$ *and the function* $f(\cdot) \in C(X \times Y)$, *then the Savage–Niehans risk function* $R_f(x, y)$ *is continuous on* $X \times Y$. (Also, see Remark A.3.3.)

Let us proceed with *the strongly-guaranteed outcome and risk* in the SCPU $\Gamma^{(1)}$. In a series of papers [122, 123], three different ways to account for uncertain factors of decision-making in conflicts under uncertainty were proposed. Our analysis below will be confined to one of them presented in [123], based on the following method. We associate with the criterion $f(x, y)$ in the problem $\Gamma^{(1)}$ its *strong guarantee* $f[x] = \min_{y \in Y} f(x, y)$. As a consequence, choosing his alternatives $x \in X$, the DM ensures an outcome $f[x] \leq f(x, y)$ $\forall y \in Y$ under any realized uncertainty $y \in Y$. Such a strongly-guaranteed outcome $f[x]$ seems natural for the *interval uncertainties* $y \in Y$ addressed in this appendix, because no additional probabilistic characteristics of y (except for information on the admissible set $Y \subseteq \mathbf{R}^m$) are available. Proposition A.3.1, in combination with Corollary A.3.1 as well as the continuity of $f(x, y)$ and $R_f(x, y)$ on $X \times Y$, leads to the following result.

Proposition A.3.2 *If in the SCPU* $\Gamma^{(1)}$ *the sets* X *and* Y *are compact and the criterion* $f(x, y)$ *is continuous on* $X \times Y$, *then the strongly-guaranteed outcome*

$$f[x] = \min_{y \in Y} f(x, y) \tag{A.3.4}$$

and the strongly-guaranteed risk

$$R_f[x] = \max_{y \in Y} R_f(x, y) \tag{A.3.5}$$

are scalar functions that are continuous on X.

Remark A.3.5 *First*, the meaning of the guaranteed outcome $f[x]$ from (A.3.4) is that, for any $y \in Y$, the realized outcome $f(x, y)$ is not smaller than $f[x]$. In other words, using his alternative $x \in X$ in the choice problem $\Gamma^{(1)}$, the DM ensures an outcome $f(x, y)$ of at least $f[x]$ under any uncertainty $y \in Y$. Therefore, the strongly-guaranteed outcome $f[x]$ gives *a lower bound* for all possible outcomes $f(x, y)$ occurring when the uncertainty y runs through all admissible values from Y. *Second*, the strongly-guaranteed risk $R_f[x]$ also gives *an upper bound* for all Savage–Niehans risks $R_f(x, y)$ that can be realized under any uncertainties $y \in Y$. Really, from (A.3.5) it immediately follows that

$$R_f[x] \geq R_f(x, y) \ \forall y \in Y.$$

Thus, adhering to his alternative $x \in X$, the DM obtains the strong guarantee in outcomes $f[x]$, and simultaneously the strong guarantee in risks $R_f[x]$.

Transition from single-criterion choice problem under uncertainty $\Gamma^{(1)}$ to bi-criteria vector optimization problem

The DM's desire to increase his outcome and simultaneously reduce his risk is described well by the new mathematical *model of a bi-criteria choice problem under uncertainty with the two-component vector criterion*

$$\Gamma_2 = \langle X, Y, \{f(x, y), -R_f(x, y)\}\rangle.$$

In this model, the sets X and Y are the same as in $\Gamma^{(1)}$. The novelty consists in the transition from the one-component criterion $f(x, y)$ to the two-component criterion $\{f(x, y), -R_f(x, y)\}$, in which $R_f(x, y)$ is the Savage–Niehans risk function for the DM. In the problem Γ_2, the DM chooses an alternative $x \in X$ in order to increase as much as possible the values of both criteria simultaneously, which explains the minus sign of $R_f(x, y)$. Moreover, the DM must expect any realization of the uncertainty $y \in Y$. Note that due to $R_f(x, y) \geq 0$, for all $(x, y) \in X \times Y$ an increase of $-R_f(x, y)$ is equivalent to a decrease of $R_f(x, y)$.

The uncertainty $y \in Y$ in the choice problem Γ_2 is of the interval type. This feature compels the DM to use the available information about the uncertainty, i.e., the limits of its range, being guided by the strongly-guaranteed outcome $f[x]$ (A.3.4) and the strongly-guaranteed risk $R_f[x]$ (A.3.5). Therefore, it seems natural to pass from $\Gamma^{(1)}$ to the two-component vector optimization problem without uncertainty

$$\Gamma_2^g = \langle X, \{f[x], -R_f[x]\}\rangle,$$

in which the DM chooses an appropriate alternative $x \in X$ for maximizing both criteria $f[x]$ and $-R_f[x]$ simultaneously.

For the practical design of the strongly-guaranteed outcome and risk in Γ_2^g, we will employ the mathematical theory of vector optimization, e.g., from [213], with its different approaches and results. Consider an optimal solution of multicriteria problems introduced in 1909 by Italian economist and sociologist V. Pareto [317]. For the problem Γ_2^g, the Pareto maximality (efficiency) of an alternative x^P is reduced to the inconsistency of the system of two inequalities $f[x] \geq f[x^P]$, $-R_f[x] \geq -R_f[x^P] \forall x \in X$, in which at least one inequality is strict. This leads to the following notion.

Definition A.3.1 *A triplet $\left(x^P, f[x^P], R_f[x^P]\right)$ is called a Pareto-maximal strongly-guaranteed solution in outcomes and risks (PSGOR) of the problem Γ_2^g if:*

(a) *the alternative x^P is Pareto-maximal in the problem Γ_2^g;*

(b) *$f[x^P]$ is the value of the strongly-guaranteed outcome $f[x] = \min_{y \in Y} f(x, y)$ in the problem Γ_2^g for $x = x^P$;*

(c) *$R_f[x^P]$ is the value of the strongly-guaranteed risk $R_f[x] = \max_{y \in Y} R_f(x, y)$ in the problem Γ_2^g for $x = x^P$.*

Remark A.3.6 Definition *A*.3.1 may also involve other optimality principles (Pareto, Geoffrion, Borwein, cone, A-optimality). All these principles as well as connections between different vector optimal solutions were considered in [365–374].

According to the definition of Pareto maximality:

(a) If x^P is a Pareto-maximal alternative, then for $\bar{x} \neq x^P, \bar{x} \in X$ an increase of value of one criterion will inevitably reduce the value of the other.

(b)   There exists no alternative $x \in X$ for which the values of both criteria will increase in comparison with their values for $x = x^P$.

Perhaps the term "Slater maximality" appeared in the Russian literature after the translation [249, 250] of a paper by Hurwitz.

If Pareto optimality is replaced by Slater maximality (weak efficiency), then Definition A.3.1 takes the following form.

Definition A.3.2 *A triplet $\left(x^S, f[x^S], R_f[x^S]\right)$ is called a Slater-strongly-guaranteed solution in outcomes and risks of the problem Γ_2^g if:*

(a)   *the alternative $x^S \in X$ is Slater-maximal in the problem Γ_2^g, i.e., for any $x \in X$ the system of two strict inequalities*

$$f[x] > f[x^S], \quad -R_f[x] > -R_f[x^S]$$

is inconsistent;

(b)   *$f[x^S]$ is the value of the strongly-guaranteed outcome in the problem Γ_2^g for $x = x^S$;*

(c)   *$R_f[x^S]$ is the value of the strongly-guaranteed risk in the problem Γ_2^g for $x = x^S$.*

Any efficient (Pareto-maximal) alternative is also weakly efficient, which follows directly from Definitions A.3.1 and A.3.2. Generally speaking, the converse is false. Also, property b) of Remark A.3.6 remains valid for the Slater-strongly-guaranteed solution in outcomes and risks of the problem $\Gamma^{(1)}$. The next result seems quite obvious.

Proposition A.3.3 *If in the problem Γ_2^g there exists an alternative $x^P \in X$ and values $\alpha, \beta \in (0, 1)$ such that x^P maximizes the scalar function $\Phi[x] = \alpha f[x] - \beta R_f[x]$, i.e.,*

$$\Phi\left[x^P\right] = \max_{x \in X} \left(\alpha f[x] - \beta R_f[x]\right), \tag{A.3.6}$$

then x^P is the Pareto-maximal alternative in the problem Γ_2^g; in other words, for any $x \in X$ the system of two inequalities

$$f[x] \geq f[x^P], \quad R_f[x] \leq R_f[x^P], \tag{A.3.7}$$

with at least one strict inequality, is inconsistent. (Here $\alpha = \beta = 1$.)

Remark A.3.7 The combination of the criteria (A.3.4) and (A.3.5) in the form $\Phi[x] = \alpha f[x] - \beta R_f[x]$ is of interest for two reasons. *First*, even if for $\bar{x} \neq x^P$ we have an increase of the guaranteed outcome $f[\bar{x}] > f[x^P]$, then due to the Pareto maximality of x^P and the fact that $R_f[\bar{x}] \geq 0$ such an improvement of the guaranteed outcome $f[\bar{x}] > f[x^P]$ will inevitably lead to an increase of the guaranteed risk $R_f[\bar{x}] > R_f[x^P]$; conversely, for the same reasons, a reduction of the guaranteed risk $R_f[\bar{x}] < R_f[x^P]$ will lead to a reduction of the guaranteed outcome $f[\bar{x}] < f[x^P]$ (both cases are undesirable for the DM). Therefore, the replacement of the bi-criteria choice problem Γ_2^g with the single-criterion choice problem $\langle X, \Phi[x] = \alpha f[x] - \beta R_f[x]\rangle$ matches well the DM's desire to increase $f[x]$ and simultaneously reduce $R_f[x]$. *Second*, since $R_f[x] \geq 0$ and $\alpha, \beta \in (0, 1)$, an increase of the difference $\alpha f[x] - \beta R_f[x]$ also matches the DM's desire to increase the guaranteed outcome $f[x]$ and simultaneously reduce the guaranteed risk $R_f[x]$.

Now, let us answer the second question from Subsection A.3.2.1: how can we combine both objectives of the DM in a single alternative taking into account the existing interval uncertainty? To do this, from the problem $\Gamma^{(1)}$ we will pass sequentially to choice problems Γ_1, Γ_2, and Γ_3:

$$\Gamma_1 = \langle X, Y, \{f(x, y), -R_f(x, y)\}\rangle,$$
$$\Gamma_2 = \langle X, \{f[x], -R_f[x]\}\rangle, \tag{A.3.8}$$
$$\Gamma_3 = \langle X, \{\Phi[x] = f[x] - R_f[x]\}\rangle.$$

In all the three choice problems, $x \in X \subseteq \mathbf{R}^n$ denotes the alternative chosen by the DM; $y \in Y \subseteq \mathbf{R}^m$ are uncertainties; the DM's criterion $f(x, y)$ is defined on the pairs $(x, y) \in X \times Y$; in (A.3.2), $R_f(x, y)$ means the Savage–Niehans risk function. In the choice problem Γ_1, the criterion has two components—the original criterion $f(x, y)$ of the problem $\Gamma^{(1)}$ and the risk function $R_f(x, y)$ of (A.3.2). In the choice problem Γ_2, the original criterion $f(x, y)$ and the risk function $R_f(x, y)$ are replaced by their guarantees $f[x] = \min_{y \in Y} f(x, y)$ and $R_f[x] = \max_{y \in Y} R_f(x, y)$, respectively. Finally, in the choice problem Γ_3, the linear convolution of the guarantees $f[x]$ and $-R_f[x]$ (see Proposition A.3.3) is used instead of the two-component criterion.

Remark A.3.8 Let us discuss the advantages of the solution formalized by Definitions A.3.1 and A.3.2. *First*, recall that economists divide all decision-makers into three categories: *risk-averse*, *risk-neutral*, and *risk-seeking*. In Definitions A.3.1 and A.3.2, the DM is assumed to be *a risk-neutral person*, who simultaneously considers the outcome and associated risk. *Second*, this solution imposes a lower bound on the outcomes and also an upper bound on the risks, $f[x] \leq f(x^P, y) \; \forall y \in Y$ and $R_f[x] \geq R_f(x^P, y) \; \forall y \in Y$, respectively. Note that the existence and continuity of the guarantees $f[x]$ and $R_f[x]$ are based on the hypotheses $X \in comp \; \mathbf{R}^n$, $Y \in comp \; \mathbf{R}^m$, and $f(\cdot) \in C(X \times Y)$; see Proposition A.3.1. *Third*, an improvement of the Pareto-maximal guaranteed outcome (in comparison with $f[x^P]$) will inevitably increase the guaranteed risk (in comparison with $R_f[x^P]$); conversely, a reduction of the risk will inevitably decrease the guaranteed payoff.

Remark A.3.9 Definitions A.3.1 and A.3.2 suggest a *constructive* method of SGPOR design. It consists of four steps as follows.

Step I. Using $f(x, y)$, find $f[y] = \max_{x \in X} f(x, y)$ and construct the Savage–Niehans risk function $R_f(x, y) = f[y] - f(x, y)$ for the criterion $f(x, y)$.

Step II. Evaluate the strong guarantee in outcomes $f[x] = \min_{y \in Y} f(x, y)$ and also the strong guarantee in risks $R_f[x] = \max_{y \in Y} R_f(x, y)$.

Step III. For the auxiliary choice problem Γ_2, calculate the Pareto-maximal alternative x^P. At this step, Proposition A.3.3 is of assistance.

Then the Pareto-maximal alternative in the auxiliary choice problem Γ_3 is x^P for which

$$\max_{x \in X} \left(f[x] - R_f[x]\right) = f[x^P] - R_f[x^P]. \tag{A.3.9}$$

Step IV. Using x^P, evaluate the strong guarantees $f[x^P]$ and $R_f[x^P]$.

The resulting triplet $\left(x^P, f[x^P], R_f[x^P]\right)$ is the requisite SGPOR, which complies with Definition A.3.1, i.e., for the original criterion $f(x, y)$ the alternative x^P leads to a guaranteed outcome $f[x^P]$ with a guaranteed Savage–Niehans risk $R_f[x^P]$.

In the next subsection of Appendix 3, this design procedure of SGPOR is applied to the linear-quadratic single-criterion choice problem under uncertainty of a general form.

A.3.3   EXPLICIT FORM OF SAVAGE–NIEHANS RISK FOR LINEAR-QUADRATIC SCPU

A.3.3.I   Problem statement

Consider the linear-quadratic single-criterion choice problem under uncertainty

$$\Gamma_{lq} = \langle \mathbf{R}^n, \mathbf{R}^m, f(x, y) \rangle,$$

in which the set of alternatives x coincides with the n-dimensional Euclidean space $\mathbf{R}^n$, the set of uncertainties y is $\mathbf{R}^m$, and the linear-quadratic criterion is given by

$$f(x, y) = x'Ax + 2x'By + y'Cy + 2a'x + 2c'y + d.$$

Here A and C are constant and symmetric matrices of dimensions $n \times n$ and $m \times m$, respectively; B is rectangular constant matrix of dimensions $n \times m$; a and c are constant vectors of dimensions n and m, respectively; finally, d is a constant. As before, the prime denotes transposition. In the problem Γ_{lq}, the DM chooses an appropriate alternative $x \in \mathbf{R}^n$ in order to maximize the linear-quadratic criterion $f(x, y)$ and simultaneously minimize a risk function under any possible realizations of the uncertainty $y \in \mathbf{R}^m$.

The problem is to design an explicit form of the Savage–Niehans risk function for the linear-quadratic choice problem Γ_{lq} (see Remark A.3.9) and then to obtain the SGPOR.

Hereinafter, for a square constant matrix A of dimensions $n \times n$, the inequality $A > 0$ ($A < 0$) means that the quadratic form with the matrix A is positive definite (negative definite, respectively).

Also, the following notations will be used below:

0_n as a zero vector of dimension n;

$$\frac{\partial f}{\partial x} = \begin{bmatrix} \frac{\partial f}{\partial x_1} \\ \vdots \\ \frac{\partial f}{\partial x_n} \end{bmatrix}$$ as the gradient of a scalar function $f(x, y)$ with respect to x under a

fixed vector y;

$$\frac{\partial^2 f}{\partial x^2} = \begin{bmatrix} \frac{\partial^2 f}{\partial x_1 \partial x_1} & \cdots & \frac{\partial^2 f}{\partial x_1 \partial x_n} \\ \vdots & \ddots & \vdots \\ \frac{\partial^2 f}{\partial x_n \partial x_1} & \cdots & \frac{\partial^2 f}{\partial x_n \partial x_n} \end{bmatrix}$$ as the Hessian of a scalar function $f(x, y)$ with

respect to x under a fixed vector y;

$\det A$ as the determinant of a matrix A;

E_n as an identity matrix of dimensions $n \times n$.

Direct calculations show that

$$\frac{\partial}{\partial x}(x'Ax) = 2Ax, \quad \frac{\partial}{\partial x}(2x'By) = 2By, \quad \frac{\partial}{\partial x}(2a'x) = 2a, \quad \frac{\partial^2}{\partial x^2}(x'Ax) = 2A.$$

A.3.3.2   Explicit form of Savage–Niehans risk function

Well, let us construct an explicit form of the Savage–Niehans risk function $R_f(x, y)$ for the linear-quadratic choice problem Γ_{lq}; see Stage I from Remark A.3.9.

Step I. Explicit-form design of the Savage–Niehans risk function $R_f(x, y)$ for the problem Γ_{lq}.

Proposition A.3.4 *In the linear-quadratic choice problem Γ_{lq} with a matrix $A < 0$, the Savage–Niehans risk function has the form*

$$R_f(x, y) = - \left(x'A + y'B' + a'\right) A^{-1} (Ax + By + a).$$

Proof An n-dimensional vector function $x(y)$ with the domain of definition $\mathbf{R}^m$ and the codomain $\mathbf{R}^n$ such that

$$\max_{z \in \mathbf{R}^n} f(z, y) = f(x(y), y) \ \forall y \in \mathbf{R}^m,$$

exists under the sufficient conditions

$$\frac{\partial f(x, y)}{\partial x}\Big|_{x=x(y)} = 2Ax(y) + 2By + 2a = 0_n \ \forall y \in \mathbf{R}^m,$$

$$\frac{\partial^2 f(x, y)}{\partial x^2}\Big|_{x=x(y)} = 2A < 0.$$

The second condition (inequality) holds due to $A < 0$; from the first condition (identity) it follows that

$$x(y) = -A^{-1}(By + a).$$

Substituting $x = x(y)$ into $f(x, y)$ gives

$$\max_{z \in \mathbf{R}^n} f(z, y) = f(x(y), y) = (y'B' + a')A^{-1}(By + a) - 2(y'B' + a')A^{-1}By + y'Cy$$

$$- 2a'A^{-1}(By + a) + 2c'y + d = -(y'B' + a')A^{-1}(By + a) + y'Cy + 2c'y + d$$

$$= y'[C - B'A^{-1}B]y + 2(c' - a'A^{-1}B)y + (d - a'A^{-1}a).$$

As a result, the Savage–Niehans risk function can be written as

$$R_f(x, y) = f(x(y), y) - f(x, y) = -x'Ax - 2x'By - 2a'x - y'B'A^{-1}By$$

$$- 2a'A^{-1}By - a'A^{-1}a = -(x'A + y'B' + a')A^{-1}(Ax + By + a).$$

The proof of this proposition is complete.

A.3.3.3   Evaluation of strong guarantee in risks

Step II. Construct the function $R_f[x] = \max_{y \in \mathbf{R}^m} R_f(x, y)$.

Proposition A.3.5 *In the linear-quadratic choice problem Γ_{lq} with matrices*

$$A < 0, \ \det B \neq 0,$$

the strong guarantee in risks is

$$R_f[x] = \max_{y \in \mathbf{R}^m} R_f(x, y) \equiv 0 \ \forall x \in \mathbf{R}^n.$$

Proof First of all, the condition $\det B \neq 0$ implies that B is a square matrix, i.e., $n = m$. For finding $R_f[x]$, define an n-dimensional vector function $y(x) : \mathbf{R}^n \to \mathbf{R}^n$ such that

$$\max_{y \in \mathbf{R}^m} R_f(x, y) = R_f(x, y(x)) = R_f[x] \ \forall x \in \mathbf{R}^n.$$

Recall the sufficient conditions of maximum for $y = y(x) : \mathbf{R}^n \to \mathbf{R}^n$:

$$\frac{\partial R_f(x, y)}{\partial y}\Big|_{y=y(x)} = -2B'x - 2B'A^{-1}By(x) - 2B'A^{-1}a = 0_m \ \forall x \in \mathbf{R}^n,$$

$$\frac{\partial^2 R_f(x, y)}{\partial y^2}\Big|_{y=y(x)} = -2B'A^{-1}B > 0. \tag{A.3.10}$$

Since $A < 0$ and $\det B \neq 0$, the following chain of implications is the case:

$$A^{-1} < 0 \Longrightarrow B'A^{-1}B < 0 \Longrightarrow -B'A^{-1}B > 0 \Longrightarrow -2B'A^{-1}B > 0.$$

(In other words, the second condition of (A.3.10) is satisfied.)
In view of

$$\left(B'A^{-1}B\right)^{-1} = B^{-1}A(B')^{-1},$$

the first condition of (A.3.10) gives

$$y(x) = -\left(B'A^{-1}B\right)^{-1}\left(B'x + B'A^{-1}a\right) = -B^{-1}A\left(x + A^{-1}a\right) = -B^{-1}(Ax + a).$$

Then, substituting $y = y(x)$ into $R_f[x]$ yields

$$R_f[x] = R_f(x, y(x)) = -\left(x'A - x'A - a' + a'\right)A^{-1}(Ax - Ax - a + a) \equiv 0 \ \forall x \in \mathbf{R}^n,$$

which finally establishes the identity $R_f[x] \equiv 0 \ \forall x \in \mathbf{R}^n$.

Continuing Step II (from Remark A.3.9), we find the strong guarantee in outcomes $\min_{y \in Y} f(x, y)$ for

$$f(x, y) = x'Ax + 2x'By + y'Cy + 2a'x + 2c'y + d,$$

that is, $f[x] = \min_{y \in \mathbf{R}^m} f(x, y)$, in the case $A < 0$, $C > 0$.

Lemma A.3.2. [38, p. 89] *For any positive definite matrix C of dimensions $n \times n$, there exists a unique positive definite matrix S of dimensions $n \times n$ such that $S^2 = C$. The matrix S is called the square root of the matrix C and denoted by $C^{\frac{1}{2}}$. Moreover, the eigenvalues of the matrix C are the squares of the eigenvalues of the matrix $C^{\frac{1}{2}}$.*

Lemma A.3.3 *For a symmetric matrix $C > 0$ of dimensions $n \times n$, $C^{-1} = [S^2]^{-1} = [S^{-1}]^2$.*

Proof Indeed, for $S = C^{\frac{1}{2}}$ it follows that

$$C = S \cdot S = S^2 \Longrightarrow C^{-1} = [S \cdot S]^{-1} = S^{-1} \cdot S^{-1} = [S^{-1}]^2.$$

Lemma A.3.4
$$A < 0 \wedge C > 0 \Longrightarrow (A - BCB') < 0 \ \forall B \in \mathbf{R}^{n \times m},$$

where $\mathbf{R}^{n \times m}$ is the set of constant matrices of dimensions $n \times m$.

Proof Really,

$$C > 0 \Longrightarrow C^{-1} > 0 \Longrightarrow BC^{-1}B' \geq 0 \ \forall B \in \mathbf{R}^{n \times m} \Longrightarrow -BC^{-1}B' \leq 0 \ \forall B \in \mathbf{R}^{n \times m}$$
$$\Longrightarrow A - BC^{-1}B' < 0 \ \forall B \in \mathbf{R}^{n \times m}.$$

Proposition A.3.6 *If* $A < 0$ *and* $C > 0$, *then*

$$f[x] = \min_{y \in \mathbf{R}^m} f(x, y) = x'[A - BC^{-1}B']x + 2x'[a - BC^{-1}c] + d - c'C^{-1}c. \quad (A.3.11)$$

Proof According to Lemma A.3.2, there exists a matrix S such that $C = S^2$; moreover, $C > 0 \implies S > 0 \wedge S = S'$. Due to $S^{-1}S^{-1} = C^{-1}$ (Lemma A.3.3), $SS = C$, and $S^{-1}S = E_n$, it follows that

$$f(x, y) = x'Ax + 2x'By + y'Cy + 2a'x + 2c'y + d = \left\| S^{-1}B'x + Sy + S^{-1}c \right\|^2$$
$$- x'BC^{-1}B'x - 2x'BC^{-1}c - c'C^{-1}c - y'Cy - 2x'By - 2c'y + x'Ax + 2x'By$$
$$+ 2a'x + d \geq x'[A - BC^{-1}B']x + 2x'[a - BC^{-1}c] + d - c'C^{-1}c = f[x]$$

for all $x \in \mathbf{R}^n$ and $y \in \mathbf{R}^m$, because $\|\cdot\| \geq 0$ by the properties of the Euclidean norm. Using the definition of the strong guarantee in outcomes,

$$f(x, y) \geq f[x] \ \forall x \in \mathbf{R}^n, y \in \mathbf{R}^m,$$

we finally arrive in (A.3.11).

Steps III–IV (construction of the Pareto-maximal alternative x^P in the problem Γ_2 (A.3.8) and calculation of $f[x^P]$).

As it has been established (see Proposition A.3.5), in the linear-quadratic problem Γ_{lq} with

$$A < 0, \ m = n, \ \det B \neq 0, \quad (A.3.12)$$

the strong guarantee in risks is $R_f[x] = 0$ for all $x \in \mathbf{R}^n$. Hence, this is also the case for the Pareto-maximal alternative x^P in the problem Γ_3 (A.3.8). Therefore, the Pareto-maximal alternative in the linear-quadratic problem Γ_{lq} with the matrices (A.3.12) and $C < 0$ can be reduced to the maximization of $f[x]$, i.e.,

$$\max_{x \in \mathbf{R}^n} f[x] = f[x^P]. \quad (A.3.13)$$

A.3.3.4   Explicit form of Pareto-maximal strongly-guaranteed solution of problem Γ_{lq}

Proposition A.3.7 *In the linear-quadratic problem* Γ_{lq} *with*

$$A < 0, \ C > 0, \ m = n, \ \det B \neq 0,$$

the Pareto-maximal strongly-guaranteed solution is given by

$$x^P = -[A - BC^{-1}B']^{-1}(a - BC^{-1}c), \quad (A.3.14)$$

$$f[x^P] = -(a' - c'C^{-1}B')[A - BC^{-1}B']^{-1}(a - BC^{-1}c) + d - c'C^{-1}c. \quad (A.3.15)$$

Proof The alternative x^P defined by (A.3.13) exists under the sufficient conditions

$$\frac{\partial f[x]}{\partial x}\Big|_{x=x^P} = 2[A - BC^{-1}B']x^P + 2(a - BC^{-1}c) = 0_n, \quad (A.3.16)$$

$$\frac{\partial^2 f[x]}{\partial x^2}\Big|_{x=x^P} = 2[A - BC^{-1}B']^{-1} < 0. \quad (A.3.17)$$

Note that (A.3.17) is satisfied due to Lemma A.3.4 and $A < 0, C > 0$. In view of $A - BC^{-1}B' < 0$, equality (A.3.16) implies

$$x^P = -\left[A - BC^{-1}B'\right]^{-1}\left(a - BC^{-1}c\right).$$

Substituting this alternative x^P into (A.3.11) gives

$$\begin{aligned}
f\left[x^P\right] &= \left(a' - c'C^{-1}B'\right)\left[A - BC^{-1}B'\right]^{-1}\left[A - BC^{-1}B'\right] \cdot \left[A - BC^{-1}B'\right]^{-1} \\
&\quad \times \left(a - BC^{-1}c\right) - 2\left(a' - c'C^{-1}B'\right)\left[A - BC^{-1}B'\right]^{-1}\left(a - BC^{-1}c\right) + d - c'C^{-1}c \\
&= -(a' - c'C^{-1}B')[A - BC^{-1}B']^{-1}(a - BC^{-1}c) + d - c'C^{-1}c.
\end{aligned}$$

Remark A.3.10 Thus, the following result has been obtained for the class of linear-quadratic SCPUs Γ_{lq}: *if the criterion in the linear-quadratic problem*

$$\Gamma_{lq} = \langle \mathbf{R}^n, \mathbf{R}^m, f(x, y) = x'Ax + 2x'By + y'Cy + 2a'x + 2c'y + d\rangle$$

satisfies the conditions $A < 0, \ C > 0,$ *and* $\det B \neq 0,$ *then the triplet* $\left(x^P, f[x^P], R_f[x^P]\right),$ *where*

$$x^P = -[A - BC^{-1}B']^{-1}(a - BC^{-1}c),$$

$$f[x^P] = -(a' - c'C^{-1}B')[A - BC^{-1}B']^{-1}(a - BC^{-1}c) + d - c'C^{-1}c, \qquad \text{(A.3.18)}$$

and

$$R_f[x^P] = 0,$$

is the Pareto-maximal strongly-guaranteed solution of Γ_{lq}.

This result has the following interpretation in terms of game theory: choosing the alternative x^P (A.3.18) in the linear-quadratic SCPU Γ_{lq}, the DM obtains the strongly-guaranteed outcome $f[x^P]$ (A.3.18) with the (minimum possible) zero risk $R_f[x^P] = 0$ (i.e., surely!). Note that by Lemma A.3.4 a considerable part of this outcome is

$$-(a' - c'C^{-1}B')[A - BC^{-1}B']^{-1}(a - BC^{-1}c) > 0.$$

A.3.4   CONCLUSIONS

The simplest conflict under uncertainty is "the game with nature," where a person (player) has to choose an optimal action (strategy) for a given criterion (e.g., profit). Moreover, each action is accompanied by incomplete or inaccurate information (uncertainty) about the results (outcome) of such an action.

This raises the question of risk associated with the results. Here an area of intensive research is focused on a special type of uncertainties (interval), for which the only available information is the ranges of their admissible values, without any probabilistic characteristics. An example of such uncertainties is the diversification problem of a deposit into sub-deposits in different currencies [398].

In Russia, interval uncertainties were called "bad uncertainties" due to the unpredictability of their realizations [33–35]. The effect of such uncertainties can be assessed using the Savage–Niehans risk function; the value of this function for a particular alternative or strategy is a measure of risk.

In Appendix 3 of the book, a solution of the single-criterion choice problem under uncertainty (SCPU) that takes into account, *first*, the effect of such uncertainties and, *second*, the DM's desire to increase the outcome and simultaneously reduce the associated risk has been presented. More specifically, the concept of a strong guarantee from [120–123] has been adopted for introducing a new approach that considers all the three factors of decision-making (uncertainty, outcome, and risk). This approach has been reduced to the construction of the game of guarantees, which contains no uncertainties. For the game of guarantees, a corresponding bi-criteria optimization problem has been designed and solved. In the future, a different approach based on vector guarantees [120–123] can be used. For a fairly general class of linear-quadratic SCPUs, the new approach proposed above has anyway yielded an explicit form of the strongly-guaranteed solution in outcomes and risks in which the guaranteed risk (and hence any Savage–Niehans risk) is 0.

The concept of strong coalitional equilibrium

> A compromise is the art of dividing a cake in such a way
> that everyone believes he has the biggest piece.
> —Erhard[42]

In this appendix the concept of strong coalitional equilibrium (SCE) for normal-form games under uncertainty is introduced. This concept integrates the notions of individual and collective rationality in normal-form games without side payments with a proposed coalitional rationality. For the sake of simplicity, the SCE is presented for the four-player games under uncertainty. Sufficient conditions for the existence of an SCE in pure strategies are established via calculating a saddle point for the Germeier convolution. Finally, following the approach of E. Borel, J. von Neumann, and J. Nash, an existence theorem of an SCE in mixed strategies is proved under the standard assumptions of normal-form games (i.e., compact and convex strategy sets of all players, a compact set of uncertainties, and continuous payoff functions of all players).

A.4.I INTRODUCTION

The theory of cooperative games has been developing in three directions as follows. The first direction involves the introduction of equilibrium solutions for normal-form games and their analysis. It is an extension of Nash's theory [310, 311]. The second direction is based on the characteristic function approach. In a characteristic function game, each coalition (subset of players) is associated with a value it can afford. The third and most recent direction considers coalition formation as a dynamic process. Since this appendix contributes to the first direction of research, we will briefly discuss the second and third ones.

In characteristic function games, the most prominent solution concept is the core proposed in [55]. The core rests on the idea of blocking: a coalition can block an imputation if it can improve the payoffs of its members by deviating from the current imputation. An imputation is in the core if it cannot be blocked by any coalition. Many other concepts were also introduced, such as the nucleolus, the kernel, and the Shapley value, to name a few. The main drawback of the characteristic function game and its solution concepts is that they do not incorporate the strategic interaction of players.

---

[42] Ludwig Erhard, (1897–1977), economist and statesman who, as economics minister (1949–63), was the chief architect of West Germany's post-World War II economic recovery.

The obvious limitations of the characteristic form games and their solution concepts led to the appearance of the third direction of research, which considers coalition formation in cooperative games as a dynamic process. The pioneering works in this direction were the publications [403], where players' farsightedness was incorporated into game-theoretic analysis (i.e., the players were assumed to care about long-term outcomes of the game); [411], where "coalition strategies" were introduced to account for the coalitional behavior of players during the game; and [408], where coalition formation was described by a Markov process. For more details, the interested reader is referred to [412].

Now, let us discuss the first direction. Many coalition-related concepts of equilibrium or solutions were introduced for n-player normal-form games. The main motivation for the inception of such investigations was to overcome a well-known drawback of Nash equilibrium (NE): NE is unstable against the deviations of coalitions. A coalition may improve the payoff of all its members by collectively deviating from NE. R. Aumann [399] introduced the strong equilibrium (SE) that is stable against such deviations. As it however turned out, the set of SE is empty for most of the games. Later, Aumann [400] suggested the α-core and β-core for relaxing the conditions of SE. A strategy profile belongs to the core of a game if no group of players has an incentive to form a coalition and choose a different strategy profile in which each of its members are made better-off, i.e., the strategy profile cannot be blocked by any coalition. The α-core and β-core differ by the definition of blocking: the α-core requires a blocking coalition to choose a specific strategy independently of the complementary coalition's choice, whereas the β-core allows a blocking coalition to vary its blocking strategy as a function of the complementary coalition's choice. C. Berge [260] introduced a very strong equilibrium, called the strong Berge equilibrium (SBE), in the sense that if one of the players chooses his strategy from an SBE, the other players have no choice but to play their strategies from the SBE. The Berge equilibrium (BE), put forward by V. Zhukovskiy [354], is an equilibrium that reflects altruism and mutual support among the players. A BE is a strategy profile in which the payoff function of each player is maximized by all the other players. Recently, research on BE has gained some momentum [417], as more empirical research showed that (besides noncooperative behavior) cooperation, mutual support, reciprocity, and caring about fairness may take place in interactions between individuals; see [404, 405, 407, 413]. Bernheim noticed that in an SE some deviations might not be self-enforcing [402], and therefore cannot be treated as credible threats. This led to the introduction of coalition-proof Nash equilibrium (CPNE). In a CPNE, only self-enforcing deviations are credible threats. A deviation by a coalition is self-enforcing if no subcoalition has an incentive to initiate a new deviation. Finally, some works combined different solution concepts, e.g., the hybrid solution of [415], which assumed a coalition structure to be formed, and the game itself to be noncooperative among coalitions but cooperative within coalitions. As a result, Nash equilibrium was adopted for the former and the core for the latter as solutions.

A common drawback of the coalition equilibria and solutions mentioned above is that their set is often empty; they do not exist under standard assumptions such as the compactness and convexity of strategy sets and the continuity and quasiconcavity of payoff functions [314, 315, 402, 409, 416], except for the α-core and Zhao's hybrid solution. Using the notion of balancedness, Scarf [413] established the non-emptiness of the α-core in the case of compact and convex strategy sets and continuous and quasiconcave payoff functions. However, Scarf's theorem suggests no method for determining an α-core element. Zhao's hybrid solution was obtained under similar hypotheses.

As most of these equilibrium concepts and solutions do not exist in the class of pure strategies under standard assumptions of continuous games, a natural question arises: Do these concepts and solutions exist in mixed strategies? Unfortunately, there are no works dealing with this question and related topics in the existing literature. Moreover, the existence of the concepts and solutions has not been considered in games under uncertainty.

In this appendix, we introduce a rational coalitional equilibrium for a game under nonprobabilistic uncertainty as well as establish its existence in mixed strategies. As a matter of fact, this concept generalizes many of the concepts mentioned above.

The mathematical model of cooperation described below is a four-player normal-form game with indeterminate parameters (interval uncertainty). The analysis has been limited to the class of four-player games for the sake of simplicity. Regarding the indeterminate parameters, it is assumed that the players know their range of admissible values only; no probabilistic characteristics are available (for some reasons). The models of game phenomena with a proper consideration of uncertainties yield more adequate results and decisions, which is supported by the numerous publications related to this field of research. (For example, a Google search on the topic "mathematical modeling under uncertainty" returns more than one million links to related works.) The uncertainty appears because of incomplete information about the players' strategy sets, the strategies being chosen by each player, and the related payoffs: "Although our intellect always longs for clarity and certainty, our nature often finds uncertainty fascinating." (C. von Clausewitz[43]). For example, different types of uncertainty in economic systems were described in Section 1.3 of this book.

One more question arises: How a player can simultaneously consider the game's strategic and cooperation aspects, and the presence of uncertainty when choosing his strategy? In this appendix, the following approach to formalize the cooperation aspect of the game is adopted. It is assumed that any non-empty subset of players has the possibility to form a coalition through communication and coordination by agreeing to choose a bundle of strategies to achieve the best possible payoff for all its members. This assumption means that the interests of all possible coalitions are considered. Further, it is also assumed that the game is without side payments or non-transferable utility (NTU). The concept of strong coalitional equilibrium (SCE) is introduced for the game described. A sufficient condition for its existence in pure strategies is provided and its existence in mixed strategies is established under standard assumptions (compact and convex strategy sets of all players, a compact set of uncertainties, and continuous payoff functions of all players).

A.4.2   GAME UNDER UNCERTAINTY

In this section we present the normal-form game under uncertainty. For the sake of simplicity, further presentation will be confined to the class of four-player games only. All the results and definitions below can be easily generalized to n-player games in a straightforward way.

Consider the four-player normal-form game under uncertainty

$$\Gamma = \langle N = \{1, 2, 3, 4\}, \{X_i\}_{i \in N}, Y, \{f_i(x, y)\}_{i \in N} \rangle,$$

---

[43]Carl von Clausewitz, in full Carl Philipp Gottlieb von Clausewitz, (1780–1831), was a Prussian general and military thinker, whose work *Vom Kriege* (1832; On War) has become one of the most respected classics on military strategy.

where $N = \{1, 2, 3, 4\}$ is the set of players; each player $i \in N$ chooses his strategy x_i from his strategy set $X_i \subset \mathbf{R}^{n_i}$, thereby forming a strategy profile $x = (x_1, x_2, x_3, x_4) \in X = \prod_{i=1}^4 X_i \subset \mathbf{R}^n$; $n = \sum_{i \in N} n_i$; an *interval uncertainty* $y \in Y \subset \mathbf{R}^m$ occurs independently of the players' actions; the payoff function of player $i \in N$ is a real-valued function $f_i(x, y)$ that depends on the pair $(x, y) \in X \times Y$. The goal of each player $i \in N$ in the game Γ is to choose a strategy x_i yielding the greatest possible payoff for him. This includes choosing strategies that maximize other players' payoffs if they are beneficial for player i. With this goal in view, the players should consider the possible formation of any coalition and also the possible realization of any uncertainty $y \in Y$. Considering the uncertainty $y \in Y$ leads to a multivalued payoff function of the form $x \to f_i(x, Y) = \bigcup_{y \in Y} f_i(x, y)$. Such multivalued payoff functions complicate further study of the cooperative games Γ. To consider the effect of uncertainty on their payoffs, the players need to adopt a principle of decision-making under uncertainty [189], such as the maximin principle [352], the principle of minimax regret [321, 322], etc. Moreover, a reasonable solution concept for the game Γ must reflect the uncertainty's effect on the players. As uncertainty is considered in equilibria of cooperative games for the first time, we assume that the players adopt a conservative (maximin or risk-averse) approach [352]. Other principles of decision-making under uncertainty in the game Γ can be studied in future works. Thus, the payoff function of each player $i \in N$ will be estimated not by its value $f_i(x, y)$, but by its guaranteed level $f_i[x]$. A guarantee over the values $f_i(x, y)$, $y \in Y$, can be defined as follows:

$$f_i[x] = \min_{y \in Y} f_i(x, y).$$

Really, we have $f_i[x] \leq f_i(x, y)$, $y \in Y$; therefore, a lower bound on the payoff function of player i can be given by $f_i[x]$. As it will be demonstrated below, under common conditions the function $x \to f_i[x]$ is well-defined and continuous on X. In this section and also in Sections A.4.3 and A.4.4, the functions $x \to f_i[x]$, $i \in N$, are assumed to be well-defined and continuous on X. This leads to the (conservative) game of guarantees

$$\Gamma^g = \langle N = \{1, 2, 3, 4\}, \{X_i\}_{i \in N}, \{f_i[x]\}_{i \in N} \rangle.$$

In the next section we will introduce the SCE of the game Γ via the game Γ^g.

A.4.3   COALITIONAL RATIONALITY AND STRONG COALITIONAL EQUILIBRIUM

First, let us present the main properties of SCE, introducing the concept itself later. To define coalitional rationality, the following notations are convenient. For any non-empty subset K of the set N, denote by $-K$ the complement of K, that is, $N \setminus K$. In particular, for each $i \in N$, denote by $-i$ the set $N \setminus \{i\}$; for each $i, j \in N$, $i \neq j$, denote by $-(i, j)$ the set $N \setminus \{i, j\}$. The notion of partitions of a set will be used as well. A partition of a set A is a family of disjoint subsets of A, the union of which equals A. In game theory, a partition of the set of players is called a coalition structure. For a strategy profile $x \in X$ and $i \in N$, denote $x = (x_i, x_{-i})$ and $X_{-i} = \prod_{j \in N \setminus \{i\}} X_j$.

In the game Γ, fifteen coalition structures can be formed as follows: $\{\{1\}, \{2\}, \{3\}, \{4\}\}$, $\{1, 2, 3, 4\}$, $K_{\{i\}} = \{\{i\}, \{-i\}\}$, $K_{\{i\}, \{j\}} = \{\{i\}, \{j\}, \{-(i, j)\}\}$, $K_{\{i,j\}} = \{\{i, j\}, \{-(i, j)\}\}$, for all $i, j \in N$, $i \neq j$. Recall some results from the theory of cooperative games without side payments [189]. For a strategy profile $x^* \in X$ in the game Γ^g, the following properties are considered:

a) x^* satisfies the *individual rationality condition* (IRC) if for all $i \in N$,

$$f_i[x^*] \geq f_i^0 = \max_{x_i \in X_i} \min_{x_{-i} \in X_{-i}} f_i[x_i, x_{-i}] = \min_{x_{-i} \in X_{-i}} f_i[x_i^0, x_{-i}].$$

The value f_i^0 is the guaranteed payoff of player $i \in N$. If player i chooses his *maximin* strategy x_i^0, then his payoff satisfies $f_i[x_i^0, x_{-i}] \geq f_i^0$ for all $x_{-i} \in X_{-i}$.

b) x^* satisfies the *collective rationality condition* (ColRC) if x^* is a Pareto-maximal alternative in the multicriteria choice problem $\Gamma^P = \langle X, f_i[x]_{i \in N} \rangle$, i.e., for all $x \in X$ the system of inequalities $f_i[x] \geq f_i[x^*]$, $i \in N$, with at least one strict inequality, is inconsistent. Note that if $\sum_{i \in N} f_i[x] \leq \sum_{i \in N} f_i[x^*]$ for all $x \in X$, then x^* is a Pareto-maximal alternative in the choice problem Γ^P.

c) x^* satisfies the *coalitional rationality condition* (CoalRC) if

$$f_k[x^*] \geq f_k[x_i^*, x_{-i}] \quad \text{for all} \quad x_{-i} \in X_{-i},$$

$$f_k[x^*] \geq f_k[x_i^*, x_j^*, x_{-(i,j)}] \quad \text{for all} \quad x_{-(i,j)} \in X_{-(i,j)},$$

$$f_k[x^*] \geq f_k[x_i, x_{-i}^*] \quad \text{for all} \quad x_i \in X_i,$$

all the three inequalities holding for all $i, j, k \in N$, $i \neq j$, where $x = (x_i, x_j, x_{-(i,j)})$ and $X_{-(i,j)} = \prod_{s \in N \setminus \{i,j\}} X_s$. This condition means that when a coalition K chooses its strategy profile from x^*, then no player can improve his payoff if the counter-coalition $-K$ deviates from its strategy profile in x^*.

Definition A.4.1 *A strategy profile $x^* \in X$ is called a strong coalitional equilibrium (SCE) for the game Γ if it satisfies IRC, ColRC, and CoalRC for the game of guarantees Γ^g.*

Remark A.4.1 According to IRC, it makes sense for a player to form coalitions with other players if he gets a payoff not less than what he can guarantee by choosing his maximin strategy. ColRC leads the players to a non-dominated strategy profile in terms of Pareto maximality. Finally, CoalRC means that the payoff of each player is stable against any deviations of individual players or coalitions from a strategy profile satisfying CoalRC. In other words, no player's payoff is increased when any coalition deviates from an SCE. Thus, it is rational for all coalitions not to deviate from x^*, because no player in a deviating coalition or outside it will benefit.

   By Definition A.4.1 a SCE must satisfy all the extremal constraints defining IRC, ColRC, and CoalRC. However, all these constraints can be easily derived from the following seventeen of them:

$$f_i[x_1^*, x_2, x_3, x_4] \leq f_i[x^*] \text{ for all } x_k \in X_k, \ k = 2, 3, 4 \text{ and } i = 1, 2, 3, 4;$$

$$f_i[x_1, x_2^*, x_3, x_4] \leq f_i[x^*] \text{ for all } x_k \in X_k, \ k = 1, 3, 4 \text{ and } i = 1, 2, 3, 4;$$

$$f_i[x_1, x_2, x_3^*, x_4] \leq f_i[x^*] \text{ for all } x_k \in X_k, \ k = 1, 2, 4 \text{ and } i = 1, 2, 3, 4; \quad \text{(A.4.1)}$$

$$f_i[x_1, x_2, x_3, x_4^*] \leq f_i[x^*] \text{ for all } x_k \in X_k, \ k = 1, 2, 3 \text{ and } i = 1, 2, 3, 4;$$

$$\sum_{i \in N} f_i[x] \leq \sum_{i \in N} f_i[x^*] \text{ for all } x \in X. \quad \text{Here } x^* = (x_1^*, x_2^*, x_3^*, x_4^*).$$

From this point onwards, we will use the system of inequalities (A.4.1) to establish that a strategy profile is an SCE of the game Γ instead of the system of inequalities involved in the definitions of IRC, ColRC, and CoalRC (see items (a)–(c) above). From (A.4.1) it can be observed that the SCE has two interesting features. First, once the players are in an SCE, they do not have incentive to deviate from it individually, collectively, or in coalitions. Second, if the players are not in an SCE, as soon as one player (or coalition) declares that he (it) will choose his (its) strategy (profile) from an SCE, the other players have no choice but to choose their strategies from the SCE. In other words, any player or coalition can enforce an SCE.

Although SCE does not exist in pure strategies in most of continuous games, in finite games it is not the case. The following example, adapted from [311], shows that an SCE exists in a class of games.

Example A.4.1 Consider a three-player game in which players 1, 2 and 3 choose rows, columns and boxes, respectively, and are named accordingly. Let $\epsilon \in [0, 1]$ and also let α, β, and γ be nonnegative numbers such that $\alpha + \beta + \gamma < 9$. Each of the players has two strategies, $\{T, B\}$ for player 1 and $\{T, L\}$ for players 2 and 3. The minimum payoffs, i.e., $f_i[x_1, x_2, x_3]$ $(i = 1, 2, 3)$, where $x_1 = T, B$ and $x_j = T, L, j = 2, 3$, are given below.

	L	R	L	R
T	2, 2, 2	0, 0, ϵ	0, 0, 0	4, 4, 1
B	α, β, γ	0, 0, ϵ	0, 0, 0	3, 3, 1
		L		R

The strategy profile (T, R, R) is an SCE. Really, this strategy profile satisfies the last inequality of the system (A.4.1): the sum of the payoffs in (T, R, R) is higher than the sum of the payoffs in any other strategy profile, including (B, L, L) due to the inequality $\alpha + \beta + \gamma < 9$ and the fact that α, β, γ are nonnegative numbers. Next, the possible deviations corresponding to the inequalities in (A.4.1) are (T, L, R), (T, R, L), and (T, L, L) when player 1 chooses the SCE strategy T; (B, R, R), (T, R, L), and (B, R, L) when player 2 chooses the SCE strategy R; (B, R, R), (T, L, R), and (B, L, R) when player 3 chooses the SCE strategy R. In all the strategy profiles mentioned, the payoffs of players 1, 2, and 3 are smaller than or equal to their payoffs in (T, R, R), which are 4, 4, and 1, respectively.

A.4.3.1 Related concepts

In this section, we recall the most prominent cooperative solutions of NTU games in normal form and compare them with the SCE. Also, we compare the SCE with the solution concepts defined in dynamic context with respect to coalition deviations.

a)   [399] A strategy profile $x^* \in X$ is a strong equilibrium (SNE) of the game Γ^g if, for all $S \subset N$ and for all $y_{-S} \in X_{-S}$, the system of inequalities $f_i[x^*] < f_i[y_S, x^*_{-S}]$ is inconsistent for all $i \in S$.

   This definition means that no coalition can improve the payoff of all its members by deviating from an SNE when the other players adhere to the SNE.

b)   [400] A strategy profile $x^* \in X$ is in the α-core of the game Γ^g if, for any coalition $S \subset N$ and for each $y_S \in X_S$, there exists $z_{-S} \in X_{-S}$ such that the system of inequalities $f_i[x^*] < f_i[y_S, z_{-S}]$ is inconsistent for all $i \in S$.

   In other words, if a coalition deviates from a strategy profile x^* belonging to the α-core, then the other players have a counterstrategy profile to punish it in such a way that not all members of the coalition are better-off.

c)   [400] A strategy profile $x^* \in X$ is in the β-core of the game Γ^g if, for each coalition $S \subset N$, there exists $z_{-S} \in Z_{-S}$ such that for all $y_S \in X_S$ the system of inequalities $f_i[x^*] < f_i[y_S, z_{-S}]$ is inconsistent for all $i \in S$.

In other words, for each coalition the other players can use a special strategy profile to punish it for any deviation from a strategy profile x^* belonging to the β-core in such a way that not all members of the coalition are better-off.

d)   [260] A strategy profile $x^* \in X$ is a strong Berge equilibrium (SBE) of the game Γ^g if for all $i \in N$, the inequality $f_j[x_i^*, z_{-i}] \leq f_j[x^*]$ holds for all $z_{-i} \in Z_{-i}$ and $j \in -i$.

In other words, no coalition of the form $-i$ can make any of its members better-off by deviating from an SBE. When a player uses his strategy from an SBE, the other players have no choice but to follow him, simply using their strategies from the SBE.

e)   [354] A strategy profile $x^* \in X$ is a Berge equilibrium (BE) of the game Γ^g if for all $i \in N$, the inequality $f_i[x_i^*, z_{-i}] \leq f_i[x^*]$ holds for all $z_{-i} \in Z_{-i}$.

In other words, in a BE the players maximize the payoff functions of each other. This equilibrium reflects mutual support and altruism among the players [417].

Using (A.4.1), we can easily verify that an SCE is also an SNE, an SBE, and a BE. As is well known, an SNE is a CPNE; then, an SCE is also a CPNE. Next, an SCE is an element of the α-core and the β-core. The SCE has similarities with the SBE. However, there are two important differences between these solution concepts. First, in an SBE for each $i \in N$ the system of inequalities $f_j[x_i^*, z_{-i}] \leq f_j[x^*]$ holding for all $z_{-i} \in Z_{-i}$ and $j \in -i$ does not include the inequality corresponding to player i, $f_i[x_i^*, z_{-i}] \leq f_i[x^*]$: the other players do not care about the payoff of player i when choosing their strategies from x^*, and his payoff is not maximized. In an SCE, the inequality $f_i[x_i^*, z_{-i}] \leq f_i[x^*]$ is included, which means that the payoff function of player i is maximized by the other players. This shows that the SCE involves mutual support, whereas the SBE does not. Second, an SBE is generally not Pareto-optimal. The Pareto-maximal SBE was investigated in [417]. The SCE has also some similarities with the BE. However, there are important differences between the two equilibria. The BE expresses mutual support and altruism and ignores the individual interests of players; it is not a refinement of Nash equilibrium as a BE may not satisfy IRC [417]. The SCE differs from the hybrid solution (HS) of [415]: in the latter, it is assumed that a coalition structure is formed and there is no cooperation among the coalitions of this structure; in the former, such assumptions are not made.

Although the coalitional equilibrium [411], the equilibrium binding agreement [412], the equilibrium process of coalition formation [408] and the consistent set [403] were defined in a dynamic context, they can be compared to the SCE based on when a coalition can deviate. In the concepts mentioned, a coalition deviates to another state or strategy profile if and only if all its members are better-off, whereas in the SCE a coalition can deviate if and only if all players of the game are better-off (not only its members).

Moreover, the concepts listed in this section do not consider uncertainty as an exogenous factor, unlike the SCE.

A.4.4   SUFFICIENT CONDITIONS FOR EXISTENCE OF SCE IN PURE STRATEGIES

In the previous section we have seen that an SCE is also an SNE and an SBE. Since these equilibria do not exist in pure strategies in most of the continuous games (see the

Introduction), the SCE suffers from this drawback too. Nevertheless, we will formulate sufficient conditions for its existence using the approach developed in [374]. The approach used in this section paves the way to the next section, where the main result of this paper will be presented. First, we introduce the convolution [54] related to the SCE

$$\varphi_1(x, z) = \max_{i \in N} \{f_i[z_1, x_2, x_3, x_4] - f_i[z]\},$$

$$\varphi_2(x, z) = \max_{i \in N} \{f_i[x_1, z_2, x_3, x_4] - f_i[z]\},$$

$$\varphi_3(x, z) = \max_{i \in N} \{f_i[x_1, x_2, z_3, x_4] - f_i[z]\},$$ (A.4.2)

$$\varphi_4(x, z) = \max_{i \in N} \{f_i[x_1, x_2, x_3, z_4] - f_i[z]\},$$

$$\varphi_5(x, z) = \sum_{i \in N} f_i[x] - \sum_{i \in N} f_i[z],$$

$$\varphi(x, z) = \max_{r=1,...,5} \{\varphi_r(x, z)\},$$

where $x = (x_1, x_2, x_3, x_4)$ and $z = (z_1, z_2, z_3, z_4) \in X = \prod_{i \in N} X_i$.

A saddle point $(x^0, z^*) \in X \times X$ of the real-valued function $\varphi(x, z)$ in (A.4.2) is defined by the chain of inequalities

$$\varphi(x, z^*) \le \varphi(x^0, z^*) \le \varphi(x^0, z), \quad \text{for all } x, z \in X.$$ (A.4.3)

Proposition A.4.1 *If $(x^0, z^*) \in X \times X$ is a saddle point of the function $\varphi(x, z)$, then the minimax strategy z^* is an SCE of the game Γ.*

Proof Letting $z = x^0$ in (A.4.3), from (A.4.2) we obtain $\varphi(x^0, x^0) = 0$. Then by transitivity, from (A.4.3) it follows that

$$\varphi\left(x^0, z^*\right) \le \varphi\left(x^0, x^0\right) = 0 \Rightarrow \varphi\left(x, z^*\right) \le 0, \quad \text{for all } x \in X,$$

which implies (A.4.1).

Remark A.4.2 According to Theorem A.4.1, the determination of an SCE reduces to the determination of a saddle point (x^0, z^*) of the Germeier convolution $\varphi(x, z)$ from (A.4.2). We obtain the following procedure for calculating an SCE in the game Γ.

Step 1. Construct the function $\varphi(x, z)$ by (A.4.2).

Step 2. Find a saddle point $(x^0, z^*) \in X \times X$ of the function $\varphi(x, z)$.

Step 3. Compute the four values $f_i[z^*]$, $i \in N$.

Then the pair $(z^*, f[z^*] = (f_1[z^*], f_2[z^*], f_3[z^*], f_4[z^*])) \in X \times \mathbb{R}^4$ consists of the SCE z^* and the corresponding payoffs of the four players. When the players choose their strategies from the SCE z^*, they gain the payoffs $f_i[z^*]$, $i \in N$, respectively.

Thus, if the function of two variables $\varphi(x, z)$ has a saddle point, the well-known numerical methods can be used for computing saddle points.

A.4.5 EXISTENCE OF SCE IN MIXED STRATEGIES

Like the SNE and SBE, the SCE does not exist in pure strategies in the majority of continuous games. Hence, we can naturally employ the strategy randomization approach, which was used in [266–268, 310–312] to establish the existence of Nash equilibrium in mixed strategies. Following these great scholars, we will establish the existence of an SCE in mixed strategies. For this purpose, some preliminary results are needed, which will help in proving the main existence theorem.

A.4.5.I Preliminaries

First, we introduce some auxiliary notations. Denote by comp R^{n_i} and cocomp R^{n_i} the set of compact subsets of R^{n_i} and the set of convex and compact subsets of R^{n_i}, respectively, and also by $C(X \times Y)$ the set of real-valued and continuous functions with a domain of definition $X \times Y$.

Assume that the elements of the game Γ satisfy the following condition.

Condition A.4.1

$$X_i \in \text{cocomp } R^{n_i}, \quad Y \in \text{cocomp } R^m, \quad f_i(\cdot) \in C(X \times Y), \quad \text{for all } i \in N. \qquad (A.4.4)$$

Then, according to Berge' maximum theorem [204], the function $x \rightarrow f_i[x]$ is well-defined and continuous on X for all $i \in N$.

Next, we construct the mixed extension of the game Γ^g, which includes the sets of mixed strategies and mixed strategy profiles as well as the expected value of the players' payoff functions.

First, we associate with each strategy set $X_i \in \text{cocomp } R^{n_i}$ the Borel σ-algebra $B(X_i)$, which consists of subsets $Q^{(i)}$ of X_i such that the intersection and union of a countable set of elements of $B(X_i)$ belong to $B(X_i)$; moreover, $B(X_i)$ is the minimal σ-algebra that contains all closed subsets of X_i. In game theory, a mixed strategy $v_i(\cdot)$ of player i can be identified with a *probability measure* on the compact set of pure strategies X_i. A probability measure is a nonnegative function $v_i(\cdot)$ defined on the Borel σ-algebra $B(X_i)$ that satisfies the two conditions:

C.1 $\quad v_i \left(\bigcup_k Q_k^{(i)} \right) = \sum_k v_i \left(Q_k^{(i)} \right)$ for any sequence of disjoint elements $\left\{ Q_k^{(i)} \right\}$ of $B(X_i)$

(countable additivity);

C.2 $\quad v_i(X_i) = 1$ (normality).

Note that (C.2) implies the inequality $v_i \left(Q^{(i)} \right) \leq 1$ for all $Q^{(i)} \in B(X_i)$.

Denote by $\{v_i\}$ the set of mixed strategies of player $i \in N$. Then a mixed strategy profile of the game Γ^g can be formulated as a product-measure

$$v(dx) = v_1(dx_1)v_2(dx_2)v_3(dx_3)v_4(dx_4);$$

the set of such measures will be denoted by $\{v\}$. The payoff of player i corresponding to his payoff function in the game Γ^g is defined by $f_i[v] = \int_X f_i[x] v(dx)$. Then the mixed extension of the game Γ^g has the form

$$\widetilde{\Gamma}^g = \langle N = \{1, 2, 3, 4\}, \{v_i\}_{i \in N}, \{f_i[v]\}_{i \in N} \rangle. \qquad (A.4.5)$$

Here we have committed an abuse of notations, denoting the expected value of the function $f_i[x]$ by $f_i[v]$. The reader can distinguish between the two functions by the variable involved.

Now, we suggest the following definition of equilibrium, using Definition A.4.1 and (A.4.1).

Definition A.4.2 *A mixed strategy profile $v^*(\cdot) \in \{v\}$ is called a mixed-strategy coalitional equilibrium (MSCE) of the game Γ if it is an SCE of the mixed-extension game (A.4.5), that is,*

(i) $v^*(\cdot)$ *satisfies individual rationality and coalitional rationality (IRC and CoalRC), which can be derived from the inequalities*

$$f_i\left[v_1^*, v_2, v_3, v_4\right] \le f_i[v^*], \text{ for all } v_k(\cdot) \in \{v_k\}, \ k = 2, 3, 4 \text{ and } i = 1, 2, 3, 4,$$

$$f_i\left[v_1, v_2^*, v_3, v_4\right] \le f_i[v^*], \text{ for all } v_k(\cdot) \in \{v_k\}, \ k = 1, 3, 4 \text{ and } i = 1, 2, 3, 4,$$

$$f_i\left[v_1, v_2, v_3^*, v_4\right] \le f_i[v^*], \text{ for all } v_k(\cdot) \in \{v_k\}, \ k = 1, 2, 4 \text{ and } i = 1, 2, 3, 4,$$

$$\text{(A.4.6)}$$

$$f_i\left[v_1, v_2, v_3, v_4^*\right] \le f_i[v^*], \text{ for all } v_k(\cdot) \in \{v_k\}, \ k = 1, 2, 3 \text{ and } i = 1, 2, 3, 4,$$

where $v^* = (v_1^*, v_2^*, v_3^*, v_4^*)$;

(ii) $v^*(\cdot)$ *satisfies collective rationality (ColRC), or it is a Pareto-maximal alternative in the quad-criteria choice problem*

$$\widetilde{\Gamma}_v^g = \langle \{v\}, \{f_i[v]\}_{i \in N} \rangle,$$

that is, for all $v(\cdot) \in \{v\}$ the system of inequalities

$$f_i[v] \ge f_i[v^*], \quad i = 1, 2, 3, 4,$$

with at least one strict inequality, is inconsistent.

A sufficient condition for Pareto optimality (see item (ii)) is as follows.

Remark A.4.3 A mixed strategy profile (alternative) $v^*(\cdot) \in \{v\}$ is Pareto-optimal in the multicriteria choice problem $\widetilde{\Gamma}_v^g = \langle \{v\}, \{f_i[v]\}_{i \in N} \rangle$ if

$$\max_{v(\cdot) \in \{v\}} \sum_{i \in N} f_i[v] = \sum_{i \in N} f_i\left[v^*\right].$$

Consider the function $\varphi_i(x, z)$, $i = 1, 2, 3, 4$ and also the function

$$\varphi(x, z) = \max_{r=1,\dots,5} \{\varphi_r(x, z)\} \tag{A.4.7}$$

introduced in (A.4.2).

Proposition A.4.1 *The inequality*

$$\max_{r=1,\dots,5} \int_{X \times X} \varphi_r(x, z) \mu(dx) v(dz) \le \int_{X \times X} \max_{r=1,\dots,5} \varphi_r(x, z) \mu(dx) v(dz) \tag{A.4.8}$$

holds for all $v(\cdot) \in \{v\}$ and $\mu(\cdot) \in \{v\}$.

Proof From (A.4.7), for all $x, z \in X$, we obtain the five inequalities

$$\varphi_r(x, z) \le \varphi(x, z) = \max_{r=1,\dots,5} \varphi_r(x, z), \quad r = 1, \dots, 5.$$

Integrating both sides of these inequalities with an arbitrary product-measure $\mu(dx)\nu(dz)$ yields

$$\int_{X \times X} \varphi_r(x, z)\mu(dx)\nu(dz) \leq \int_{X \times X} \max_{r=1,\dots,5} \varphi_r(x, z)\mu(dx)\nu(dz),$$

for all $\mu(\cdot), \nu(\cdot) \in \{\nu\}$ and $r = 1, \dots, 5$. Therefore,

$$\max_{r=1,\dots,5} \int_{X \times X} \varphi_r(x, z)\mu(dx)\nu(dz) \leq \int_{X \times X} \max_{r=1,\dots,5} \varphi_r(x, z)\mu(dx)\nu(dz),$$

for all $\mu(\cdot), \nu(\cdot) \in \{\nu\}$. Hence, (A.4.8) is satisfied.

Remark A.4.4 In fact, Proposition A.4.1 generalizes the well-known property of maximization: the maximum of a sum of some functions does not exceed the sum of their maxima.

Proposition A.4.2 *The function $\varphi(x, z)$ defined in (A.4.7) is continuous on $X \times Z$, where $Z = X$.*

The proof of a more general result (the continuity of the maximum of a finite number of continuous functions on a compact set) can be found in many textbooks, e.g., in [204].

A.4.5.2   Existence theorem

In this subsection, we prove the main result of Appendix 4, that is, the existence of an MSCE in the game Γ.

Theorem A.4.2 *Under Condition A.4.1, the game Γ has an MSCE.*

Proof Consider the two-player zero-sum game

$$\Gamma^a = \langle \{1, 2\}, X, Z, \varphi(x, z) \rangle,$$

where $X = Z$. In the game Γ^a, the maximizing and minimizing players choose their strategies from the same set X; $\varphi(x, z)$ is the payoff function of the maximizing player and $-\varphi(x, z)$ is the payoff function of the minimizing player. Any saddle point (x^0, z^*) of the function $\varphi(x, z)$ is an NE in the game Γ^a. Really, by the saddle point definition

$$\varphi(x, z^*) \leq \varphi\left(x^0, z^*\right) \leq \varphi\left(x^0, z\right), \text{ for all } (x, z) \in X \times Z,$$

the strategy profile (x^0, z^*) is an NE in the game Γ^a. Now, we associate with the game Γ^a its mixed strategy extension

$$\widetilde{\Gamma}^a = \langle \{1, 2\}, \{\mu\}, \{\nu\}, \varphi(\mu, \nu) \rangle,$$

where $\{\mu\}$ is the set of all strategies of the maximizing player; $\{\nu\} = \{\mu\}$ is the set of all strategies of the minimizing player; $\varphi(\mu, \nu)$ is the payoff (expected utility) of the maximizing player,

$$\varphi(\mu, \nu) = \int_{X \times X} \varphi(x, z)\mu(dx)\nu(dz). \tag{A.4.9}$$

Here, we have committed another abuse of notations, denoting the expected value of the function $\varphi(x, z)$ by $\varphi(\mu, \nu)$. The reader can distinguish between the two by the variables involved. In a similar fashion, any saddle point (μ^0, ν^*) of the function $\varphi(\mu, \nu)$ is an NE in the game $\widetilde{\Gamma}^a$. Really, by the saddle point definition

$$\varphi(\mu, \nu^*) \le \varphi(\mu^0, \nu^*) \le \varphi(\mu^0, \nu), \quad \text{for all } (\mu, \nu) \in \{\nu\} \times \{\nu\}, \tag{A.4.10}$$

the strategy profile (μ^0, ν^*) is an NE in the game $\widetilde{\Gamma}^a$.

In 1952 I. Gliksberg established the existence of an NE in mixed strategies for the N-player games with $N > 1$; see the original paper [55]. Using his result for the two-player zero-sum game Γ^a as a special case, we obtain the following statement. Since the set of all strategy profiles $X \subset \mathbb{R}^n$ is convex and compact and the function $\varphi(x, z)$ is continuous on $X \times X$ (Proposition A.4.2), the game Γ^a has a mixed-strategy NE (μ^0, ν^*) satisfying (A.4.10).

In view of (A.4.7) and (A.4.9), inequalities (A.4.10) take the form

$$\int_{X \times X} \max_{r=1,\ldots,5} \varphi_r(x, z) \mu(dx) \nu^*(dz) \le \int_{X \times X} \max_{r=1,\ldots,5} \varphi_r(x, z) \mu^0(dx) \nu^*(dz)$$

$$\le \int_{X \times X} \max_{r=1,\ldots,5} \varphi_r(x, z) \mu^0(dx) \nu(dz),$$

for all $(\mu, \nu) \in \{\nu\} \times \{\nu\}$. Letting $\nu_i(dz_i) = \mu_i^0(dx_i)$, $i \in N$, in

$$\varphi(\mu^0, \nu) = \int_{X \times X} \max_{r=1,\ldots,5} \varphi_r(x, z) \mu^0(dx) \nu(dz),$$

we obtain

$$\varphi(\mu^0, \mu^0) = \int_{X \times X} \max_{r=1,\ldots,5} \varphi_r(x, x) \mu^0(dx) \mu^0(dx).$$

(In this case, $\nu(dz) = \mu^0(dx)$.) From (A.4.2) it follows that $\varphi_r(x, x)$, $r = 1, \ldots, 5$, for all $x \in X$. Then the previous integral yields $\varphi(\mu^0, \mu^0) = 0$. A similar reasoning leads to $\varphi(\nu^*, \nu^*) = 0$. From (A.4.10) we obtain

$$\varphi(\mu^0, \nu^*) = 0. \tag{A.4.11}$$

Using (A.4.11) and inequalities (A.4.10), by transitivity we write

$$\varphi(\mu, \nu^*) = \int_{X \times X} \max_{r=1,\ldots,5} \varphi_r(x, z) \mu(dx) \nu^*(dz) \le 0, \quad \text{for all } \mu \in \{\nu\}.$$

According to Proposition A.4.1,

$$\max_{r=1,\ldots,5} \int_{X \times X} \varphi_r(x, z) \mu(dx) \nu^*(dz) \le \int_{X \times X} \max_{r=1,\ldots,5} \varphi_r(x, z) \mu(dx) \nu^*(dz) \le 0,$$

for all $\mu \in \{\nu\}$. Therefore,

$$\int_{X \times X} \varphi_r(x, z) \mu(dx) \nu^*(dz) \le 0, \quad \text{for all } \mu \in \{\nu\} \text{ and for all } r = 1, \ldots, 5. \tag{A.4.12}$$

We will distinguish two cases as follows.

Case 1 ($r = 1, \ldots, 4$). Due to (A.4.2), (A.4.12) and the fact that $\mu(\cdot)$ is normalized $\left(\text{i.e., } \int\limits_X \mu(dx) = 1\right)$, for example, for $r = 1$ we write

$$f_i[\nu_1^*, \mu_2, \mu_3, \mu_4] - f_i[\nu^*] = \int\limits_{X \times X} f_i[z_1, x_2, x_3, x_4]\mu(dx)\nu^*(dz)$$

$$- \int\limits_X f_i[z]\nu^*(dz) \int\limits_X \mu(dx) = \int\limits_{X \times X} f_i[z_1, x_2, x_3, x_4]\mu(dx)\nu^*(dz)$$

$$- \int\limits_{X \times X} f_i[z]\mu(dx)\nu^*(dz) = \int\limits_{X \times X} (f_i[z_1, x_2, x_3, x_4] - f_i[z])\mu(dx)\nu^*(dz)$$

$$\leq \int\limits_{X \times X} \max_{i \in N}\{f_i[z_1, x_2, x_3, x_4] - f_i[z]\}\mu(dx)\nu^*(dz) = \int\limits_{X \times X} \varphi_1(x, z)\mu(dx)\nu^*(dz) \leq 0,$$

which holds for all $i \in N$. Thus, $f_i[\nu_1^*, \mu_2, \mu_3, \mu_4] - f_i[\nu^*] \leq 0$ for all $i \in N$ and $\mu_k \in \{\nu_k(\cdot)\}$, $k = 2, 3, 4$.

Similar considerations can be used to establish the following three inequalities for $r = 2, 3, 4$:

$$f_i\left[\mu_1, \nu_2^*, \mu_3, \mu_4\right] - f_i[\nu^*] \leq 0, \text{ for all } i \in N \text{ and } \mu_k \in \{\nu_k(\cdot)\}, k = 1, 3, 4;$$

$$f_i\left[\mu_1, \mu_2, \nu_3^*, \mu_4\right] - f_i[\nu^*] \leq 0, \text{ for all } i \in N \text{ and } \mu_k \in \{\nu_k(\cdot)\}, k = 1, 2, 4;$$

$$f_i\left[\mu_1, \mu_2, \mu_3, \nu_4^*\right] - f_i[\nu^*] \leq 0, \text{ for all } i \in N \text{ and } \mu_k \in \{\nu_k(\cdot)\}, k = 1, 2, 3.$$

Therefore, the mixed strategy profile $\nu^*(\cdot)$ satisfies the four inequalities (A.4.6) from condition (i) of Definition A.4.2. It remains to prove that $\nu^*(\cdot)$ also satisfies condition (ii) of Definition A.4.2, i.e., that it has the property of Pareto optimality or collective rationality. For this purpose, we will use Remark A.4.3.

Case 2 ($r = 5$). Due to (A.4.2), (A.4.12) and the fact that $\mu(\cdot)$ and $\nu^*(\cdot)$ are both normalized $\left(\int\limits_X \mu(dx) = \int\limits_X \nu^*(dx) = 1\right)$, we write

$$\sum_{i \in N} f_i[\mu] - \sum_{i \in N} f_i[\nu^*] = \sum_{i \in N} \int\limits_X f_i[x]\mu(dx) - \sum_{i \in N} \int\limits_X f_i[z]\nu^*(dz)$$

$$= \int\limits_X \sum_{i \in N} f_i[x]\mu(dx) \int\limits_X \nu^*(dz) - \int\limits_X \sum_{i \in N} f_i[z]\nu^*(dz) \int\limits_X \mu(dx)$$

$$= \int\limits_{X \times X} \left[\sum_{i \in N} f_i[x] - \sum_{i \in N} f_i[z]\right] \mu(dx)\nu^*(dz) = \int\limits_{X \times X} \varphi_5(x, z)\mu(dx)\nu^*(dz) \leq 0.$$

Therefore, $\sum_{i \in N} f_i[\mu] - \sum_{i \in N} f_i[\nu^*] \leq 0$ for all $\mu \in \{\nu\}$. Then, according to Remark A.4.3, the mixed strategy profile $\nu^*(\cdot)$ is a Pareto-optimal alternative in the multicriteria choice problem

$$\widetilde{\Gamma}_\nu^g = \langle \{\nu\}, \{f_i[\nu]\}_{i \in N}\rangle.$$

Thus, we have established that the mixed strategy profile $v^*(\cdot)$ is an SCE in the game $\tilde{\Gamma}^g$. By Definition A.4.2, $v^*(\cdot)$ is an MSCE in the game Γ and $f[v^*]$ is the players' payoff vector.

A.4.6   CONCLUSIONS

Appendix 4 has contributed to the theory of cooperative normal-form games in the following way. First, the concept of the strong coalitional equilibrium (SCE) in normal-form games under uncertainty has been formalized. This concept considers the interests of all coalitions. Second, a constructive procedure for determining a pure-strategy SCE has been provided; this procedure reduces to saddle point calculation for a function of two variables. Third, the existence of SCE in mixed strategies has been proved under standard assumptions of cooperative game theory (continuous payoff functions of the players, compact and convex strategy sets of the players, and a compact set of uncertainty).

In our view, the following qualitative results of Appendix 4 are important.

1) The approach presented here can be extended to the games with any finite number of players (more than four players).

2) An SCE $x^* \in X$ is stable against any deviation of any coalition of players and is attractive, because when a coalition chooses its strategies from x^*, all other players will have incentive to choose their strategies from x^* as well.

3) The SCE could be applied even if the coalition structure changes over time.

4) The SCE could be used for the formation of stable alliances.

5) Game theory has been focusing on individual rationality and collective rationality so far. On the one hand, the individual interests of players are represented by the prominent Nash equilibrium with its selfish character (each player acts for himself only). On the other, the collective interests of players are represented by the concept of Berge equilibrium with its altruism (each player helps others, neglecting his own interests). Such an omission is not rooted in the human nature of players. The SCE partially addresses the incomplete representation of human behavior in the two concepts mentioned. In the game Γ, when player 1 chooses his SCE strategy, he does not neglect his own interests as an SCE is also a Nash equilibrium; moreover, according to (A.4.1), he also helps (maximizes the payoffs of) all other players, which is the inherent property of a Berge equilibrium. The other players act in a similar way. Thus, the SCE fills the gap between the concepts of Nash equilibrium and Berge equilibrium, completing the former by "caring about others" and the latter by "caring about oneself."

Finally, we suggest two possible ways of extending this research. The first is to investigate the SCE in finite games. The second is to consider other approaches to manage uncertainty, such as the maximin regret principle.

Bibliography

1. Isaacs, R., *Differential Games: A Mathematical Theory with Applications to Warfare and Pursuit, Control and Optimization*, Mineola, NY: Dover, 1999.

2. Altunin, A.E. and Semukhin, N.V., *Modeli i algoritmy prinyatiya reshenii v nechetkikh usloviyakh: Monografiya* (Models and Algorithms of Decision-Making under Fuzzy Conditions: A Monograph), Tyumen: Tyumensk. Gos. Univ., 2000.

3. Antipin, A.S., *Gradientnyi i ekstragradientnyi podkhody v bilineinom ravnovesnom programmirovanii* (Gradient and Extra-gradient Approaches in Bilinear Equilibrium Programming), Moscow: Vychisl. Tsentr Ross. Akad. Nauk, 2002.

4. Aris, R., *Discrete Dynamic Programming*, New York: Blaisdell, 1964.

5. Apresyan, R.G., The Golden Rule, in *Etika: novye starye problemy* (Ethics: New Old Problems), Moscow: Gardariki, 1999.

6. Apresyan, R., The Golden Rule of Ethics. *Genesis*. http://tvkultura.ru/video/show/brand_id/ 20898/video_id/270913/viewtype/picture

7. *Aristotle's Nicomachean Ethics*, translated by Bartlett, R.C. and Collins, S.D., University of Chicago Press, 2012.

8. Aumann, R.J. and Serrano, R, An Economic Index of Riskiness, *Journal of Political Economy*, 2008, vol. 116, no. 5, pp. 810–836.

9. Akhrameev, P.K. and Zhukovskiy, V.I., Guaranteed Solution in Outcomes and Risks of Linear-Quadratic Control Problem, in *Spektral'nye i evolyutsionnye zadachi* (Spectral and Evolutionary Problems), 2013, vol. 23, pp. 104–109.

10. Ashmanov, S.A. and Timokhov, A.V., *Teoriya optimizatsii v zadachakh i uprazhneniyakh* (Optimization Theory in Problems and Exercises), Moscow: Nauka, 1991.

11. Bardin, A.E., *Riski v igrovykh modelyakh konfliktov* (Risks in Game-Theoretic Models of Conflicts), Orekhovo-Zuevo: Mosk. Gos. Oblast. Gumanit. Inst., 2014.

12. Bel'skikh, Yu.A., Zhukovskiy, V.I., and Samsonov, S.P., Altruistic Berge Equilibrium in Bertrand Duopoly Model, *Vestn. Udmurt. Univ. Ser. Math. Mech. Comp. Sci.*, 2016, vol. 26, no. 1, pp. 27–45.

13. Boribekova, K.A. and Zharkynbaev, S., Berge Equilibrium in a Differential Game, *Trudy mezhdunarodnoi konferentsii "Mnogokriterial'nye dinamicheskie zadachi pri neopredelennosti"* (Proc. Int. Conf. "Multicriteria Dynamic Problems under Uncertainty"), 1991, pp. 83–86.

14. Bellman, R., *Adaptive Control Processes. A Guided Tour*, Princeton: Princeton Univ. Press, 1961.

15. Bellman, R.E. and Dreyfus, S.E., *Applied Dynamic Programming*, Princeton: Princeton Univ. Press, 1962.

16. Bel'skikh, Yu.A. and Zhukovskiy, V.I., Nash and Berge Equilibria in a Two-Player Linear-Quadratic Game, *Trudy XIII Mezhdunarodnoi nauchno-prakticheskoi konferentsii "Otechestvennaya nauka v epokhu izmenenii: postulaty proshlogo i teorii novogo vremeni"*

(Proc. XIII Int. Scientific-Practical Conf. "National Science in the Era of Change: Postulates of Past and Theories of New Time"), St. Petersburg, National Association of Scientists, 2015, vol. 13, pp. 41–44.

17. Berge, C., *Obshchaya teoriya igr neskol'kikh lits*, Moscow: Fizmatgiz, 1961.

18. Berezin, S.A., Lavrovskii, B.L., Rybakova, T.A., and Satanova, E.A., *Faktor neopredelennosti v mezhotraslevykh modelyakh* (The Uncertain Factor in Interdisciplinary Models), Novosibirsk: Nauka, 1983.

19. Boltyanskii, V.G., *Optimal'noe upravlenie diskretnymi sistemami* (Optimal Control of Discrete Systems), Moscow: Nauka, 1973.

20. *Bol'shaya kniga aforizmov* (A Big Book of Aphorisms), Moscow: Eksmo, 2005.

21. *Bol'shoi tolkovyi slovar' russkogo yazyka* (The Big Explanatory Dictionary of the Russian Language), St. Peterburg: Norint, 2003.

22. Borisovich, Yu.G., Gel'man, B.D., Myshkis, A.D., and Obukhovskii, V.V., *Vvedenie v teoriyu mnogoznachnykh otobrazhenii* (An Introduction to Theory of Multivalued Maps), Voronezh: Voronezh. Univ., 1986.

23. *The Essential Talmud: Tractates Ta'Anit and Bava Metzia*, translated by Steinsaltz, A., Basic Books, 2006.

24. Vaisman, K.S., Berge Equilibrium, *Cand. Sci. (Phys.-Math.) Dissertation*, St. Petersburg: Sankt-Peterb. Gos. Univ., 1995.

25. Vaisman, K.S., Berge Equilibrium, in *Lineino-kvadratichnye differentsial'nye igry* (Linear-Quadratic Differential Games), Zhukovskiy, V.I. and Chikrii, A.A., Eds., Kiev: Naukova Dumka, 1994, pp. 119–143.

26. Vaisman, K.S., The Existence of Guaranteed Berge Equilibrium in a Differential Game, *Tezisy dokladov Vesennei Matematicheskoi Shkoly "Pontryaginskie chteniya–VI"* (Abstr. Proc. Spring Mathematical School "Pontryagin's Readings–VI"), Voronezh, 1995, p. 19.

27. Vaisman, K.S., On a Solution to Strictly Convex Noncooperative Game, in *Slozhnye upravlyaemye sistemy* (Complex Controlled Systems), Moscow: Ross. Zaochn. Inst. Tekstil. Legk. Promyshl., 1996, pp. 13–15.

28. Vaisman, K.S., Berge Equilibrium in a Differential Game, in *Slozhnye dinamicheskie sistemy* (Complex Dynamic Systems), Pskov: Pskov. Ped. Inst., 1994, pp. 58–63.

29. Vaisman, K.S., Nash Bargaining Solution under Uncertainty, in *Kooperativnye igry pri neopredelennosti i ikh primenenie* (Cooperative Games under Uncertainty and Their Application), Moscow: Editorial URSS, 2010, pp. 213–249.

30. Vaisman, K.S. and Aimukhanov, N.Zh., Berge Equilibrium in a Differential-Difference Game, in *Slozhnye upravlyaemye sistemy* (Complex Controlled Systems), Moscow: Ross. Zaochn. Inst. Tekstil. Legk. Promyshl., 1996, pp. 90–93.

31. Vaisman, K.S., and Zhukovskiy, V.I., The Properties of Berge Equilibrium, *Tezisy dokladov V shkoly "Matematicheskie problemy ekologii"* (Abstr. Proc. V school "Mathematical Problems of Ecology"), Chita, 1994, pp. 27–28.

32. Vaisman, K.S., and Zhukovskiy, V.I., The Structure of Berge Equilibria, *Tezisy dokladov Vesennei Matematicheskoi Shkoly "Pontryaginskie chteniya–V"* (Abstr. Proc. Spring Mathematical School "Pontryagin's Readings–V"), Voronezh, 1994, p. 29.

33. Vasil'ev, F.P., *Metody optimizatsii* (Optimization Methods), Moscow: Faktorial Press, 2002.

34. Wentzel, E.S., *Issledovanie operatsii* (Operations Research), Moscow: Znanie, 1976.

35. Wentzel, E.S., *Elementy dinamicheskogo programmirovaniya* (Elements of Dynamic Programming), Moscow: Nauka, 1964.

36. Wentzel, E.S., *Issledovanie operatsii: zadachi, printsipy, metodologiya* (Operations Research: Problems, Principles, Methodology), Moscow: Nauka, 1980.

37. Vilkas, E.I., *Optimal'nost' v igrakh i resheniyakh* (Optimality in Games and Decisions), Moscow: Nauka, 1990.

38. Voevodin, V.V. and Kuznetsov, Yu.A., *Matritsy i vychisleniya* (Matrices and Calculations), Moscow: Nauka, 1984.

39. Vorobiev, N.N., Noncooperative Games, in *Problemy kibernetiki* (Problems of Cybernetics), 1978, vol. 33, pp. 69–90.

40. Vorobiev, N.N., *Issledovanie operatsii*, (Operations Research), Moscow: Nauka, 1972.

41. Vorobiev, N.N., *Osnovy teorii igr. Beskoalitsionnye igry* (Fundamentals of Game Theory. Noncooperative Games), Moscow: Nauka, 1984.

42. Vorobiev, N.N., The Present State of the Theory of Games, *Russian Mathematical Surveys*, 1970, vol. 25, no. 2, pp. 77–136

43. Vorobiev, N.N., *Teoriya igr dlya ekonomistov-kibernetikov* (Game Theory for Economists-Cyberneticians), Moscow: Nauka, 1985.

44. Vorontsovskii, A.V., *Upravlenie riskami* (Risk Management), St. Petersburg: Sankt-Peterb. Gos. Univ., 2000.

45. Voshchinin, A.P. and Sotirov, G.R., *Optimizatsiya v usloviyakh neopredelennosti* (Optimization under Uncertainty), Moscow: Mosk. Energ. Inst., 1989.

46. Vysokos, M.I. and Zhukovskiy, V.I., The Diversification Problem from the View of a Risk-Neutral Player, *Analiz, Modelirovanie, Upravlenie, Razvitie Sotsial'no-Ekonomicheskikh Sistem: Trudy IX Mezhdunarodnoi Shkoly-Simpoziuma (AMUR-2015)* (Analysis, Modeling, Management, Development of Socio-economic Systems: Proc. IX Int. School-Symposium (AMMD-2015)), Krymskii Fed. Univ., Simferopol, 2015, p. 84.

47. Vysokos, M.I. and Zhukovskiy, V.I., Berge Equilibrium—A Mathematical Model of the Golden Rule of Ethics, *Trudy mezhdunarodnoi nauchno-prakticheskoi konferentsii* (Proc. Int. Scientific and Practical Conference), Volgograd, October 11, 2016, pp. 30–33.

48. Vysokos, M.I. and Zhukovskiy V.I., The Golden Rule in Cournot Duopoly Model, *Taurida J. Comp. Sci. Theory and Math.*, 2015, vol. 27, no. 2, pp. 46–54.

49. Gabasov, R. and Kirillova, F.M., *Osnovy dinamicheskogo programmirovaniya* (Fundamentals of Dynamic Programming), Minsk: Belaruss. Gos. Univ., 1975.

50. *Poisk*, 2004, nos. 34–35, p. 7.

51. Gantmacher, F.R., *The Theory of Matrices. Vols. 1 and 2*, New York: Chelsea Publishing, 1959.

52. Gelfond, A.O., *Ischislenie konechnykh raznostei* (Calculus of Finite Differences), Moscow: URSS, KomKniga, 2006.

53. Germeier, Yu.B., *Vvedenie v teoriyu issledovaniya operatsii* (Introduction to Operations Research), Moscow: Nauka, 1971.

54. Germeier, Yu.B., *Non-antagonistic Games*, Reidel, 1986.

55. Glicksberg, I.L., A Further Generalization of the Kakutani Fixed Point Theorem, with Application to Nash Equilibrium Points, *Proc. Am. Math. Soc.*, 1952, vol. 3, no. 1, pp. 170–174.

56. Gorbatov, A.S. and Zhukovskiy, V.I., On a Guaranteed Equilibrium, in *Spektral'nye i evolyutsionnye zadachi* (Spectral and Evolutionary Problems), 2013, vol. 23, pp. 88–103.

57. Gorbatov, A.S. and Zhukovskiy, V.I., A New Guaranteed Equilibrium in Conflict under Uncertainty, *Analiz, Modelirovanie, Upravlenie, Razvitie Sotsial'no-Ekonomicheskikh Sistem: Trudy IX Mezhdunarodnoi Shkoly-Simpoziuma (AMUR-2013)* (Analysis, Modeling, Management, Development of Socio-economic Systems: Proc. VII Int. School-Symposium (AMMD-2013)), Krymskii Fed. Univ., Simferopol, 2013, pp. 134–138.

58. Gorbatov, A.S. and Zhukovskiy V.I., Berge Equilibrium in Bertrand Oligopoly with Import, *Taurida J. Comp. Sci. Theory and Math.*, 2015, vol. 27, no. 2, pp. 55–64.

59. Gorelik, V.A. and Kononenko, A.F., *Teoretiko–igrovye modeli prinyatiya reshenii v ekologo-ekonomicheskikh sistemakh* (Game-Theoretic Models of Decision-Making in Ecological-Economic Systems), Moscow: Radio i Svyaz', 1982.

60. Gorobets, B.S., *Pedagogi shutyat tozhe...Tol'ko strozhe* (Teachers Are Joking Too...But in a Stricter Way), Moscow: Librokom, 2011.

61. Granatulov, V.N., *Ekonomicheskii risk* (Economic Risk), Moscow: Delo i Servis, 1999.

62. Gracheva, S.S. and Pershin, M.A., Discrete Optimization Problem of Company's Advertising Policy in the Case of Linear Demand Dynamics Model, *Upravlen. Ekon. Sist.*, 2013, vol. 3, no. 51, p. 26.

63. Gracheva, S.S., Optimization of Company's Advertising Strategy in the Case of Nonlinear Demand Function, *Vestn. Samarsk. Gos. Univ.*, 2014, no. 2(113), pp. 180–185.

64. Hurwicz, L., Programming in Linear Topological Spaces, in *Issledovaniya po lineinomu programmirovaniyu* (Studies in Linear Programming), Arrow, K.J., Hurwicz, L., and Uzawa, H., Moscow: Inostrannaya Literatura, 1962, pp. 65–155.

65. Guseinov, A.A., *Zolotoe pravilo nravstvennosti* (The Golden Rule of Ethics), Moscow: Molodaya Gvardiya, 1988.

66. Guseinov, A.A., *Velikie proroki i mysliteli. Nravstvennye ucheniya ot Moiseya do nashikh dnei* (Great Prophets and Thinkers. Ethical Teachings from Moses to Present), Moscow: Veche, 2009.

67. Guseinov, A.A., *Antichnaya etika* (Antique Ethics), Moscow: Librokom, 2011.

68. Guseinov, A.A., *Filosofiya - mysl' i postupok: stat'i, doklady, lektsii, interv'yu* (Philosophy—Thought and Deed: Papers, Presentations, Lectures, Interviews), St. Petersburg: Gumanit. Univ. Profsoyuzov, 2012.

69. Guseinov, A.A., Philosophy as an Ethical Project, *Voprosy Filosofii*, 2014, no. 4, pp. 16–26.

70. Guseinov, A.A., *Sotsial'naya priroda nravstvennosti* (The Social Nature of Ethics), Moscow: Mosk. Gos. Univ., 1974.

71. Guseinov, A.A., Zhukovskii, V.I., and Kudryavtsev, K.N., *Matematicheskie osnovy Zolotogo pravila nravstvennosti: teoriya novogo al'truisticheskogo uravnoveshivaniya konfliktov v protivoves "egoistichnomu" ravnovesiyu po Neshu* (Mathematical Foundations of the Golden Rule of Ethics: The Theory of a New Altruistic Equilibration of Conflicts to Counterbalance the Selfish Nash Equilibrium), Moscow: LENAND, 2016.

72. Dunford, N. and Schwartz, J.T., *Linear Operators*, New York: Interscience, 1958. Translated under the title *Lineinye operatory. Obshchaya teoriya*, Moscow: Inostrannaya Literatura, 1962, vol. 1.

73. Degtyarev, D.A., *Vvedenie v teoriyu igr dlya politologov i mezhdunarodnikov* (Introduction to Game Theory for Political Scientists and Experts in International Relations), Moscow: Mosk. Gos. Inst. Mezh. Otnosh., 2010.

74. Diev, V.S., *Upravlencheskie resheniya: neopredelennost', modeli, intuitsiya* (Managerial Decisions: Uncertainty, Models, Intuition), Novosibirsk: Novosibirsk. Gos. Univ., 2001.

75. *Differentsial'nye igry. Ukazatel' literatury za 1968–1976* (Differential Games. Index of Literature for 1968–1976), Ushakov, V.N., Ed., Sverdlovsk: Ural. Nauchn. Tsentr Akad. Nauk SSSR, 1978.

76. *Differentsial'nye igry neskol'kikh lits. Ukazatel' literatury za 1968–1983* (Differential Multiperson Games. Index of Literature for 1968–1983), Zhukovskiy, V.I. and Dochev, D.T., Eds., Ruse, Bulgaria: Center for Mathematics, 1985.

77. *Differentsial'nye igry neskol'kikh lits. Ukazatel' literatury za 1984–1988* (Differential Multiperson Games. Index of Literature for 1984–1988), Zhukovskiy, V.I. and Ushakov, V.N., Eds., Sverdlovsk: Ural. Otd. Akad. Nauk SSSR, 1990.

78. *Differentsial'nye igry neskol'kikh lits. Ukazatel' literatury za 1989–1994* (Differential Multiperson Games. Index of Literature for 1989–1994), Zhukovskiy, V.I. and Ukhobotov, V.I., Eds., Chelyabinsk: Gos. Univ., 1995.

79. Dmitruk, A.V., *Vypuklyi analiz. Elementarnyi vvodnyi kurs* (Convex Analysis. An Elementary Introductory Course), Moscow: Makspress, 2012.

80. Dubina, I.N., *Osnovy teorii ekonomicheskikh igr* (Theoretical Foundations of Economic Games), Moscow: Knorus, 2010.

81. Egorov, A.I., *Osnovy teorii upravleniya* (Foundations of Control Theory), Moscow: Fizmatlit, 2004.

82. Emelyanov, S.V., Kostyleva, I.E., Matiĉ, Ozernoi, V.M., and Zimokha, V.A., *Mnogokriterial'naya otsenka lokal'nykh sistem upravleniya tekhnologicheskimi protsessami* (Multicriteria Assessment of Local Industrial Control Systems), Moscow: Inst. Probl. Upravlen., 1974.

83. Zhautykov, O.A., Zhukovskiy, V.I., and Zharkynbaev, S.Zh, *Differentsial'nye igry neskol'kikh lits (s zapazdyvaniem vremeni)* (Differential Multiperson Games with Time Delay), Almaty: Nauka, 1988.

84. Zhitomirskii, G.I., Dynamic Problems with Conflict, *Extended Abstract of Cand. Sci. Dissertation (Phys.-Math.)*, Leningrad State Univ., Leningrad, 1989.

85. Zhitomirskii, G.I. and Vaisman, K.S., On Berge Equilibrium, in *Slozhnye dinamicheskie sistemy* (Complex Dynamic Systems), Pskov: Pskovsk. Pedagog. Inst., 1994, pp. 52–57.

86. Zhukovskiy, V.I., *Vvedenie v differentsial'nye igry pri neopredelennosti* (Introduction to Differential Games under Uncertainty), Moscow: Mezhd. Nauchno-Issled. Inst. Probl. Upravlen., 1997.

87. Zhukovskiy, V.I., *Kooperativnye igry pri neopredelennosti i ikh prilozheniya* (Cooperative Games under Uncertainty and Their Applications), Moscow: URSS, 2010, 2nd ed.

88. Zhukovskiy, V.I., *Konflikty i riski* (Conflicts and Risks), Moscow: Ross. Zaochn. Inst. Tekstil. Legk. Promysh., 2007.

89. Zhukovskiy, V.I., *Vvedenie v differentsial'nye igry pri neopredelennosti. Ravnovesie po Neshu* (Introduction to Differential Games under Uncertainty. Nash Equilibrium), Moscow: URSS, 2010.

90. Zhukovskiy, V.I., *Vvedenie v differentsial'nye igry pri neopredelennosti. Ravnovesie ugroz i kontrugroz* (Introduction to Differential Games under Uncertainty. Equilibrium in Threats and Counter-Threats), Moscow: URSS, 2010.

91. Zhukovskiy, V.I., *Vvedenie v differentsial'nye igry pri neopredelennosti. Ravnovesie po Berzhu–Vaismanu* (Introduction to Differential Games under Uncertainty. Berge–Vaisman Equilibrium), Moscow: URSS, 2010.

92. Zhukovskiy, V.I., *Riski pri konfliktnykh situatsiyakh* (Risks in Conflict Situations), Moscow: URSS, LENAND, 2011.

93. Zhukovskiy, V.I., Mathematical Theory of the Golden Rule, *Trudy XXVII Krymskoi Osennei Matematicheskoi Shkoly-simpoziuma po spektral'nym i evolyutsionnym zadacham (KROMSH-2016)* (Proc. XXVII Crimean Autumn Mathematical School-Symposium on Spectral and Evolutionary Problems (CAMSS-2016)), Simferopol, 2016, pp. 72–73.

94. Zhukovskiy, V.I., Berge Equilibrium: Sufficient Conditions and Existence, *Tezisy dokladov konferentsii "Lomonosovskie chteniya"* (Abstr. Proc. Conf. "Lomonosov's Readings"), Moscow: Maks PRESS, 2014, p. 22.

95. Zhukovskiy, V.I., Berge Equilibrium under Uncertainty, *Tezisy dokladov Mezhdunarodnoi konferentsii "Dinamika sistem i protsessy upravleniya" posvyashchennoi 90-letiyu akademika N.N. Krasovskogo* (Abstr. Int. Conf. "Dynamics of Systems and Control Processes' dedicated to the 90th Anniversary of Academician N.N. Krasovskii), September 15–20, 2014, Yekaterinburg: Inst. Mat. Mekh. Ural. Otd. Ross. Akad. Nauk, pp. 85–86.

96. Zhukovskiy, V.I., The Existence of Berge Equilibrium in the Class of Mixed Strategies, *Trudy XXV Krymskoi Osennei Matematicheskoi Shkoly-simpoziuma po spektral'nym i evolyutsionnym zadacham (KROMSH-2014)* (Proc. XXV Crimean Autumn Mathematical School-Symposium on Spectral and Evolutionary Problems (CAMSS-2014)), Simferopol, 2014, pp. 84–85.

97. Zhukovskiy, V.I., Conflict Equilibration: Guaranteed Equilibria, *Tezisy Mezhdunarodnoi letnei matematicheskoi shkoly pamyati V.A. Plotnikova* (Abstr. Int. Summer Mathematical School in memory of V.A. Plotnikov), Odessa: Astroprint, 2013, pp. 51–52.

98. Zhukovskiy, V.I. and Akhrameev, P.K., Guaranteed Solution in Risks of Deposit Diversification Problem with Three Currencies (RUB, USD and Euro), *Uchen. Zapiski Tavrich. Natsion. Univ. Ser. Fiz.-Mat. Nauki*, 2014, vol. 29, no. 1, pp. 177–197.

99. Zhukovskiy, V.I., Bel'skikh, Yu.A., and Samsonov, S.P., Coefficient Criteria for Equilibrium Choice (An Example of a Linear-Quadratic Two-Player Game), *Vestn. South Ural State Univ. Ser. Math. Mech. Phys.*, 2015, vol. 7, no. 4, pp. 20–26.

100. Zhukovskiy, V.I. and Vysokos, M.I., Guaranteed Solution in Outcomes and Risks of Single-Criterion Problem, *Uchen. Zapiski Tavrich. Natsion. Univ. Ser. Fiz.-Mat. Nauki*, 2014, vol. 29, no. 1, pp. 198–210.

101. Zhukovskiy, V.I. and Vysokos, M.I., Outcome and Risk in Multistage Positional Problem under Uncertainty, *Trudy XXVII Krymskoi Osennei Matematicheskoi Shkoly-simpoziuma po spektral'nym i evolyutsionnym zadacham (KROMSH-2016)* (Proc. XXVII Crimean Autumn Mathematical School-Symposium on Spectral and Evolutionary Problems (CAMSS-2016)), Simferopol, 2016, pp. 73–74.

102. Zhukovskiy, V.I. and Vysokos, M.I., Multistage Setup of Controlled Bertrand Duopoly, in *Spektral'nye i evolyutsionnye zadachi* (Spectral and Evolutionary Problems), 2013, vol. 23, pp. 118–123.

103. Zhukovskiy, V.I., Vysokos, M.I., and Gorbatov, A.S., Guaranteed Outcome in a Discrete Positional Problem under Strategic Uncertainty, *Analiz, Modelirovanie, Upravlenie, Razvitie Sotsial'no-Ekonomicheskikh Sistem: Trudy IX Mezhdunarodnoi Shkoly-Simpoziuma (AMUR-2016)* (Analysis, Modeling, Management, Development of Socio-economic Systems: Proc. X Int. School-Symposium (AMMD-2016)), Krymskii Fed. Univ., Simferopol, 2016, pp. 109–114.

104. Zhukovskiy, V.I., Vysokos, M.I., and Gorbatov, A.S., Berge Equilibrium in the Single-Stage Setup of Controlled Cournot Duopoly, *Analiz, Modelirovanie, Upravlenie, Razvitie Sotsial'no-Ekonomicheskikh Sistem: Trudy IX Mezhdunarodnoi Shkoly-Simpoziuma (AMUR-2015)* (Analysis, Modeling, Management, Development of Socio-economic Systems: Proc. IX Int. School-Symposium (AMMD-2015)), Krymskii Fed. Univ., Simferopol, 2015, pp. 105–109.

105. Zhukovskiy, V.I., Vysokov, M.I., and Gorbatov, A.S., Outcome and Risk in Multistage Positional Problem under Uncertainty, *Taurida J. Comp. Sci. Theory and Math.*, 2016, vol. 30, no. 2, pp. 65–77.

106. Zhukovskiy, V.I., Vysokos, M.I., and Zharkynbaev, S., Equilibrium in Difference Model of Bertrand Duopoly, *Tezisy dokladov Krymskoi Mezhdunarodnoi Matematicheskoi Konferentsii (KMMK-2013)* (Abstr. Crimean Int. Mathematical Conf. (CIMC-2013), Sudak, September 22–October 4, 2013, Simferopol: Tavrich. Natsion. Univ., 2013, p. 27.

107. Zhukovskiy, V.I., Vysokos, M.I., and Zharkynbaev, S., To the Theory of Bertrand Duopoly, *Analiz, Modelirovanie, Upravlenie, Razvitie Sotsial'no-Ekonomicheskikh Sistem: Trudy VII Mezhdunarodnoi Shkoly-Simpoziuma (AMUR-2013)* (Analysis, Modeling, Management, Development of Socio-economic Systems: Proc. VII Int. School-Symposium (AMMD-2013)), Krymskii Fed. Univ., Simferopol, 2013, pp. 186–191.

108. Zhukovskiy, V.I. and Gorbatov, A.S., Mathematical Model of the Golden Rule of Ethics, *Tezisy dokladov Mezhdunarodnoi konferentsii "Dinamicheskie sistemy: obratnye zadachi, ustoichivost' i protsessy upravleniya", posvyashchennaya 80-letiyu akademika Yu.S. Osipova*

(Abstr. Int. Conf. "Dynamic Systems: Inverse Problems, Stability and Control Processes" dedicated to the 80th Anniversary of Academician Yu.S. Osipov), Moscow, September 22–23, 2016, Moscow: Mat. Inst. Ross. Akad. Nauk, 2016, pp. 51–53.

109. Zhukovskiy, V.I. and Gorbatov, A.S., Mathematical Foundations of the Golden Rule of Ethics, *Trudy VIII Moskovskoi mezhdunarodnoi konferentsii po issledovaniyu operatsii* (Abstr. VIII Moscow Int. Conf. on Operations Research (ORM2016)), Moscow, October 17–22, 2016, vol. 2, Moscow: Feder. Issled. Tsentr Inf. Upravlen. Ross. Akad. Nauk, 2016, pp. 179–180.

110. Zhukovskiy, V.I. and Gorbatov, A.S., Zero Risk in Single-Criterion Choice Problems, *Upravlen. Risk.*, 2015, vol. 74, no. 2, pp. 29–36.

111. Zhukovskiy, V.I. and Gorbatov, A.S., Connection between the Golden Rule and Berge Equilibrium, *Control Theory and Theory of Generalized Solutions of Hamilton–Jacobi Equations (CGS'2015): Abstr. II Int. Seminar dedicated to the 70th Anniversary of Academician A.I. Subbotin*, Yekaterinburg, Inst. Mat. Mekh. Ural. Otd. Ross. Akad. Nauk, 2015, pp. 58–60.

112. Zhukovskiy, V.I. and Dochev, D.T., *Vektornaya optimizatsiya dinamicheskikh sistem* (Vector Optimization of Dynamic Systems), Ruse, Bulgaria: Center for Mathematics, 1981.

113. Zhukovskiy, V.I. and Zhiteneva, Yu.M., Bertrand Difference–Differential Game Model in the Framework of Berge Equilibrium, *Analiz, Modelirovanie, Upravlenie, Razvitie Sotsial'no-Ekonomicheskikh Sistem: Tr. IX Mezhd. Shkoly-Simpoziuma (AMUR-2015)* (Analysis, Modeling, Management, Development of Socio-economic Systems: Proc. IX Int. School-Symposium (AMMD-2015)), Krymskii Fed. Univ., Simferopol, 2015, pp. 109–114.

114. Zhukovskiy, V.I. and Zhukovskaya, L.V., *Risk v mnogokriterial'nykh i konfliktnykh sistemakh pri neopredelennosti* (Risk in Multicriteria Choice and Conflict Systems under Uncertainty), Moscow: URSS, 2004.

115. Zhukovskiy, V.I. and Kirichenko, M.M., Risks and Outcomes in Multicriteria Problem under Uncertainty, *Upravlenie Riskom*, 2016, vol. 78, no. 2, pp. 17–25.

116. Zhukovskiy, V.I., Kirichenko, M.M., and Boldyrev, M.V., Risks and Outcomes in Multicriteria Problem under Uncertainty, *Trudy XXVII Krymskoi Osennei Matematicheskoi Shkoly-simpoziuma po spektral'nym i ekolyutsionnym zadacham (KROMSH-2016)* (Proc. XXVII Crimean Autumn Mathematical School-Symposium on Spectral and Evolutionary Problems (CAMSS-2016)), Simferopol, 2016, pp. 74–75.

117. Zhukovskiy, V.I., Kirichenko, M.M., and Boldyrev, M.V., Risks and Outcomes in a Multicriteria Problem, *Trudy VIII Moskovskoi mezhdunarodnoi konferentsii po issledovaniyu operatsii* (Abstr. VIII Moscow Int. Conf. on Operations Research (ORM2016)), Moscow, October 17–22, 2016, vol. 2, Moscow: Feder. Issled. Tsentr Inf. Upravlen. Ross. Akad. Nauk, 2016, pp. 72–73.

118. Zhukovskiy, V.I. and Kudryavtsev, K.N., Pareto-Optimal Nash Equilibrium: Sufficient Conditions and Existence in Mixed Strategies, *Autom. Remote Control*, 2016, vol. 77, no. 8, pp. 1500–1510.

119. Zhukovskiy, V.I., Kudryavtsev, K.N., and Smirnova, L.V., *Garantirovannye resheniya konfliktov i prilozheniya* (Guaranteed Solutions of Conflicts and Applications), Moscow: URSS, 2013.

120. Zhukovskiy, V.I. and Kudryavtsev, K.N., Mathematical Foundations of the Golden Rule. I. Static Case, *Autom. Remote Control*, 2017, vol. 78, no. 10, pp. 1920–1940.

121. Zhukovskiy, V.I. and Kudryavtsev, K.N., *Uravnoveshivanie konfliktov i prilozheniya* (Equilibrating Conflicts and Applications), Moscow: URSS, 2012.

122. Zhukovskiy, V.I. and Kudryavtsev, K.N., Equilibrating Conflicts under Uncertainty. I. Analog of Saddle-Point, *Mat. Teor. Igr Prilozh.*, 2013, vol. 5, no. 1, pp. 27–44.

123. Zhukovskiy, V.I. and Kudryavtsev, K.N., Equilibrating Conflicts under Uncertainty. II. Analog of Maximin, *Mat. Teor. Igr Prilozh.*, 2013, vol. 5, no. 2, pp. 3–45.

124. Zhukovskiy, V.I., Kudryavtsev, K.N., and Gorbatov, A.S., The Berge Equilibrium in Cournot Oligopoly Model, *Vestn. Udmurt. Univ. Ser. Math. Mech. Comp. Sci.*, 2015, vol. 25, no. 2, pp. 147–156.

125. Zhukovskiy, V.I. and Makarkina, T.V., Altruistic Equilibrium in Competitive Bertrand Oligopoly Model, *Evrop. Soyuz Uchenykh*, 2015, vol. 17, no. 8, pp. 48–52.

126. Zhukovskiy, V.I. and Makarkina, T.V., Which Type of Equilibrium (Nash or Berge) Should Be Used in Competitive Bertrand Duopoly Depending on Maximum Produce Price?, *Analiz, Modelirovanie, Upravlenie, Razvitie Sotsial'no-Ekonomicheskikh Sistem: Trudy X Mezhdunarodnoi Shkoly-Simpoziuma (AMUR-2016)* (Analysis, Modeling, Management, Development of Socio-economic Systems: Proc. X Int. School-Symposium (AMMD-2016)), Krymskii Fed. Univ., Simferopol, 2016, pp. 114–119.

127. Zhukovskiy, V.I. and Makarkina, T.V., Pareto-Maximin Guaranteed Solution in Outcomes and Risks of Linear-Quadratic Problem, *Upravlenie Riskom*, 2016, vol. 78, no. 3, pp. 24–29.

128. Zhukovskiy, V.I. and Makarkina, T.V., Comparison of Nash and Berge Equilibria, *Trudy XXVII Krymskoi Osennei Matematicheskoi Shkoly-simpoziuma po spektral'nym i ekolyutsionnym zadacham (KROMSH-2016)* (Proc. XXVII Crimean Autumn Mathematical School-Symposium on Spectral and Evolutionary Problems (CAMSS-2016)), Simferopol, 2016, pp. 75–76.

129. Zhukovskiy, V.I. and Makarkina, T.V., Comparison of Nash and Berge Equilibria in Bertrand Duopoly Model, *Taurida J. Comp. Sci. Theory and Math.*, 2016, vol. 30, no. 1, pp. 78–88.

130. Zhukovskiy, V.I., Makarkina, T.V., Samsonov, S.P., and Gorbatov, A.S., Berge Equilibrium in Competition Model, *Natsion. Assots. Uchenykh*, 2015, vol. 1, no. 7(12), pp. 94–98.

131. Zhukovskiy, V.I., Makarkina, T.V., Samsonov, S.P., and Gorbatov, A.S., Berge Equilibrium for a Price Competition Model in a Duopoly Market, *Tr. XII Mezhd. Nauchno-Praktich. Konf. "Otechestvennaya nauka v epokhu izmenenii: postulaty proshlogo i teorii novogo vremeni"* (Proc. XII Int. Scientific-Practical Conf. "National Science in the Era of Change: Postulates of Past and Theories of New Time"), Yekaterinburg, National Association of Scientists, 2015, vol. 7, pp. 94–98.

132. Zhukovskiy, V.I. and Makarkina, T.V., Consideration of Import in Cournot Duopoly, *Taurida J. Comp. Sci. Theory and Math.*, 2015, vol. 27, no. 2, pp. 65–76.

133. Zhukovskiy, V.I. and Molostvov, V.S., *Mnogokriterial'naya optimizatsiya sistem v usloviyakh nepolnoi informatsii* (Multicriteria Optimization of Systems under Incomplete Information), Moscow: Mezhd. Nauchno-Issled. Inst. Probl. Upravlen., 1990.

134. Zhukovskiy, V.I. and Molostvov, V.S., *Mnogokriterial'noe prinyatie reshenii v usloviyakh neopredelennosti* (Multicriteria Decision-Making under Uncertainty), Moscow: Mezhd. Nauchno-Issled. Inst. Probl. Upravlen., 1988.

135. Zhukovskiy, V.I. and Salukvadze, M.E., *Mnogokriterial'nye zadachi upravleniya v usloviyakh neopredelennosti* (Multicriteria Control Problems under Uncertainty), Tbilisi: Metsniereba, 1991.

136. Zhukovskiy, V.I. and Salukvadze, M.E., *Nekotorye igrovye zadachi upravleniya i ikh prilozheniya* (Some Game-Theoretic Problems of Control and Their Applications), Tbilisi: Metsniereba, 1998.

137. Zhukovskiy, V.I. and Salukvadze, M.E., *Optimizatsiya garantii v mnogokriterial'nykh zadachakh upravleniya* (Optimization of Guarantees in Multicriteria Control Problems), Tbilisi: Metsniereba, 1996.

138. Zhukovskiy, V.I. and Salukvadze, M.E., *Riski i iskhody v mnogokriterial'nykh zadachakh upravleniya* (Risks and Outcomes in Multicriteria Control Problems), Tbilisi: Intelekti, 2004.

139. Zhukovskiy, V.I. and Salukvadze, M.E., *Garantii i riski v konfliktakh i prilozheniya* (Guarantees and Risks in Conflicts and Applications), Moscow–Tbilisi: Izd. Natsion. Akad. Nauk Gruzii, 2014.

140. Zhukovskiy, V.I. and Salukvadze, M.E., *Dinamika Zolotogo Pravila nravstvennosti* (Dynamics of the Golden Rule of Ethics), Tbilisi: Natsion. Akad. Nauk Gruzii, 2018.

141. Zhukovskiy, V.I. and Sachkov, S.N., An Unusual Way of Resolving Conflicts, *Tr. Konf. "Sovremennye tekhnologii upravleniya-2014"* (Proc. Int. Conf. "Contemporary Technology of Management-2014"), Moscow, 2014, pp. 337–355.

142. Zhukovskiy, V.I. and Smirnova, L.V., Application of the Golden Rule in a Positional Differential Game, *Analiz, Modelirovanie, Upravlenie, Razvitie Sotsial'no-Ekonomicheskikh Sistem: Tr. IX Mezhd. Shkoly-Simpoziuma (AMUR-2015)* (Analysis, Modeling, Management, Development of Socio-economic Systems: Proc. IX Int. School-Symposium (AMMD-2015)), Krymskii Fed. Univ., Simferopol, 2015, pp. 115–120.

143. Zhukovskiy, V.I. and Smirnova, L.V., The Existence of Berge Equilibrium in a Differential Game with Separated Dynamics, *Taurida J. Comp. Sci. Theory and Math.*, 2015, vol. 27, no. 2, pp. 77–86.

144. Zhukovskiy, V.I. and Smirnova, L.S., The Existence of Berge Equilibrium in Linear-Quadratic Differential Game with Small Parameter, *Analiz, Modelirovanie, Upravlenie, Razvitie Sotsial'no-Ekonomicheskikh Sistem: Trudy X Mezhdunarodnoi Shkoly-Simpoziuma (AMUR-2016)* (Analysis, Modeling, Management, Development of Socio-economic Systems: Proc. X Int. School-Symposium (AMMD-2016)), Krymskii Fed. Univ., Simferopol, 2016, pp. 119–123.

145. Zhukovskiy, V.I., Smirnova, L.V., and Molostvov, V.S., Maximin in a Positional Control Problem, *Analiz, Modelirovanie, Upravlenie, Razvitie Sotsial'no-Ekonomicheskikh Sistem: Trudy VII Mezhdunarodnoi Shkoly-Simpoziuma (AMUR-2013)* (Analysis, Modeling, Management, Development of Socio-economic Systems: Proc. VII Int. School-Symposium (AMMD-2013)), Krymskii Fed. Univ., Simferopol, 2013, pp. 119–123.

146. Zhukovskiy, V.I., Smirnova, L.V., and Gorbatov, A.S., Mathematical Foundations of the Golden Rule. II. Dynamic Case, *Autom. Remote Control*, 2018, vol. 79, no. 10, pp. 1929–1952.

147. Zhukovskiy, V.I. and Soldatova, N.G., Guaranteed Risks and Outcomes in the Game with Nature, *Probl. Upravlen.*, 2014, no. 1, pp. 14–26.

148. Zhukovskiy, V.I. and Soldatova, N.G., To the Deposit Diversification Problem with Three Currencies, *Vestn. Udmurt. Univ. Ser. Mat. Mekh. Komp. Nauki*, 2013, no. 4, pp. 55–61.

149. Zhukovskiy, V.I. and Soldatova, N.G., On the Choice of Guaranteed Solution in Outcomes and Risks for a Single-Criterion Problem, *Trudy XXI Mezhdunarodnoi zaochnoi nauchno-prakticheskoi konferentsii "Nauchnaya diskussiya: voprosy ekonomiki i upravleniya"* (Proc. XXI Int. Extramural Scientific and Practical Conf. "Scholarly Dispute: Issues of Economics and Control"), Moscow: Int. Center of Science and Education, 2013, vol. 12, pp. 31–35.

150. Zhukovskiy, V.I. and Soldatova, N.G., On Guaranteed Solution in Outcomes and Risks, *Trudy XXI Mezhdunarodnoi zaochnoi nauchno-prakticheskoi konferentsii "Nauchnaya diskussiya: voprosy matematiki, fiziki, khimii, biologii"* (Proc. XXI Int. Extramural Scientific and Practical Conf. "Scholarly Dispute: Issues of Mathematics, Physics, Chemistry and Biology"), Moscow: Int. Center of Science and Education, 2013, vol. 12, pp. 20–24.

151. Zhukovskiy, V.I. and Soldatova, N.G., Risks and Outcomes in Linear-Quadratic Problem with Bounded Uncertainty, *Trudy XXI Mezhdunarodnoi zaochnoi nauchno-prakticheskoi konferentsii "Nauchnaya diskussiya: voprosy matematiki, fiziki, khimii, biologii"* (Proc. XXI Int. Extramural Scientific and Practical Conf. "Scholarly Dispute: Issues of Mathematics, Physics, Chemistry and Biology"), Moscow: Int. Center of Science and Education, 2013, vol. 12, pp. 216–222.

152. Zhukovskiy, V.I. and Soldatova, N.G., Risks Connected with Deposit Diversification, *Upravlenie Riskom*, 2014, vol. 69, no. 1, pp. 15–24.

153. Zhukovskiy, V.I. and Soldatova, N.G., Risks Connected with Deposit Diversification, *Upravlenie Riskom*, 2015, vol. 73, no. 1, pp. 3–13.

154. Zhukovskiy, V.I. and Soldatova, N.G., A Method to Balance Conflicts under Uncertainty, *Vestn. Udmurt. Univ. Ser. Mat. Mekh. Komp. Nauki*, 2013, no. 3, pp. 28–33.

155. Zhukovskiy, V.I. and Tynyanskii, N.T., *Ravnovesnye upravleniya mnogokriterial'nykh dinamicheskikh sistem* (Equilibrium Controls of Multicriteria Dynamic Systems), Moscow: Mosk. Gos. Univ., 1984.

156. Zhukovskiy, V.I. and Chikrii, A.A., *Lineino-kvadratichnye differentsial'nye igry* (Linear-Quadratic Differential Games), Kiev: Naukova Dumka, 1994.

157. Zhukovskiy, V.I., Chikrii, A.A., and Soldatova, N.G., The Berge Equilibrium in the Conflicts under Uncertainty, *XII Vserossiiskoe soveshchanie po problemam upravleniya (VSPU-2014)* (Proc. XII All-Russian Meeting on Control Problems (AMCP-2014)), Moscow: Inst. Probl. Upravlen., 2014, pp. 8290–8302.

158. Zhukovskiy, V.I., Chikrii, A.A., and Soldatova, N.G., Existence of Berge Equilibrium in Conflicts under Uncertainty, *Automation and Remote Control*, 2016, vol. 77, no. 4, pp. 640–655.

159. Zhukovskiy, V.I., Chikrii, A.A., and Soldatova, N.G., Diversification Problem of Deposits, in *Analiz, modelirovanie, upravlenie, razvitie* (Analysis, Modeling, Management, Development), Vernadskii Gos. Univ., Sevastopol, 2015, pp. 134–142.

160. Zhukovskiy, V.I. and Shershekov, M.I., Import in Two-Stage Duopoly Model, *Analiz, Modelirovanie, Upravlenie, Razvitie Sotsial'no-Ekonomicheskikh Sistem: Trudy VII Mezhdunarodnoi Shkoly-Simpoziuma (AMUR-2013)* (Analysis, Modeling, Management, Development of Socio-economic Systems: Proc. VII Int. School-Symposium (AMMD-2013)), Krymskii Fed. Univ., Simferopol, 2013, pp. 126–129.

161. Zhukovskiy, V.I. and Shershekov, M.I., Multistage Bertrand Duopoly Model with Import, in *Spektral'nye i evolyutsionnye zadachi* (Spectral and Evolutionary Problems), 2013, vol. 23, pp. 72–77.

162. Zhuravlev, S.G. and Anikovskii, V.I., *Differentsial'nye uravneniya. Sbornik zadach* (Differential Equations. A Problem Book), Moscow: Ekzamen, 2005.

163. *Zabavnye anekdoty* (Funny Stories), St. Petersburg: Dilya, 1994.

164. Zadeh, L.A., The Concept of a Linguistic Variable and Its Application to Approximate Reasoning, *Information Sciences*, 1975, vol. 8, no. 4, pp. 301–357.

165. Zolotarev, V.V., Hybrid equilibrium under uncertainty, *Cand. Sci. (Phys.-Math.) Dissertation*, Moscow: Pedagogical State University, 2002.

166. *Muhammad's Sayings Not Included in the Holy Quran*, in *L.N. Tolstoi: Polnoe sobranie sochinenii. T. 40* (L.N. Tolstoy: Full Collection of Works, Vol. 40), Moscow, 1956.

167. *Issledovanie operatsii* (Operations Research), Moscow: Nauka, 1972.

168. Kyburg, H.E., *Probability and Inductive Logic*, Macmillan, 1970.

169. *I. Kant: Osnovy metafiziki nravstvennosti. Razd. 2. Perekhod ot populyarnoi nravstvennoi filosofii k metafizike nravstvennosti* (I. Kant: Foundations of the Metaphysics of Morals. Sect. 2. Transition from Popular Philosophy of Morals to the Metaphysics of Morals), Asmus, V.F., Gulyga, A.V., and Oizerman, T.I., Eds., Moscow: Mysl', 1999.

170. *I. Kant: Osnovy metafiziki nravstvennosti* (I. Kant: Foundations of the Metaphysics of Morals), Asmus, V.F., Gulyga, A.V., and Oizerman, T.I., Eds., Moscow: Mysl', 1999.

171. Kapitonenko, V.V., *Finansovaya matematika i ee prilozheniya* (Financial Mathematics and Its Applications), Moscow: Prior, 2000.

172. Karlin, S., *Mathematical Methods and Theory in Games, Programming and Economics*, London-Paris: Pergamon Press, 1959.

173. Kleimenov, A.F., *Neantagonisticheskie pozitsionnye differentsial'nye igry* (Nonantagonistic Positional Differential Games), Yekaterinburg: Nauka, 1993.

174. Kolemaev, V.A., *Matematicheskaya ekonomika* (Mathematical Economics), Moscow: Yuniti, 2002.

175. Kolmogorov, A.N. and Fomin, S.V., *Elementy teorii funktsii i funktsional'nogo analiza* (Elements of Theory of Functions and Functional Analysis), Moscow: Nauka, 1976.

176. *Confucius: Discussions/Conversations, or The Analects (Lun-yu)*, translated by Schiller, D.R., Saga Virtual Publishers, 2011.

177. Kononenko, A.F., The Structure of the Optimal Strategy in Controlled Dynamic Systems, *USSR Comp. Math. Math. Phys.*, 1980, vol. 20, no. 5, pp. 13–24.

178. Krasovskii, N.N., *Upravlenie dinamicheskoi sistemoi* (Control of Dynamic System), Moscow: Nauka, 1985.

179. Krasovskii, N.N. and Subbotin, A.I., *Pozitsionnye differentsial'nye igry* (Positional Differential Games), Moscow: Nauka, 1985.

180. *Plato: Crito*, translated by Jowett, B., CreateSpace Independent Publishing Platform, 2015.

181. Kudryavtsev, K.N., Coordinated Solutions in Multiagent Information Environment, *Extended Abstract of Cand. Sci. Dissertation (Phys.-Math.)*, South-Ural Federal University, Chelyabinsk, 2011.

182. Kudryavtsev, K.N., On the Existence of Guaranteed Solutions in Payoffs and Risks in Cooperative Games under Uncertainty, *Sist. Upravlen. Inform. Tekh.*, 2010, no. 1.1(39), pp. 148–152.

183. Kudryavtsev, K.N., Controlled Cournot Duopoly with Import, in *Spektral'nye i evolyutsionnye zadachi* (Spectral and Evolutionary Problems), 2013, vol. 23, pp. 109–115.

184. Kudryavtsev, K.N. and Meshkov, V.M., About One Planning Model of Advertising Budget in Duopoly, *Trudy XV Mezhdunarodnoi Konferentsii "Sistemy komp'yuternoy matematiki i ikh prilozheniya"* (Proc. XV Int. Conf. "Systems of Computer Mathematics and Their Applications"), Smolensk: Gos. Univ., 2014, vol. 15, pp. 173–175.

185. Lagunov, V.N., *Vvedenie v differentsial'nye igry* (Introduction to Differential Games), Vilnius: Inst. Mat. Kibern. Akad. Nauk LSSR, 1979.

186. *Larets ostroslovov* (Casket of Wisecrackers), Moscow: Izdatel'stvo Politicheskoi Literatury, 1991.

187. *Diogen Laertskii: O zhizni, ucheniyakh i izrecheniyakh znamenitykh filosofov* (Diogenes Laërtius: About the Life, Doctrines and Aphorisms of Famous Philosophers), Losev, A.F., Ed., Moscow: Akad. Nauk. SSSR, 1979.

188. Lezhnev, A.V., *Dinamicheskoe programmirovanie v ekonomicheskikh zadachakh* (Dynamic Programming in Economic Problems), Moscow: BINOM, Laboratoriya Znanii, 2006.

189. Duncan Luce, R. and Raiffa, H., *Games and Decisions: Introduction and Critical Survey*, Dover Books on Mathematics, 1989.

190. Lyusternik, L.A. and Sobolev, V.I., *Elementy funktsional'nogo analiza* (Elements of Functional Analysis), Moscow: Nauka, 1969.

191. Makarov, V.A. and Rubinov, A.M., *Matematicheskaya teoriya ekonomicheskoi dinamiki i ravnovesie* (Mathematical Theory of Economic Dynamics and Equilibrium), Moscow: Nauka, 1973.

192. McConnell, C.R., Brue, S.L., and Flynn, S.M., *Economics: Principles, Problems, and Policies*, McGraw-Hill, 2011.

193. Malkin, G.E., *Bol'shaya kniga aforizmov dlya ochen' umnykh* (A Big Book of Aphorisms for Really Smart Persons), Moscow: RIPOL Klassik, 2005.

194. Malkin, I.G., *Teoriya ustoychivosti dvizheniya* (Theory of Motion Stability), Moscow: Nauka, 1996.

195. Mamedov, M.B., About Pareto-optimal Nash Equilibrium, *Izv. Akad. Nauk Azerbaidzh. Ser. Fiz.-Mat. Nauk*, 1983, vol. 4, no. 2, pp. 11–17.

196. Matveev, V.A., The Multicriteria Positional Problem for a Parabolic System, *Extended Abstract of Cand. Sci. (Phys.-Math.) Dissertation*, Ural State Univ., Yekaterinburg, 1992.

197. *Matematicheskaya entsiklopediya. Tom 3* (Encyclopedia of Mathematics. Vol. 3), Moscow: Sovetsk. Entsiklop., 1982.

198. *The Complete Mahabharata Volume 4: Udyoga Parva*, translated by Kumar, J., and edited by Menon, R., Rupa, 2013.

199. Mashchenko, S.O., Concept of Nash Equilibrium and Its Development, *Zh. Obchisloval. Prikladn. Mat.*, 2012, no. 1(107), pp. 40–65.

200. Mishchenko, E.F. and Rozov, N.Kh., *Differentsial'nye uravneniya s malym parametrom i relaksatsionnye kolebaniya* (Small-Parameter Differential Equations and Relaxation Oscillations), Moscow: Nauka, 1975.

201. Moiseev, N.N., *Elementy teorii optimal'nykh sistem* (Elements of the Theory of Optimal Systems), Moscow: Nauka, 1975.

202. Morozov, V.V., *Osnovy teorii igr* (Foundations of Game Theory), Moscow: Mosk. Gos. Univ., 2002.

203. Morozov, V.V., Mixed Strategies in a Game with Vector Payoffs, *Vestn. Mosk. Gos. Univ. Ser. Vychisl. Mat. Kibern.*, 1978, no. 4, pp. 44–49.

204. Morozov, V.V., Sukharev, A.G., and Fedorov, V.V., *Issledovanie operatsii v zadachakh i uprazhneniyakh* (Operations Research in Problems and Exercises), Moscow: Vysshaya Shkola, 1986.

205. Muschick, E. and Müller, P., *Metody prinyatiya tekhnicheskikh reshenii* (Methods of Technical Decision-Making), Moscow: Mir, 1990.

206. Knight, F.H., *Risk, Uncertainty, and Profit*, Boston: Houghton Mifflin, 1921.

207. Noghin, V.D., *Prinyatie reshenii v mnogokriterial'noi srede* (Decision-Making in Multicriteria Environment), Moscow: Fizmatlit, 2005.

208. Petrosjan, L.A. and Danilov, N.N., *Kooperativnye differentsial'nye igry i ikh prilozheniya* (Cooperative Differential Games and Their Applications), Tomsk: Gos. Univ., 1985.

209. Petrosjan, L.A. and Zakharov, V.V., *Matematicheskie modeli v ekologii* (Mathematical Models in Ecology), St. Petersburg: Sankt-Peterb. Gos. Univ., 1997.

210. Petrosjan, L.A. and Murzov, N.V., The n-person Tug-of-War Game, *Vestn. Leningrad. Univ. Ser. Mat.*, 1967, vol. 22, no. 13, pp. 125–129.

211. Petrosjan, L.A. and Tomskii, G.V., *Dinamicheskie igry i ikh prilozheniya* (Dynamic Games and Their Applications), Leningrad: Leningrad. Univ., 1982.

212. Pecherskii, S.L. and Belyaeva, A.A., *Teoriya igr dlya ekonomistov. Vvodnyy kurs* (Game Theory for Economists. An Introductory Course), St. Petersburg: Evrop. Univ., 2004.

213. Podinovskii, V.V. and Noghin, V.D., *Pareto-optimal'nye resheniya mnogokriterial'nykh zadach* (Pareto-ptimal Solutions of Multicriteria Problems), Moscow: Fizmatlit, 2007.

214. Pontryagin, L.S., *Ordinary Differential Equations*, Adiwes International Series in Mathematics, Addison-Wesley, 1962.

215. Pontryagin, L.S., Boltyanskii, V.G., Gamkrelidze, R.V., and Mishchenko, E.F., *The Mathematical Theory of Optimal Processes*, Interscience Publishers, 1962.

216. Propoi, A.I., *Elementy teorii optimal'nykh diskretnykh protsessov* (Elements of the Theory of Optimal Discrete Processes), Moscow: Nauka, 1973.

217. Raiffa, H., *Decision Analysis. Introductory Lectures on Choices under Uncertainty*, Longman Higher Education Division, 1968.

218. Ris, A., *Davaite melochit'sya* (Let's Niggle), Moscow: Ripol Klassik, 2006.

219. Rogov, M.A., *Risk-menedzhment* (Risk Management), Moscow: Finansy i Statistika, 2001.

220. Rozen, V.V., *Modeli prinyatiya reshenii v ekonomike* (Models of Decision-Making in Economics), Moscow: Vysshaya Shkola, 2002.

221. Rozen, V.V., Equilibria in Games with Ordered Outcomes, in *Sovremennye napravleniya teorii igr* (Modern Directions of Game Theory), Vilnius: Moklas, 1976, pp. 115–118.

222. Rozen, V.V., Properties of Outcomes in Equilibria, in *Matematicheskie modeli povedeniya* (Mathematical Models of Behavior), Saratov: Gos. Univ., 1975, vol. 2, pp. 45–49.

223. Rosenmüller, J., *The Theory of Games and Markets*, Elsevier, 1981.

224. Romanenko, V.K., *Raznostnye uravneniya* (Difference Equations), Moscow: BINOM, 2006.

225. *Russko–anglo–nemetskii tolkovyi slovar' po biznesu* (Russian–English–German Glossary on Business Science), Kuznetsova, N.N., Novikova, E.V., Plekhanov, S.V., and Chekmezov, N.A., Eds., Moscow: Gorizont, 1992.

226. Sartre, J.-P., *Existentialism Is a Humanism*, World Publishing, 1956.

227. Sirazetdinov, T.K. and Sirazetdinov, R.T., The Problem of Risk and Its Modeling, *Problema Chelovecheskogo Riska*, 2007, no. 1, pp. 31–43.

228. *Slovar' inostrannykh slov* (Dictionary of Foreign Words), Moscow: Sovetsk. Entsiklop., 1964.

229. Smirnova, L.V., Hurwitz Principle in Complex Controlled Systems, *Extended Abstract of Cand. Sci. Dissertation (Phys.-Math.)*, Moscow Pedagogical Stat University, Moscow, 1998.

230. Smol'yakov, E.R., *Ravnovesnye modeli pri nesovpadayushchikh interesakh uchastnikov* (Equilibrium Models with Noncoinciding Interests of Parties), Moscow: Nauka, 1986.

231. Smol'yakov, E.R., *Teoriya konfliktnykh ravnovesii* (Theory of Conflict Equilibria), Moscow: URSS, 2004.

232. *Sochineniya Koz'my Prutkova* (Works of Kozma Prutkov), Moscow: Sovetskaya Rossiya, 1981.

233. Subbotin, A.I. and Chentsov, A.G., *Optimizatsiya garantii v zadachakh upravleniya* (Optimization of Guarantees in Control Problems), Moscow: Nauka, 1981.

234. Feldbaum, A.A., *Osnovy teorii avtomaticheskikh sistem* (Foundations of Automatic Control Systems), Moscow: Nauka, 1966.

235. *Filosofia* (Philosophy), Zotov, A.F., Mironov, V.V., and Razin, A.V., Eds., Moscow: Akademicheskii Proekt, 2009.

236. Fischer, S., Dornbusch, R., and Schmalensee, R., *Economics*, McGraw-Hill, 1988.

237. Harsanyi J.C. and Selten R., *A General Theory of Equilibrium Selection in Games*, Cambridge: MIT Press, 1988.

238. Hedley, G., *Nonlinear and Dynamic Programming*, Reading, Massachusets, London: Addison–Wesley, 1964.

239. Hille, E. and Phillips, R.S., *Functional Analysis and Semi-Groups*, American Mathematical Society Colloquium Publications, vol. 31, Providence: Am. Math. Soc., 1957.

240. Khoroshilova, E.V., *Sovremennye anekdoty i vyskazyvaniya* (Modern Funny Stories and Sayings), Moscow: Maks-Press, 2009.

241. Tsvetkova, E.V. and Arlyukova, I.O., *Riski v ekonomicheskoi deyatel'nosti* (Risks in Economic Activity), St.-Petersburg: Inst. Vneshneekon. Svyav. Ekon. Prav., 2002.

242. Tsel'mer, G., Consideration of Risk in Management Decisions, *Problemy Mezh. Sovet. Nauchn. Tekh. Inf.*, 1980, no. 3, pp. 94–105.

243. Cheremnykh, Yu.N., *Mikroekonomika. Prodvinutyi uroven'* (Microeconomics: Advance Level), Moscow: Info-M, 2008.

244. Cherkasov, V.V., *Delovoi risk v predprinimatel'skoi deyatel'nosti* (Business Risk in Entrepreneurship), Kiev: Libra, 1996.

245. Shakhov, V.V., *Vvedenie v strakhovanie. Ekonomicheskii aspekt* (Introduction to Insurance. Economic Aspect), Moscow: Finansy i Statistika, 1994.

246. Shokin, Yu.I., *Interval'nyi analiz* (Interval Analysis), Novosibirsk: Sib. Otd. Akad. Nauk SSSR, 1981.

247. Euler, L., *Metod nakhozhdeniya krivykh linii, obladayushchikh svoistvami maksimuma ili minimuma, ili reshenie izoperimetricheskoi zadachi, vzyatoi v samom shirokom smysle* (A Method

to Find Curves with the Properties of Maximum or Minimum, or a Solution to the Isoparametric Problem in the Widest Sense), Moscow: Finansy i Statistika, 1994.

248. *Entsiklopediya kibernetiki. Tom 2* (Encyclopeadia of Cybernetics. Vol. 2), Kiev: Glavn. Redakts. Ukrainsk. Sovetsk. Entsiklop., 1974.

249. *Entsiklopediya matematiki. Tom 1–5* (Encyclopeadia of Mathematics. Vols. 1–5), Moscow: Sovetsk. Entsiklop., 1977–1985.

250. Arrow, K.J., Hurwicz, L., and Uzawa, H., *Studies in Linear and Non-linear Programming*, Stanford Mathematical Studies in the Social Sciences, Stanford Univ. Press, 1958.

251. Abalo, K.Y. and Kostreva, M., Equi–Well–Posed Games, *J. Optim. Theory Appl.*, 1996, vol. 89, no. 1, pp. 89–99.

252. Abalo, K.Y. and Kostreva, M., Fixed Points, Nash Games and Their Organization, *Topol. Methods Nonlinear Anal*, 1996, no. 8, pp. 205–215.

253. Abalo, K.Y. and Kostreva, M., Intersection Theorems and Their Applications to Berge Equilibria, *Appl. Math. Comput.*, 2006, vol. 182, pp. 1840–1848.

254. Abalo, K.Y. and Kostreva, M.M., Berge Equilibrium: Some Recent Results from Fixed-Point Theorems, *Appl. Math. Comput.*, 2005, vol. 169, pp. 624–638.

255. Abalo, K.Y. and Kostreva, M.M., Some Existence Theorems of Nash and Berge Equilibria, *Appl. Math. Lett.*, 2004, vol. 17, pp. 569–573.

256. Basar, T., A Contrexample in Linear-Quadratic Game: Existence of Nonlinear Nash Solutions, *J. Optimiz. Theory Appl.*, 1974, vol. 14, no. 4, pp. 425–430.

257. Basar, T. and Olsder, G.J., *Dynamic Noncooperative Game Theory*, London: Academic Press, 1982.

258. Bellman, R., Dynamic Programming and a New Formalism in the Calculus of Variations, *Proc. Nat. Acad. Sci. USA*, 1953, vol. 39, pp. 1077–1082.

259. Bellman, R., On the Application of Dynamic Programming to Variational Problems Arising in Mathematical Economics, *Proc. Symposium in Calculus of Variations and Applications*, New York: McGraw–Hill, 1956, pp. 115–138.

260. Berge, C., *Théorie générale des jeux á n personnes games*, Paris: Gauthier Villars, 1957. Translated under the title *Obshchaya teoriya igr neskol'kikh lits*, Moscow: Fizmatgiz, 1961.

261. Bertrand, J., *Caleul des probabilities*, Paris: Gauthier-Villars, 1888.

262. Bertrand, J., Book review of theorie mathematique de la richesse sociale and of recherches sur les principles mathematiques de la theorie des richesses, *Journal de Savants*, 1883, vol. 67, pp. 499–508.

263. Bertrand, J., Review of theory mathematique de la richesse sociale recherches sur les principes mathematique de la theorie des richesses, *Journal des Savants*, 1883, vol. 68, pp. 449–508.

264. Blackwell, O., An Analog of the Minimax Theorem for Vector-Payoffs, *Pacific J. Math.*, 1956, no. 6, pp. 1–8.

265. Bohnenblust, H.F. and Karlin, S., On a Theorem of Ville, in *Contributions to the Theory of Games*, Princeton: University Press, 1953, pp. 155–160.

266. Borel, E., La théorie du jeu et les equations intégrales a noyau symétrique, *Compes Rendus de l'Académic des Sciences*, 1921, vol. 173, pp. 1304–1308.

267. Borel, E., Sur les jeux an interviennent l'hasard et l'abilité des joueurs, in *Théorie des probabilite's*, Paris, 1924, pp. 204–224.

268. Borel, E., Sur les systemes de formes lineares a determinant symetrique gauche et la theorie generale du jeu, *Comptes Rendus de l'Academie des Sciences*, 1927, vol. 184, pp. 52–53.

269. Born, P., Tijs, S., and van der Aarssen, J., Pareto–Equilibrium in Multi-Objective Games, *Methods Oper. Res.*, 1988, vol. 60, pp. 302–312.

270. Borwein, J., Proper Efficient Points for Maximization with Respect to Cones, *SIAM J. Control Optimiz.*, 1977, vol. 15, no. 1, pp. 57–63.

271. Case, J.H. and Kimeldorf, G., On Nash Equilibrium Points and Games of Imperfect Information, *J. Optimiz. Theory Appl.*, 1972, vol. 9, no. 5, pp. 302–323.

272. Categorical Imperative. http://www.qee.cuny.edu/socialSciences/ppecorino/INTRO_TEXT/Chapter%208%20Ethics/Categorical_Imperative.htm

273. Chan, W.L. and Lau, W.T., Vector Saddle-Point and Distributed Parameter Differential Games, *Comput. Math. Appl.*, 1989, vol. 18, nos. 1–3, pp. 195–207.

274. Chernoff, H., Rational Selection of Decision Function, *Econometrica*, 1954, vol. 22, pp. 422–443.

275. Chiarella, C. and Szidarovszky, F., Discrete Dynamic Oligopolies with Intertemporal Demand Interactions, *Mathematica Pannonica*, 2008, vol. 19, no. 1, pp. 107–115.

276. Cochrane, J. and Zeleny, M., *Multiple Criteria Decision Making*, Columbia: University of South Carolina Press, 1973.

277. Cohon, J.L., *Multiobjective Programming and Planning*, New York: Academic Press, 1978.

278. Colman, A.M., Körner, T.W., Musy, O., and Tazdaït, T., Mutual Support in Games: Some Properties of Berge Equilibria, *J. Math. Psychol.*, 2011, vol. 55, no. 2, pp. 166–175.

279. Colman, A.M., *Game Theory and Its Applications in the Social and Biological Sciences*, 2nd ed., Oxford: Butterworth–Heinemann, 1995.

280. Corley, H.W. and Kwain, P., Algorithm for Computing All Berge Equilibria, *Game Theory*, Hindawi Publishing Corporation, 2015, vol. 2015, article ID 862842.

281. Cournot, A., *Recherches sur les principes mathématiques de la théorie de richesses*, Paris: Chez L. Hachette, 1838.

282. Courtois, P., Nessah, R. and Tazdait, T., How to Play Games? Nash Versus Berge Behaviour Rules, *Econ. Philo.*, 2015, vol. 3, no. 31, pp. 123–139.

283. Demyanov, V.F. and Rubinov, A.M., *Constructive Nonsmooth Analysis*, Frankfurt a/M: erl. Peter Lang, 1995.

284. Dinh, T.L. and Vargas, C., A Saddle-Point Theorem for Set-Valued Maps, *Nonlinear Anal.-Theor.*, 1992, vol. 18, no. 1, pp. 1–7.

285. Dolezal, J., Some Properties of Nonzero-Sum Multistage Games, *Lect. Notes Comput. Sci.*, 1975, vol. 27, pp. 451–459.

286. Drâgustin, C., Min-max pour des criteres multiples. Recherche operationelle, *Oper. Res.*, 1979, vol. 12, no. 2, pp. 169–180.

287. Erickson, G.M., Advertising Strategies in a Dynamic Oligopoly, *J. Market. Res.*, 1995, vol. 32, no. 2, pp. 233–237.

288. Ferro, F., Minimax Theorem for Vector-Valued Functions, *J. Optimiz. Theory Appl.*, 1989, vol. 60, pp. 19–31.

289. Gaidov, S.D., Optimal Strategies in Two-Player Stochastic Differential Games, *C.R. Acad. Bulgare Sci.*, 1986, vol. 39, pp. 33–36.

290. Gaidov, S.D., Berge Equilibrium in Stochastic Differential Games, *Math. Balkanica*, 1987, vol. 1, pp. 25–32.

291. Gintchev, I.A., Method to Obtain Berge Equilibrium in Bimatrix Games, *Multicriteria Dinamical Problems under Uncertainty*, Orekhovo-Zuevo, 1991, pp. 78–82.

292. Geoffrion, A.M., Proper Efficiency and the Theory of Vector Maximization, *J. Math. Anal. Appl.*, 1968, vol. 22, no. 3, pp. 618–630.

293. Germeier, J.B., *Einfuhrung in die Theorie der Operationsforschung*, Berlin: Academie-Verlag, 1974.

294. Glicksberg, I.L., A Further Generalization of Kakutani's Fixed Point Theorem with Application to Nash Equilibrium Points, *Proc. Amer. Math. Soc.*, 1952, vol. 3, no. 1, pp. 170–174.

295. Kakutani, S.A., Generalization of Brouwer's Fixed Point Theorem, *Duke Math. J.*, 1941, vol. 8, pp. 457–459.

296. Keskin, K. and Saglam, H.C., On the Existence of Berge Equilibrium: An Order Theoretic Approach, *International Game Theory Review*, 2015, vol. 17, no. 3, pp. 1550007. DOI:10.1142/S0219198915500073

297. Krim, F., An In-Depth Study of Berge Equilibrium, *Master of Science Thesis*, Faculty of Exact Sciences, University of Tizi-Ouzou, Algeria, 2001.

298. Koopmans, T.C., Analysis of Production as an Efficient Combination of Activities, in *Activity Analysis of Production and Allocation*, New York: Wiley, 1951, pp. 33–97.

299. Kudryavtsev, K., Zhukovskiy, V., and Stabulit, I., One Method for Constructing Pareto-Optimal Nash Equilibrium, *Proc. DOOR-2016. CEUR-WS.*, vol. 1623, pp. 618–623.

300. Larbani, M., Sur l'existence de l'equilibre de Berge pour un jeu a *n*-personnes, in *Optimisation et decision. Actes des deuxiemes journees de recherche operationnelle Francoro. II Sousse*, 6–8 Avril, 1998, Ben Abdelaziz, F., Haouari, M., and Mellouli, K., Eds., Centre de Publication Universitaire de Tunisie. 2000, pp. 291–300.

301. Larbani, M. and Nessah, R., A Note on the Existence of Berge and Berge–Nash Equilibria, *Math. Soc. Sci.*, 2008, vol. 55, pp. 258–271.

302. Ledyard, J.O., Public Goods: A Survey of Experimental Research, in *The Handbook of Experimental Economics*, Kagel, J.H. and Roth, A.E., Eds., Princeton: Princeton University Press, 1994, pp. 111–194.

303. Leitmann, G., *Cooperative and Non-Cooperative Many Player Differential Games*, Vienna: Springer Verlag, 1974.

304. *Lexicon fur Theologie und Kirche*, Bd. 4, 1960.

305. Little, J.D.C., Aggregate Advertising Models: The state of the Art, *Oper. Res.*, 1979, vol. 27, no. 4, pp. 629–667.

306. Markovitz, H.M., Portfolio Selection, *J. Finance*, 1952, vol. 7, no. 1, pp. 77–89.

307. Van Megen, F., Born, P., and Tijs, S., A Preference Concept for Multicriteria Game, *Math. Methods Oper*, 1999, vol. 49, no. 3, pp. 401–412.

308. Milnor, J., Games Against Nature, in *Decision Processes*, New York: Wiley, 1954, pp. 49–60.

309. Moore, R.E., *Interval Analysis*, New York: Prentice-Hall, 1966.

310. Nash, J.F., Non-Cooperative Games, *Ann. Math.*, 1951, vol. 54, pp. 286–295.

311. Nash, J.F., Equilibrium Points in *N*-Person Games, *Proc. Nat. Academ. Sci. USA*, 1950, vol. 36, pp. 48–49.

312. Von Neumann, J., Zur Theorie der Gesellschaftspiele, *Math. Ann.*, 1928, vol. 100, pp. 295–320.

313. Von Neumann, J. and Morgenstern, O., *Theory of Games and Economic Behavior*, Princeton Univ. Press, 1944.

314. Nessah, R., Larbani, M., and Tazdait, T., A Note on Berge Equilibrium, *Appl. Math. Lett.*, 2007, vol. 20, no. 8, pp. 926–932.

315. Nessah, R., Larbani, M., and Tazdait, T., On Berge Equilibrium, HAL, halshs-00271452. Archives ouvertes.

316. Niehans, J., Zur Preisbildung bei Unterwissen, *Schweizerische Zeitschrift Association*, 1951, vol. 46, no. 3, pp. 55–67.

317. Pareto, V., *Manuel d'économie politique*, Paris: Geard, 1909.

318. Petrosian, L.A., *Differential Games of Pursuit*, London, Singapore: World Scientific, 1993.

319. Reiner, H., Die "Golden Regel": Die Bedeutung einer sittlichen Grundformel der Menschheit, *Zeitschrift fur philosophische Forschung*, Bd. 3, 1948.

320. Radjef, M., Sur l'existence d'un équilibre de Berge pour un jeu différential n-personnes (On the Existence of a Berge Equilibrium for an *n*-Person Differential Game), *Cahiers Mathématiques de l'Université d'Oran*, 1998, no. 1, pp. 89–93.

321. Savage, L.Y., *The Foundation of Statistics*, New York: Wiley, 1954.

322. Savage, L.Y., The Theory of Statistical Decision, *J. Amer. Stat. Assoc.*, 1951, no. 46, pp. 55–67.

323. Seneca, *Epistulae ad Lucilium*, 94:43.

324. Seneca, *Epistulae ad Lucilium*, 94:67.

325. Shapley, L.S., Equilibrium Point in Game with Vector Payoffs, *Naval Res. Logistics Quart.*, 1959, vol. 1, pp. 57–61.

326. Shubik, M., Review of C. Berge "General theory of n-person games," *Econometrica*, 1961, vol. 29, no. 4, p. 821.

327. Sethi, S.P., Deterministic and Stochastic Optimization of Dynamic Advertising Model, *Optim. Control Appl. Methods*, 1983, vol. 4, no. 2, pp. 179–184.

328. Sethi, S.P., Prasad, A., and He, X., Optimal Advertising and Pricing in a New-Product Adoption Model, *J. Optim. Theory Appl.*, 2008, vol. 139, no. 2, pp. 351–360.

329. Sethi, S.P. and Thompson, G.L., *Optimal Control Theory: Applications to Management Science and Economics*, 2nd ed., New York: Springer, 2000.

330. Shi, D.S. and Ling, C., Minimax Theorems and Cone Saddle-Points of Uniformly Some Order Vector-Valued Functions, *J. Optim. Theory Appl.*, 1995, vol. 84, no. 3, pp. 575–587.

331. Slade, M., Product Rivalry with Multiple Strategic Weapons: An Analysis of Price and Advertising Competition, *J. Econ. Manage. Strat.*, 1995, vol. 4, no. 3, pp. 445–476.

332. Simaan, M. and Cruz, I., Sampled–Data Nash Controls in Non-Zero-Sum Differential Games, *Int. J. Control*, 1973, no. 17, pp. 1201–1209.

333. Sorger, G., Competitive Dynamic Advertising: A Modification of the Case Game, *J. Econ. Dyn. Control*, 1989, vol. 13, pp. 55–80.

334. Steuer, R., *Multiple Criteria Optimization: Theory, Computation and Application*, New York: John Wiley and Sons, 1986.

335. Al-Suhrawardy, A., *The Sayings of Muhammad*, New York, 1990.

336. Syrus Publilitis, *Sententiae*, sententia 2.

337. Tanaka, T., Existence Theorems for Cone Saddle-Points of Vector-Valued Functions in Infinite—Dimensional Space, *J. Optimiz. Theory Appl.*, 1989, vol. 62, no. 1, pp. 127–138.

338. Tanaka, T., Generalized Quasiconvexities Cone Saddle-Points and Minimax Theorem for Vector—Valued Functions, *J. Optimiz. Theory Appl.*, 1994, vol. 81, no. 2, pp. 355–377.

339. Tanaka, T., Generalized Semicontinuity and Existence Theorems for Cone Saddle—Points, *Appl. Math. Opt.*, 1997, vol. 36, no. 3, pp. 313–322.

340. Tan, K.K., Yu, J., and Yuan, X.Z., Existence Theorems for Saddle-Points of Vector-Valued Maps, *J. Optimiz. Theory Appl.*, 1996, vol. 89, no. 3, pp. 731–747.

341. Okeja, U.B., *Normative Justification of a Global ethic: A Perspective from African Philosophy*, New York: Lexington Books, 2013.

342. Vaisbord, E.M. and Zhukovskiy, V.I., *Introduction to Multi Player Differential Games and Their Applications*, New York: Gordon and Breach, 1988.

343. Vaisman, K.S., About Differential Game under Uncertainty, *Abstr. 3rd Int. Workshop "Nonsmooth and Discontinuous Problems of Control and Optimization"*, St. Petersburg, 1995, pp. 45–48.

344. Vaisman, K.S., The Berge Equilibrium for Linear–Quadratic Differential Game, *The 3-rd Intern. Workshop on Multiple Criteria Problems under Uncertainty*, Orekhovo-Zuevo, Russia, 1994, p. 96.

345. Vaisman, K.S., Nash Equilibrium Routing and Ring Networks, *Game Theory Appl.*, vol. III, 1997, pp. 147–160.

346. Vaisman, K.S. and Zhukovskiy, V.I., The Berge Equilibrium under Uncertainty, *The 3-rd Intern. Workshop on Multiple Criteria Problems under Uncertainty*, Orekhovo-Zuevo, Russia, 1994, pp. 97–98.

347. Vidale, M.L. and Wolfe, H.B., An Operations-Research Study of Sales Response to Advertising, in *Mathematical Models in Marketing. Lecture Notes in Economics and Mathematical Systems (Operations Research)*, vol. 132, Berlin–Heidelberg: Springer, 1976, pp. 223–225.

348. Vidura–Niti, *Mhb*, vols. 31–41; vols. 39–47.

349. Barletta, V., *Death in Babylon: Alexander the Great and Iberian Empire in the Muslim Orient*, University of Chicago Press, 2010.

350. Vinckle, P., *Multicriteria Decision – Aid*, Chichester: John Wiley and Sons, 1992.

351. Wald, A., Contribution to the Theory of Statistical Estimation and Testing Hypothesis, *Annu. Math. Statist.*, 1939, vol. 10, pp. 299–326.

352. Wald, A., *Statistical Decision Functions*, New York: Wiley, 1950.

353. Zeleny, M., Games with Multiple Payoffs, *International Journal of Game Theory*, 1976, vol. 4, pp. 179–191.

354. Zhukovskiy, V.I., Some Problems of Non-Antagonistic Differential Games, in *Mathematical Methods in Operations Research*, Kendrov, P., Ed., Bulgarian Academy of Sciences, Sofia, 1985, pp. 103–195.

355. Zhukovskiy, V.I., *Lyapunov Functions in Differential Games*, London and New York: Taylor & Francis, 2003.

356. Zhukovskiy, V.I., Chikrii, A.A., and Soldatova, N.G., Existence of Berge Equilibrium in Conflicts under Uncertainty, *Autom. Remote Control*, 2016, vol. 77, no. 4, pp. 607–622.

357. Zhukovskiy, V.I. and Gorbatov, A.S., About Deposit Diversification Problem, *Uchen. Zapiski Tavrich. Natsion. Univ. Ser. Fiz.-Mat. Nauki*, 2014, vol. 27, no. 1, pp. 222–233.

358. Zhukovskiy, V.I., Kirichenko, M.M., and Boldyrev, M.V., Guaranteed Outcomes and Risks in Multicriteria Problem, *Taurida J. Comput. Sci. Theory Math.*, 2016, no. 1, pp. 7–18.

359. Zhukovskiy, V.I. and Kudryavtsev, K.N., Pareto-Optimal Nash equilibrium: Sufficient Conditions and Existence in Mixed Strategies, *Autom. Remote Control*, 2016, vol. 77, no. 8, pp. 1500–1510.

360. Zhukovskiy, V.I., Molostvov, V.S., and Topchishvili, A.L., Problem of Multicurrency Deposit Diversification - Three Possible Approaches to Risk Accounting, *Int. J. Oper. Quant. Manage.*, 2014, vol. 20, no. 1, pp. 1–15.

361. Zhukovskiy, V.I., Molostvov, V.S., and Vaisman, K.S., Non-cooperative Games under Uncertainty, *Game Theory Appl.*, vol. III, 1997, pp. 189–222.

362. Zhukovskiy, V.I., Molostvov, V.S., and Zhukovskaya, L.V., On Existence of Guaranteed Risk, *Abstr. Int. Conf. Appl. Math.*, Kyiv: Glushkov Inst. Cybern., 2002, p. 36.

363. Zhukovskiy, V.I. and Sachkov, S.N., Bilanciamento conflitti Friendly, *Ital. Sci. Rev.*, 2014, vol. 9(18), pp. 169–179.

364. Zhukovskiy, V.I., Sachkov, S.N., and Gorbatov, A.S., Mathematical Model of the Golden Rule, *SCIENCE, TECHNOLOGY AND LIFE - 2014: Proc. Int. Scientific Conf.*, Czech Republic, Karlovy Vary, December 27–28, 2014, pp. 17–23.

365. Zhukovskiy, V.I. and Salukvadze, M.E., *The Vector-Valued Maximin*, New York: Academic Press, 1994.

366. Zhukovskiy, V.I. and Salukvadze, M.E., Sufficient Conditions in Vector-Valued Maximin Problems, *J. Optimiz. Theory Appl.*, 1996, vol. 90, no. 3, pp. 523–534.

367. Zhukovskiy, V.I., Salukvadze, M.E., and Vaisman, K.S., *The Berge Equilibrium under Uncertainty*, Preprint. Tbilisi: Institute of Control Systems, 1995.

368. Zhukovskiy, V.I., Salukvadze, M.E., and Vaisman, K.S., *The Berge Equilibrium*, Preprint. Tbilisi: Institute of Control Systems, 1994.

369. Zhukovskiy, V.I., Sachkov, S.N., and Smirnova, L.V., Berge Equilibrium, *Analiz, Modelirovanie, Upravlenie, Razvitie Sotsial'no-Ekonomicheskikh Sistem: Tr. VIII Mezhd. Shkoly-Simpoziuma (AMUR-2014)* (Analysis, Modeling, Management, Development of Socioeconomic Systems: Proc. VIII Int. School-Symposium (AMMD-2014)), Krymskii Fed. Univ., Simferopol, 2014, pp. 124–133.

370. Zhukovskiy, V.I., Sachkov, S.N., and Smirnova, L.V., Existence of Berge Equilibrium in Mixed Strategies, *Uchenye Zapiski Tavrich. Natsional. Univ.*, 2014, pp. 261–279.

371. Zhukovskiy, V.I. and Smirnova, L.V., Berge–Vaisman Equilibrium for One Linear-Quadratic Differential Game, *Taurida J. Comput. Sci. Theory Mathematics*, 2016, no. 1, pp. 19–28.

372. Zhukovskiy, V.I. and Topchishvili, A.T., The Explicit Form of Berge-Strongly-Guaranteed Equilibrium in Linear-Quadratic Non-Cooperative Game, *Int. J. Oper. Quant. Manage.*, 2015, vol. 21, no. 4, pp. 265–273.

373. Zhukovskiy, V.I. and Topchishvili, A.T., Mathematical Model of the Golden Rule in Form of Differential Positional Game of Many Persons, *Int. J. Oper. Quant. Manage.*, 2016, vol. 22, no. 3, pp. 203–229.

374. Zhukovskiy, V., Topchishvili, A., and Sachkov, S., Application of Probability Measures to the Existence Problem of Berge–Vaisman Guaranteed Equilibrium, *Model Assist. Stat. Appl.*, 2014, vol. 9, no. 3, pp. 223–239.

375. Zhukovskiy, V.I. and Vaisman, K.S., Solution in Non-cooperative Games, *Abstracts of N.N. Vorob'ev's Memorial Conference on Game Theory and Economics*, St. Petersburg, 1996, p. 77.

376. Zhukovskiy, V.I. and Vaisman, K.S., To a Problem of Berge Equilibrium, *Vestn. Pskovsk. Vol'nog. Univ. Ser. Mat. Inform.*, 1997, vol. 1, pp. 49–70.

377. Zhukovskiy, V.I. and Larbani, M., Berge Equilibrium in Normal Form Static Games: A Literature Review, *Izv. Inst. Mat. Inform. Udmurt. Gos. Univ.*, 2018, vol. 49, no. 3 (40), pp. 47–70.

378. Zhukovskiy, V.I., Sachkov, S.N., and Sachkova, E.N., Guaranteed Solution for Risk-Neutrality: An Analog of Maximin in Single-Criterion Choice Problem, *Taurida J. Comput. Sci. Theory Math.*, 2018, no. 3(40), pp. 46–70.

379. Zhukovskiy, V.I., Smirnova, L.V., and Vysokos, M.I., One Variable Problem in Matrix Ordinary Differential Equations, *Taurida J. Comput. Sci. Theory Math.*, 2019, no. 1(42), pp. 62–73.

380. Zhukovskiy, V.I. and Boldyrev, M.V., Guaranteed Risks and Payoffs in a Single-Criterion Problem, *Taurida J. Comput. Sci. Theory Math.*, 2019, no. 1(42), pp. 7–23.

381. Vaisbord, E.M., Coalitional Differential Games, *Diff. Uravn.*, 1974, vol. 10, no. 4, pp. 613–623.

382. Vaisbord, E.M. and Zhukovskiy, V.I., *Vvedenie v differentsial'nye igry neskol'kikh lits i ikh prilozheniya* (Introduction to Differential Games of Many Players and Their Applications), Moscow: Sovetskoe Radio, 1980.

383. Vilkas, E.I., Formalization of Choice Problem of Game-Theoretic Optimality Criterion, in *Matematicheskie metody v sotsial'nykh naukakh* (Mathematical Methods in Social Sciences), Vilnius: Inst. Mat. Kibern. Akad. Nauk LSSR, 1972, vol. 2, pp. 9–55.

384. Vilkas, E.I. and Maiminas, E.Z., *Resheniya: Teoriya, Informatsiya, Modelirovanie* (Decisions: Theory, Information, Modeling), Moscow: Radio i Svyaz', 1981.

385. Zhukovskiy, V.I., Gorbatov, A.S., and Kudryavtsev, K.N., Nash and Berge Equilibria in a Linear-Quadratic Game, *Mat. Teor. Igr Prilozh.*, 2017, vol. 9, no. 1, pp. 62–94.

386. Zhukovskiy, V.I., Kudryavtsev, K.N., Samsonov, S.P., Vysokov, M.I., and Bel'skikh, Yu.A., A Class of Differential Games in Which Nash Equilibrium is Absent But There Exists a Berge Equilibrium, *Vestn. South Ural State Univ. Ser. Math. Mech. Phys.*, 2018, vol. 10, no. 2, pp. 5–21.

387. Owen, *Game Theory*, Saunders, 1968.

388. Biltchev, S.V., $\varepsilon - Z$-Equilibrium in a Differential Game Described by a Parabolic System, in *Many Players Differential Game*, Bulgaria, Rousse: Technical Univ., 1984, pp. 47–52.

389. Case, J.H., A Class of Games Having Pareto Optimal Nash Equilibrium, *J. Optimiz. Theory Appl.*, 1974, vol. 13, no. 3, pp. 378–385.

390. Dochev, D.T. and Stojanov, N.V., Existence of Z-Equilibrium in a Differential Game with Delay, in *Many Players Differential Game*, Bulgaria, Rousse: Technical Univ., 1984, pp. 64–72.

391. Gaidov, S.D., Z-Equilibrium in Stochastic Differential Game, in *Many Players Differential Game*, Bulgaria, Rousse: Technical Univ., 1984, pp. 53–63.

392. Rashkov, P.I., Sufficient Conditions for Z-Equilibrium in a Differential Game in Banach Space, *Many Players Differential Game*, Bulgaria, Rousse: Technical Univ., 1984, pp. 91–99.

393. Tersian, St.A., On the Z-Equilibrium Points in a Differential Game, *Many Players Differential Game*, Bulgaria, Rousse: Technical Univ., 1984, pp. 106–111.

394. Zhukovskiy, V.I. and Kudryavtsev, K.N., Coalition Equilibrium in a Three-Person Game, *2017 Constructive Nonsmooth Analysis and Related Topics (dedicated to the memory of V.F. Demyanov) (CNSA)*, St. Petersburg, May 22–27, 2017, pp. 1–4.

395. Vatel', I.A. and Ereshko, F.I., A Game with a Hierarchical Structure, in *Matematicheskaya entsiklopediya* (Mathematical Encyclopedia), Moscow, 1979, vol. 2, pp. 477–481.

396. Kukushkin, I.S. and Morozov, V.V., *Teoriya neantagonisticheskikh igr* (Theory of Non-antagonistic Games), Moscow: Mosk. Gos. Univ., 1984.

397. Zhukovskiy, V.I., Makarkina, T.V., and Vysokos, M.I., A New Approach to Noncooperative Game under Uncertainty, *Int. Game Theory Rev.*, 2007, vol. 2, no. 19, pp. 1750024-1-1750024-19.

398. Zhukovskiy, V.I., Molostvov, V.S., and Topchishvili, A.L., Problem of Multicurrency Deposit Diversification—Three Possible Approaches to Risk Accounting, *Int. J. Oper. Quant. Manage.*, 2014, vol. 20, no. 1, pp. 1–14.

399. Aumann, R.J., Acceptable Points in General Cooperative n-Person Games, in *Contributions to the Theory of Games IV, Annals of Mathematics Study*, Tucker, A.W. and Duncan Luce, R., Eds., Princeton: Princeton University Press, 1959, vol. 40, pp. 287–324.

400. Aumann, R.J., The Core of a Cooperative Game without Side Payments, *Trans. Am. Math. Soc.*, 1961, vol. 98, pp. 539–552.

401. Berge, C., *Espace Topologiques*, Paris: Dunod, 1963.

402. Douglas, B.B., Peleg, B., and Whinston, M.D., Coalition-Proof Nash Equilibria I. Concepts, *J. Econ. Theory*, 1987, vol. 42, no. 1, pp. 1–12.

403. Chwe, M.S.Y., Farsighted Coalitional Stability, *J. Econ. Theory*, 1994, vol. 63, pp. 299–325.

404. Engel, C., Dictator Games: A Meta Study, *Exp. Econ.*, 2011, vol. 14, pp. 583–610.

405. Fehr, E. and Schmidt, K. M., The Economics of Fairness, Reciprocity and Altruism: Experimental Evidence and New Theories, in *Handbook of the Economics of Giving, Altruism and Reciprocity*, 2006, vol. 1, pp. 615–691.

406. Gillies, D.B., Solutions to General Non-Zero-Sum Games, in *Contributions to the Theory of Games IV, Annals of Mathematics Study*, Tucker, A.W. and Duncan Luce, R., Eds., Princeton: Princeton University Press, 1959, vol. 40, pp. 47–85.

407. Kahneman, D., Knetsch, J.L., and Thaler, R.H., Fairness and the Assumptions of Economics, *J. Business*, 1986, vol. 59, pp. 285–300.

408. Kornishi, H. and Ray, D., Coalition Formation as a Dynamic Process, *J. Econ. Theory*, 2003, vol. 110, pp. 1–41.

409. Larbani, M. and Nessah, R., Sur l'quilibre fort selon Berge, *RAIRO Oper. Res.*, 2001, vol. 35, pp. 439–451.

410. Mariotti, M., A Model of Agreement in Strategic Form Games, *J. Econ. Theory*, 1997, vol. 74, pp. 196–217.

411. Ray, D. and Vohra, R., Equilibrium Binding Agreement, *J. Econ. Theory*, 1997, vol. 73, pp. 30–78.

412. Sally, D., Conversation and Cooperation in Social Dilemmas: A Meta-Analysis of Experiments from 1958 to 1992, *Ration. Soc.*, 1995, vol. 7, pp. 58–92.

413. Scarf, H.E., On the Existence of a Cooperative Solution for a General n-Person Game, *J. Econ. Theory*, 1971, vol. 32, pp. 169–181.

414. Zhao, J., The Hybrid Solutions of an N-Person Game, *Games Econ. Behav.*, 1992, vol. 4, pp. 145–160.

415. Zhao, J., A Cooperative Analysis of Covert Collusion in Oligopolistic Industries, *Int. J. Game Theory*, 1997, vol. 26, pp. 249–266.

416. Zhukovskiy V.I. and Larbani M., Alliance in Three Person Games, *Res. Math. Mechan.*, 2017, vol. 22, no. 1(29), pp. 105–119.

417. Zhukovskiy, V.I. and Salukvadze, M.E., *The Berge Equilibrium: A Game-Theoretic Framework for the Golden Rule of Ethics*, Springer, 2020.

Subject Index

Author Index

Communications in Cybernetics, Systems Science and Engineering

Book Series Editor: Jeffrey 'Yi-Lin' Forrest

ISSN: 2164-9693

Publisher: CRC Press/Balkema, Taylor & Francis Group

1. A Systemic Perspective on Cognition and Mathematics
   Jeffrey Yi-Lin Forrest
   ISBN: 978-1-138-00016-2 (Hb)

2. Control of Fluid-Containing Rotating Rigid Bodies
   Anatoly A. Gurchenkov, Mikhail V. Nosov & Vladimir I. Tsurkov
   ISBN: 978-1-138-00021-6 (Hb)

3. Research Methodology: From Philosophy of Science to Research Design
   Alexander M. Novikov & Dmitry A. Novikov
   ISBN: 978-1-138-00030-8 (Hb)

4. Fast Fashion Systems: Theories and Applications
   Tsan-Ming Choi (ed.)
   ISBN: 978-1-138-00029-2 (Hb)

5. Reflexion and Control: Mathematical Models
   Dmitry A. Novikov & Alexander G. Chkhartishvili
   ISBN: 978-1-138-02473-1 (Hb)

6. A Systems Perspective on Financial Systems
   Jeffrey Yi-Lin Forrest
   ISBN: 978-1-138-02628-5 (Hb)

7. Fashion Retail Supply Chain Management: A Systems Optimization Approach
   Tsan-Ming Choi
   ISBN: 978-1-138-00028-5 (Hb)

8. Service Supply Chain Systems: A Systems Engineering Approach
   Tsan-Ming Choi (ed.)
   ISBN: 978-1-138-02829-6 (Hb)

9. The Golden Rule of Ethics: A Dynamic Game-Theoretic Framework Based on Berge Equilibrium
   Vladislav I. Zhukovskiy & Mindia E. Salukvadze
   ISBN: 978-0-367-68179-1 (Hb)